SPECKLE METROLOGY

OPTICAL ENGINEERING

Series Editor

Brian J. Thompson
Provost
University of Rochester
Rochester, New York

1. Electron and Ion Microscopy and Microanalysis: Principles and Applications, *Lawrence E. Murr*
2. Acousto-Optic Signal Processing: Theory and Implementation, *edited by Norman J. Berg and John N. Lee*
3. Electro-Optic and Acousto-Optic Scanning and Deflection, *Milton Gottlieb, Clive L. M. Ireland, and John Martin Ley*
4. Single-Mode Fiber Optics: Principles and Applications, *Luc B. Jeunhomme*
5. Pulse Code Formats for Fiber Optical Data Communication: Basic Principles and Applications, *David J. Morris*
6. Optical Materials: An Introduction to Selection and Application, *Solomon Musikant*
7. Infrared Methods for Gaseous Measurements: Theory and Practice, *edited by Joda Wormhoudt*
8. Laser Beam Scanning: Opto-Mechanical Devices, Systems, and Data Storage Optics, *edited by Gerald F. Marshall*
9. Opto-Mechanical Systems Design, *Paul R. Yoder, Jr.*
10. Optical Fiber Splices and Connectors: Theory and Methods, *Calvin M. Miller with Stephen C. Mettler and Ian A. White*
11. Laser Spectroscopy and Its Applications, *edited by Leon J. Radziemski, Richard W. Solarz and Jeffrey A. Paisner*
12. Infrared Optoelectronics: Devices and Applications, *William Nunley and J. Scott Bechtel*
13. Integrated Optical Circuits and Components: Design and Applications, *edited by Lynn D. Hutcheson*
14. Handbook of Molecular Lasers, *edited by Peter K. Cheo*
15. Handbook of Optical Fibers and Cables, *Hiroshi Murata*
16. Acousto-Optics, *Adrian Korpel*
17. Procedures in Applied Optics, *John Strong*
18. Handbook of Solid-State Lasers, *edited by Peter K. Cheo*
19. Optical Computing: Digital and Symbolic, *edited by Raymond Arrathoon*
20. Laser Applications in Physical Chemistry, *edited by D. K. Evans*

21. Laser-Induced Plasmas and Applications, *edited by Leon J. Radziemski and David A. Cremers*
22. Infrared Technology Fundamentals, *Irving J. Spiro and Monroe Schlessinger*
23. Single-Mode Fiber Optics: Principles and Applications, Second Edition, Revised and Expanded, *Luc B. Jeunhomme*
24. Image Analysis Applications, *edited by Rangachar Kasturi and Mohan M. Trivedi*
25. Photoconductivity: Art, Science, and Technology, *N. V. Joshi*
26. Principles of Optical Circuit Engineering, *Mark A. Mentzer*
27. Lens Design, *Milton Laikin*
28. Optical Components, Systems, and Measurement Techniques, *Rajpal S. Sirohi and M. P. Kothiyal*
29. Electron and Ion Microscopy and Microanalysis: Principles and Applications, Second Edition, Revised and Expanded, *Lawrence E. Murr*
30. Handbook of Infrared Optical Materials, *edited by Paul Klocek*
31. Optical Scanning, *edited by Gerald F. Marshall*
32. Polymers for Lightwave and Integrated Optics: Technology and Applications, *edited by Lawrence A. Hornak*
33. Electro-Optical Displays, *edited by Mohammad A. Karim*
34. Mathematical Morphology in Image Processing, *edited by Edward R. Dougherty*
35. Opto-Mechanical Systems Design: Second Edition, Revised and Expanded, *Paul R. Yoder, Jr.*
36. Polarized Light: Fundamentals and Applications, *Edward Collett*
37. Rare Earth Doped Fiber Lasers and Amplifiers, *edited by Michel J. F. Digonnet*
38. Speckle Metrology, *edited by Rajpal S. Sirohi*

Additional Volumes in Preparation

Organic Photoreceptors for Imaging Systems, *Paul Borsenberger and David Weiss*

Photonic Switching and Interconnects, *edited by Abdellatif Marrakchi*

SPECKLE METROLOGY

EDITED BY

RAJPAL S. SIROHI
Indian Institute of Technology
Madras, India

Marcel Dekker, Inc. New York • Basel • Hong Kong

Library of Congress Cataloging-in-Publication Data

Speckle metrology / edited by Rajpal S. Sirohi.
p. cm. -- (Optical engineering ; 38)
ISBN 0-8247-8932-6 (acid-free paper)
1. Non-destructive testing. 2. Speckle metrology. I. Sirohi, R. S. II. Series: Optical engineering (Marcel Dekker, Inc.) ; v. 38.
TA417.2.S66 1993
620.1'127--dc20 93-734
CIP

The publisher offers discounts on this book when ordered in bulk quantities. For more information, write to Special Sales/Professional Marketing at the address below.

This book is printed on acid-free paper.

Marcel Dekker, Inc.
270 Madison Avenue, New York, New York 10016

Current printing (last digit):
10 9 8 7 6 5 4 3 2 1

PRINTED IN THE UNITED STATES OF AMERICA

About the Series

The series came of age with the publication of our twenty-first volume in 1989. The twenty-first volume was entitled *Laser-Induced Plasmas and Applications* and was a multi-authored work involving some twenty contributors and two editors: as such it represents one end of the spectrum of books that range from single-authored texts to multi-authored volumes. However, the philosophy of the series has remained the same: to discuss topics in optical engineering at the level that will be useful to those working in the field or attempting to design subsystems that are based on optical techniques or that have significant optical subsystems. The concept is not to provide detailed monographs on narrow subject areas but to deal with the material at a level that makes it immediately useful to the practicing scientist and engineer. These are not research monographs, although we expect that workers in optical research will find them extremely valuable.

There is no doubt that optical engineering is now established as an important discipline in its own right. The range of topics that can and should be included continues to grow. In the "About the Series" that I wrote for earlier volumes, I noted that the series covers "the topics that have been part of the rapid expansion of optical engineering." I then followed this with a list of such topics which we have already outgrown. I will not repeat that mistake this time! Since the series now exists, the topics that are appropriate are best exemplified by the titles of the volumes listed in the front of this book. More topics and volumes are forthcoming.

Brian J. Thompson
University of Rochester
Rochester, New York

Preface

The speckle effect has the rare distinction of being both a nuisance and an asset to the optics community engaged in the development of laser-based methods to monitor displacements and deformations of objects. Its remarkable information-carrying ability was recognized around 1968 and put to use for measurement purposes. Since then the application areas of the speckle phenomenon have multiplied. The first book on the subject (*Speckle Metrology*, ed. R. K. Erf) appeared in 1978. Over the last decade a better understanding has emerged of the measurement capabilities of the speckle phenomenon and its new applications coupled with new techniques. Fringe analysis and particle image velocimetry (PIV) are two of the new areas that have developed during the last decade. To fill the need for a book covering various aspects of speckles, newer topics, and measurement and analysis techniques, *Speckle Metrology* was conceived. Its 11 chapters provide a timely, comprehensive treatment.

First, the chapter by Yamaguchi on theory and applications of speckle displacement and decorrelation gives a general survey of the origin of speckle displacement and decorrelation. The theory of cross-correlation of intensity and amplitude is described. The derivation of cross-correlation of intensity reduces to that of the amplitude cross-correlation. The cross-correlation function of the amplitude is derived using Fresnel-Kirchhoff diffraction theory for both free-space propagation and imaging geometry. The cross-correlation shows a peak whose position corresponds to speckle displacement and whose height indicates speckle decorrelation accompanying the displacement. Dependences of the speckle displacement on individual deformation parameters such as translation, rotation, and strain, as well as on optical configuration, are presented explicitly. The speckle displacement is detected by either speckle photography or opto-electronic methods. In opto-electronic methods the signals from the video camera are cross-correlated before and after object deformation. The real-time correlators can also use a bipolar algorithm. A serious disadvantage of correlation analysis is its time consumption. A spatial filtering detector is a kind of parallel image processor for detecting speckle displacement in real time. The working of spatial filtering detectors is described and a variety of spatial filtering detectors are mentioned; some of them can be used with simpler electronics. A number of applications of electronic speckle correlation are discussed. Speckle encoders that use the surface of the object as a random grating can measure the displacement of the surface itself. The speckle strain gauge is another example; laser speckle strain gauges directly detect elongation or contraction of the microscopic structure of solid surfaces and are free from any kind of surface contact or preparation. By using image sensors in two orthogonal directions, Poisson's ratio of the material can also be determined quickly. The velocity of speckle is measured by speckle velocimeters; it can also be sensed by spatial filtering detectors. Some recent developments, particularly fast detection of speckle displacement and two-dimensional cross-correlation, are described.

Chapter 2 by Rastogi, on techniques of displacement and deformation measurements, presents various procedures of deformation analysis and shape measurement of rough objects. Emphasis is placed upon analysis of observed fringes. These procedures are grouped under three categories: photography, interferometry, and speckle-shear interferometry. The theory underlying these techniques, as well as the fringe formation analysis and various procedures adopted to obtain components of deformation vectors and also for contouring, have been detailed.

My chapter on speckle methods in experimental mechanics covers various techniques of speckle metrology and discusses new features not covered in Chapter 2. The influence of lens aperturing on data recording and retrieval, use of multiple apertures in front of the lens, and consequent isolation of deformation components, slope, and curvature information during filtering operation are highlighted. A combination of hologram interferometry and speckle photography, called holospeckle interferometry, is discussed. These techniques are applied to the study of various problems in fracture mechanics, vibration analysis, thermal stress analysis, and nondestructive testing.

Chapter 4 by Løkberg introduces video speckle interferometry and gives details of its early development. The term video speckle interferometry is used in the broader sense, and all techniques for deformation measurements, vibration analysis, and derivative measurements are included. The fringe functions for two cases, namely data acquisition in a single frame or a combination of frames, are detailed. Technical details for system configuration are given. The influence of frame subtraction and addition for detecting very low amplitude vibrations is shown. The averaging of many interferograms results in noise reduction. This procedure is different for vibration analysis and deformation measurement. Fringe analysis can be performed using phase-shifting interferometry. A routine for detecting defects under unstable conditions of the object is presented. A technique based on time-average recordings using sinusoidal phase modulation to measure vibration amplitudes and phases at very low excitation is also described, and phase-shifting interferometry to obtain phase of vibration is discussed. The chapter concludes with a note on the problems in automatic analysis of fringes and gives some examples. It is stressed that shear methods will be the inspection methods of the future.

Chapter 5 by Vikram gives some novel applications of speckle metrology. Use of speckles for measuring ultralow thermal expansion, study of large structures, measurement of large deflections and small motions, use of fibers in speckle metrology, and speckle methods of measurement in daylight are some of the topics explored. Other topics include heterodyne speckle photography, speckle photography with electron microscope, liquid crystals used for speckle photography, and use of photorefractive crystals for quasi-real time applications.

Chapter 6 on particle image velocimetry (PIV), by Hinsch, begins with the limits of flow visualization and surveys developments in PIV. Then follows the theoretical analysis of PIV in space domain and spatial frequency domain, and by auto-correlation methods. Fringe formulas thus derived have been further analyzed for two cases: constant displacement and nonuniform displacement within the interrogation region. The analysis of noise in PIV is important as the information is extracted from doubly/multiply exposed specklegrams. The analysis of PIV records by various methods—e.g., the well-known Young's fringe formation method—and the attendant problems with the processing of Young's fringe systems, follows. The space-dependent visibility of fringes in large-area processing has been exploited to obtain statistical information about a flow field. Various suggestions for future analysis of PIV records have been put forward. The practical problems associated with the design of an experimental setup for faithful mapping of the flow field are outlined. The problem of direction ambiguity can be solved by introducing a device that shifts the image of the interrogation region between two records. The solution to the problem of measuring mean velocity in the presence of large turbulence is presented. Also, problems associated with three-dimensional flow field recordings are discussed. Some examples are presented to demonstrate the state of the art of PIV

Speckle metrology need not be restricted to the use of laser-produced speckles. In fact, large deformations and tilts introduce decorrelations, making the laser speckle–based methods unsuitable. The artificially created speckles, for example, by painting the object with retro-reflecting paint, are used to advantage for measuring large deformations of large objects and obtaining the derivatives of the displacement. Asundi's chapter on white light speckle metrology presents an account of the methods of production of artificial speckles and their properties. The procedures for extracting information from white light specklegrams are identical to those used for laser specklegrams. The methods of obtaining components of a displacement vector, measuring strain fields, and contouring with white light speckles are presented. White light speckle methods can also be used to measure displacements inside a transparent object, as well as for vibration analysis.

The speckle pattern is caused by reflection from a rough surface and carries information about the roughness. Chapter 8 on surface roughness evaluation by Briers gives a brief account of the statistics of rough surfaces followed by the origin of the speckle effect and speckle statistics. The contrast of partially developed speckles from weak diffusers depends on the roughness of the surface. A few practical systems based on contrast measurements are described. There are also methods of evaluating surface roughness that use polychromatic/dichromatic speckles. The range of surface roughness measurement is increased manyfold by employing polychromatic illumination. Decorrelation sets in when either the wavelength or the angle of incidence is varied; and the degree of decorrelation depends on the surface roughness. The techniques based on correlation of speckles are described. Apart from speckle techniques, there are many contact and noncontact methods of surface roughness measurement. The techniques based on scattering and surface profilometry are presented. Finally, a comparison between laser speckle technique and other optical techniques is given.

The specklegram, when interrogated with either a narrow or an expanded beam, results in the formation of fringes. Fringe analysis is carried out to obtain information on the object deformation. The speckle fringes are very noisy, and hence special considerations apply. Chapter 9 on automatic fringe analysis procedures by Kaufmann deals with various aspects of fringe analysis. Starting with the evaluation of Young's fringes from double-exposure specklegrams with semiautomatic procedures, it goes on to describe the automatic procedures for the measurement of displacement fields and fluid velocity fields, and then compares the various methods. The analysis of fringes generated by dual-field speckle photography is presented using heterodyning and phase-shift techniques. The use of photorefractive crystals for fringe analysis is outlined. The fringes formed in ESPI can also be analyzed using phase-shift techniques; the details are presented.

All the speckle techniques for deformation measurement use imaging geometries for recording specklegrams. A good-quality imaging lens is then invariably used. An imaging system can be made holographically and employed for speckle metrology. Chapter 10 by Shakher discusses hololenses, in particular, a dual hololens system with optimized aberration correction. Using the hololens for imaging, displacements around the crack tip are measured and results are compared with those given by theory. Further, the design of a compact hololens system with better performance is given. Both speckle photography and speckle interferometries may be performed with or without aperturing the hololens system. However, due to the system's skewed viewing, isolation of in-plane and out-of-plane components is a bit tricky.

Chapter 11 on correlation speckle interferometry in the mechanics of contact interaction, by Ostrovsky and Shchepinov, works out the contrast of Young's fringes. The variation in visibility arises due to the change of microrelief of the surface, and hence the decorrelation of speckle patterns. This has been worked out for two cases: when recording is performed with full and apertured lenses. Experimental verification of the variation of fringe contrast is given. The technique is applied to the measurement of contact pressure. I am extremely grieved to report the demise of Prof. Ostrovsky; this chapter was completed just before his death.

It is appropriate to record my appreciation for the support and cooperation I received from the contributors. I am thankful to many others who have contributed directly or indirectly to the completion of the book. Eric Stannard and Ruth Dawe from Marcel Dekker, Inc., are to be congratulated for their prompt handling of the book. Finally, I acknowledge the support and encouragement given by my wife, Vijayalaxmi Sirohi, which made this work a pleasure.

Rajpal S. Sirohi

Contents

Contributors

Anand K. Asundi *University of Hong Kong, Hong Kong*

J. D. Briers *Kingston University, Kingston upon Thames, England*

K. D. Hinsch *University of Oldenburg, Oldenburg, Germany*

G. H. Kaufmann *Rosario Institute of Physics, Rosario, Argentina*

Ole J. Løkberg *The Norwegian Institute of Technology, Trondheim, Norway*

Yuri I. Ostrovsky† *A. F. Ioffe Physical Technical Institute, St. Petersburg, Russia*

Pramod K. Rastogi *Swiss Federal Institute of Technology, Lausanne, Switzerland*

V. P. Shchepinov *Physical Engineering Institute, Moscow, Russia*

Chandra Shakher *Indian Institute of Technology, New Delhi, India*

Rajpal S. Sirohi *Indian Institute of Technology, Madras, India*

Chandra S. Vikram *The University of Alabama in Huntsville, Huntsville, Alabama*

I. Yamaguchi *Institute of Physical and Chemical Research (RIKEN), Wako, Japan*

---

†Deceased.

SPECKLE METROLOGY

1

Theory and Applications of Speckle Displacement and Decorrelation

I. Yamaguchi

Institute of Physical and Chemical Research (RIKEN)
Wako, Japan

1.1 INTRODUCTION

When a laser-illuminated diffuse surface undergoes displacement and/or deformation, the speckles appearing in the scattered field or in its image show displacement accompanied by change in structure. Because of the high contrast of laser speckles, speckle displacement can be easily observed everywhere in space. It is not, however, uniform in space but shows remarkable dependence on observation position and on the optical system as well as on object deformation. The purpose of this chapter is to discuss these dynamic behaviors theoretically and to describe several methods for detection of speckle displacement. Then we discuss the application of speckle displacement to the measurement of surface displacement, strain, and velocity.

1.2 GENERAL SURVEY

1.2.1 Definition and Origin of Speckle Displacement and Decorrelation

Speckle displacement is quantitatively represented by the peak position of the cross-correlation function between the intensity distributions of speckle patterns before and after object displacement or deformation [1]. Figure 1.1 shows speckle patterns taken by a video camera before and after in-

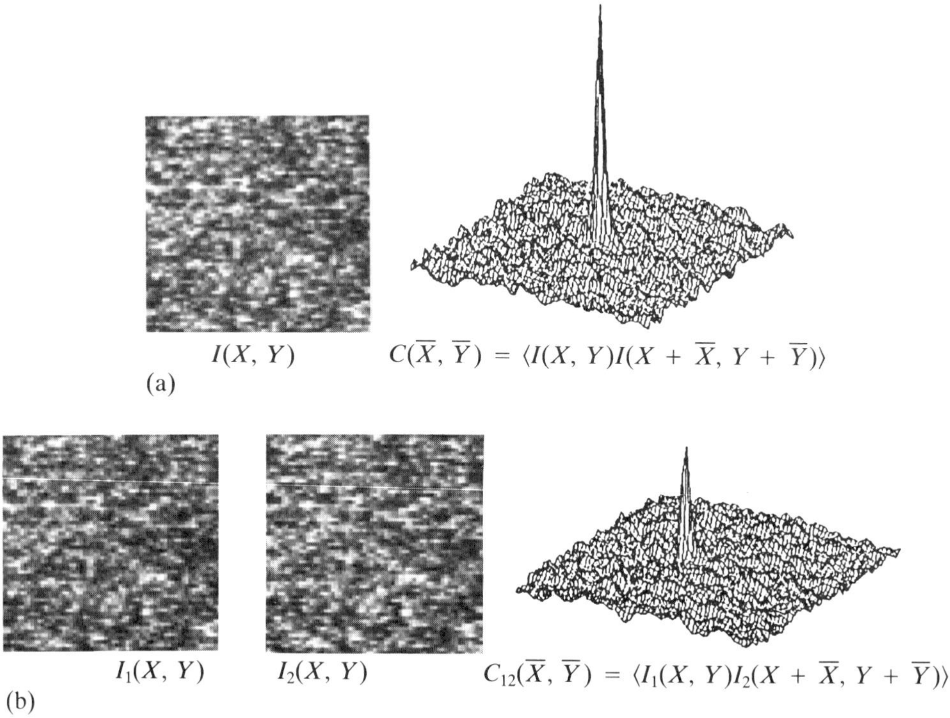

Figure 1.1 Speckle patterns appearing in video images of a paper sheet before and after in-plane translation and their autocorrelation (a) and cross-correlation (b).

plane translation of a paper. The two-dimensional cross-correlation was computed by a digital computer. For comparison, the autocorrelation of the pattern before the translation is also displayed. The peak of the autocorrelation is always located at zero. On the other hand, the peak position of the cross-correlation, which generally deviates from zero, corresponds to the speckle displacement whereas the decrease in peak height is associated with change in structure, so-called decorrelation.

Origins of speckle displacement and decorrelation are illustrated in Fig. 1.2 for the diffraction field (a) and the image field (b). In the diffraction field, speckle displacement $\boldsymbol{A}$ is caused by the change in phase relationship of interfering light waves by the summation of randomly directed vectors as shown in Fig. 1.2a. The phase change that results from the variation in optical path lengths due to object deformation brings about changes in the vector directions. Therefore, speckle displacement can be derived by cal-

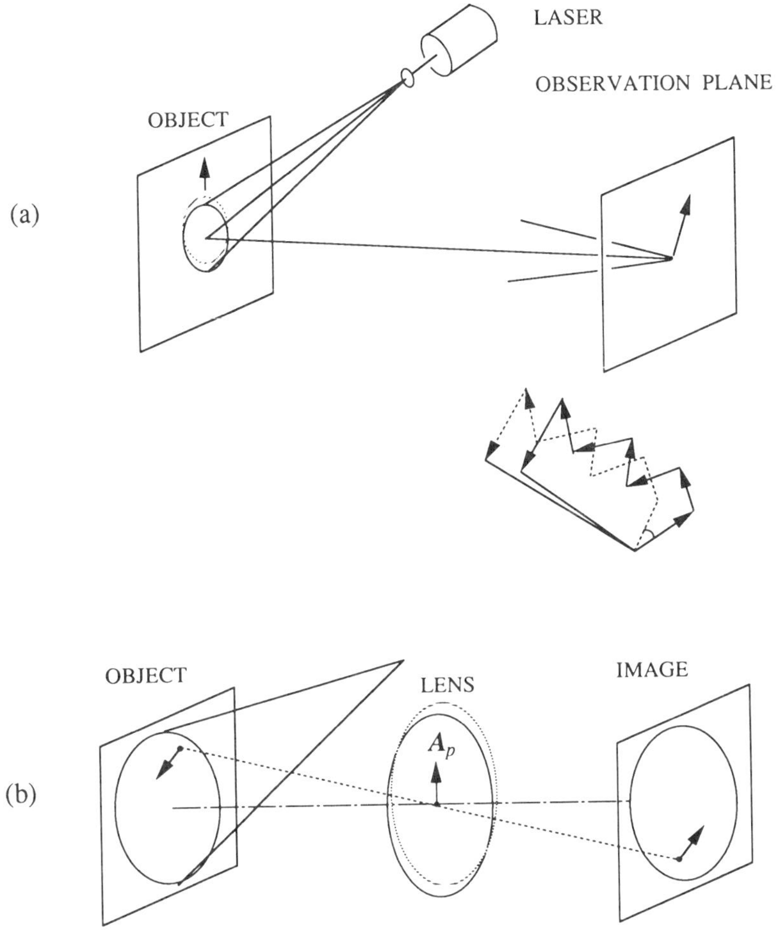

Figure 1.2 Origins of speckle displacement and decorrelation due to object displacement and deformation. (a) Diffraction field; (b) image field.

culating the change in optical path lengths appearing in the diffraction integral. In a configuration it can also be regarded as the displacement of diffraction orders from a random grating because speckle can be regarded as its randomly distributed diffraction orders. On the other hand, speckle decorrelation takes place owing to change in the membership of the interfering light waves. This effect depends on the diameter of an incident beam. In the case of speckle patterns in the image field, the decorrelation depends on the speckle displacement $\boldsymbol{A}p$ at an aperture of an imaging lens and its size as shown in Fig. 1.2b.

1.2.2 Roles of Speckle Displacement

Since speckle displacement is proportional to the amount of object deformation, speckles can be used as random markings on the surface. We should notice, however, the following points. First, displacement sensitivity depends strongly on optical configuration and type of deformation, and second the accompanying decorrelation deteriorates the conservation of the markings. Speckle displacement also indicates the shift of the spatial coherence area and decreases the coherence of the scattered fields that are superposed before and after object deformation in holographic interferometry and speckle interferometry as shown in Fig. 1.3. Thus visibility of the fringes observed in these methods of diffuse surface interferometry vanishes as speckle displacement $|\boldsymbol{A}|$ exceeds the mean speckle size δ [1].

1.2.3 Detection and Applications of Speckle Displacement

As mentioned above, by detecting speckle displacement we can measure displacement or deformation of diffusely reflecting surfaces. Methods of detection are classified into a photographic method and optoelectronic methods. Speckle photography is illustrated by Fig. 1.4, where a double-exposed negative (specklegram) recording the speckle patterns before and after speckle displacement is optically Fourier transformed [2–9]. Young's fringes appearing at the Fourier plane represent the direction and magnitude of speckle displacement. The first of the optoelectronic methods is the speckle correlation technique in which speckle patterns are detected by one- or two-dimensional image sensors whose output signals are cross-correlated with displacement in between by using a video memory and computers or a hardware correlator [10, 11]. The one-dimensional speckle correlation method is illustrated in Fig. 1.5. This detection scheme is used for measuring displacement, rotation, and strain of solid surfaces. The unique advantage of this method is simplicity in optical configurations and direct delivery of the desired quantities, for example, surface strain without differentiation of displacements [12, 13]. If we detect speckle displacement for a fixed time interval, surface velocity can also be determined. The second of the optoelectronic techniques, which is often used for velocity measurement, is a spatial filtering method in which a speckle pattern is detected by an array detector whose periodic elements are mutually connected as shown in Fig. 1.6, where a differential type is illustrated suppressing the dc component [14–16]. As the pattern moves, the detector delivers a signal that is periodic as a function of the pattern displacement

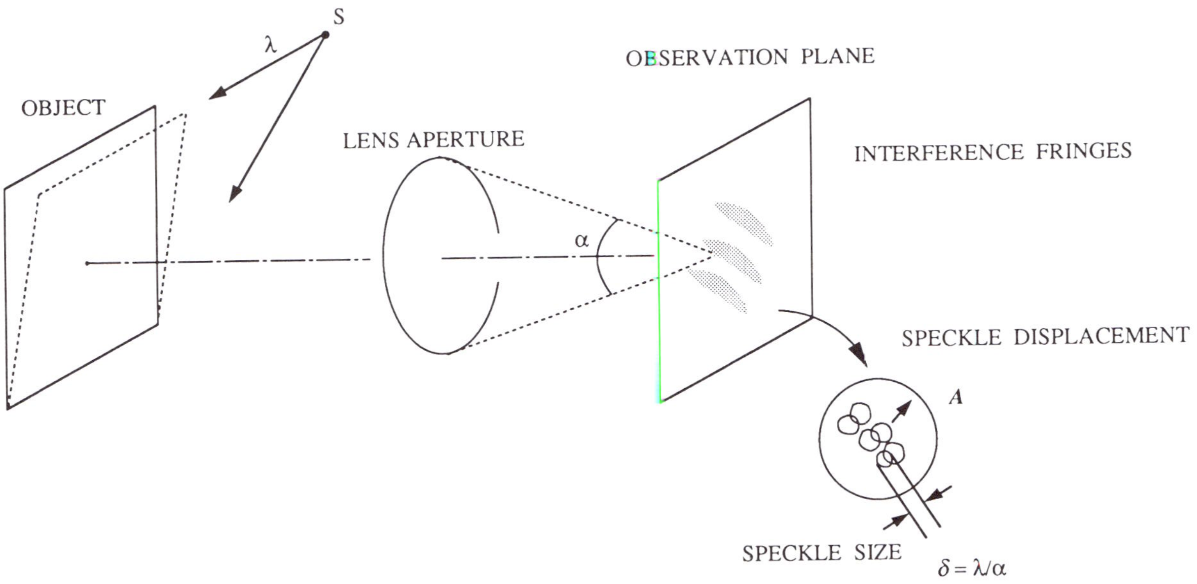

Figure 1.3 Role of speckle displacement in formation of interference fringes in holographic and speckle interferometry.

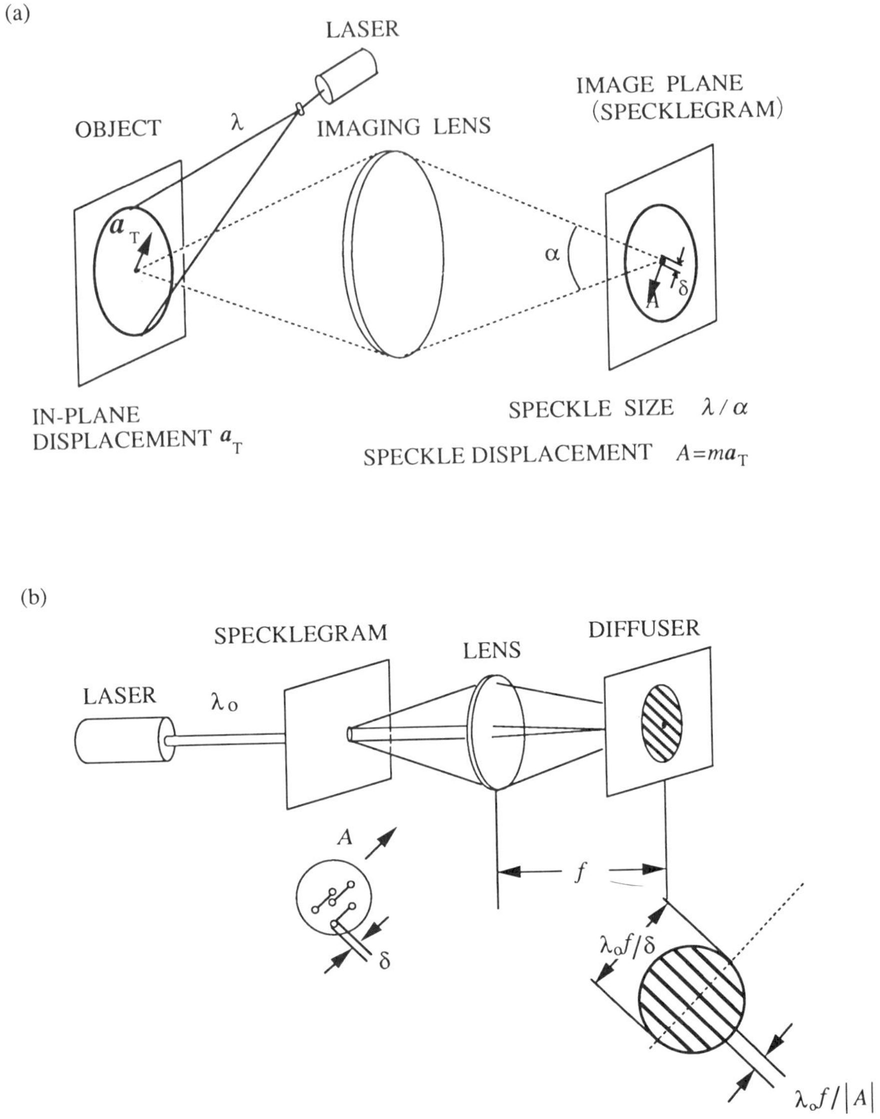

Figure 1.4 Principles of speckle photography. (a) Recording a specklegram; (b) observation of Young's fringes.

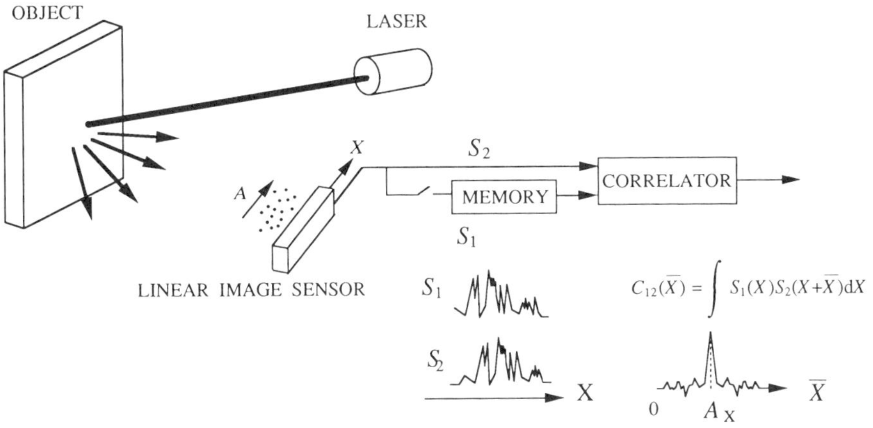

Figure 1.5 Detection of speckle displacement by the one-dimensional electronic speckle correlation method.

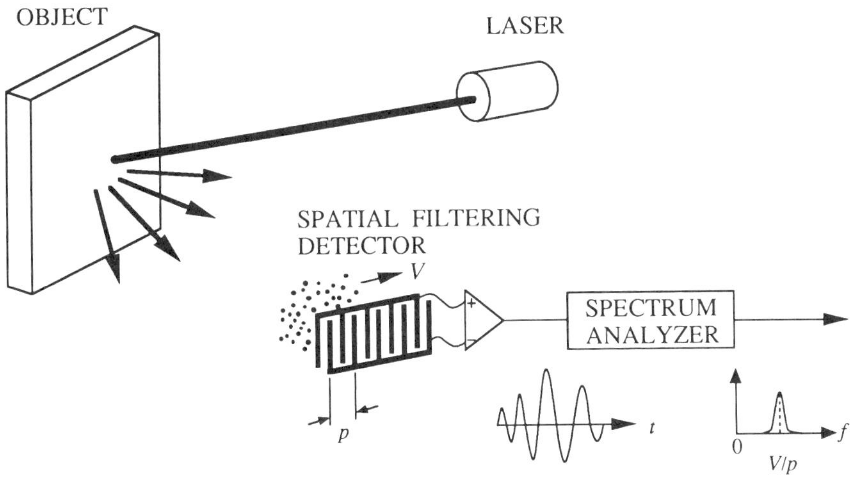

Figure 1.6 Velocity measurement by a spatial filtering detector. The differential setup is shown for eliminating the bias component.

normal to the array. The output signal is frequency analyzed, and the position of the spectrum peak is proportional to speckle velocity.

1.3 THEORY OF SPECKLE DISPLACEMENT AND DECORRELATION

1.3.1 Cross-Correlations of Intensity and Amplitude

Speckle displacement caused by surface displacement and/or deformation is theoretically derived from the peak position of the cross-correlation function $\langle I_1(X, Y)I_2(X + \overline{X}, Y + \overline{Y})\rangle$ between the intensity distributions at an observation plane before and after the displacement and/or deformation. The averaging procedure $\langle\ \rangle$ is performed over a statistical ensemble of the microscopic structures of a diffuse surface. In practice the averaging is carried out over an area including a sufficient number of speckles. When the correlation patch of the microstructure of the diffuse surface is much smaller than the beam diameter or resolution limit of the imaging system, and when the rms surface roughness is larger than the wavelength, speckles are generated by coherent superposition of a sufficient number of wavelets with random phase differences uniformly distributed between 0 and 2π. Then the complex amplitude distribution of the speckle field can be represented by the Gaussian random process, and the intensity cross-correlation function is expressed in terms of the cross-correlation of the complex amplitudes $U_1(X, Y)$ and $U_2(X, Y)$ through

$$\begin{aligned}
&\langle I_1(X, Y)I_2(X + \overline{X}, Y + \overline{Y})\rangle \\
&\quad = \langle I_1(X, Y)\rangle\langle I_2(X + \overline{X}, Y + \overline{Y})\rangle \\
&\qquad + |\langle U_1(X, Y)U_2^*(X + \overline{X}, Y + \overline{Y})\rangle|^2
\end{aligned} \tag{1.1}$$

The second term on the right-hand side is equal to the cross-correlation of the intensity fluctuation $\Delta I = I - \langle I\rangle$, that is,

$$|\langle U_1(\boldsymbol{R})U_2^*(\boldsymbol{R} + \overline{\boldsymbol{R}})\rangle|^2 = \langle \Delta I_1\,(\boldsymbol{R})\,\Delta I_2\,(\boldsymbol{R} + \overline{\boldsymbol{R}})\rangle \tag{1.2}$$

Thus derivation of the intensity cross-correlation reduces to that of the amplitude cross-correlation. The cross-correlation function of the complex amplitude was derived by applying the Fresnel–Kirchhoff diffraction integral to the wave propagation from the source point to the observation plane through the object. In this way speckle displacement and decorrelation were investigated for a general optical system including both diffraction and the image fields [17]. In the following the derivation that is based on the Fresnel–Kirchhoff diffraction theory is briefly traced.

1.3.2 Free-Space Geometry

We first study the simplest case of free-space geometry, shown in Fig. 1.7, where a position on the mean surface of the object is represented by $\boldsymbol{r}(x, y)$, and that on an observation plane is represented by $\boldsymbol{R}(X, Y)$. The original study [17] was later simplified and extended [18]. The essence is presented here. We denote the intensity distribution of the incident beam by $I_O(\boldsymbol{r})$, the random phase added to the reflected light by the surface roughness by $\phi(\boldsymbol{r})$, the optical path from the center of curvature of the wave-front S to the object point by $L_A(\boldsymbol{r})$, and the path length from the

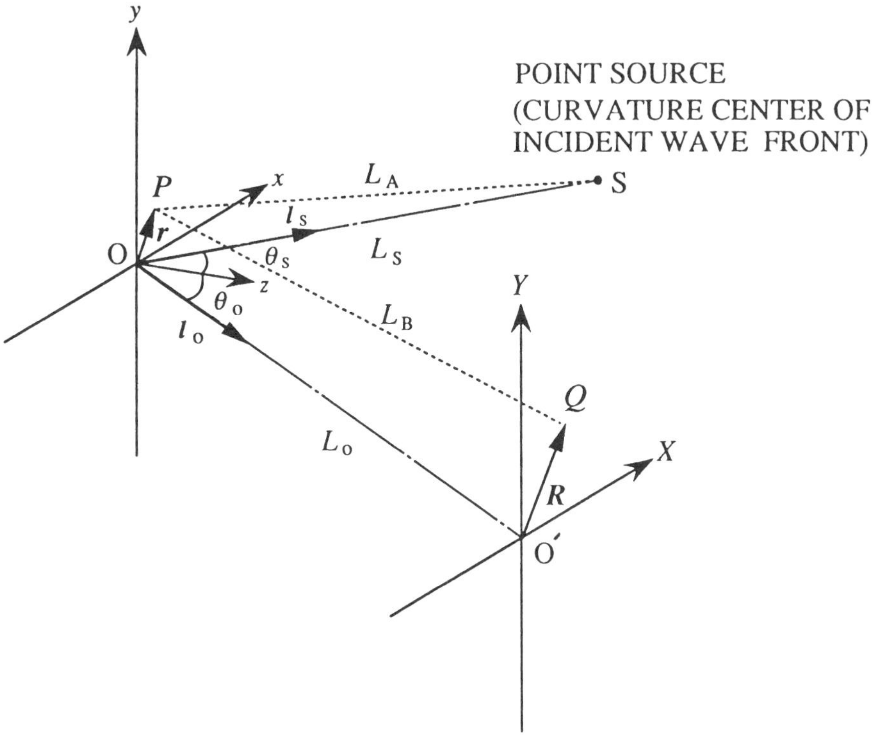

Figure 1.7 Coordinate systems for derivation of speckle displacement in the diffraction field.

object to the observation point $\boldsymbol{r}$ by $L_B(\boldsymbol{r}, \boldsymbol{R})$. Then the complex amplitude at an observation point before object deformation is given by

$$U_1(\boldsymbol{R}) = \int \sqrt{I_O(\boldsymbol{r})} \exp[i\phi(\boldsymbol{r})] \exp[ik\{L_A(\boldsymbol{r}) + L_B(\boldsymbol{r}, \boldsymbol{R})\}]\, d^2\boldsymbol{r} \qquad (1.3)$$

with the wave number $k = 2\pi/\lambda$ (λ wavelength). Equation (1.3) gives the superposition of the spherical waves scattered by the object point $\boldsymbol{r}$. After object movement or deformation, the point $\boldsymbol{r}$ is displaced to $\boldsymbol{r} + \boldsymbol{a}(\boldsymbol{r})$ by the displacement vector $\boldsymbol{a}(\boldsymbol{r})$ of the point $\boldsymbol{r}$. Then the complex amplitude is represented by

$$U_2(\boldsymbol{R}) = \int \sqrt{I_O(\boldsymbol{r} + \boldsymbol{a}_T)} \exp[i\phi(\boldsymbol{r})] \times \exp[ik\{L_A(\boldsymbol{r} + \boldsymbol{a}) + L_B(\boldsymbol{r} + \boldsymbol{a}, \boldsymbol{R})\}]\, d^2\boldsymbol{r} \qquad (1.4)$$

where $\boldsymbol{a}_T$ is the displacement component parallel to the mean object surface. After substitution of Eqs. (1.3) and (1.4), the amplitude cross-correlation becomes

$$\begin{aligned}\langle U_1(\boldsymbol{R})U_2^*(\boldsymbol{R} + \overline{\boldsymbol{R}})\rangle &= \int\int \sqrt{I_O(\boldsymbol{r})I_O(\boldsymbol{r}' + \boldsymbol{a}_T)}\langle \exp[i\{\phi(\boldsymbol{r}) - \phi(\boldsymbol{r}')\}]\rangle \\ &\quad \times \exp[ik\{L_A(\boldsymbol{r}) - L_A(\boldsymbol{r}' + \boldsymbol{a}) + L_B(\boldsymbol{r}, \boldsymbol{R}) \\ &\quad - L_B(\boldsymbol{r}' + \boldsymbol{a}, \boldsymbol{R} + \overline{\boldsymbol{R}})\}]\, d^2\boldsymbol{r}\, d^2\boldsymbol{r}' \qquad (1.5)\end{aligned}$$

Supposing that the surface roughness is much finer than the diameter of the incident spot and that the rms surface roughness is larger than the laser wavelength, we use the approximation with the two-dimensional delta function, that is,

$$\langle \exp[i\{\phi(\boldsymbol{r}) - \phi(\boldsymbol{r}')\}]\rangle = \delta^{(2)}(\boldsymbol{r} - \boldsymbol{r}') \qquad (1.6)$$

which leads to

$$\begin{aligned}\langle U_1(\boldsymbol{R})U_2^*(\boldsymbol{R} + \overline{\boldsymbol{R}})\rangle &= \int \sqrt{I_O(\boldsymbol{r})I_O(\boldsymbol{r} + \boldsymbol{a}_T)} \exp[ik\{L_A(\boldsymbol{r}) - L_A(\boldsymbol{r} + \boldsymbol{a}) + L_B(\boldsymbol{r}, \boldsymbol{R}) \\ &\quad - L_B(\boldsymbol{r} + \boldsymbol{a}, \boldsymbol{R} + \overline{\boldsymbol{R}})\}]\, d^2\boldsymbol{r} \qquad (1.7)\end{aligned}$$

When the displacement $\boldsymbol{a}_T$ is much smaller than the spot diameter, the peak position of $|\langle U_1(\boldsymbol{R})U_2^*(\boldsymbol{R} + \overline{\boldsymbol{R}})\rangle|$ with respect to $\overline{\boldsymbol{R}}$ depends on the exponent of Eq. (1.7), where we adopt the approximation based on the assumption $|\boldsymbol{a}| \ll L_A(\boldsymbol{r}) = \overline{\mathrm{SP}}$,

$$L_A(\boldsymbol{r} + \boldsymbol{a}) = L_A(\boldsymbol{r}) - \boldsymbol{l}_A(\boldsymbol{r})\cdot\boldsymbol{a}(\boldsymbol{r}) \qquad (1.8)$$

with the unit vector $\boldsymbol{l}_A(\boldsymbol{r})$ along $\overrightarrow{\mathrm{PS}}$. Similarly, we also approximate the distance from the displaced point to the observation plane such as

$$L_B(\boldsymbol{r} + \boldsymbol{a}, \boldsymbol{R} + \overline{\boldsymbol{R}}) = L_B(\boldsymbol{r}, \boldsymbol{R} + \overline{\boldsymbol{R}}) - \boldsymbol{l}_B(\boldsymbol{r}, \boldsymbol{R} + \overline{\boldsymbol{R}}) \cdot \boldsymbol{a}(\boldsymbol{r})$$

where we have introduced $L_B(\boldsymbol{r}, \boldsymbol{R} + \overline{\boldsymbol{R}}) = \overline{\mathrm{PQ}}$ and the unit vector $\boldsymbol{l}_B(\boldsymbol{r}, \boldsymbol{R} + \overline{\boldsymbol{R}})$ along $\overrightarrow{\mathrm{PQ}}$. Therefore, the exponent ϕ in Eq. (1.7) divided by the constant becomes

$$\begin{aligned}\Phi/k &= [\boldsymbol{l}_A(\boldsymbol{r}) + \boldsymbol{l}_B(\boldsymbol{r}, \boldsymbol{R} + \overline{\boldsymbol{R}})] \cdot \boldsymbol{a}(\boldsymbol{r}) + L_B(\boldsymbol{r}, \boldsymbol{R}) - L_B(\boldsymbol{r}, \boldsymbol{R} + \overline{\boldsymbol{R}}) \\ &\approx [\boldsymbol{l}_A(\boldsymbol{r}) + \boldsymbol{l}_B(\boldsymbol{r}, \boldsymbol{R})] \cdot \boldsymbol{a}(\boldsymbol{r}) + L_B(\boldsymbol{r}, \boldsymbol{R}) - L_B(\boldsymbol{r}, \boldsymbol{R} + \overline{\boldsymbol{R}})\end{aligned} \tag{1.9}$$

In the coordinate systems shown in Fig. 1.7, $\boldsymbol{l}_O$ and L_O are the direction and the distance of the center O′ of the observation plane (also assumed to be in the xz plane) from the center of the spot O. We adopt the approximation based on $L_O \gg |\boldsymbol{r}|, |\boldsymbol{R}|$,

$$\begin{aligned}L_B(\boldsymbol{r}, \boldsymbol{R}) &= \begin{matrix}[L_O^2 + X^2 + Y^2 + x^2 + y^2 - 2L_O(x \sin\theta_O + y) \\ - 2(Xx\cos\theta_O + Yy)]^{1/2}\end{matrix} \\ &\approx L_O - x\sin\theta_O - y - Xx\cos\theta_O/L_O - Yy/L_O \\ &\quad + (x^2 + y^2 + X^2 + Y^2)/2L_O\end{aligned} \tag{1.10}$$

where we assumed the vector $\boldsymbol{l}_O(\sin\theta_O, 0, \cos\theta_O) = \boldsymbol{l}_B(0, 0)$ to be in the xz plane and denoted the angle between $\boldsymbol{l}_O$ and the z axis by θ_O. Using Eq. (1.10), we get

$$L_B(\boldsymbol{r}, \boldsymbol{R}) - L_B(\boldsymbol{r}, \boldsymbol{R} + \overline{\boldsymbol{R}}) = \overline{X}x\cos\theta_O/L_O + \overline{Y}y/L_O \tag{1.11}$$

where we retained only the terms containing x and y. We further suppose that the deformation of the illuminated region is homogeneous; that is, the displacement distribution of each surface point is expressed by a linear function of position. Thus we expand the first term in Eq. (1.9) at the center of the illuminated region,

$$\begin{aligned}[\boldsymbol{l}_A(\boldsymbol{r}) + \boldsymbol{l}_B(\boldsymbol{r}, \boldsymbol{R})] \cdot \boldsymbol{a}(\boldsymbol{r}) &\approx [\boldsymbol{l}_A(\boldsymbol{r}) + \boldsymbol{l}_B(\boldsymbol{r}, 0)] \cdot \boldsymbol{a}(\boldsymbol{r}) \\ &= [\boldsymbol{l}_A(0) + \boldsymbol{l}_B(0, 0)] \cdot \boldsymbol{a}(0) \\ &\quad + \frac{\partial}{\partial x}\{[\boldsymbol{l}_A(\boldsymbol{r}) + \boldsymbol{l}_B(\boldsymbol{r}, 0)] \cdot \boldsymbol{a}(\boldsymbol{r})\}_{\boldsymbol{r}=0}x \\ &\quad + \frac{\partial}{\partial y}\{[\boldsymbol{l}_A(\boldsymbol{r}) + \boldsymbol{l}_B(\boldsymbol{r}, 0)] \cdot \boldsymbol{a}(\boldsymbol{r})\}_{\boldsymbol{r}=0}y\end{aligned} \tag{1.12}$$

Substitution of Eqs. (1.11) and (1.12) into Eq. (1.9) yields a result that can be summarized as

$$\frac{\Phi}{k} = (\boldsymbol{l}_S + \boldsymbol{l}_O)\cdot\boldsymbol{a}(0) + \frac{x\cos\theta_O}{L_O}(\overline{X} - A_X) + \frac{y}{L_O}(\overline{Y} - A_Y) \tag{1.13}$$

where $\boldsymbol{l}_S = \boldsymbol{l}_A(0)$ is the unit vector directed from O to S. In Eq. (1.13) the vector (A_X, A_Y) is given by

$$\begin{aligned} A_X &= -\frac{L_O}{\cos\theta_O}\left[\frac{\partial}{\partial x}\{[\boldsymbol{l}_A(\boldsymbol{r}) + \boldsymbol{l}_B(\boldsymbol{r}, 0)]\cdot\boldsymbol{a}(\boldsymbol{r})\}\right]_{x=y=0} \\ &= a_x\left(\frac{L_O\cos^2\theta_S}{L_S\cos\theta_O} + \cos\theta_O\right) - a_z\left(\frac{L_O\cos\theta_S\sin\theta_S}{L_S\cos\theta_O} + \sin\theta_O\right) \\ &\quad - L_O\left[\varepsilon_{xx}\left(\frac{\sin\theta_S}{\cos\theta_O} + \tan\theta_O\right) - \Omega_y\left(\frac{\cos\theta_S}{\cos\theta_O} + 1\right)\right] \end{aligned} \tag{1.14}$$

and

$$\begin{aligned} A_Y &= -L_O\left[\frac{\partial}{\partial y}\{[\boldsymbol{l}_A(\boldsymbol{r}) + \boldsymbol{l}_B(\boldsymbol{r}, 0)]\cdot\boldsymbol{a}(\boldsymbol{r})\}\right]_{x=y=0} \\ &= a_y\left(\frac{L_O}{L_S} + 1\right) - L_O[\varepsilon_{xy}(\sin\theta_S + \sin\theta_O) \\ &\quad - \Omega_x(\cos\theta_S + \cos\theta_O) - \Omega_z(\sin\theta_S + \sin\theta_O)] \end{aligned} \tag{1.15}$$

where (a_x, a_y, a_z), $(\Omega_x, \Omega_y, \Omega_z)$, and $(\varepsilon_{xx}, \varepsilon_{xy}, \varepsilon_{yy})$ are the translation, rotation, and strain coefficients of the illuminated object region, respectively. These coefficients are related to the displacement vector through

$$(a_x, a_y, a_z) = [a_x(0), a_y(0), a_z(0)] \tag{1.16}$$

$$(\Omega_x, \Omega_y, \Omega_z) = \left\{\left(\frac{\partial a_z}{\partial y}\right)_0, -\left(\frac{\partial a_z}{\partial x}\right)_0, \frac{1}{2}\left[\left(\frac{\partial a_y}{\partial x}\right)_0 - \left(\frac{\partial a_x}{\partial y}\right)_0\right]\right\} \tag{1.17}$$

and

$$\begin{aligned} \varepsilon_{xx} &= \left(\frac{\partial a_x}{\partial x}\right)_0, \qquad \varepsilon_{yy} = \left(\frac{\partial a_y}{\partial y}\right)_0, \\ \varepsilon_{xy} &= \varepsilon_{yx} = \frac{1}{2}\left[\left(\frac{\partial a_x}{\partial y}\right)_0 + \left(\frac{\partial a_y}{\partial x}\right)_0\right] \end{aligned} \tag{1.18}$$

Furthermore, θ_S and L_S are the incident angle and the curvature of radius of the wavefront of the incident laser beam. By substituting Eq. (1.13) into Eq. (1.7) and using Eq. (1.2), we reach

$$\langle \Delta I_1 (X, Y)\, \Delta I_2 (X + \overline{X}, Y + \overline{Y})\rangle$$
$$= \left| \iint [I_O(x, y) I_O(x + a_x, y + a_y)]^{1/2} \times \exp\left\{ -\frac{ik[x(\overline{X} - A_X)\cos\theta_O + y(\overline{Y} - A_Y)]}{L_O} \right\} dx\, dy \right|^2 \tag{1.19}$$

For small values of a_x and a_y compared with the spot diameter, the cross-correlation given by Eq. (1.19) takes its maximum at $\overline{X} = A_X$ and $\overline{Y} = A_Y$, and thus the vector (A_X, A_Y) is the speckle displacement caused by the object deformation. Speckle decorrelation is related to the maximum value

$$\langle \Delta I_1 (X, Y)\, \Delta I_2 (X + \overline{X}, Y + \overline{Y})\rangle$$
$$= \left| \iint [I_O(x, y) I_O(x + a_x, y + a_y)]^{1/2}\, dx\, dy \right|^2 \tag{1.20}$$

which is dependent on the in-plane translation compared with the beam diameter. Thus the degree of correlation is proportional to the overlapping area of the laser spots shifted by the object translation.

1.3.4 Generalization

In general configurations that contain lenses forming diffraction patterns or images as shown in Fig. 1.8, it is convenient to use the linear transmission formulation of the complex amplitude to the observation plane [17]. The complex amplitude at the observation point $\boldsymbol{R}$ can be represented as a linear superposition of the complex amplitude arising from each point of the object. Hence the complex amplitudes before and after object deformation are given by

$$U_1(\boldsymbol{R}) = \int \sqrt{I_O(\boldsymbol{r})} \exp[i\phi(\boldsymbol{r})] \exp[ikL_A(\boldsymbol{r})] K(\boldsymbol{R}; \boldsymbol{r})\, d^2\boldsymbol{r} \tag{1.21}$$

and

$$U_2(\boldsymbol{R}) = \int \sqrt{I_O(\boldsymbol{r} + \boldsymbol{a}_T)} \exp[i\phi(\boldsymbol{r})] \times \exp[ikL_A(\boldsymbol{r} + \boldsymbol{a})] K(\boldsymbol{R}; \boldsymbol{r} + \boldsymbol{a})\, d^2\boldsymbol{r} \tag{1.22}$$

where $K(\boldsymbol{R}; \boldsymbol{r})$ denotes the system transmission function that is the distribution of the complex amplitude caused by a point source positioned at

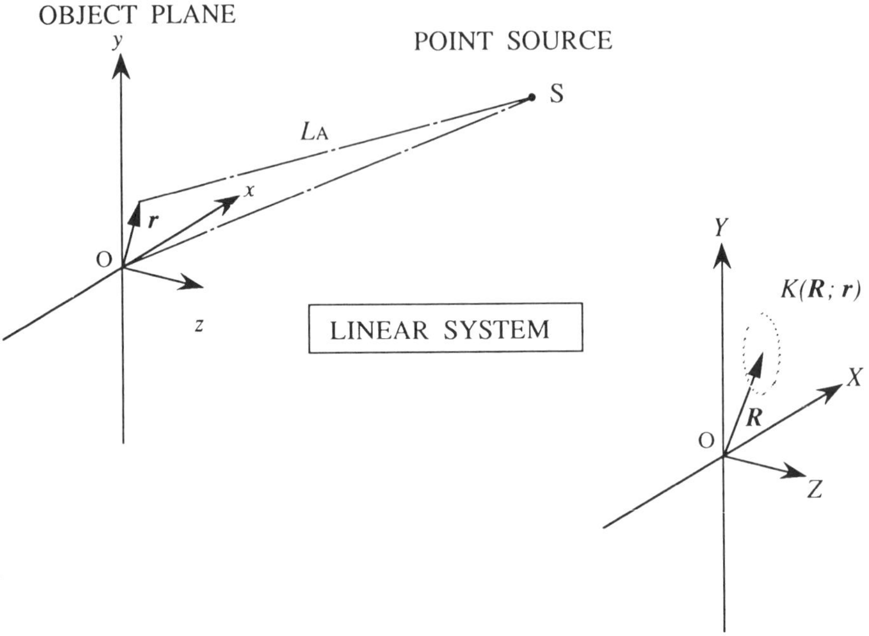

Figure 1.8 Linear transmission system for derivation of speckle displacement in general optical systems.

the object point $\boldsymbol{r}$. With these expressions, the cross-correlation function of the complex amplitude becomes

$$\begin{aligned}&\langle U_1(\boldsymbol{R})U_2^*(\boldsymbol{R} + \overline{\boldsymbol{R}})\rangle \\ &\quad = \int \sqrt{I_O(\boldsymbol{r})I_O(\boldsymbol{r} + \boldsymbol{a}_T)} \\ &\qquad \times \exp\,[ik\boldsymbol{l}_A(\boldsymbol{r})\cdot\boldsymbol{a}(\boldsymbol{r})]K(\boldsymbol{R};\boldsymbol{r})K^*(\boldsymbol{R} + \overline{\boldsymbol{R}};\boldsymbol{r} + \boldsymbol{a})\,d^2\boldsymbol{r} \qquad (1.23)\end{aligned}$$

where we assumed that the correlation patch of the microscopic reflection function $\exp\,[i\phi(\boldsymbol{r})]$ is much smaller than that of the system transmission function and also used the approximation of Eq. (1.8).

1.3.5 Image Field

For studying speckle displacement and decorrelation in the image field shown in Fig. 1.9, we only consider the region with uniform average in-

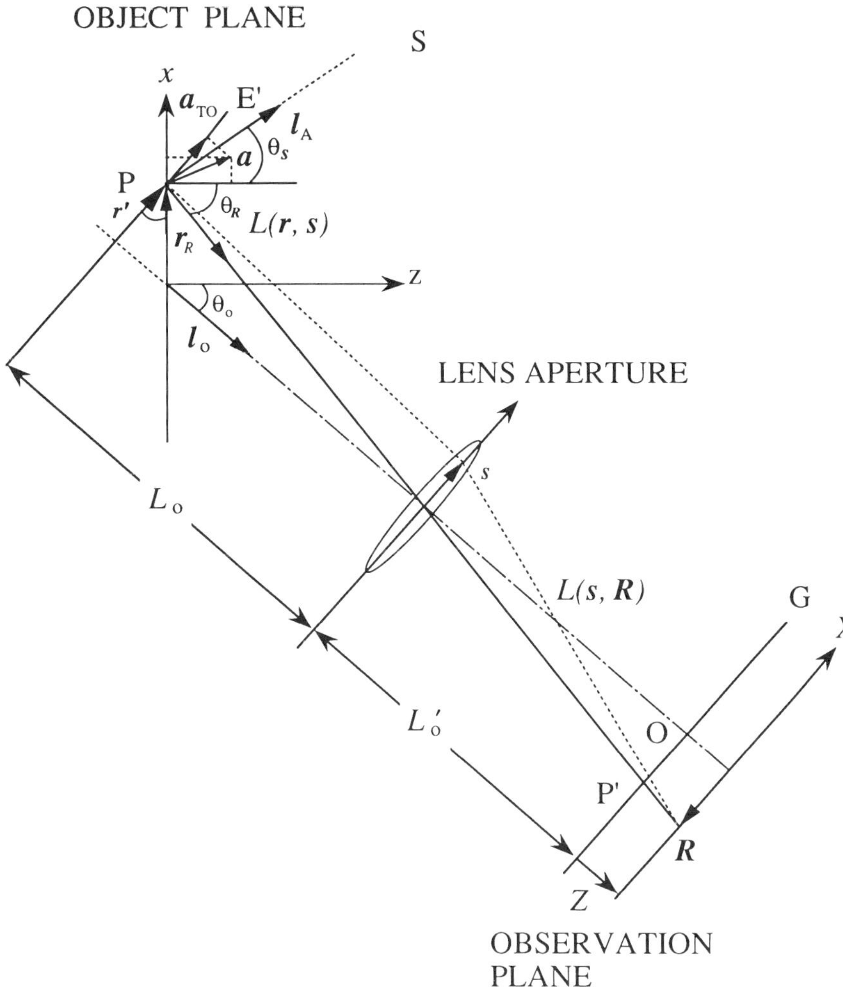

Figure 1.9 Coordinate systems for derivation of speckle displacement in the image field.

tensity that leads to $I_O(\boldsymbol{r}) = I_O =$ constant. Therefore, Eq. (1.23) becomes

$$\langle U_1(\boldsymbol{R})U_2^*(\boldsymbol{R} + \overline{\boldsymbol{R}})\rangle = I_O \int \exp\,[ik\boldsymbol{l}_A(\boldsymbol{r})\cdot\boldsymbol{a}(\boldsymbol{r})]K(\boldsymbol{R};\,\boldsymbol{r})K^*(\boldsymbol{R} + \overline{\boldsymbol{R}};\,\boldsymbol{r} + \boldsymbol{a})\,d^2\boldsymbol{r} \qquad (1.24)$$

This is an extension of an earlier analysis that was restricted to an imaging system in the normal direction [17]. The optical axis of the imaging system is inclined at an angle θ_O with respect to the z axis. The position of the observation plane is represented by an amount of defocus Z from the Gaussian plane G of the object point P, which is located at the intersection of the principal ray through the observation point $(X, Y, Z) = (\boldsymbol{R}, Z)$ with the object plane. The system transmission function can be represented by the Fresnel–Kirchhoff diffraction integral for the lens aperture and is thus given by

$$\begin{aligned} &\boldsymbol{K}[\boldsymbol{R}, Z; \boldsymbol{r} + \boldsymbol{a}(\boldsymbol{r})] \\ &\quad = \int \exp\,[ik\,|\boldsymbol{L}(\boldsymbol{r}, \mathbf{s}) - \boldsymbol{a}|]P(\mathbf{s}) \\ &\qquad \times \exp\left(-ik\frac{|\mathbf{s}|^2}{2F}\right) \exp\,[ik\,|(L'_O + Z)\boldsymbol{l}_O + \boldsymbol{R} - \mathbf{s}|]\,d^2\mathbf{s} \end{aligned} \tag{1.25}$$

where $P(\mathbf{s})$ and F are the pupil function and the focal length of the lens, respectively. We use the approximation

$$|\boldsymbol{L}(\boldsymbol{r}, \mathbf{s}) - \boldsymbol{a}(\boldsymbol{r})| = L(\boldsymbol{r}, \mathbf{s}) - \boldsymbol{l}(\boldsymbol{r}, \mathbf{s})\cdot\boldsymbol{a}(\boldsymbol{r}) \approx L(\boldsymbol{r}, \mathbf{s}) - \boldsymbol{l}(\boldsymbol{r}, 0)\cdot\boldsymbol{a}(\boldsymbol{r}) \tag{1.26}$$

with $L(\boldsymbol{r}, \mathbf{s}) = |\boldsymbol{L}(\boldsymbol{r}, \mathbf{s})|$ and $\boldsymbol{l}(\boldsymbol{r}, \mathbf{s}) = \boldsymbol{L}(\boldsymbol{r}, \mathbf{s})/L(\boldsymbol{r}, \mathbf{s})$. We adopt a parabolic approximation

$$L(\boldsymbol{r}, \mathbf{s}) = L_O + |\boldsymbol{r}' - \mathbf{s}|^2/2L_O \tag{1.27}$$

with the projection $\boldsymbol{r}'$ of $\boldsymbol{r}$ to the plane E' normal to the optical axis and the axial distance L_O from the point P to the aperture. The same approximation is applied to the distance between the aperture and the observation point,

$$\begin{aligned} L(\mathbf{s}, \boldsymbol{R}) &= |(L'_O + Z)\boldsymbol{l}_O + \boldsymbol{R} - \mathbf{s}| \\ &= L'_O + Z + |\boldsymbol{R} - \mathbf{s}|^2/2(L'_O + Z) \end{aligned} \tag{1.28}$$

where $\boldsymbol{l}_O$ is the unit vector along the optical axis and L'_O is the distance between the lens aperture and the Gaussian image plane G. Substituting these expressions into Eq. (1.21) results in

$$\begin{aligned} &\boldsymbol{K}[\boldsymbol{R}, Z; \boldsymbol{r} + \boldsymbol{a}(\boldsymbol{r})] \\ &\quad = \exp\left\{ik\left[L_O + L'_O + Z + \frac{|\boldsymbol{r}'|^2}{2L_O} + \frac{|\boldsymbol{R}|^2}{2(L'_O + Z)}\right]\right\} \\ &\qquad \times \int P(\mathbf{s}) \exp\left[-ik\frac{|\mathbf{s}|^2 Z}{2L'_O(L'_O + Z)}\right] \\ &\qquad \times \exp\left\{-ik\left[\mathbf{s}\cdot\left(\frac{\boldsymbol{r}'}{L_O} + \frac{\boldsymbol{R}}{L'_O + Z}\right) - \boldsymbol{l}(\boldsymbol{r}, \mathbf{s})\cdot\boldsymbol{a}(\boldsymbol{r})\right]\right\} d^2\mathbf{s} \end{aligned} \tag{1.29}$$

where we considered the imaging relation $1/L_O + 1/L'_O = 1/F$. We calculate the cross-correlation function of complex amplitude by substituting Eq. (1.29) into Eq. (1.24). Initially we have

$$\begin{aligned}&\langle U_1(\boldsymbol{R}, Z)U_2^*(\boldsymbol{R} + \overline{\boldsymbol{R}}, Z)\rangle \\ &\quad = \exp\left[ik\frac{\boldsymbol{R}\cdot\overline{\boldsymbol{R}}}{L'_O + Z}\right]\int\int\int P(\boldsymbol{s}_a)P^*(\boldsymbol{s}_b)\exp\left[-ik\frac{Z(|\boldsymbol{s}_a|^2 - |\boldsymbol{s}_b|^2)}{2L'_O(L'_O + Z)}\right] \\ &\qquad \times \exp\left\{-ik\left[\frac{\boldsymbol{r}'\cdot(\boldsymbol{s}_a - \boldsymbol{s}_b)}{L_O} + \frac{\boldsymbol{R}\cdot\boldsymbol{s}_a - (\boldsymbol{R} + \overline{\boldsymbol{R}})\cdot\boldsymbol{s}_b}{L'_O + Z}\right.\right. \\ &\qquad \left.\left. - \boldsymbol{m}(\boldsymbol{r}, \boldsymbol{s}_b)\cdot\boldsymbol{a}(\boldsymbol{r})\right]\right\} d^2\boldsymbol{s}_a\, d^2\boldsymbol{s}_b\, d^2\boldsymbol{r}' \end{aligned} \tag{1.30}$$

where

$$\boldsymbol{m}(\boldsymbol{r}, \boldsymbol{s}) = \boldsymbol{l}_A(\boldsymbol{r}) + \boldsymbol{l}(\boldsymbol{r}, \boldsymbol{s}) \tag{1.31}$$

is the vector along the bisector between $\boldsymbol{l}_A$ and $\boldsymbol{l}$.

To simplify Eq. (1.30), we consider that both $U_1(\boldsymbol{R}, Z)$ and $U_2(\boldsymbol{R}, \overline{\boldsymbol{R}}, Z)$ receive contributions solely from limited object regions. These regions coincide with those covered by the diffraction images of the lens aperture that are formed by point sources positioned at $(\boldsymbol{R}, Z)$ and $(\boldsymbol{R} + \overline{\boldsymbol{R}}, Z)$. As long as $|\overline{\boldsymbol{R}}|$ is small enough and Z is much less than L'_O, the region has such a small extent that within it the following linear approximation applies:

$$\boldsymbol{m}(\boldsymbol{r}, \boldsymbol{s})\cdot\boldsymbol{a}(\boldsymbol{r}) = \boldsymbol{m}(\boldsymbol{r}_R, \boldsymbol{s})\cdot\boldsymbol{a}(\boldsymbol{r}_R) + [\nabla\boldsymbol{m}(\boldsymbol{r}, \boldsymbol{s})\cdot\boldsymbol{a}(\boldsymbol{r})]_{\boldsymbol{r}_R}\cdot(\boldsymbol{r} - \boldsymbol{r}_R) \tag{1.32}$$

where $\boldsymbol{r}_R = -L_O\boldsymbol{R}/(L'_O + Z)$ is the object point conjugate to the observation point. This approximation is valid when the deformation in the contributing region is homogeneous and when the wave front incident at the region can be approximated by a paraboloid. Considering that the exponent in the integrand of Eq. (1.30) contains only the terms less than the third order of $\boldsymbol{r}$, $\boldsymbol{s}$, and $\boldsymbol{R}$, we deform the first term on the right-hand side of Eq. (1.31) such that

$$\boldsymbol{m}(\boldsymbol{r}_R, \boldsymbol{s})\cdot\boldsymbol{a}(\boldsymbol{r}_R) = \boldsymbol{l}_A(\boldsymbol{r}_R)\cdot\boldsymbol{a}(\boldsymbol{r}_R) + \frac{(\boldsymbol{s} - \boldsymbol{r}'_R)}{L_O}\cdot\boldsymbol{a}_{\mathrm{TO}}(\boldsymbol{r}_R) \tag{1.33}$$

where $\boldsymbol{r}'_R$ is the position vector of $\boldsymbol{r}_R$ in the plane E' and $\boldsymbol{a}_{\mathrm{TO}}$ means the displacement component transverse to the optic axis of the lens. The second term of Eq. (1.32) becomes

$$[\nabla\boldsymbol{m}(\boldsymbol{r}, \boldsymbol{s})\cdot\boldsymbol{a}(\boldsymbol{r})]_{\boldsymbol{r}_R}\cdot(\boldsymbol{r} - \boldsymbol{r}_R) \approx [\nabla\boldsymbol{m}(\boldsymbol{r}, 0)\cdot\boldsymbol{a}(\boldsymbol{r})]_{\boldsymbol{r}_R}\cdot(\boldsymbol{r} - \boldsymbol{r}_R) \tag{1.34}$$

We substitute Eq. (1.32) with Eqs. (1.33) and (1.34) into Eq. (1.30). The integration over $\boldsymbol{r}'$ yields the two-dimensional delta function

$$\delta^{(2)}\{\boldsymbol{s}_a - \boldsymbol{s}_b - L_O[\nabla \boldsymbol{m}(\boldsymbol{r}, 0)\cdot\boldsymbol{a}(\boldsymbol{r})]_{\boldsymbol{r}_R}\} \tag{1.35}$$

which vanishes except for

$$\boldsymbol{s}_a = \boldsymbol{s}_b + L_O[\nabla \boldsymbol{m}(\boldsymbol{r}, 0)\cdot\boldsymbol{a}(\boldsymbol{r})]_{\boldsymbol{r}_R} \tag{1.36}$$

After integration over $\boldsymbol{s}_a$ and neglecting the second-order terms of $\boldsymbol{a}$ and constant factors of proportionality, we finally obtain

$$\begin{aligned}\langle U_1(\boldsymbol{R}, Z)U_2^*(\boldsymbol{R} + \overline{\boldsymbol{R}}, Z)\rangle &= \exp\left\{ik\left[\boldsymbol{m}(\boldsymbol{r}_R, 0)\cdot\boldsymbol{a}(\boldsymbol{r}_R) + \frac{\boldsymbol{R}\cdot\overline{\boldsymbol{R}}}{L_O' + Z}\right]\right\} \\ &\times \int P^*(\boldsymbol{s})P(\boldsymbol{s} + L_O[\nabla \boldsymbol{m}(\boldsymbol{r}, 0)\cdot\boldsymbol{a}]_{\boldsymbol{r}_R}) \\ &\times \exp\left\{ik\,\frac{\boldsymbol{s}}{L_O' + Z}\cdot[\overline{\boldsymbol{R}} - \boldsymbol{A}(\boldsymbol{R}, Z)]\right\} d^2\boldsymbol{s}\end{aligned} \tag{1.37}$$

where the correlation peak is located at $\overline{\boldsymbol{R}} = \boldsymbol{A}$ with

$$\boldsymbol{A} = -\frac{L_O' + Z}{L_O}\,\boldsymbol{a}_{\mathrm{TO}} + \frac{L_O Z}{L_O'}\,[\nabla \boldsymbol{m}(\boldsymbol{r}, 0)\cdot\boldsymbol{a}]_{\boldsymbol{r}_R} \tag{1.38}$$

which is the speckle displacement. In Ref. 17 the second-order term of $\boldsymbol{s}$ was also derived, however, for the configuration where the optical axis is normal to the object.

The displacement vector given by Eq. (1.38) consists of two components, one independent of defocus Z and the other proportional to the amount of defocus. It can also be written as

$$\boldsymbol{A} = -\frac{L_O'}{L_O}\,\boldsymbol{a}_{\mathrm{TO}} - \frac{Z}{L_O'}\left\{\frac{L_O'}{L_O}\,\boldsymbol{a}_{\mathrm{TO}} - L_O[\nabla \boldsymbol{m}(\boldsymbol{r}, 0)\cdot\boldsymbol{a}]_{\boldsymbol{r}_R}\right\} \tag{1.39}$$

In the coordinate systems shown in Fig. 1.9 we can easily get the relations for the displacement component in the plane E',

$$a_{\mathrm{TO}X} = a_x \cos\theta_R + a_z \sin\theta_R \tag{1.40}$$

and

$$a_{\mathrm{TO}Y} = a_y \tag{1.41}$$

We also use Eqs. (1.14) and (1.15) for $[\nabla \boldsymbol{m}(\boldsymbol{r}, 0)\cdot\boldsymbol{a}]_{r_R}$ to reach the components of the speckle displacement vector,

$$A_X = m(a_x \cos\theta_R + a_z \sin\theta_R) - \frac{Z}{L'_O}\Bigg\{m(a_x \cos\theta_R + a_z \sin\theta_R) - a_x\left(\frac{L_O \cos^2\theta_S}{L_S \cos\theta_R} + \cos\theta_R\right) + a_z\left(\frac{L_O \cos\theta_S \sin\theta_S}{L_S \cos\theta_R} + \sin\theta_R\right) + L_O\left[\varepsilon_{xx}\left(\frac{\sin\theta_S}{\cos\theta_R} + \tan\theta_R\right) - \Omega_y\left(\frac{\cos\theta_S}{\cos\theta_R} + 1\right)\right]\Bigg\} \tag{1.42}$$

and

$$A_Y = ma_y - \frac{Z}{L'_O}\Bigg\{ma_y - a_y\left(\frac{L_O}{L_S} + 1\right) + L_O[\varepsilon_{xy}(\sin\theta_S + \sin\theta_R) - \Omega_x(\cos\theta_S + \cos\theta_R) - \Omega_z(\sin\theta_S + \sin\theta_R)]\Bigg\} \tag{1.43}$$

where m is the imaging magnification $m = -L'_O/L_O$ for the Gaussian plane and the variables representing the incident angle and the curvature radius of the illuminating beam as well as object deformation denote their values at $x_R = -L_O X/(L'_O + Z)$, $y_R = -L_O Y/(L'_O + Z)$.

The peak value of the cross-correlation that governs the change in speckle structure accompanying the displacement is given by

$$\langle \Delta I_1(\boldsymbol{R}, Z)\Delta I_2(\boldsymbol{R} + \boldsymbol{A}, Z)\rangle = \left|\int P^*(\mathbf{s})P(\mathbf{s} + \boldsymbol{A}_P)\, d^2\mathbf{s}\right|^2 \tag{1.44}$$

where

$$\boldsymbol{A}_p = L_O[\boldsymbol{m}(\boldsymbol{r}, 0)\cdot\boldsymbol{a}]_{r_R} \tag{1.45}$$

is just equal to the speckle displacement on the pupil plane that is located in the diffraction field of the object. This is easily seen from reference to Eqs. (1.14) and (1.15). Equation (1.44) says that the correlation peak height is proportional to the squared MTF (modulation transfer function) of the imaging system where the spatial frequency is proportional to the speckle displacement at the pupil plane. Hence the speckle decorrelation effect in the image field depends on the speckle displacement at the pupil plane compared with the aperture diameter. It is also accelerated by the existence of lens aberrations [19–21].

1.4 DETECTION OF SPECKLE DISPLACEMENT

The displacement of speckle, derived above, can be detected by various methods, either photographic or optoelectronic. In speckle photography, speckle displacement is evaluated from the Young's fringes that result from the optical Fourier transform of a double-exposed negative, called a specklegram. The optoelectronic methods consist of electronic correlation where the peak position of the cross-correlation function of the speckle image is detected and the spatial filtering method where the change in the total amount of light is utilized as a signal that is transmitted through a periodic slit array or detected by a periodic detector array.

1.4.1 Speckle Photography

The basic principle of speckle photography is shown in Fig. 1.10, where measurement of in-plane displacement of a plane surface is depicted. Here we discuss formation of Young's fringes whose direction is normal to that of the speckle displacement and whose spacing is inversely proportional

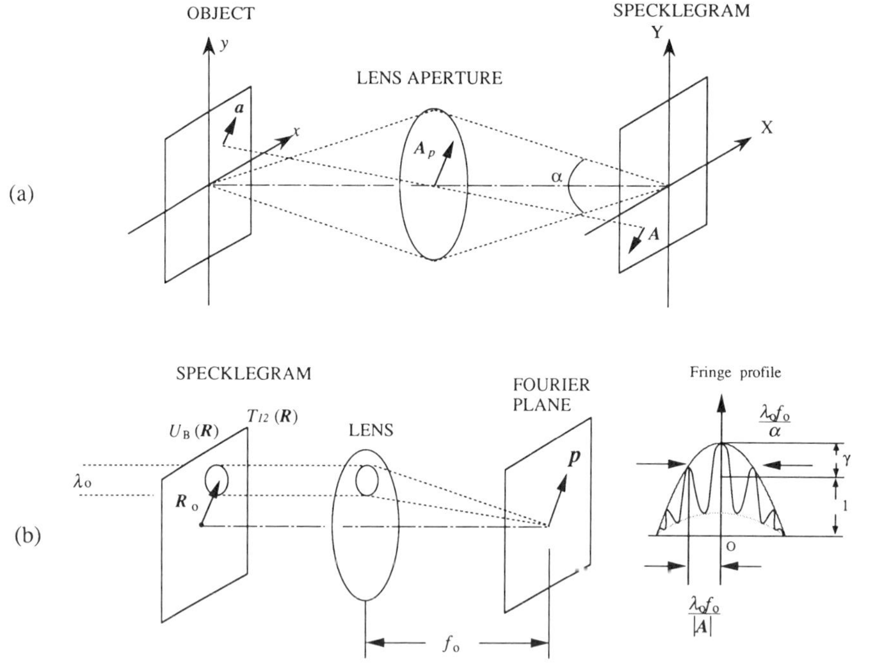

Figure 1.10 Coordinate systems for analyzing formation of Young's fringes. (a) Specklegram recording; (b) formation of Young's fringes.

to the magnitude of displacement. This has been studied in terms of the correlation properties of speckle [21]. The intensity distribution of the Young's fringes is expressed as the statistical average of the intensity at the Fourier plane,

$$J_Y(\boldsymbol{p}) = \left\langle \left| \int U_B(\boldsymbol{R} - \boldsymbol{R}_O) T_{12}(\boldsymbol{R}) \exp\left(-i2\pi \frac{\boldsymbol{R}\cdot\boldsymbol{p}}{\lambda_O f_O} \right) d^2\boldsymbol{R} \right|^2 \right\rangle \tag{1.46}$$

where $U_B(\boldsymbol{R})$ is the distribution of the complex amplitude of the reading beam, $\boldsymbol{R}_O$ the center of the beam incidence, and $T_{12}(\boldsymbol{R})$ the transmission function of the specklegram to be represented by

$$T_{12}(\boldsymbol{R}) = T_O - \eta[I_1(\boldsymbol{R}, Z) + I_2(\boldsymbol{R}, Z)] \tag{1.47}$$

where we assume that the specklegram is recorded at image or at the defocus plane. We substitute Eq. (1.47) into Eq. (1.46) and use the relationship valid for small $|\overline{\boldsymbol{R}}|$

$$C_{12}(\overline{\boldsymbol{R}}) = \langle I_1(\boldsymbol{R}, Z) I_2(\boldsymbol{R} + \overline{\boldsymbol{R}}, Z) \rangle = C(0) + \gamma C(\overline{\boldsymbol{R}} - \boldsymbol{A}) \tag{1.48}$$

where $C(\overline{\boldsymbol{R}})$ is the autocorrelation of intensity fluctuation and is given by

$$\begin{aligned} C(\overline{\boldsymbol{R}}) &= \langle \Delta I\,(\boldsymbol{R}, Z)\, \Delta I\,(\boldsymbol{R} + \overline{\boldsymbol{R}}, Z) \rangle \\ &= \left| \int |P(\boldsymbol{s})|^2 \exp\left(-ik \frac{\boldsymbol{s}\cdot\overline{\boldsymbol{R}}}{L'_O + Z} \right) d^2\boldsymbol{s} \right|^2 \end{aligned} \tag{1.49}$$

Relation (1.49) was derived by letting $\boldsymbol{a} = \boldsymbol{A} = 0$ in Eq. (1.37). The coefficient γ represents the cross-correlation peak normalized by the autocorrelation and takes the form

$$\gamma = \frac{\langle \Delta I_1\,(\boldsymbol{R}, Z)\, \Delta I_2\,(\boldsymbol{R} + \boldsymbol{A}, Z) \rangle}{\langle [\Delta I\,(\boldsymbol{R}, Z)]^2 \rangle} = \frac{\left| \int P(\boldsymbol{s}) P^*(\boldsymbol{s} + \boldsymbol{A}_p)\, d^2\boldsymbol{s} \right|^2}{\left[\int |P(\boldsymbol{s})|^2\, d^2\boldsymbol{s} \right]^2} \tag{1.50}$$

which indicates the decorrelation factor. Finally we obtain for the intensity distribution at the Fourier plane

$$\begin{aligned} J_Y(\boldsymbol{p}) = {} & (T_O - 2\eta\langle I \rangle)^2 \left| \int U_B(\boldsymbol{R}) \exp\left(-i2\pi \frac{\boldsymbol{R}\cdot\boldsymbol{p}}{\lambda_O f_O} \right) d^2\boldsymbol{R} \right|^2 \\ & + 2\eta^2 \int |U_B(\boldsymbol{R})|^2\, d^2\boldsymbol{R}\; \tilde{C}\left(\frac{\boldsymbol{p}}{\lambda_O f_O} \right) \left[1 + \gamma \cos\left(2\pi \frac{\boldsymbol{p}\cdot\boldsymbol{A}}{\lambda_O f_O} \right) \right] \end{aligned} \tag{1.51}$$

where $\tilde{C}(\boldsymbol{S})$ is the Fourier transform of the autocorrelation and is given by

$$\tilde{C}\left(\frac{\mathbf{P}}{\lambda_O f_O}\right)$$

$$\begin{aligned}\tilde{C}(\boldsymbol{S}) &= \int C(\boldsymbol{R}) \exp(i2\pi \boldsymbol{q}\cdot\boldsymbol{R})\, d^2\boldsymbol{R} \\ &= \int |P(\boldsymbol{s})|^2\, |P[\boldsymbol{s} + \lambda(L_O' + D)\boldsymbol{S}]|^2\, d^2\boldsymbol{s} \qquad (1.52)\end{aligned}$$

The first term on the right-hand side of Eq. (1.51) refers to the bright spot appearing at the center, while the second term corresponds to Young's fringes. Hence the orientation of the Young's fringes is transverse to the speckle displacement $\boldsymbol{A}$ and their spacing is equal to $\lambda_O f_O/|\boldsymbol{A}|$. If the specklegram is recorded at the Gaussian plane, $\boldsymbol{A}$ is proportional to in-plane displacement of the object, whereas the recording at the defocused plane causes the contributions from surface tilt and strain to appear in the speckle displacement.

The envelope of Young's fringes, the so-called diffraction halo, corresponds to the Fraunhofer diffraction pattern of the individual speckle patterns. Equation (1.51) also implies that the contrast γ of the Young's fringes depends on the speckle decorrelation, which is caused by the speckle displacement $\boldsymbol{A}_p$ at the lens aperture. The speckle decorrelation is especially prominent when the surface deformation contains the tilt component Ω_x or Ω_y. The Young's fringes are especially suited for automatic analysis using video cameras followed by computers because of their simple structures. Various algorithms have been proposed in combination with automatic scanning of the reading beam on a specklegram [22–29].

1.4.2 Electronic Correlation

Speckle patterns can also be detected by a video camera whose output signals are cross-correlated before and after object deformation. The first proposal was the one-dimensional correlation technique where speckle patterns were detected by a linear photodiode array and its output was stored and correlated by a microcomputer [10, 15, 30] to avoid the time needed for two-dimensional correlation computation. The optical system used in the 1-D correlation is very simple, as shown in Fig. 1.5. A narrow laser beam is, either directly or after divergence by a lens, incident on a surface subject to motion, and the speckle pattern in the scattered field is detected by a linear photodiode array as shown in Fig. 1.11. The speckle displacement component along the sensor axis causes its output signal to shift and change as shown in Fig. 1.12a. This shift can be detected as a peak of the cross-correlation of the outputs before and after object motion. By adjusting the beam diameter and the sensor distance we can easily

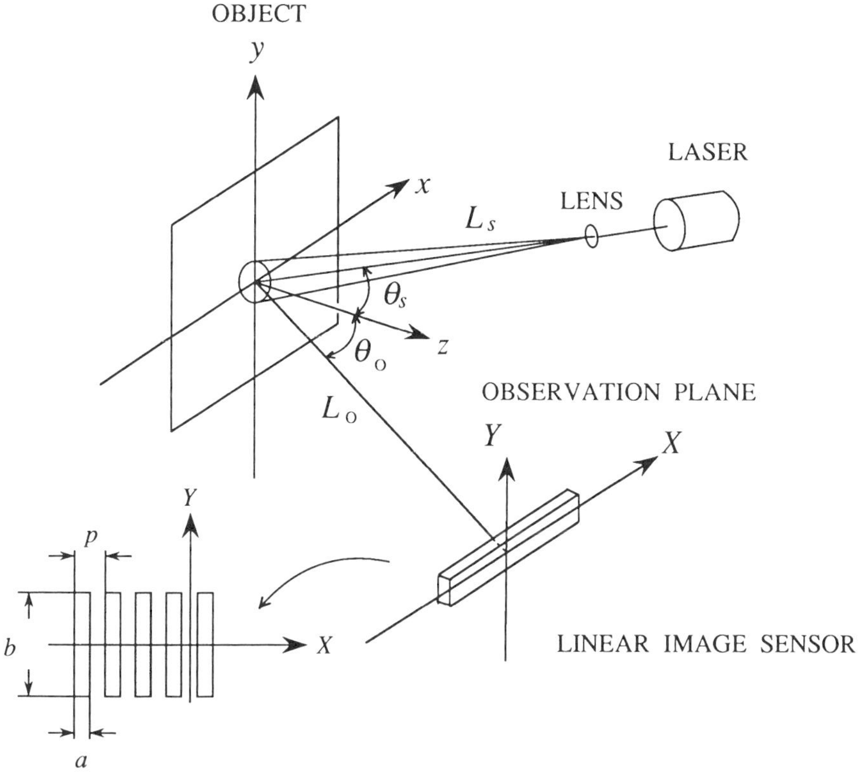

Figure 1.11 Optical system for one-dimensional electronic speckle correlation.

match the speckle size to the sensor resolution. The resultant cross-correlation of the sensor outputs corresponds to

$$\langle \Delta I_1 (X, 0)\, \Delta I_2 (X + \overline{X}, 0) \rangle = \left| \int\int [I_O(x, y) I_O(x + a_x, y + a_y)]^{1/2} \times \exp\{-ik[x(\overline{X} - A_X) \cos \theta_O + yA_Y]/L_O\}\, dx\, dy \right|^2 \tag{1.53}$$

For $\overline{X} = A_X$ this function has a peak given by

$$\langle \Delta I_1 (X, 0)\, \Delta I_2 (X + A_X, 0) \rangle = \left| \int\int [I_O(x, y) I_O(x + a_x, y + a_y)]^{1/2} \times \exp(-ikyA_Y/L_O)\, dx\, dy \right|^2 \tag{1.54}$$

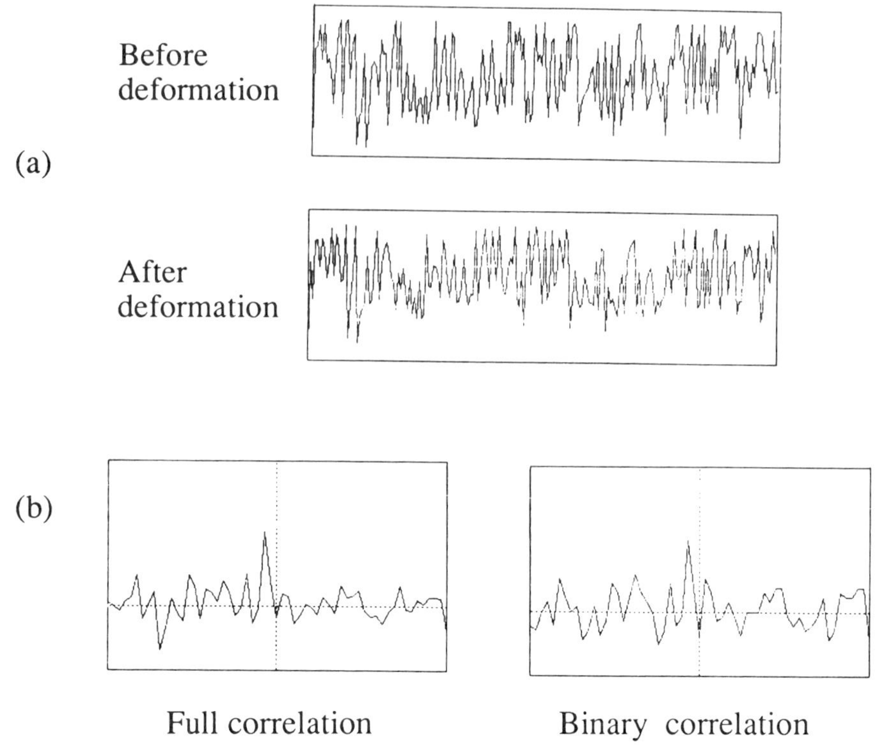

Figure 1.12 Waveforms used in the one-dimensional speckle correlation. (a) Sensor outputs before and after object movement; (b) full and binary correlations.

Therefore, the height of the correlation peak depends not only on the in-plane object translation but also on the speckle displacement component transverse to the image sensor. Actually, the signals from the jth pixel of the image sensor (pitch p, pixel length a, pixel width b) are given by

$$S_1(j) = \alpha \int_{-b/2}^{b/2} \int_{jp-a/2}^{jp+a/2} I_1(X, Y)\, dX\, dY \tag{1.55}$$

and

$$S_2(j) = \alpha \int_{-b/2}^{b/2} \int_{jp-a/2}^{jp+a/2} I_2(X, Y)\, dX\, dY \tag{1.56}$$

where $1 \leq j \leq N$ (N, total number of pixels) and α is a constant representing the sensitivity of the sensor. To reduce computation time, these signals are converted into binary signals by using the relation

$$B_{1,2}(j) = \begin{cases} 1 & \text{if } S_{1,2}(j) \geq \dfrac{1}{N} \displaystyle\sum_{i=1}^{N} S_{1,2}(i) \\ -1 & \text{otherwise} \end{cases} \tag{1.57}$$

to yield the bipolar cross-correlation defined by

$$B_{1,2}(i) = \frac{1}{N - j} \sum_{i=1}^{N} B_1(j) B_2(i + j) \tag{1.58}$$

Our experimental results show that the peak position of the bipolar cross-correlation coincides with that of the full correlation function defined by

$$C_{12}(j) = \frac{1}{N - j} \sum_{i=1}^{N} S_1(i) S_2(i + j) \tag{1.59}$$

which corresponds to Eq. (1.53). This fact is attributed to the extremely high contrast of speckle patterns. Examples of both the full and binary correlations are shown in Fig. 1.12b. The bipolar correlation algorithm is also easily implemented by real-time correlators, as will be described in Section 1.6.1.

1.4.3 Spatial Filtering Detector

A serious disadvantage of the correlation analysis is its time consumption. A spatial filtering detector is a kind of parallel image processor for detecting speckle displacement in real time. For real-time detection of speckle displacement with simpler electronics, a spatial filter made from a slit array is also available. Its first version is a combination of parallel-slit reticles and a photodetector whose output shows repetition due to movement of patterns on the reticles [14,31]. This construction has been advanced in various forms using optical fiber bundles or reticular gratings [32, 33]. The reticles and the detector were unified to a spatial filtering detector, which is an interconnected photodiode array as shown earlier in Fig. 1.6. It has been used in speckle velocimeters [34]. Its output shows the frequency, which is equal to the product of the velocity and the spatial frequency of the array. This detector, however, cannot deliver the sign of velocity. To determine the sign, we need to move the detector at constant velocity. Instead of actually moving the detector, electronic scanning methods have been proposed for cyclic opening of a liquid crystal shutter array [35] and for changing the interconnection of a photodiode array [13]. The second type of detector was applied to strain measurement and improved its time

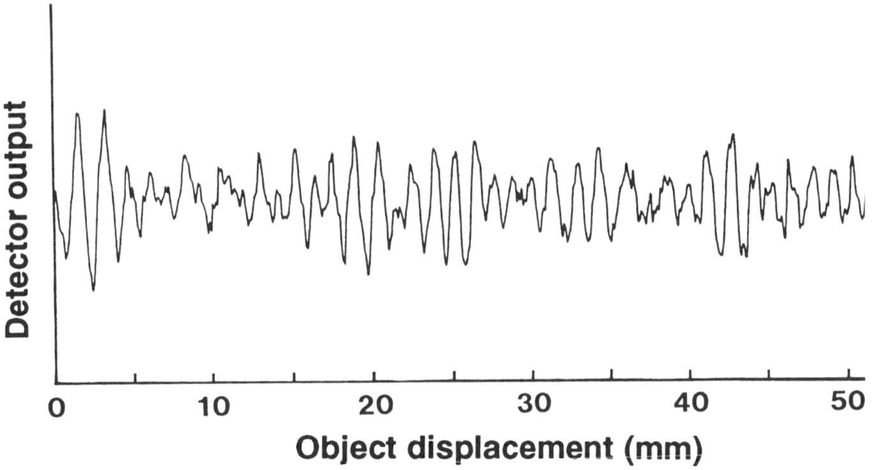

Figure 1.13 Signal from a spatial filtering detector due to the in-plane movement of a metal plate.

response from a few tenths of a hertz that had been realized by correlation analysis to a few tens of hertz [36, 37]. This detector provides a sinusoidal signal of constant frequency whose phase is shifted in proportion to displacement of a pattern on a photodiode array.

The output of the detector is given by the integral of speckled intensity over position and time,

$$V = \int\int I(X, Y)G(X, Y)\, dX\, dY \tag{1.60}$$

where $G(X, Y)$ represents the spatial response of the detector and for time dependence we assumed an ideal temporal response. In the presence of speckle movement the detector output shows dynamic behavior as shown in Fig. 1.13. It is a random signal with periodic modulation where the period is equal to the displacement component transverse to the detector. Hence we consider the autocorrelation of the output given by

$$C_V(\boldsymbol{A}) = \langle V_1 V_2 \rangle = \int C_{12}(\boldsymbol{R}) \int |\tilde{G}(\boldsymbol{S})|^2 \exp(-i2\pi \boldsymbol{S}\cdot\boldsymbol{R})\, d^2\boldsymbol{S}\, d^2\boldsymbol{R} \tag{1.61}$$

with the cross-correlation of intensity $C_{12}(\boldsymbol{R})$ and the Fourier transform $\tilde{G}(\boldsymbol{S})$ of the detector response. If we substitute Eq. (1.48) into Eq. (1.61), the following simple relation is obtained.

$$C_V(\boldsymbol{A}) = C(0)|\tilde{G}(0)|^2 + \gamma \int |\tilde{G}(\boldsymbol{S})|^2 \tilde{C}(\boldsymbol{S}) \exp (i2\pi \boldsymbol{S}\cdot\boldsymbol{A}) \, d^2\boldsymbol{S} \qquad (1.62)$$

where $\tilde{C}(\boldsymbol{S})$ is the Fourier transform of $\tilde{C}(\boldsymbol{R})$, that is, the Wiener spectrum of the speckle pattern, and is given by Eq. (1.52) for the image field. In the case of a slit or detector array along the y axis with pitch p and the unit cell $G_X(X)$ $(0 \leq X \leq p)$, its response becomes

$$|\tilde{G}(\boldsymbol{S})|^2 = |\tilde{G}_X(S_X)|^2 \, |\tilde{G}_Y(S_Y)|^2 \frac{\sin^2 \pi N p S_X}{\sin^2 \pi p S_X} \qquad (1.63)$$

with the total number of periods N. The response along the spatial frequency S_X is illustrated in Fig. 1.14. The frequency selectivity depends on the number N of array periods. In the differential scheme shown in Fig. 1.6 the response vanishes at zero frequency. If we pass only the first-order spectrum, Eq. (1.62) becomes, except for a constant,

$$C_V(\boldsymbol{A}) = \gamma \tilde{C}\left(\frac{1}{p}, 0\right) \cos 2\pi \frac{A_X}{p} \qquad (1.64)$$

which carries modulation with period p as well as attenuation due to γ for increasing $|\boldsymbol{A}|$. The modulation is due to the displacement signal, while the attenuation is concerned with speckle decorrelation and causes the signal drop off mentioned below. The properties of the signal produced from speckle movement in the diffraction field have been discussed in detail elsewhere [38].

Although the spatial filter has a quick time response in principle, there is a serious disadvantage. This is the extinction of the output signal, which occurs owing to disappearance of the spatial frequency component of speckle patterns corresponding to that of the detector.

1.5 APPLICATIONS OF ELECTRONIC SPECKLE CORRELATION

1.5.1 Encoders

In-plane displacement of a diffuse surface is difficult to measure by conventional techniques. Normally we attach a linear encoder to the surface to detect the displacement—either a magnetic grating or an optical grating. It is, however, troublesome, and moreover the maximum amount of displacement is limited by the length of the encoder. Speckle encoders, which use the surface itself as a random grating, can measure the displacement of the surface with a simple optical setup.

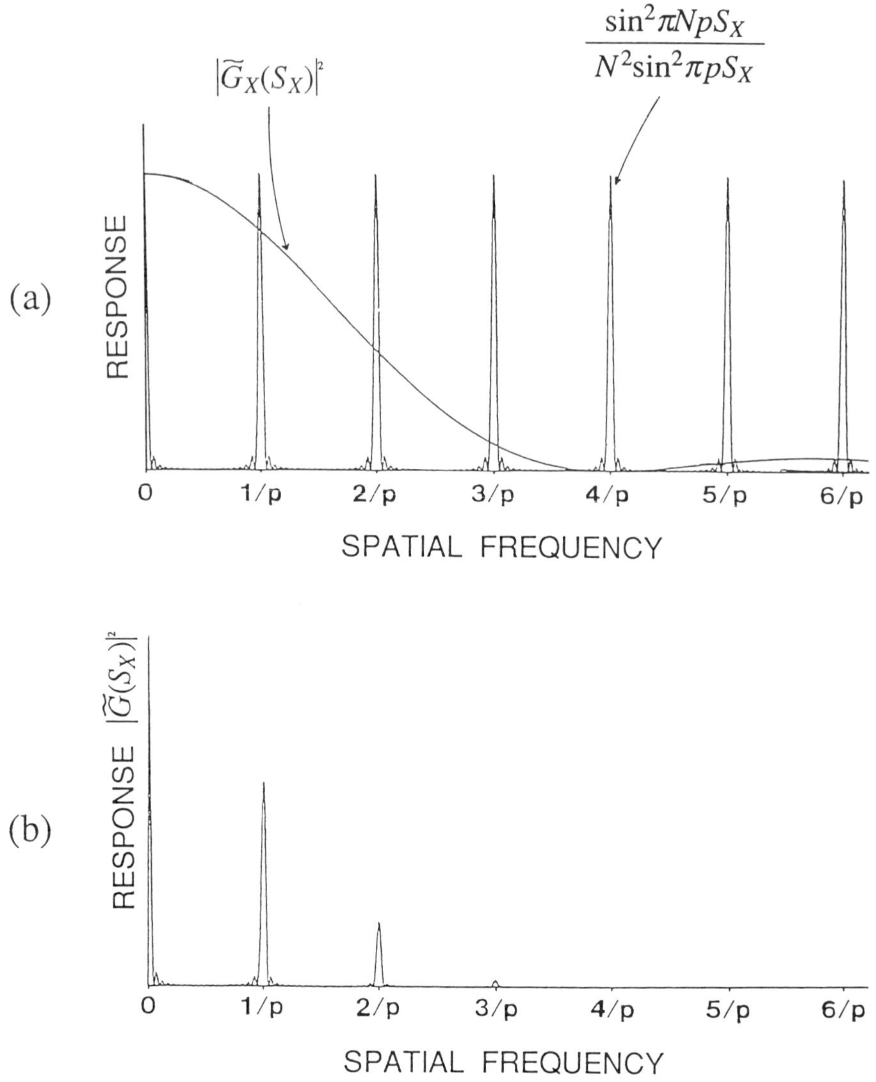

Figure 1.14 Response of a spatial filtering detector. (a) Responses of a unit cell (slit width $p/4$) and its repetition ($N = 20$); (b) total response.

The experimental setup for measuring in-plane translation is shown in Fig. 1.15. This in-plane translation has been measured by the dual-beam interferometers that are used in the laser-Doppler velocimeter [39, 40]. However, its accuracy is limited by the speckle noise. The present system is free of noise and requires a much simpler setup [18]. The laser beam is

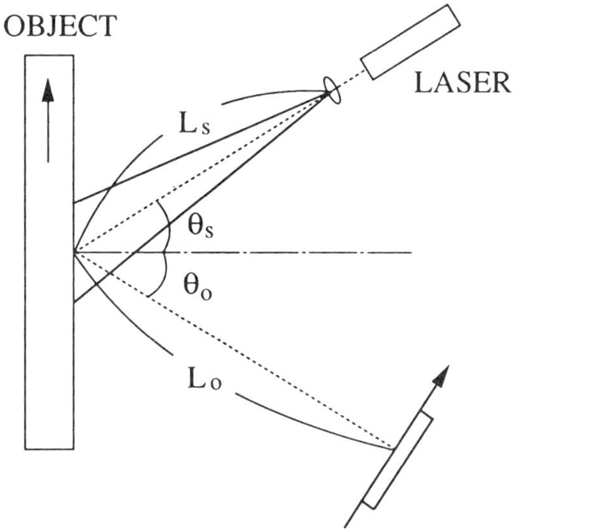

Figure 1.15 Setup of the speckle linear encoder.

either directly incident on the object or diverged by a lens to realize the finite value of the radius of the wave-front curvature L_S. A linear image sensor is placed along the plane of incidence, which is parallel to the direction of surface movement a_x. The detected speckle displacement is given by

$$A_X = a_x \left(\frac{L_O \cos^2 \theta_S}{L_S \cos \theta_O} + \cos \theta_O \right) = \sigma a_x \tag{1.65}$$

Thus we can amplify the object translation by adjusting L_S to be smaller than L_O. If we adopt the collimated illumination, the speckle displacement becomes

$$A_X = a_x \cos \theta_O \tag{1.66}$$

which is independent of both the object distance L_O and the incident angle θ_S. Sensitivity of the speckle encoder depends on the minimum speckle displacement that can be detected by the image sensor and also on speckle displacement sensitivity σ. The former is made as small as one-tenth of the pitch of the sensor by use of an interpolation algorithm for detection of the peak position of the correlation peak. σ can easily be made as much as 10 by adopting divergent illumination of the surface. The range of measurement can be extended arbitrarily if we integrate small incremental speckle displacements for which the decorrelation effect remains small. In

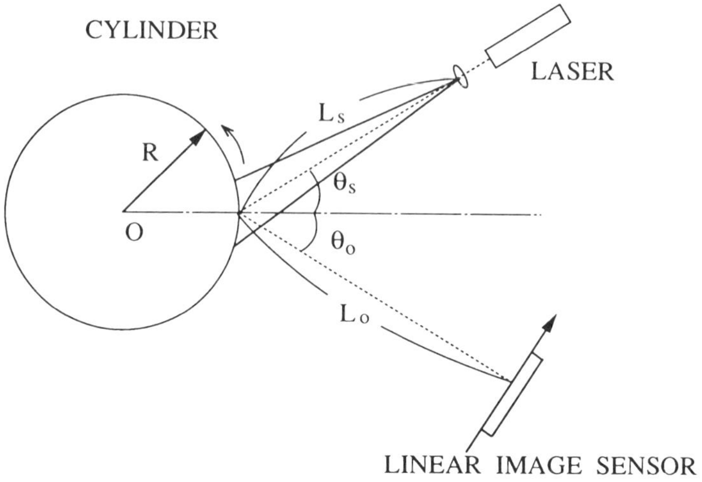

Figure 1.16 Setup of the speckle rotary encoder.

actual implementation this can be accomplished as follows. We monitor not only the position of the correlation peak but also its height and automatically exchange the reference signal of the correlation with the latest signal when the peak becomes lower than a fixed level of threshold. Then the peak is recovered to give speckle displacement clearly again. The peak position is summed up at each exchange of the reference to yield total speckle displacement.

The same arrangement can also be used for out-of-plane translation a_z, but the sensitivity is much lower [18].

The rotation angle of a cylinder can also be measured by employing the setup shown in Fig. 1.16 [18, 41].

1.5.2 Strain Gauge

Speckle displacement is also proportional to surface strain, which is equal to the derivative of in-plane displacement. In general, strain is superposed by rigid-body motion consisting of translation and rotation. The rigid-body motion can be canceled out if we compare the displacements of speckles appearing at opposite scattering angles. The two configurations are available as shown in Fig. 1.17. In the dual-beam system [15] the displacements of speckles generated by each of the symmetrical incident beams are de-

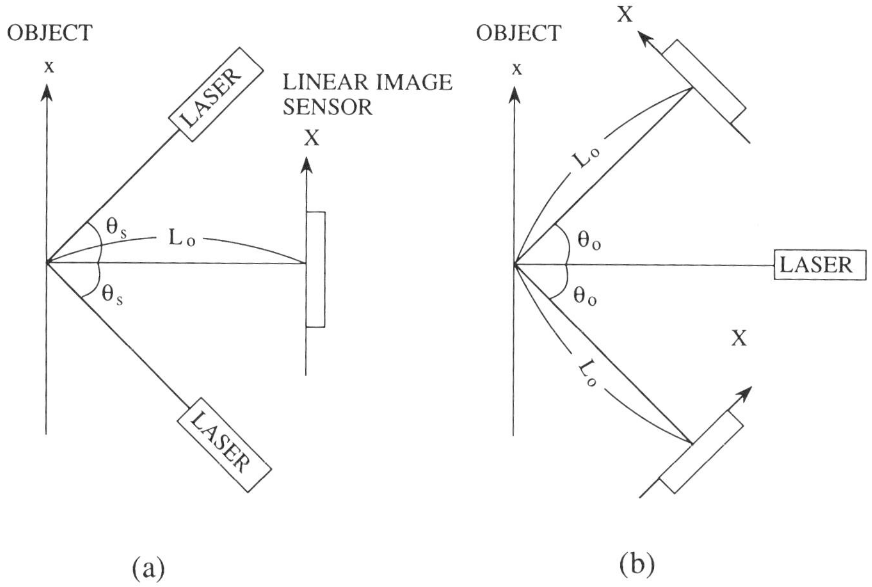

Figure 1.17 Two configurations of speckle strain gauge. (a) Dual beams; (b) double sensors.

tected at the normal direction. The difference of the speckle displacements is derived from Eq. (1.14) with $\theta_O = 0$ and $L_S = \infty$ (collimated beam) as

$$\Delta A_X = A_X(\theta_S, 0) - A_X(-\theta_S, 0) = -2\varepsilon_{xx} L_O \sin \theta_S \tag{1.67}$$

which is proportional to the strain ε_{xx} along the plane of incidence. In this case the incident beams have to be exchanged one after another each time the image sensor signal is stored in memory. After the complete storage of the signals, cross-correlation functions are computed between the signals belonging to the same incident beam.

In the double-sensor system [16, 37, 42], displacements of speckles produced by a single incident beam are detected simultaneously by the sensors symmetrically positioned, and therefore measurement can be done more quickly. The difference of speckle displacement is given by

$$\Delta A_X = A_X(\theta_S, \theta) - A_X(\theta_S, -\theta) = -2\varepsilon_{xx} L_O \tan \theta_O - 2a_z \sin \theta_O \tag{1.68}$$

which also depends on out-of-plane translation a_z. This effect can be neglected if $|a_z|$ is much smaller than $|\varepsilon_{xx}| L_O / \cos \theta_O$ as realized in normal extension loading.

These laser speckle strain gauges directly detect elongation or contraction of the microscopic structure of solid surfaces and are free of any kind of surface contact or preparation. The minimum detectable strain magnitude is on the order of a few tens of microstrains, that is, 10^{-5}, when we use image sensors with the pitch of 10 μm and L_O of a few tens of centimeters. There is no upper limit of measurement because incremental speckle displacements are integrated for which decorrelation remains small. The range of measurement covers both elastic and plastic strain. Its gauge length, which is given by the spot diameter, can be made as small as 1 mm. Both elastic and plastic strains of various materials such as metal, plastic, paper, and film sheet have been measured on loading by commercial testing instruments. Using a biaxial configuration with two orthogonal pairs of image sensors, Poisson's ratio of a material can be quickly determined. Thermal expansion has also been successfully measured [42].

1.5.3 Velocity Measurement

Speckle velocimeters are based on detection of the velocity of speckle displacement. When we apply Doppler velocimeters to diffusely reflecting objects, the resultant Doppler beat signal is deteriorated by the speckle noise and dropout of the signal takes place to reduce accuracy of measurement. Since speckle velocimeters detect movement of this speckle noise itself, they are free of this signal dropout. In addition, speckle velocimeters require simpler optics than Doppler velocimeters.

In Fig. 1.18 basic configurations of speckle velocimeters applied to in-plane motion are shown, where velocity V of speckle displacement is detected. In the two-detector setup (Fig. 1.18a), speckle velocity V_X along the X axis is determined from the peak position of the temporal cross-correlation of the temporal intensity fluctuations at a pair of points $\boldsymbol{R}_1$ and $\boldsymbol{R}_2$ [20, 43],

$$C_{12}(\tau) = \lim_{T\to\infty} \frac{1}{T} \int_0^T I(\boldsymbol{R}_1, t) I(\boldsymbol{R}_2, t + \tau)\, dt \tag{1.69}$$

We consider that the object displacement during the delay time is equal to $a_x = \tau v$ and that speckle displacement for the time interval is given by $A_X = \tau V_X$. We denote the sensor distance by $\overline{X}$ and further replace the time average by the ensemble average to arrive at

$$\begin{aligned} C_{12}(\tau) &= \langle I_1(X, Y) I_2(X + \overline{X}, Y) \rangle \\ &= \left| \iint [I_O(x, y) I_O(x + v\tau, y)]^{1/2} \right. \\ &\qquad \left. \times \exp\,[-ikx(\overline{X} - \tau V_X)]\, dx\, dy \right|^2 \end{aligned} \tag{1.70}$$

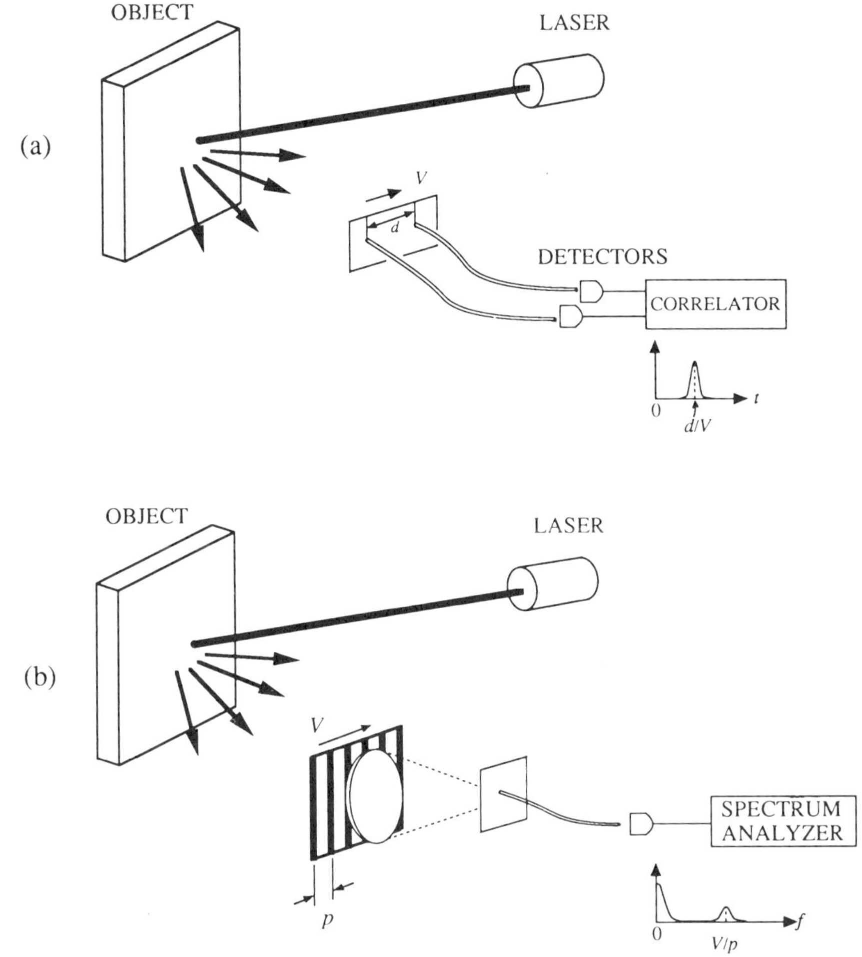

Figure 1.18 Basic configurations of speckle velocimeters. (a) Two-point correlation; (b) reticle aperture.

where the speckle velocity is proportional to object velocity such that

$$V_X = \sigma v \tag{1.71}$$

Therefore, the correlation peak position is equal to

$$\tau_O = \overline{X}/V_X \tag{1.72}$$

from which the speckle velocity and hence the surface velocity can be derived. Very low velocity, which requires too long a delay time to be implemented, can also be detected by the spatial correlation technique used in the speckle encoders.

Speckle velocity can also be sensed by the spatial filtering detector shown in Fig. 1.6. The output of the spatial filtering detector is represented by

$$V(t) = \int\int I(X, Y; t)G(X, Y)\, dX\, dY \tag{1.73}$$

whose output has the autocorrelation related to Eq. (1.64)

$$C_V(\tau) = \gamma(\tau)\tilde{C}\left(\frac{1}{p}, 0\right) \cos 2\pi \frac{\tau V_X}{p} \tag{1.74}$$

with the decorrelation factor

$$\gamma(\tau) = \frac{\left| \int\int [I_O(x, y)I_O(x + \tau v, y)]^{1/2}\, dx\, dy \right|^2}{\left[\int\int I_O(x, y)\, dx\, dy \right]^2} \tag{1.75}$$

The output shows the spectrum, which has a peak at the frequency that is equal to the product V_X/p of speckle velocity and the spatial frequency of the array. Its width is increased by the decorrelation effect.

In the reticle arrangement shown in Fig. 1.18b, speckle velocity V across the reticle produces periodic modulation of the detector output, which is to be analyzed [45, 46].

1.6 RECENT DEVELOPMENTS

1.6.1 Acceleration

The cross-correlation function of the image sensor signals was first computed by a microcomputer. For sensors of 1024 pixels it took a few seconds. Moreover, the frame rate of the sensor was limited to less than 25 Hz owing to the capacity of the computer. Use of the spatial filtering detector improved the response up to 10 kHz, but a problem with the signal dropoff prevented stable and reliable measurement [37, 38]. These problems were overcome after a real-time correlator was developed that was as fast as 1 kHz and capable of pursuing an arbitrarily large amount of speckle displacement by automatic exchange of reference signals and integration of incremental displacements [18, 42, 47]. A system diagram of the correlator is shown in Fig. 1.19.

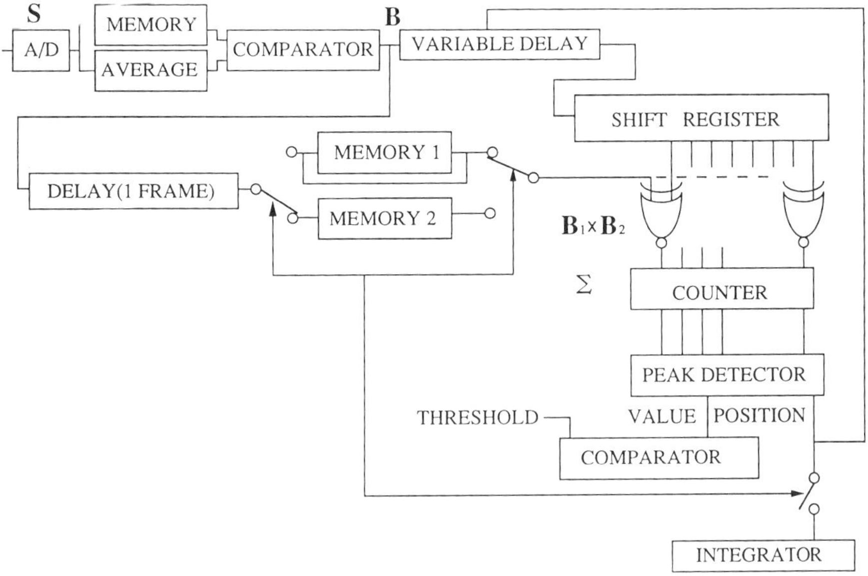

Figure 1.19 System diagram of the real-time correlator, which automatically changes the reference signal when the correlation peak is lower than a threshold level to extend the range of speckle displacement.

1.6.2 Whole-Field Analysis

For measuring distributions of displacement and strain, which are important for evaluation of stress distribution and prediction of destruction, we can employ two-dimensional cross-correlation of the image field speckle. In this case, however, reduction of computation time is strongly desired for practical applications [48]. For this purpose spatial filtering detectors might be suited in which parallel detection and processing of images are performed. Optical flow analysis can also be used for the distribution measurement [49]. These techniques could also be applied to real-time measurement of strain distribution and real-time measurement of flow fields [50].

1.6.3 Improvement in Resolution

Resolution of speckle displacement detection depends fundamentally on the width of the cross-correlation function, which is equal to that of the autocorrelation, that is, the speckle size. Finally, the resolution is governed by the pixels of the imaging detector. For improving the resolution, fitting methods have been proposed. The spatial frequency of the Young's fringes is determined by fitting a cosinusoidal curve to a one-dimensional projection of the fringe along their direction. Consequently, even the fringe spacing, which is twice as wide as the diffraction halo, can be correctly analyzed [28]. The fringe direction was determined by comparing the integrals of fringe intensity over bands of various directions. In the electronic speckle correlation technique, resolution that is smaller than the pixel size of the detector has also been attained by fitting a parabola to the cross-correlation peak [47]. Several other fitting algorithms have also been examined.

Spatial filtering detectors deliver continuous output with automatic introduction of the interpolation function. A serious problem encountered in this kind of detector, however, is the signal dropoff that arises from speckle decorrelation and causes random variation of the spatial frequency spectrum of speckle on the detector. This dropoff can be partially eliminated by using a pair of detectors with different frequency response and selecting signals of larger amplitude from their outputs [51].

1.6.4 Applications of Nonoptical Speckles

Acoustic speckle and electronic speckle have a similar appearance when coherent sources and suitable imaging devices are employed. Movement of acoustic speckle generated by an acoustic diffuser was detected by double exposure of small random ripple patterns appearing at a water surface [52]. The electronic speckle correlation technique will find broad applications with nonoptical speckles if suitable detector arrays can be used.

1.7 CONCLUSIONS

We have discussed the dynamic properties of laser speckles that are caused by displacement or deformation of diffusely reflecting surfaces. They can be quantitatively described in terms of the cross-correlation function of the intensity distributions at an observation plane before and after the surface movement. This function has been derived for the diffraction field (free-space geometry) and the image field including the defocused region by using the Fresnel–Kirchhoff diffraction formula. The cross-correlation shows a peak whose position corresponds to speckle displacement and whose height indicates speckle decorrelation accompanying the displacement. The dependence of the speckle displacement on individual deformation parameters such as translation, rotation, and strain as well as on optical configurations has been explicitly presented. Displacement is caused by the change in the phase relationship of the superposed wavelets, while decorrelation is related to the walk-off of the surface region out of the incident laser spot or of the speckle out of the imaging area of the lens.

Next we discussed how to detect the speckle displacement. Speckle photography and the electronic correlation method were discussed. After the formation of Young's fringes in the first technique was discussed, the one- and two-dimensional correlation techniques were described. For accelerating the detection of speckle displacement, spatial filters and spatial filtering detectors based on parallel detection of speckle patterns were mentioned. Properties of images or output signals obtained with these methods were quantitatively evaluated from the speckle correlation functions.

As applications of speckle displacement detection we mentioned the measurement of displacement, strain, and velocity. General advantages of these methods consist in simplicity in optics and direct delivery of the desired quantities. The latter advantage, however, is strongly dependent on the performance of the signal processing system. Hence both recent progress in these systems and steadily increasing industrial demand for simple and versatile optical metrology would promote the further development and use of the techniques presented in this chapter.

REFERENCES

1. I. Yamaguchi, Fringe formations in deformation and vibration measurements using laser light, in *Progress in Optics*, Vol. 22, E. Wolf, Ed., North-Holland, Amsterdam, 1985, p. 271.
2. J. M. Burch and M. J. Tokarski, Production of multiple beam fringes from photographic scatterers, *Opt. Acta*, *15*, 101 (1968).
3. W. Fink and P. Büger, Eine Methode zur kontaktlosen Messung kleiner Verschiebungen rauher Oberflächen, *Z. Angew. Phys.*, *30*, 176 (1970).
4. E. Archbold, J. M. Burch, and A. E. Ennos, Recording of in-plane surface displacement by double-exposure speckle photography, *Opt. Acta*, *17*, 883 (1970).
5. H. J. Tiziani, Applications of speckling for in-plane vibration analysis, *Opt. Acta*, *18*, 891 (1971).
6. U. Köpf, Ein kohärent-optisches Verfahren zur berührungslosen Messung mechanischer Beformungen und Schwingungen, *Optik*, *35*, 144 (1972).
7. H. J. Tiziani, A study of the use of laser speckles to measure small tilts of optically rough surfaces, *Opt. Commun.*, *5*, 271 (1972).
8. E. Archbold, J. M. Burch, and A. E. Ennos, Displacement measurement from double-exposure laser photographs, *Opt. Acta*, *19*, 253 (1972).
9. H. J. Tiziani, Analysis of mechanical oscillations by speckling, *Appl. Opt.*, *11*, 2911 (1972).
10. I. Yamaguchi, Real-time measurement of in-plane translation and tilt by electronic speckle correlation, *Jpn. J. Appl. Phys.*, *19*, L133 (1980).

11. W. H. Peters and W. F. Ranson, Digital imaging techniques in experimental stress analysis, *Opt. Eng.*, *21*(3), 427 (1982).
12. I. Yamaguchi, A laser speckle strain gauge, *J. Phys. E: Sci. Instrum.*, *14*, 1270 (1981).
13. I. Yamaguchi, Simplified laser-speckle strain gauge, *Opt. Eng.*, *21*, 436 (1982).
14. G. Stavis, Optical diffraction velocimeter, *Instrum. Control Syst.*, *39*, 99 (1966).
15. H. Ogiwara and H. Ukita, A speckle pattern velocimeter using a periodical differential detector, *Jpn. J. Appl. Phys.*, *14*(Suppl. 14-1), 307 (1975).
16. E. Ogita, T. Ueda, and I. Yamazaki, Optical three-dimensional displacement meter, *Proc. SPIE*, *556*, 139 (1985).
17. I. Yamaguchi, Speckle displacement and decorrelation in the diffraction and image fields for small object deformation, *Opt. Acta*, *28*, 1359 (1981).
18. I. Yamaguchi and T. Fujita, Linear and rotary encoders using electronic speckle correlation, *Opt. Eng.*, *30*, 1862 (1991).
19. I. Yamaguchi, S. Komatsu, and H. Saito, Dynamics of speckles produced by a moving object and its applications, *Jpn. J. Appl. Phys.*, *14*(Suppl. 1), 301 (1975).
20. I. Yamaguchi and S. Komatsu, Theory and applications of dynamic laser speckles due to in-plane object motion, *Opt. Acta*, *24*, 705 (1977).
21. I. Yamaguchi, Fringe formation in speckle photography, *J. Opt. Soc. Am.*, *A1*, 81 (1984).
22. B. Ineichen, P. Eglin, and R. Dändliker, Hybrid optical and electronic image processing for strain measurements by speckle photography, *Appl. Opt.*, *19*, 2119 (1980).
23. D. W. Robinson, Automatic fringe analysis with a computer image-processing system, *Appl. Opt.*, *22*, 2169 (1983).
24. R. Erbeck, Fast image processing with a microcomputer applied to speckle photography, *Appl. Opt.*, *24*, 3838 (1985).
25. H. D. Navone and G. H. Kaufmann, Two-dimensional digital processing of speckle photography fringes. 3. Accuracy in angular determination, *Appl. Opt.*, *26*, 154 (1987).
26. F. Ansari and G. Ciurpita, Automated fringe measurement in speckle photography, *Appl. Opt.*, *26*, 1688 (1987).
27. J. M. Huntley, Speckle photography fringe analysis: assessment of current algorithms, *Appl. Opt.*, *28*, 4316 (1989).
28. N. Deng and I. Yamaguchi, Automatic analysis of speckle photographs with extended range and improved accuracy, *Appl. Opt.*, *29*, 296 (1990).
29. J. M. Coupland and N. A. Halliwell, Automated optical analysis of Young's fringes—optical autocorrelation, *Opt. Lasers Eng.*, *14*, 351 (1991).
30. I. Yamaguchi, Automatic measurement of in-plane translation by speckle correlation using a linear image sensor, *J. Phys. E: Sci. Instrum.*, *19*, 944 (1986).
31. J. T. Ator, Image velocity sensing with parallel-slit reticles, *J. Opt. Soc. Am.*, *53*, 1416 (1963).

32. A. Hayashi and Y. Kitagawa, Image velocity sensing using an optical fiber array, *Appl. Opt.*, *21*, 1394 (1982).
33. Y. Aizu and T. Asakura, Principles and development of spatial filtering velocimetry, *Appl. Phys.*, *B43*, 209 (1987).
34. M. Naito, Y. Ohkami, and A. Kobayashi, Spatial filter and its application to industrial measurement, *Proc. IMEKO-V JA-129*, 1970.
35. Y. Itakura, A. Sugimura, and S. Tsutsumi, Amplitude-modulated reticle constructed by a liquid crystal cell array, *Appl. Opt.*, *20*, 2819 (1981).
36. I. Yamaguchi, T. Furukawa, T. Ueda, and E. Ogita, Accelerated laser speckle strain gauge, *Opt. Eng.*, *25*, 671 (1986).
37. I. Yamaguchi, Advances in the laser speckle strain gauge, *Opt. Eng.*, *27*, 214 (1988).
38. I. Yamaguchi, T. Okamoto, and H. Nagayama, Performance of a spatial filtering detector applied to a speckle displacement sensor, *Meas. Sci. Technol.*, *1*, 406 (1990).
39. D. Joyeux, Real time measurement of very small transverse displacements of diffuse objects by random moiré. 1. Theory, *Appl. Opt.*, *15*, 1241 (1976).
40. D. Joyeux, Real time measurement of very small transverse displacements of diffuse objects by random moiré. 2. Experiments, *Appl. Opt.*, *15*, 1248 (1976).
41. I. Yamaguchi and T. Fujita, Laser speckle rotary encoder, *Appl. Opt.*, *28*, 4401 (1989).
42. I. Yamaguchi and K. Kobayashi, Material testing by the laser speckle strain gauge, *Proc. SPIE*, *1554A*, 240 (1991).
43. P. N. Pusey, Photon correlation study of laser speckle produced by a moving rough surface, *J. Phys. D*, *9*, 1399 (1976).
44. T. Asakura and N. Takai, Dynamic laser speckles and their application to velocity measurements of the diffuse objects, *Appl. Phys.*, *25*, 179 (1981).
45. S. Komatsu, I. Yamaguchi, and H. Saito, *Jpn. J. Appl. Phys.*, *15*, 1715 (1976).
46. S. Komatsu, I. Yamaguchi, and H. Saito, *Opt. Commun.*, *18*, 314 (1976).
47. T. Takemori, K. Fujita, and I. Yamaguchi, Resolution improvement in speckle displacement and strain sensor by correlation interpolation, *Proc. SPIE*, *1553*, 137 (1991).
48. N. Takai and T. Asakura, Vectorial measurements of speckle displacement by the 2-D electronic correlation method, *Appl. Opt.*, *24*, 660 (1985).
49. P. Sobey and M. V. Srinivasan, Measurement of optical flow by a generalized gradient scheme, *J. Opt. Soc. Am.*, *A8*, 1488 (1991).
50. Z. H. He, M. A. Sutton, W. F. Ranson, and W. H. Peters, Two-dimensional fluid-velocity measurements by use of digital-speckle correlation techniques, *Exp. Mech.*, *24*, 117 (1984).
51. H. Nagayama, T. Okamoto, and I. Yamaguchi, Adaptive spatial filtering detector and its application to speckle displacement sensor with high resolution, *Kogaku* (*Jpn. J. Opt.*), *20*, 530 (1991).
52. D. M. He, Basic use of acoustic speckle pattern for metrology, *Proc. 15th ICO Congress on Optics in Complex Systems*, Garmisch-Partenkirchen, 1990, p. 323.

2

Techniques of Displacement and Deformation Measurements in Speckle Metrology

Pramod K. Rastogi

Swiss Federal Institute of Technology
Lausanne, Switzerland

2.1 INTRODUCTION

The speckle effect has the rare distinction of being both a nuisance and an asset to the optics community working on the development of laser-based optical methods to monitor displacements and deformations of engineering structures. Long considered to be a bane for holographers, the speckle effect grew out of its complex to shape an important niche for itself in deformation analysis [1–7]. Ever since the appearance of the pioneering works of Leendertz [8] and Archbold et al. [9] in 1970 demonstrating the use of speckles in displacement measurements, the development of speckle-based methods has gone from strength to strength backed by spates of activities and programs in different laboratories around the world. As a result of this investigation a wide range of procedures have been developed, which can be classified into three broad categories: speckle photography, speckle interferometry, and speckle shear interferometry. While the modes of operation of the different categories will open out to readers as they go through the chapter, it will suffice to say at this stage that it would be dubious, if not misleading, in many instances to make an outright distinction between the procedures belonging to the different categories.

A significant feature of the speckle methods not sufficiently stressed in the literature resides in their extreme flexibility in handling large-deformation problems. The capacity to generate reduced sensitivities makes these methods—speckle photography and speckle interferometry—very suitable for measuring relatively large displacements. Variable in-plane sensitivities in the range of 0.3–10 μm for in-plane displacements and in the range of 0.25–30 μm for out-of-plane displacements can be realized. The term "sensitivity" here refers to the increment of displacement between two adjacent fringes.

The family of techniques forming the ensemble of the three categories have been employed to determine the displacement, strain, slope change, and curvature contours of a deformed structure. These techniques have also been used to measure shape and slope variations of a three-dimensional specimen. Of the variety of techniques, each has its own particularity, advantages, and disadvantages. The goal of this chapter is to review the speckle techniques applied to deformation analysis. The emphasis will be to bring out clearly the limitations attached to these methods and propose conditions under which they can be used without introducing ambiguity into measurements.

2.2 SPECKLE PHOTOGRAPHY

2.2.1 Image Analysis and Some Limiting Aspects

Image Processing and Measurement Range

By far the simplest technique to set up and operate in speckle metrology is that of speckle photography [9–66]. This method permits one to measure in-plane displacements, derivatives of displacements, and surface slopes, is easy to handle, and requires simple optical systems. The general procedure consists of exposing a photographic plate to two displaced speckle patterns, once before and once after the object has been moved [9–12]. An alternative approach to the recording of a double-exposure specklegram resides in recording the two speckle patterns on separate photographic plates [13, 14]. Subsequent superposition of the two plates allows for eliminating rigid-body movements undergone by the object surface during its deformation.

The recorded specklegram can be analyzed in two ways. The first method, a point-by-point approach, consists of interrogating the specklegram by means of a narrow laser beam (Fig. 2.1). The laser beam diffracted by the speckles lying within the beam area gives rise to a diffraction halo. The halo is modulated by an equidistant system of fringes arising from the interference of two identical but displaced speckles. The direction of these fringes is perpendicular to the direction of displacement. The magnitude of displacement, inversely proportional to the fringe spacing p, is given by

$$d_\xi = m\sqrt{(a_x^2 + a_y^2)} = \lambda L/p \tag{2.1}$$

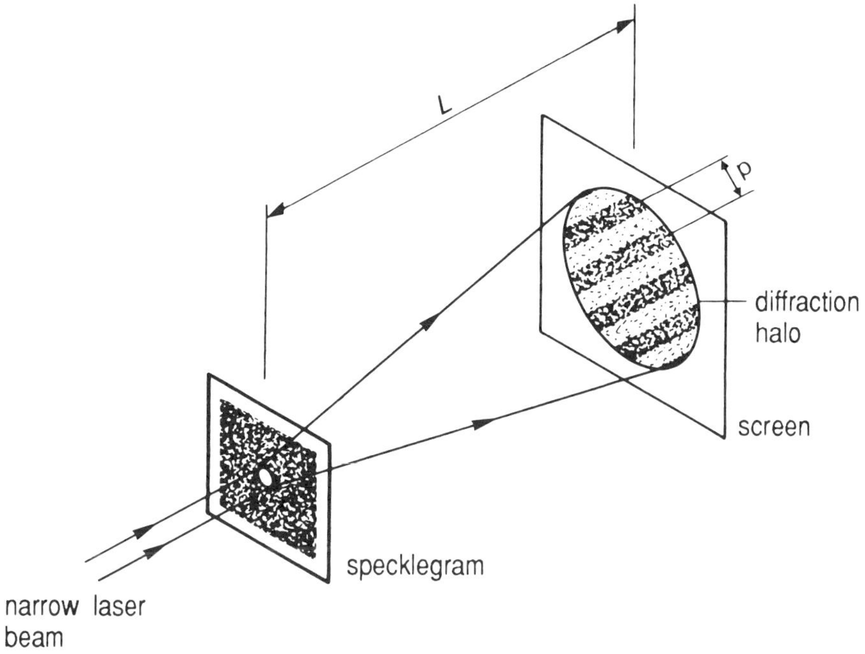

Figure 2.1 Point-by-point analysis of the recorded specklegrams.

where d_ξ is the displacement in the image plane, a_x and a_y are the components of speckle movements along the x and y directions in the object plane, m is the magnification factor of the speckle recording imaging system, and L is the distance between the interrogated point and the observation screen.

The second method, a whole-field approach, consists of optically filtering the recorded specklegram in the setup schematized in Fig. 2.2. An observer looking through a small offset aperture placed in the focal plane of the Fourier transform lens sees contours of speckle displacement along the direction defined by the position of the aperture. The fringe contours are described as the loci of points where

$$d_\xi = n\lambda f/\zeta_{0\xi} \tag{2.2}$$

f is the focal length of the Fourier transform lens. ξ and $\zeta_{0\xi}$ denote the azimuth and radial distance from the position of the zero diffraction order; ξ denotes the direction along which the object displacement is measured, and $\zeta_{0\xi}$ determines the sensitivity of the fringe contours.

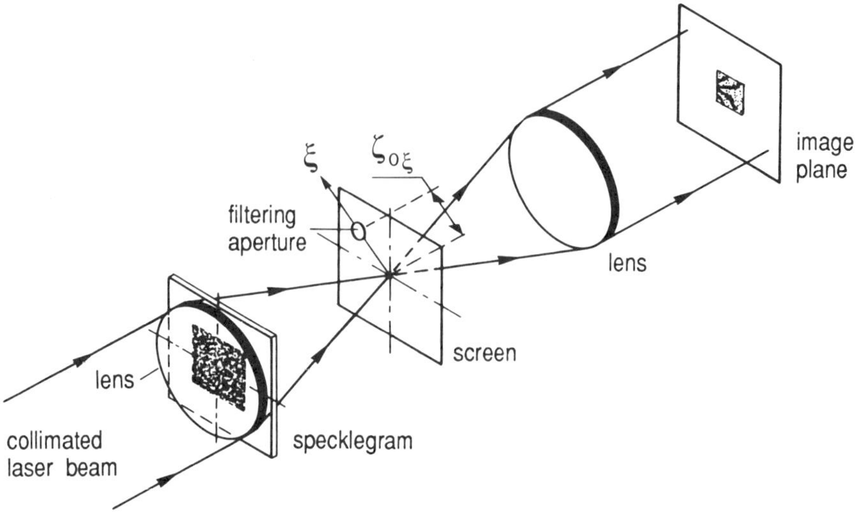

Figure 2.2 Spatial filtering setup to obtain whole-field fringe display in speckle photography.

As far as the range of the method is concerned, upper and lower limits, respectively, on measurable displacements are imposed by the size of the diffraction halo and by the mean size of the secondary speckles modulating the diffraction halo. The diameter of the diffraction halo, D_h, and the mean size of the secondary speckles in the halo, s_h, are given by

$$D_h = 2Lb/d_i, \qquad s_h = 1.22\lambda L/t \tag{2.3}$$

where b is the diameter of the imaging lens aperture, t is the width of the laser beam interrogating the specklegram, and d_i is the distance between the imaging lens and the film plane during the recording of the specklegram.

Hence one of the fundamental limitations on the measurable range arises from the presence of secondary speckles in the halo. In order to successfully carry out the measurements, the Young's interfringe distance, p, must satisfy the condition

$$\langle s_h \rangle < p \leq D_h \tag{2.4}$$

It is also possible to measure still smaller displacements by introducing between the two exposures an additional lateral displacement to the photographic plate.

Finally, for the method to work it is essential that the axial speckle shift, a_z, remain inferior to the mean longitudinal size of the speckles, s_l, or

$$a_z \ll \frac{8\lambda d_o^2}{b^2} \tag{2.5}$$

where the term on the right-hand side represents the value of s_l; the significance of d_o is understood from Fig. 2.3.

Heterodyne and Phase Shifting Applied to Fringe Analysis

Some important drawbacks preventing the widespread use of speckle photography have been (1) nonconcomitant mode of operation, (2) tedious and time-consuming processing of the Young's fringes, and (3) relatively poor quality of the constant-displacement fringes obtained in the whole-field analysis. These difficulties have prompted work toward developing means to make the technique more user-friendly. Of the three problems, the first was initially attacked by Tiziani et al. [18], who reported a quasi-real-time realization of the speckle photography technique by using photorefractive $Bi_{12}SiO_{20}$ crystals to record the specklegrams. The use of these crystal plates for recording specklegrams is being investigated by several researchers, for example, Nakagawa and coworkers [19, 20]. On the other hand, the application of heterodyne [67] and phase shifting [68, 69] to speckle photography [21–23] has helped in considerably reducing the latter two drawbacks. Finally, the development of automatic microcomputer-

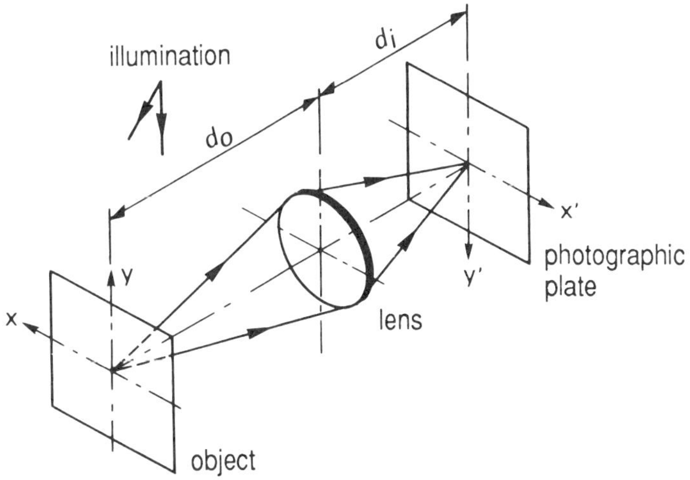

Figure 2.3 Schematic of the focused speckle photography setup.

based image processing systems has greatly facilitated the evaluation of Young's fringes [24–28].

Heterodyne readout provides an attractive approach to the processing of specklegrams. First introduced by Smith and Stetson [21], the method consists of recording the two exposures of the object, before and after deformation, on separate photographic plates. The diffraction halos issuing from the same points on the object images are subsequently combined by means of a beamsplitter in a heterodyne interferometer. A small frequency shift between the two readout beams then causes Young's fringes to scan over the halo. Two pairs of photodetectors, each oriented along the x and y axes, are placed in the halo. The phase differences between the signals from the aperture pairs along the x and y axes give information relative to the displacement components u and v, respectively. In addition to improving the precision of measurements, the heterodyne readout system removes the displacement-sign ambiguity drawback inherent in speckle photography. The analysis of error sources and a performance evaluation of a heterodyne readout system have been reported by Kim et al. [22].

Kaufmann and Jacquot [23] have reported a phase-shifting readout system to analyze whole-field fringes obtained by optical filtering of dual-plate specklegrams. The two specklegrams, before and after object deformation, are placed in the two arms of a Mach–Zehnder interferometer. The Fourier transform lens is placed in the beams exiting the interferometer via a beamsplitter. The filtered image is phase-shifted by placing a glass plate that can be tilted in one of the arms of the interferometer. Three frames of intensities are recorded corresponding to phase shifts of $2\pi/3$, 0, and $-2\pi/3$. The phase of the measured displacement component is then calculated from the set of three intensity equations. This calculation provides information on the sign of the displacement component as well.

Influence of Varying Halo Intensity on Measurements

A particularity of the halo resides in its nonuniform distribution of intensity. The intensity in the halo is not constant but varies monotonically from its center to its periphery. The effect of the varying diffraction halo intensity on the measurement of Young's fringes has been widely investigated in speckle photography [29–43] because of the errors the halo function intrinsically induces in the determination of the periodicity of these fringes. For a circular aperture, the intensity distribution perpendicular to the Young's fringes can be written as

$$I(\rho) = \left[\frac{1 - V}{1 + V} + \frac{2V}{1 + V}\cos^2(D_0\rho)\right][\cos^{-1}\rho - \rho\sqrt{(1 - \rho^2)}] \tag{2.6}$$

where V is the fringe visibility, $\rho = |r/r_m|$ is the normalized position ($0 \leq \rho \leq 1$) in the halo, r_m is the radius of the halo, and

$$D_0 = \pi d_\xi r_m / m\lambda L \tag{2.7}$$

Equation (2.6) assumes that the intensity distribution in the diffraction halo is given by the autocorrelation of the aperture function. A nonuniform distribution of intensity in the halo causes the positions of the fringes to shift from their true positions. The shift occurs only at the fringe maxima for unit visibility $\cos^2$ fringes but extends to both fringe maxima and minima for lower visibility fringes. The fringe shifts thus result in measurement errors if the intensity variation in the halo is not accounted for.

Several researchers have reported the use of the halo function for incoherent speckle patterns (given by the square of the autocorrelation of the aperture function [44]) to evaluate and correct the errors introduced in double-exposure speckle photography. This inconsistency, reported by Meynart [39], however, does not cast any prejudice on the earlier proposed approaches. Sophisticated electronic techniques [24, 35] have been used to delete the influence of intensity nonuniformity in the halo. Although simple in concept, these techniques require special equipment and sometimes specialized computer techniques to implement. Some other approaches consist of calculating the error by analytical, computer, or numerical techniques and then incorporating the corrections to fringe positions [29–37]. An approach developed by Vikram and Vedam [38] requires using a selective counting path for eliminating the halo intensity effect on fringe maxima and minima positions. Joenathan and Sirohi [40] reported the use of proper aperturing of the imaging lens to generate constant intensity over a portion of the halo. The Young's fringes over this region yield measurements free of errors from varying intensity of the halo. The technique proposed by Georgieva [41] does not require a priori knowledge of the exact profile of the halo function. Hinsch [42] has taken into account the positional variation of fringe visibility in the halo and presented a generalized theory of these effects. Finally, Joenathan [43] has reported a comparative study of the errors introduced in coherent and incoherent speckle photography due to the presence of the halo function.

Factors Affecting Fringe Quality

Several authors [45–50] have presented a detailed theoretical analysis to account for the decorrelation factors that affect the quality of fringes produced in speckle photography. The decrease in fringe visibility has been attributed mainly to the microstructural changes that occur in the displaced speckle patterns or to the decorrelation effects associated with the spatial filtering process.

When an object is deformed, speckle movements occur in the pupil plane of the imaging lens. As a result, the imaging lens samples slightly different compositions of speckles during each exposure. That the wave fronts contributing to speckle formation during the two exposures are not perfectly identical induces a state of partial decorrelation between the two speckle fields. The loss in correlation reduces the contrast of the Young's fringes. The effect of arbitrary displacement and deformation on fringe visibility has been examined in detail. The presence of tilts and slopes contributes strongly to the microstructural changes between the two speckle patterns. The effect of longitudinal speckle movements remains small as long as the displacement remains well within the limits described by Eq. (2.5).

The speckle displacement caused by object deformation is, in general, nonuniformly distributed in the image plane. If we consider pointwise filtering, each pair of speckles contained in the illuminated area produces its own set of Young's fringes of given spacing and orientation. Since several such speckle pairs coexist in the illuminated area, an incoherent superposition of these fringe patterns ensues, giving rise to a "blurring" of the overall pattern.

A specklegram is the recording of a large number of identical speckle pairs. Each speckle pair, in turn, is made up of two correlated and mutually displaced speckles. The illuminated area on the specklegram contains many identical speckle pairs but also many isolated speckles whose counterparts lie outside the illumination aperture area. The light diffracted by these isolated speckles does not participate in fringe formation and just adds to the noisy background.

2.2.2 Focused-Image Speckle Photography

The method due to Archbold et al. [9, 10] consists of recording in the image plane the irradiance distributions of a coherently lit object surface in its two states of deformation. The object surface is assumed to be sharply focused in the image plane (Fig. 2.3). The photographic plate that records two laterally displaced speckle patterns could then be analyzed in either of the two ways described in Section 2.2.1 to yield the in-plane displacement undergone by the object surface. At an arbitrary point on the scattering surface, the speckle shifts are given by

$$a_x(x, y) = u(x, y) + \frac{x}{d_o} w \tag{2.8a}$$

and

$$a_y(x, y) = v(x, y) + \frac{y}{d_o} w \tag{2.8b}$$

where $u(x, y)$, $v(x, y)$, and $w(x, y)$ are the displacements undergone by the object point along the x, y, and z directions, respectively. The speckle thus moves in a prescribed fashion governed by Eqs. (2.8).

The influence of the out-of-plane displacement component (w) including tilts on the in-plane component of speckle shift becomes progressively important as the observation point on the object surface shifts away from the optic axis [51, 52]. If one were to neglect the influence of the out-of-plane displacements, the resulting interpretation of the Young's fringes might in some cases lead to important errors. The relative errors entering into the measurement are given by

$$\frac{\delta u}{u} = \frac{w}{u} \tan \psi_x; \qquad \frac{\delta v}{v} = \frac{w}{v} \tan \psi_y \tag{2.9}$$

where ψ_x and ψ_y are the angular components of the viewing angle ψ along the x and y axes, respectively. Thus for the same measurement problem, relatively large errors would be obtained by the use of a wide-angle lens as compared to a telephoto lens.

Similarly, the whole-field display of in-plane displacement contours is also not independent of errors. These fringes are described by

$$u + \frac{x}{d_o} w = \frac{n\lambda f}{\zeta_{0x}} \tag{2.10a}$$

$$v + \frac{y}{d_o} w = \frac{n\lambda f}{\zeta_{0y}} \tag{2.10b}$$

where the filtering aperture is assumed to be placed along the x axis to obtain Eq. (2.10a) and along the y axis to obtain Eq. (2.10b). The lateral rigid-body movements of the object surface do not affect the fringe interpretation. Equation (2.10) shows that in the presence of out-of-plane movements the whole-field fringe display is subject to the same errors as the point-by-point mode of fringe analysis. Hence, depending on the metrological problem at hand, serious errors could occur during measurements. An example of a fringe pattern depicting the in-plane displacements along the x and y directions in a wooden beam structure subjected to deformation is shown in Fig. 2.4.

A single pair of equations as in expressions (2.8) or (2.10) is not sufficient to solve for the three unknown displacement components contained in the respective equation pairs. One solution would be to create an additional pair of equations similar to (2.8) or (2.10) by recording simultaneously on separate photographic plates the intensities irradiated by a point on the object in two different directions. A point-by-point analysis of the two specklegrams thus allows one to not only eliminate measurement errors

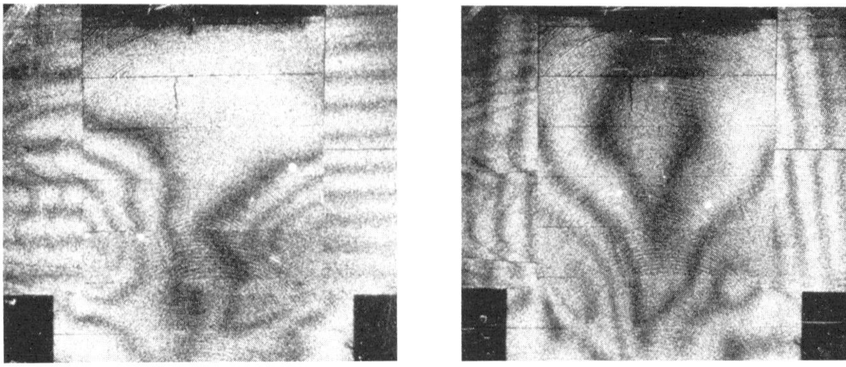

a) b)

Figure 2.4 Examples of in-plane fringe contours along the (a) x and (b) y directions in a wooden beam structure subjected to deformation.

but also to reinforce the method's measurement capacity by providing information on the complete displacement vector.

In focused-image speckle photography a purely geometrical description is sufficient to describe the speckle movements for both on- and off-axis points. Although the inclination and shape of the illumination beam do not affect the measurements, these parameters do affect the out-of-plane speckle shifts and the speckle movements in the pupil plane. These movements, by reducing the fringe quality, impose additional limitations on the measurement range of the method. The influence of lens aberration on measurements has been examined in detail by Stetson [53].

2.2.3 Defocused Speckle Photography

A broad class of methods in coherent speckle photography do not necessitate the focusing of the object surface on the film plane to obtain a speckle pattern. The coherent speckles are present all over the space surrounding the illuminated object surface. The concept of focusing the imaging lens on a plane in front of or behind the specimen surface has been developed and widely applied to separate out and measure different displacement and deformation components. The introduction of defocusing into an imaging system modifies the characteristics of speckle movements in the film plane. In such a situation, an ensemble of points contained in the cone projection on the object contribute to an observed point in the image plane. When an object is deformed, the speckles in the three-dimensional space surrounding the object also move. Aiming at speckles lying in a defocused

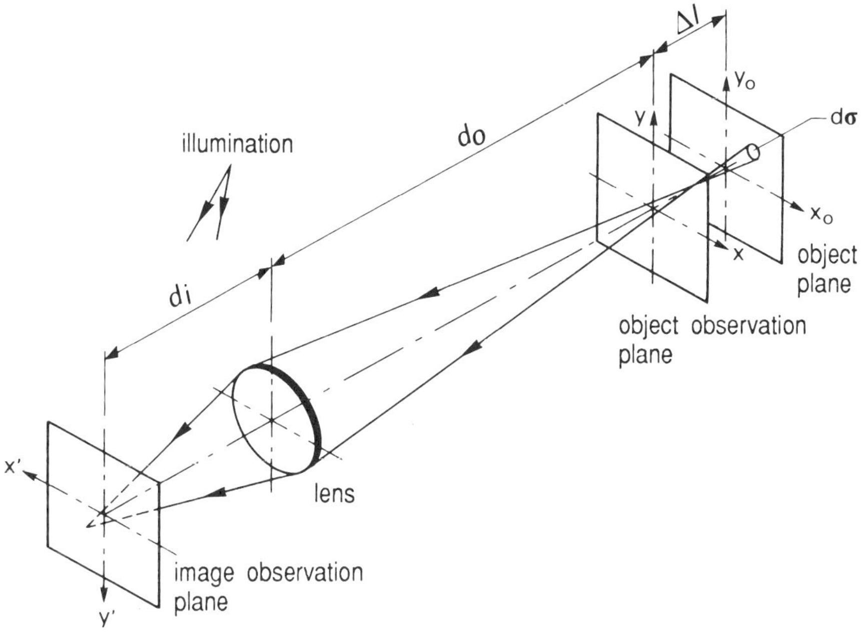

Figure 2.5 Optical system illustrating the voluntary defocusing introduced in speckle photography.

plane, Tiziani [54, 55], Gregory [56–58], and Chiang and Juang [59] have, through their different works, extended remarkably the domain of application of speckle photography.

Because of the voluntary defocusing present in the optical system, the observed speckle displacements at a point in the object-observation plane result from the movements over the element $d\sigma$ on the object surface. The scheme is shown in Fig. 2.5. A practical consequence of this averaging is that $d\sigma$ also corresponds to the smallest region that can be distinguished on the object surface during image reconstruction. These undesirable effects can be reduced by keeping the diameter κ of the surface element $d\sigma$ as small as possible,

$$\kappa = \frac{\Delta l \cdot m}{N(1 + m)} \tag{2.11}$$

where N is the f number of the lens and Δl is the distance arising due to defocusing between the object and object-observation planes as shown in Fig. 2.5. Equation (2.11) suggests that in order to minimize the difficulties associated with defocusing, it is advisable to work with as large N and as small m as possible.

The measurement of in-plane displacements assumes that the imaging lens is focused on the object, whereby a faithful correspondence between the object and its image, leaving aside the aberration effects, is maintained. This correspondence is totally lost in Tiziani's method [54], in which all points on the object contribute to a speckle in the back focal plane of the lens. Lying between the two extremes of no loss and total loss of correspondence between the object and imaged speckles are the techniques of Gregory [56] and Chiang and Juang [59]. These authors emphasize the application of speckle photography to those deformation problems that show rapid variations of tilt over the object surface. Gregory describes a theory to show the existence of transverse-motion-independent tilt-only planes, while Chiang and Juang demonstrate an optical configuration that links the speckle shifts in a defocused plane to the slope change.

Finally, general expressions relating the speckle displacement vector in a function of the optical configuration and displacement/deformation components of the object have been developed by a number of authors, notably Stetson [60], Jacquot and Rastogi [61], Jacquot [62], and Yamaguchi [63]. Throughout this section, we will employ expressions developed and reported in Refs. 61, 62, and 64.

Tiziani's Method

The two-exposure method for measuring out-of-plane rotations was conceived by Tiziani [54]. The principle is based on the fact that each speckle in the back focal plane of the lens receives contributions from all the rays scattered in a particular direction by each diffusing point on the object surface. This is schematized in Fig. 2.6. The object is assumed to be illu-

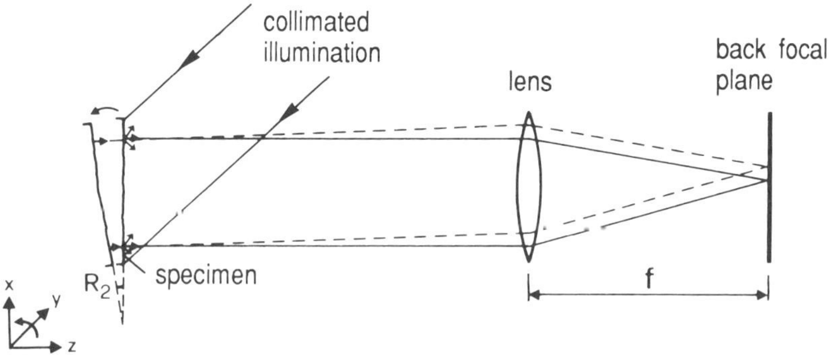

Figure 2.6 Configuration, due to Tiziani, to measure surface tilts.

minated by a collimated beam making an angle θ with respect to the surface normal. In the presence of translations, the speckle in the back focal plane remains stationary, whereas in the case of tilt, all the rays contributing to the formation of a given speckle travel in a slightly different direction, resulting in a lateral shift of the speckle in the focal plane. In the presence of out-of-plane rotations, the lateral speckle shifts along the horizontal and vertical axes in the back focal plane are given, respectively, by

$$d_x = \frac{fR_2(\cos\theta + \cos\psi_x)}{\cos^2\psi_x} \tag{2.12a}$$

and

$$d_y = -fR_1(\cos\theta + \cos\psi_x) \tag{2.12b}$$

where f is the focal length of the imaging lens; R_1, R_2, and R_3 are the rigid-body rotations around the x, y, and z axes, respectively; and the illumination source is assumed to lie in the xz plane. Assuming normal illumination and observation, Eqs. (2.12) reduce to

$$d_x = 2fR_2 \tag{2.13a}$$

$$d_y = -2fR_1 \tag{2.13b}$$

A film placed in the back focal plane records the speckle pattern before and after a tilt is given to the object. The measurement of tilt is obtained without any ambiguity as the speckles in this plane are independent of the in-plane movements and sensitive only to specimen rotations. The film after development is processed in the setup of Fig. 2.1. The speckles in the back focal plane of the lens, however, cease to be insensitive to lateral surface displacements when the object is illuminated by a divergent beam.

A disadvantage of Tiziani's system thus ensues from the use of collimated illumination, which necessarily puts an upper bound on the object size that can be studied unless large collimated optics are incorporated in the system.

Gregory's Method for the Measurement of Tilt Variations

In many engineering applications the need often arises to use a divergent illumination source to address large structures. Another need is to visualize the response of these structures to specific displacement components in the presence of complex deformations undergone by the object surface. Gregory [56, 57] recognized these problems and reported on the existence of a tilt-only plane for diverging beam illumination. He showed that the translation-independent tilt plane corresponds to the virtual plane containing the image of the point source of illumination in the object surface simulated as a mirror. The scheme for locating the image of the point source of illumination is shown in Fig. 2.7.

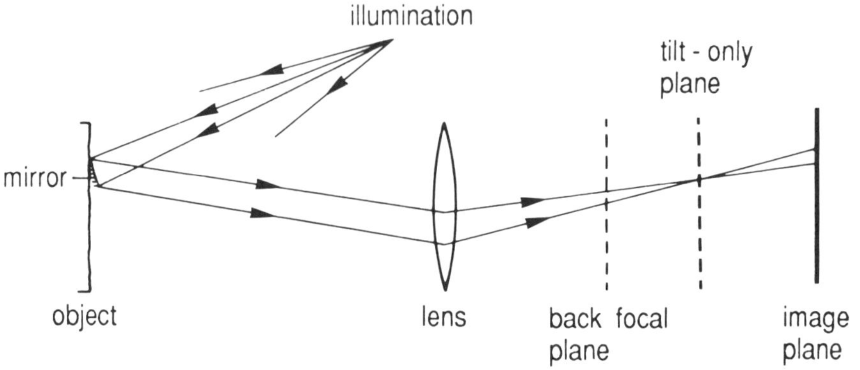

Figure 2.7 Illustration of the tilt-only plane in a configuration using divergent light to illuminate the object (Gregory).

If we assume that the illumination source lies at the center of the lens pupil, the speckle shifts caused by deformations, at a point $(x, 0)$ in the virtual plane containing the image of the point source of illumination, are given by

$$a_x = \frac{-2\,\Delta l\cdot(\beta - \varepsilon_x)\tan\psi_x}{\cos^2\psi_x} \tag{2.14a}$$

and

$$a_y = 2\,\Delta l[\alpha + (\varepsilon_y - \gamma)\tan\psi_x] \tag{2.14b}$$

where Δl refers to the distance between the object and virtual planes; α, β, and γ denote the components of rotation around the x, y, and z axes, respectively; and ε_x and ε_y are the components of in-plane derivatives along the x and y directions, respectively. These components are expressed as

$$\alpha = R_1 + \frac{\partial w}{\partial y}, \qquad \beta = R_2 - \frac{\partial w}{\partial x}, \qquad \gamma = R_3 + \frac{1}{2}\left(\frac{\partial v}{\partial x} - \frac{\partial u}{\partial y}\right) \tag{2.15a}$$

and

$$\varepsilon_x = \frac{\partial u}{\partial x}, \qquad \varepsilon_y = \frac{\partial v}{\partial y} \tag{2.15b}$$

Under these conditions of illumination and observation, the virtual plane corresponds to an error-free null transverse and axial translation plane as predicted by Gregory.

Chiang and Juang System

Chiang and Juang [59] described a method for the measurement of slope change in recording the speckle movements in a defocused plane. In the configuration of collimated illumination at an angle θ with respect to the optical axis, and observation at a plane in front of the object surface, the speckle shifts observed at a point $(x, 0)$ on the object-observation plane can be written as

$$a_x = u - w \tan \psi_x + \frac{\Delta l}{\cos^3 \psi_x} \times \{(\cos \theta + \cos \psi_x)\beta - (\sin \theta + \sin \psi_x)\varepsilon_x\} \quad (2.16a)$$

$$a_y = v - \frac{\Delta l}{\cos \psi_x} \times \{(\cos \theta + \cos \psi_x)\alpha + (\sin \theta + \sin \psi_x)(\varepsilon_y - \gamma)\} \quad (2.16b)$$

The extension of the formula to any point (x, y) is far more complicated, being a function of other deformation components as well. Hence, using this method, one can rely on the measurement of slopes, along the horizontal and vertical axes, only as long as the lateral and axial displacements can be considered to be negligible.

A basic problem in the application of the method to slope-change measurement arises from the presence of in-plane displacements and strains. However, in the study of the plate-bending problem, for which this system was initially conceived, the in-plane displacement field is much smaller than the out-of-plane displacement field. Rigid-body rotations and translations can be assumed to be eliminated by means of a well-built loading device. Under these assumptions, the speckle shifts for points lying in the neighborhood of the optical axis are given by

$$a_x = -\Delta l \left[\varepsilon_x \sin \theta + \frac{\partial w}{\partial x}(1 + \cos \theta) \right] \quad (2.17a)$$

$$a_y = -\Delta l \left[\varepsilon_y \sin \theta + \frac{\partial w}{\partial y}(1 + \cos \theta) \right] \quad (2.17b)$$

As we will see later in Section 2.4, expression (2.17) relating the speckle movements to the displacement derivatives is the same as that in speckle shear interferometry. Stated briefly, the method is based on the premise that by deliberately introducing defocusing into the optical system, the configuration's sensitivity characteristics are modified to be responsive to the displacement derivative components only. The sensitivity of measurements can be varied by modifying the degree of defocusing, Δl.

A case of particular interest arises when the illumination source is moved such that its apparent position seems to lie at the center of the imaging lens. The speckle shifts at a point $(x, 0)$ on the object observation plane are accordingly given by

$$a_x = \left(1 \pm \frac{\Delta l}{d_o \pm \Delta l}\right)\left(u - \frac{x}{d_o} w\right) \pm 2\,\Delta l \left(1 + \frac{x^2}{d_o^2}\right)\left(\beta - \frac{x}{d_o} \varepsilon_x\right) \tag{2.18a}$$

and

$$a_y = \left(1 \pm \frac{\Delta l}{d_o \pm \Delta l}\right) v \mp 2\,\Delta l \cdot \alpha \pm \frac{2\,\Delta l \cdot x}{d_o} (\gamma - \varepsilon_y) \tag{2.18b}$$

where the upper sign refers to observation in a plane lying in front of the object and the lower one refers to virtual observation in a plane lying behind the object.

Figure 2.8a shows some examples of the variation of the amplification factor Λ of the object displacements u and w (defined as the ratio of the speckle displacement to the object displacement producing the given speckle shift) as a function of $\Delta l/d_o$ for three different values of field angles ψ_x. Curves 1 and 1′ display the plots of the amplification factors associated

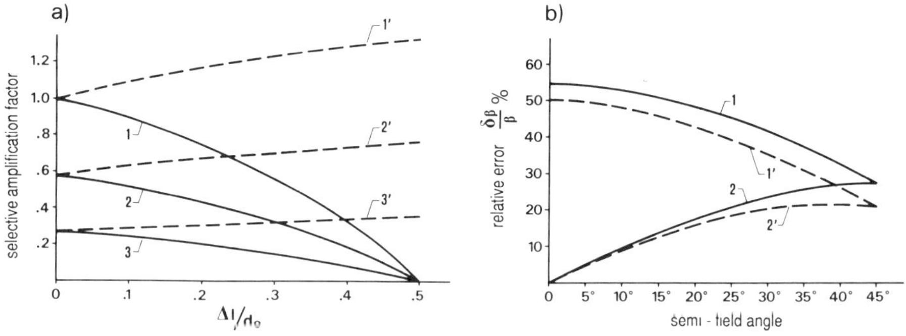

Figure 2.8 (a) Plots of the variation of amplification factors Λ_{xu} (1 and 1′) and Λ_{xw} (1 and 1′, 2 and 2′, 3 and 3′) as a function of $\Delta l/d_o$; Λ_{xu} is independent of field angles. Curves 1 and 1′, 2 and 2′, 3 and 3′ correspond to $\psi_x = 45°$, 30°, and 15°, respectively. (b) Plots of the variation of the relative error $\delta\beta/\beta$ introduced in the measurement of slope change, $\beta = \partial w/\partial x$, in the presence of u (1 and 1′) and w (2 and 2′) displacement components.

with the displacement component u for speckle shifts in the x direction (a_x) as a function of $\Delta l/d_o$. The corresponding equation is given by

$$\Lambda_{xu} = \frac{a_x(u)}{u} = 1 \pm \frac{\Delta l}{d_o \pm \Delta l} \tag{2.19}$$

Curves 1 and 1′, 2 and 2′, and 3 and 3′ display the amplification factors Λ_{xw} associated with the displacement component w for speckle shifts in the x direction, a_x, as a function of $\Delta l/d_o$,

$$\Lambda_{xw} = \frac{a_x(w)}{w} = \left(1 \pm \frac{\Delta l}{d_o \pm \Delta l}\right) \tan \psi_x \tag{2.20}$$

for semi-field angles $\Psi_x = 45°, 30°$, and 15°, respectively.

The sensitivity to translation components decreases for increasing $\Delta l/d_o$ for virtual speckles, eventually dropping to zero for $\Delta l/d_o = 0.5$, the so-called tilt-only plane of Gregory. On the other hand, for real speckle observation, the sensitivity to translation components increases with increasing $\Delta l/d_o$. Thus, the relative response of the method to slope measurements, in the presence of in-plane displacements, is better for an observation plane lying behind the object than in front of it.

As indicated by Eqs. (2.19) and (2.20) and Fig. 2.8a, the amplification factors of the displacement components u and w could be of the same order as that of deformation components, α, β, or γ. Figure 2.8b illustrates by means of a few examples the uncertainties to which the measurements can be exposed in interpreting the point-by-point fringes as being uniquely due to the local slope component. Curve 1 shows the variation of the relative error $\delta\beta/\beta$ introduced in the measurement of slope change, $\beta = \partial w/\partial x$, as a function of the semi-field angle ψ_x:

$$\frac{\delta\beta}{\beta} = \frac{u}{2\,\Delta l\cdot\beta}\left(1 + \frac{\Delta l/d_o}{1 + \Delta l/d_o}\right) \cos^2 \psi_x \tag{2.21}$$

where, apart from the slopes, the only other displacement component assumed to be present is u. Similarly, curve 2 shows the variation of the relative error $\delta\beta/\beta$ in the presence of w:

$$\frac{\delta\beta}{\beta} = \frac{w}{2\,\Delta l\cdot\beta}\left(1 + \frac{\Delta l/d_o}{1 + \Delta l/d_o}\right) \sin \psi_x \cos \psi_x \tag{2.22}$$

Curves 1 and 2 have been drawn under the assumption that the illumination source is situated at the center of the viewing system. On the other hand, curve 1′ shows the variation of the relative error $\delta\beta/\beta$ in the presence of u but in the configuration of collimated illumination at zero incidence:

$$\frac{\delta\beta}{\beta} = \frac{u \cos^3 \psi_x}{\Delta l \cdot \beta(1 + \cos \psi_x)} \tag{2.23}$$

Curve 2′ is the analog of curve 2 but in the configuration of collimated illumination at zero incidence.

$$\frac{\delta\beta}{\beta} = \frac{w \sin \psi_x \cos^2 \psi_x}{\Delta l \cdot \beta(1 + \cos \psi_x)} \tag{2.24}$$

For the purpose of plotting these curves we have assumed that the observation plane lies in front of the object, $\Delta l/d_o = 1$, and that $u/\Delta l \cdot \beta$ is constant and equal to 1. The value used for $u/\Delta l \cdot \beta$ is only illustrative, but in real situations should be chosen as a function of the deformation problem at hand in order to evaluate the measurement uncertainty associated with the method. Serious errors may arise if defocusing is arbitrary and the products of type $\Delta l \cdot \alpha$ or $\Delta l \cdot \beta$ approach the values of the displacement components u and w. Finally, curves 1 and 1′ show that it is possible to obtain configurations in which relative errors decrease with increasing field angles. Quite an unusual result when considered from the point of view of optical metrology. Likewise, the whole-field fringes are subject to errors similar to those associated with the pointwise fringes in the measurement of slopes.

This section would be incomplete without a word on the conceptually elegant technique of tandem speckle photography proposed by Stetson and Harrison [66]. Their technique requires recording a focused plane and two tandem specklegrams simultaneously for at least three different directions of illumination. Subseqent solving of the linear system of equations allows one to determine the displacements, slopes, and surface strains undergone by the object.

2.3 SPECKLE INTERFEROMETRY

2.3.1 Principles

Another interesting and attractive class of methods in speckle metrology is speckle interferometry [8, 70–94]. A significant feature of this method is that its sensitivity is comparable to that obtainable in holographic interferometry. Speckle interferometric methods are based on the coherent addition of a speckle field diffracted by an object and a reference field, which can either be a plane wave or another speckle field not necessarily originating from the object surface under study.

The introduction of a reference field brings about a significant change in the behavior of the speckle pattern when the object is deformed. As the observed speckle pattern is formed by the interference of two coherent fields, the intensity in the resultant pattern depends on the relative phase distributions of the added fields. If the object is deformed, the relative phase of the two fields will change, causing a variation in the irradiance of the resultant pattern. The deformed speckle pattern will match the initial pattern (correlated) if the variation of the path difference is an integral multiple of λ, and a contrast reversal occurs (decorrelation) for a variation of half the integral multiple of λ.

The principle of measurement based on speckle interferometry consists of observing the evolution of a speckle pattern obtained by the coherent superposition of two independent fields, as a function of the phase difference between the two fields. Let us suppose that the speckle pattern in the image plane results from the coherent combination of two uncorrelated speckle fields. Any displacement of a point on the object surface gives rise to a relative phase difference between the two fields, given by

$$\Delta\varphi = \varphi_1 - \varphi_2 \tag{2.25}$$

where φ_1 and φ_2 are the phase differences introduced in the two fields by the object deformation.

If a_1 and a_1', a_2 and a_2' represent the diffracted amplitudes at a point in the image plane before and after the deformation of the object surface, the total amplitude in each state is given by

$$A = a_1 + a_2, \qquad A' = a_1' + a_2' \tag{2.26}$$

The coefficient of correlation between the two amplitudes is defined by

$$\mu = \frac{\langle AA'^{*}\rangle}{\langle |A|^2\rangle\langle |A'|^2\rangle} \tag{2.27}$$

the mean being taken over an infinity of independent scattering surface elements. Owing to the circular complex Gaussian statistics of the speckle fields, the coefficient of correlation between the corresponding intensities,

$$I = |A|^2 \qquad \text{and} \qquad I' = |A'|^2 \tag{2.28}$$

is related to μ [80] by the equation

$$C = \frac{\langle II'\rangle - \langle I\rangle\langle I'\rangle}{[\langle (I - \langle I\rangle)^2\rangle\langle (I' - \langle I'\rangle)^2\rangle]^{1/2}} = |\mu|^2 \tag{2.29}$$

The expression for μ can be written as

$$\mu = \frac{\exp(-j\varphi_1) + \exp(-j\varphi_2)}{2} \tag{2.30}$$

where we have made use of

$$\langle a_1 a_2'^* \rangle = \langle a_2 a_1'^* \rangle = 0 \tag{2.31}$$

Consequently, the coefficient of correlation is given by

$$C = \frac{1 + \cos \Delta\varphi}{2} = \cos^2 \frac{\Delta\varphi}{2} \tag{2.32}$$

The fringes in speckle interferometry, suitably interpreted as "speckle correlation fringes," arise from the correlation of two speckle pattern intensity distributions obtained before and after the object deformation. The technique basically consists of recording the intensities I and I'. Several methods exist for comparing the two random intensity distributions to put in evidence the coefficient $C = |\mu|^2$. The regions of correlation are obtained as the loci of points that have undergone a phase change of 2π or its multiple between the two resultant speckle patterns. Finally, for the methods based on speckle and speckle shear interferometry to function successfully, the mean lateral and longitudinal dimensions of the speckle in the measurement system must necessarily be greater than the respective displacement components when the object is deformed.

Rastogi [81] and Spajer et al. [82] applied the notion of the coefficient of correlation between two interfering wave fronts to study fringe formation and visibility in speckle interferometry. The parameters affecting the visibility of the in-plane displacement fringes are analyzed in the case of an arbitrary displacement and deformation of the object surface. In the presence of strong granularity, the fringe visibility is, however, not sufficient to define adequately the quality of the fringes. The concept of signal-to-noise was taken into account for describing the fringe quality [81, 82].

Influence of Speckle Polarization

In many applications of speckle interferometry one is concerned with rough metallic surfaces. When illuminated by a polarized light such surfaces diffract a strongly polarized wave. In this case, the probability density function for speckle intensity is given by

$$p(I) = \frac{1}{\langle I \rangle} \exp\left(-\frac{I}{\langle I \rangle}\right) \tag{2.33}$$

However, to obtain a more luminous image it may sometimes be necessary to deposit on the surface a strongly diffusing paint, which is generally depolarizing. The speckle so obtained is the sum of two orthogonally polarized and uncorrelated speckles. The corresponding intensity distribution has the same statistical properties as the sum in intensity of the two speckles,

$$p(I) = \frac{4I}{\langle I \rangle^2} \exp\left(-\frac{2I}{\langle I \rangle}\right) \tag{2.34}$$

If I_v and I_h are the intensity distributions of the two orthogonally polarized speckles, it follows that

$$I = I_v + I_h, \qquad I' = I'_v + I'_h \tag{2.35a}$$

$$\langle I_v I'_h \rangle - \langle I_v \rangle \langle I'_h \rangle = 0 \tag{2.35b}$$

$$\langle I'_v I_h \rangle - \langle I'_v \rangle \langle I_h \rangle = 0 \tag{2.35c}$$

The expression for the coefficient of correlation between I and I' is thus unchanged. The difference in statistics affects only the profile of the correlation fringes.

2.3.2 Observation of Correlation Fringes

Methods

Several methods have been developed to visualize the correlation fringes. The first, though not the most employed, of these methods is the mask technique. The technique consists of recording the initial speckle intensity on a photographic plate. The plate after processing is replaced in its original position. The speckle distribution I' is observed through the mask (negative of distribution I). In the regions where I and I' are identical, the images are complementary. The flux transmitted in these regions is much weaker than in the regions where I and I' are uncorrelated. As a result, correlation fringes are seen to appear covering the image in the form of light and dark bands arising from the variation in the light transmitted through the mask. These fringes are obtained in real time.

A method quite akin to the mask technique relies on photographic subtraction to detect the positions of correlation. The speckle distribution I' is recorded on a separate plate, and its positive contact print is made on another plate. The two plates, one containing the negative image of the distribution I and the other the positive image of the distribution I', are placed in register and adjusted to yield correlation fringes to an observer looking through the pair of plates. This mode of generating correlation fringes is very interesting as it allows for compensating rigid-body movements with the result that the fringe contours are of relatively good quality.

Double-exposure methods have been used to detect the positions of correlation. Speckle patterns are recorded on a photographic plate before and after object deformation. The regions where the phase difference between the two speckle patterns remains the same or changes by $2n\pi$ give rise to a coherent addition of the two speckle fields. The regions where the phase changes by $(2n + 1)\pi$ between the two exposures give rise to an incoherent addition of the two speckle fields. Although these regions have the same average intensities, they do differ in texture. The detection can be improved by making use of the nonlinearity of the photographic process. In the regions with speckle texture (coherent combination), the flux transmitted is much higher than in the regions of incoherent combination of the speckle fields.

A useful method of generating correlation fringes uses a filtering technique that consists of recording on a photographic plate two slightly shifted images of the object. Illuminating the recorded film with a plane wave gives rise to a diffraction pattern that is modulated by fringes of spacing inversely proportional to the given shift. The light intensity at the minima of the diffraction fringes receives contributions only from the uncorrelated regions on the plate. A slit centered on a dark fringe in the diffraction halo passes light from only these regions. The dark fringes correspond to path length variations of $n\lambda$, while the bright fringes correspond to path length variations of $(2n + 1)\lambda/2$. Such a filtering system is shown in Fig. 2.9.

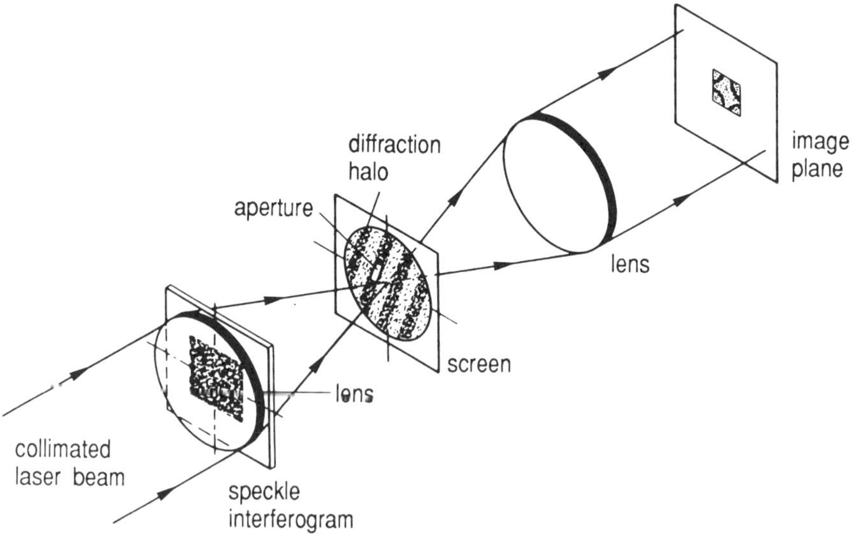

Figure 2.9 Fourier filtering arrangement to obtain the visualization of correlation fringes in speckle interferometry.

Although the information supplied by the mask or photographic subtraction techniques is similar to that obtained by methods based on double exposure, the former two methods distinguish themselves by their capacity to compensate for rigid-body movements undergone by the object during loading.

Another attractive method for observating correlation fringes is based on the electronic subtraction of intensities [71, 83–87] before and after object deformation. Supposing that the image intensity displayed on the TV monitor is proportional to $|I - I'|$, electronic speckle pattern interferometry (ESPI) presents an analogy, at least qualitative, in the imaging aspects with the mask technique. The rapidity of ESPI is best illustrated by its possibility to provide an interferogram every 1/25 sec. ESPI has been combined with the phase-shifting technique to yield a quantitative display of the phase of the object's displacement corresponding to the interference fringes [88–90, 95]. Phase steppings are introduced in one of the two arms of the interferometer, irrespective of the arm's dependence on object deformation. Finally, digital image processing techniques have been applied to speckle pattern interferometry to obtain improved quality correlation fringes [91, 92].

Profile and Contrast of Fringes as a Function of Emulsion Characteristics

We consider here for purposes of illustration the case of mask or sandwich techniques. The negative image of I' is superposed on the distribution I. If $T_n(I')$ is the transmittance of the negative emulsion, the mean intensity of the resulting image is

$$\langle E \rangle = \iint_0^\infty I T_n(I') p(I, I') \, dI \, dI' \tag{2.36}$$

where the probability density function $p(I, I')$ for polarized speckle [80] is given by

$$p(I, I') = \frac{\exp\left[-(I + I')/\langle I \rangle(1 - C)\right]}{\langle I \rangle^2(1 - C)} \cdot I_0 \left[\frac{2\sqrt{II'C}}{\langle I \rangle(1 - C)}\right] \tag{2.37}$$

where $I_0(x)$ is the modified Bessel function of zero order,

$$I_0(x) = \frac{1}{2\pi} \int_{-\pi}^{+\pi} \exp(x \cos\theta) \, d\theta \tag{2.38}$$

If we superpose a positive image of I onto a negative image of I', the mean transmittance of the superposed image becomes

$$\langle T \rangle = \iint_0^\infty T_p(I) T_n(I') p(I, I') \, dI \, dI' \tag{2.39}$$

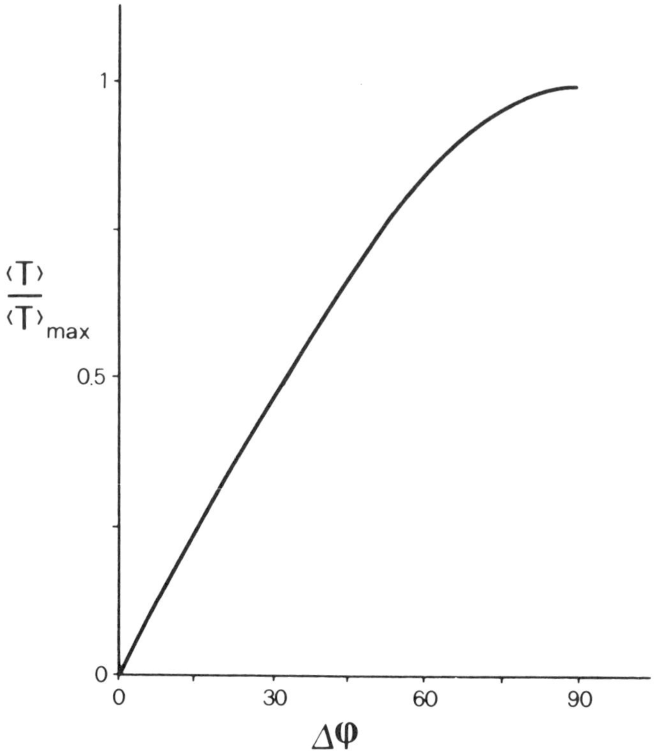

Figure 2.10 An example of the profile of correlation fringes.

where $T_p(I)$ denotes the transmittance of the positive emulsion. This expression defines the profile of the observed fringes [81, 82]. An example of the profile of the correlation fringes obtained by the sandwich method is shown in Fig. 2.10.

The mean transmittance $\langle T \rangle$ is minimum in those regions where I and I' vary identically ($C = 1$),

$$\langle T \rangle_{\min} = \int_0^\infty T_p(I) T_n(I) p(I)\, dI = \langle T_p(I) T_n(I) \rangle \tag{2.40}$$

and is maximum in those regions where I and I' are uncorrelated,

$$\begin{aligned} \langle T \rangle_{\max} &= \int_0^\infty T_p(I) p(I)\, dI \int_0^\infty T_n(I') p(I')\, dI' \\ &= \langle T_p(I) \rangle \langle T_n(I') \rangle \end{aligned} \tag{2.41}$$

where $p(I)$ depends on the degree of polarization of the granular fields. The contrast thus follows as

$$V = \frac{\langle T_p \rangle \langle T_n \rangle - \langle T_p T_n \rangle}{\langle T_p T_n \rangle + \langle T_p \rangle \langle T_n \rangle} \tag{2.42}$$

2.3.3 Measurement of Out-of-Plane Displacements

Leendertz System

The schematic of the Leendertz optical system [8] is shown in Fig. 2.11. It basically uses a setup analogous to a Michelson interferometer in which both mirrors are replaced by scatter surfaces. The resultant speckle pattern in the image plane is formed by the interference of the two speckle fields issuing independently from the two scatter surfaces. One of the surfaces is subjected to deformation, and the other serves as the source of reference speckles. If the object is illuminated by a plane parallel wave at zero

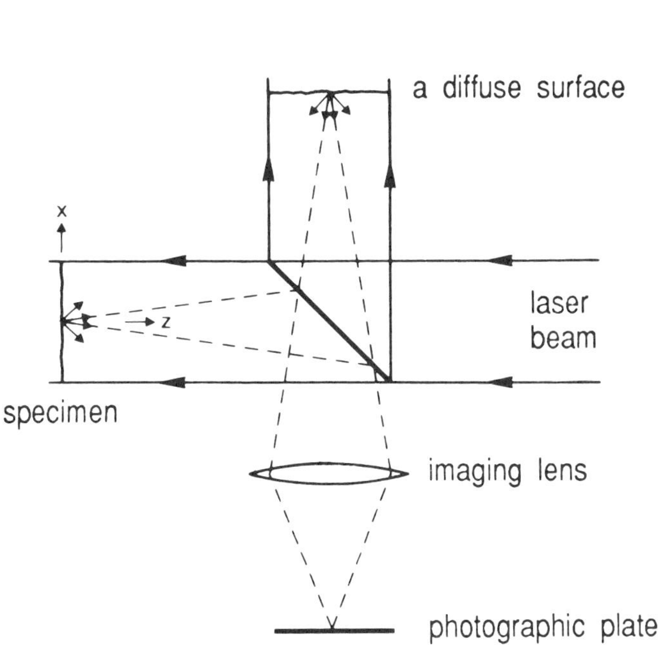

Figure 2.11 An optical setup analogous to the Michelson interferometer for measuring out-of-plane displacements.

incidence, the relative phase change between the two speckles is given by

$$\Delta\varphi\,(x, y) = \frac{2\pi}{\lambda}\left[\frac{xu}{d_o}\left(1 - \frac{x^2 + y^2}{2d_o^2}\right) + \frac{yv}{d_o}\left(1 - \frac{x^2 + y^2}{2d_o^2}\right) + 2w\left(1 - \frac{x^2 + y^2}{4d_o^2}\right)\right] \tag{2.43}$$

where we have taken the reasonable looking approximation of $x, y \ll d_o$ and the illumination beam is assumed to lie in the xz plane.

Equation (2.43) brings out clearly that in the presence of generalized deformations, and for off-axis points, one can no longer rely on the interpretation of the observed contour pattern as fringes of equal out-of-plane displacements. There are many measurement problems in which $w \gg u, v$. In such situations the fringe equation simply reduces to

$$2w\left(1 - \frac{x^2 + y^2}{4d_o^2}\right) = n\lambda \tag{2.44}$$

The interpretation of the fringes as contours of equal out-of-plane displacements is thus valid only for surface points lying in the neighborhood of the optical axis. As the fringe increment varies over the object surface, a correction at each object point (x, y) is called for when analyzing the fringe contours. Dropping the corrective quadratic terms in x and y, to facilitate fringe analysis, comes to introducing relative errors in the measurement of the out-of-plane displacement. These are of the order of

$$\frac{\delta w}{w} = \frac{x^2 + y^2}{4d_o^2} = \frac{\tan^2\psi}{4} \tag{2.45}$$

This method has been extended to determine the difference between the mechanical responses of two similarly stressed specimens [78, 86]. This difference is observed in the form of fringes that contour the change in the surface shapes of two specimens, namely, master and test, caused by deformations. The schematic of the optical system is shown in Fig. 2.12. The laser beams are incident upon the two specimen surfaces at the same angle. The speckle in the image plane results from the interference of light diffused by the master and test specimens. The two specimens are subjected to the same deformations. If φ_m and φ_t are the corresponding phase changes induced in the two speckles, the relative phase change between the two speckle fields is given by

$$\Delta\varphi = \varphi_m - \varphi_t$$

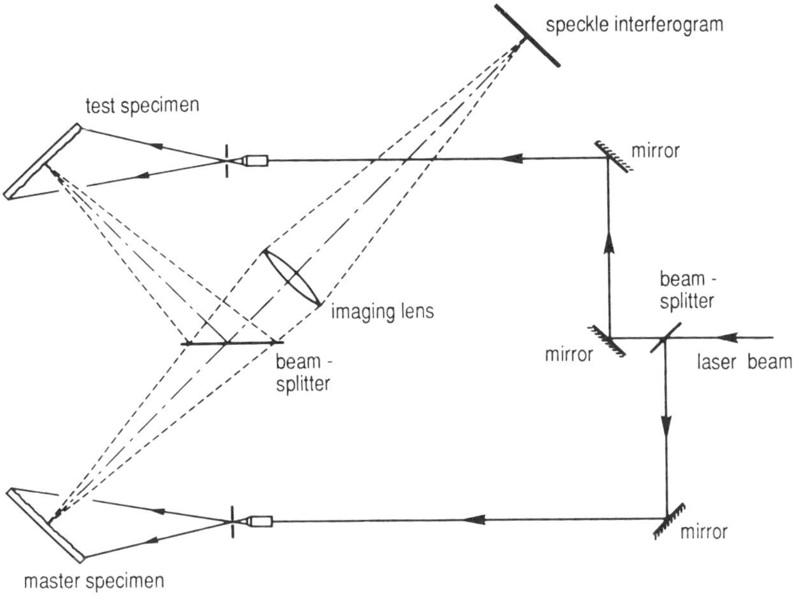

Figure 2.12 A speckle interferometric setup to measure difference of displacements of two objects subjected to the same deformations.

Under small-angle approximation and in the configuration of collimated illumination at zero incidence, the equation of the observed fringe contours is given by the relation

$$\Delta w = n\lambda/2 \tag{2.46}$$

An example of a fringe pattern obtained by the method is shown in Fig. 2.13. The fringe pattern corresponds to the difference in displacements of two square plates clamped along the edges and subjected to slightly different centrally concentrated loads.

The measurement of the difference in deformations of two macroscopically similar but different object surfaces has considerable appeal in nondestructive testing. Ganesan et al. [86] reported the use of the technique for fringe compensation and flaw detection in applications related to nondestructive testing.

Butters and Leendertz System

In the Butters and Leendertz system [71], a smooth reference wave, plane or spherical, traveling along the optical axis of the interferometer is added to the object speckle field in the image plane (Fig. 2.14). The resultant

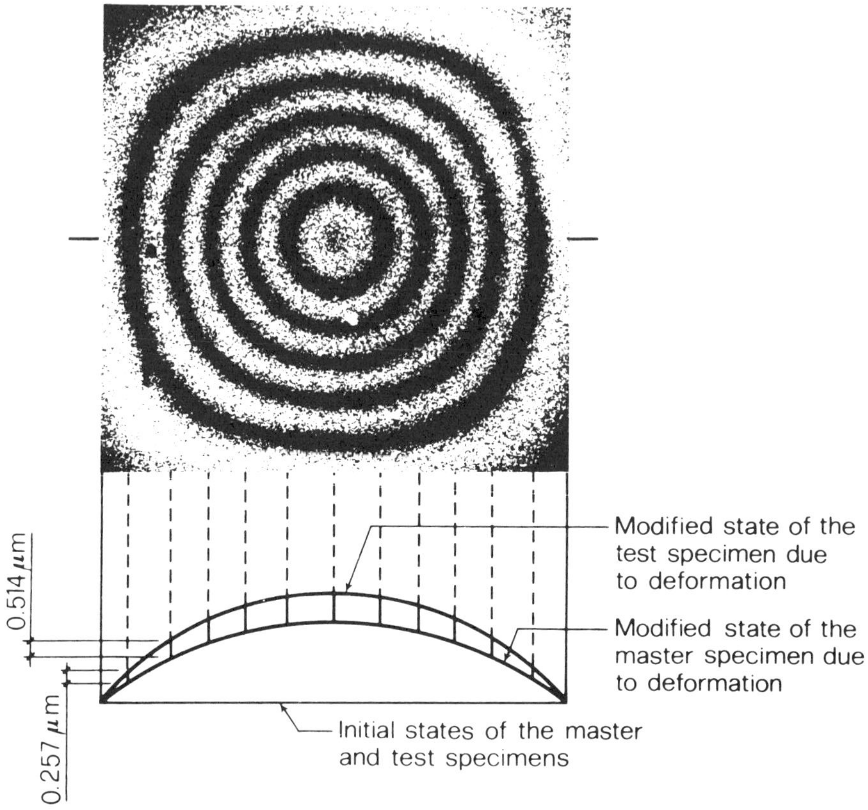

Figure 2.13 An example of the speckle correlation fringes obtained by Fourier filtering, in a setup similar to that of Fig. 2.10, of a speckle interferogram recorded in the configuration shown in Fig. 2.12. The fringes represent contours of constant difference in out-of-plane displacements Δw.

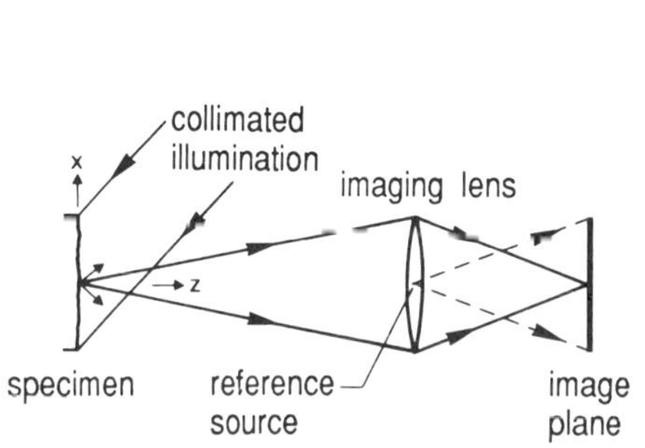

Figure 2.14 An interferometric configuration for measuring out-of-plane displacements.

speckle pattern is due to the interference of the speckle field with the smooth reference wave. The reference field is independent of the object deformation. The object is illuminated by a collimated wave lying in the xz plane and making an angle θ with respect to the surface normal. When the object is deformed, the relative path length change, $\Delta\tau$, between the two speckles is given by

$$\Delta\tau = u\left(\sin\theta + \frac{x}{d_o}\right) + \frac{vy}{d_o} + w(1 + \cos\theta) \tag{2.47}$$

In writing Eq. (2.47) we have neglected the influence of the second-order terms. This interferometer has found extensive applications in out-of-plane displacement measurements and vibration analysis. Important errors could occur if appropriate measurers are not taken to minimize the effect of parasitic displacement components during the recording of the speckle interferogram.

2.3.4 Measurement of In-Plane Displacements

Leendertz System

Figure 2.15 schematizes the basic optical system [8] used for the measurement of in-plane displacements undergone by a diffuse object surface. The object is illuminated by two collimated waves placed symmetrically and making an angle θ with respect to the surface normal. The resulting speckle pattern in the image plane is generated by the coherent superposition of the speckle fields corresponding to each of the two illumination beams.

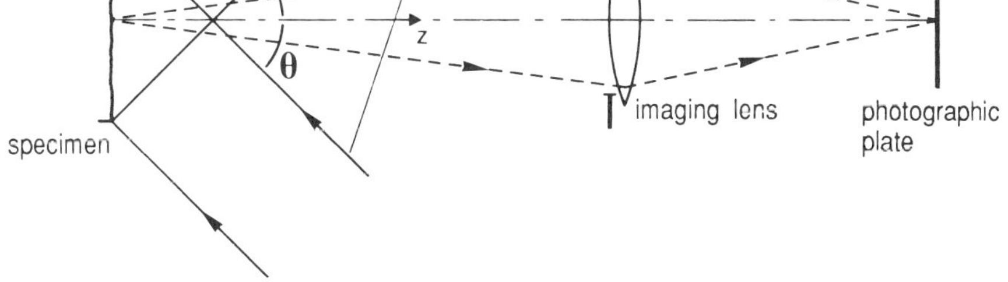

Figure 2.15 Schematic of the speckle interferometry setup to measure in-plane displacement component u along the x direction.

When the object is deformed, both speckle fields are influenced, albeit in a different manner, by the object deformation. The relative phase change between the two speckle patterns is given by

$$\Delta\varphi = \frac{4\pi}{\lambda} u \sin\theta \tag{2.48}$$

The corresponding fringes represent contours of constant in-plane displacement u. Equation (2.48) shows that the system is independent of the out-of-plane and as well as in-plane displacements in a direction perpendicular to the plane containing the two illumination beams. This configuration is also one of those rare speckle metrology methods that escape from off-axis error terms. Figure 2.16 shows an example of a fringe pattern obtained on a ring subjected to diametral compression.

Figure 2.16 An example of correlation fringes produced on a ring subjected to diametral compression. The fringe pattern depicting u was obtained using the mask technique.

On the other hand, if we suppose an error in the symmetry of the two beams with respect to the surface normal, the interpretation of the contour pattern as fringes of equal in-plane displacement induces a relative error in the measurement of the fringe order,

$$\frac{\delta n}{n} = \frac{\delta\theta}{2}\left(\cot\theta - \frac{w}{u}\right) \tag{2.49}$$

where $\delta\theta$, the difference between the two illumination beam angles, is assumed to be small. Equation (2.49) clearly shows that errors in measurement are to be expected when divergent beams are used to illuminate the object.

Duffy's Two-Aperture Method

A method of measuring in-plane displacement has been described by Duffy [75, 76]. The technique consists of imaging the object surface through a lens fitted with a double aperture located symmetrically about the lens center. The imaging system is illustrated in Fig. 2.17. The lens focused on the surface sees an arbitrary point on the object surface along two different directions. The two viewing apertures are separated by a distance D. Large speckles of characteristic size dependent on the aperture diameter are formed in the image plane. As a result of the coherent superposition of

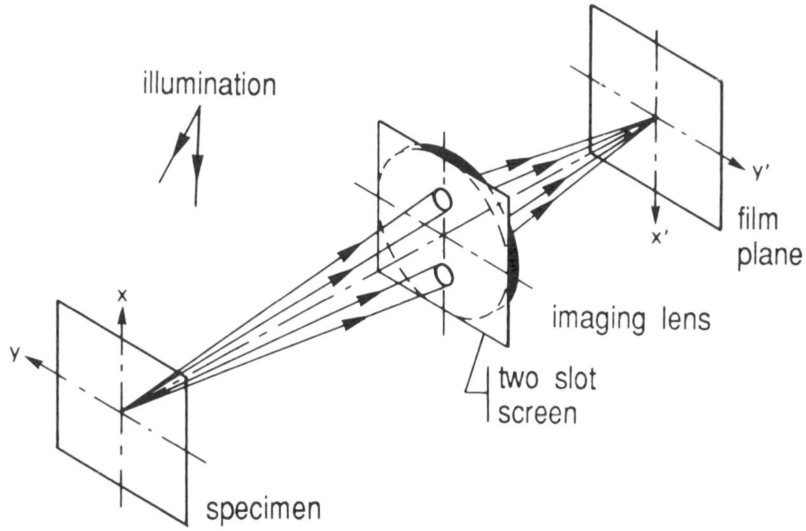

Figure 2.17 Schematic of the Duffy method for measuring in-plane displacement component u.

the beams emerging from the two apertures, each speckle is modulated by a grid structure running perpendicular to the line joining the apertures. The pitch of the fringe system embedded in each speckle depends on the separation of the two apertures.

An insight into the method's functioning is obtained as follows. For movements that are out-of-plane, or in-plane in a direction perpendicular to the line joining the apertures, the two beams diffused by a point on the object undergo equal path length changes and the speckle grid remains unchanged. In reality this statement is rigorously true for only those object points that lie on the optical axis. On the other hand, movements in a direction parallel to the line joining the two apertures cause the speckle grid to shift parallel to itself by a grating interval for every successive surface displacement equivalent to the grid spacing. For those areas that have undergone odd half-integral values of this displacement, the grid pattern becomes out of phase with the original pattern, resulting in a loss of the periodic structure of the grid within the speckle in the case of a double-exposure recording.

The relative phase change between the two beams [64, 94] before and after the object deformation can be written as

$$\Delta\varphi = \frac{2\pi D}{\lambda d_o}\left[\left(1 - \frac{3x^2}{2d_o^2} - \frac{y^2}{2d_o^2}\right) u - \frac{xy}{d_o^2} v + \frac{x}{d_o} w\right] \tag{2.50}$$

where we have used the approximation $x, y, D \ll d_o$. An examination of Eq. (2.50) reveals that if one were to interpret the fringes as due to the u component alone, large relative errors could be induced in the measurement in the presence of generalized deformations, especially at those points where u is small.

Neglecting the second-order terms, the equation of the fringe pattern reduces to

$$\frac{D}{d_o}\left(u + \frac{x}{d_o} w\right) = n\lambda \tag{2.51}$$

Fringes corresponding to Eqs. (2.50) and (2.51) can be observed and photographed in the Fourier filtering setup of Fig. 2.18. In actual deformation problems, w may not be so limited, and errors will consequently appear in the measurements. Equation (2.51) shows that the measurement of in-plane components by the present technique is prone to the same relative errors as those found in speckle photography

$$\frac{\delta u}{u} = \frac{w}{u} \tan \psi_x \tag{2.52}$$

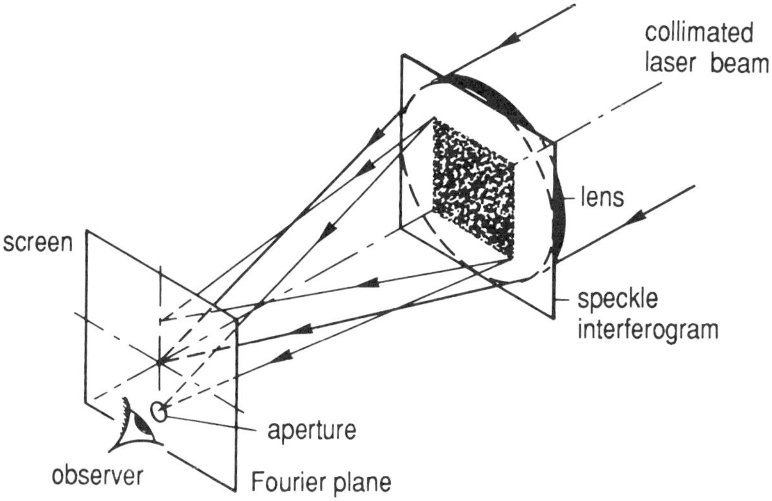

Figure 2.18 Observation of fringes in Duffy's method.

It is interesting to note that one can also arrive at the result shown in Eq. (2.50) by first writing the expression of the two-beam interference pattern from the pair of apertures. When the object is deformed, the relative path length change between the two diffused beams causes a shift in the speckle grid. The initial and shifted grid patterns can be expressed as

$$f(x, y) = n_1\lambda \tag{2.53a}$$

and

$$f(x + a_x(x, y), y + a_y(x, y)) = n_2\lambda \tag{2.53b}$$

A beat pattern is then generated between the two grid structures. The moiré pattern obtained by the superposition of the two grids follows as

$$f(x + a_x(x, y), y + a_y(x, y)) - f(x, y) = (n_2 - n_1)\lambda \tag{2.53c}$$

Computation of Eq. (2.53c) leads to results identical to those of Eq. (2.50). On the other hand, for the aperture pair placed in the yz plane, the equation of fringe contours becomes

$$\frac{D}{d_o}\left[\left(1 - \frac{x^2}{2d_o^2} - \frac{3y^2}{2d_o^2}\right) v - \frac{xy}{d_o^2} u + \frac{y}{d_o} w\right] = n\lambda \tag{2.54}$$

In a first-order approximation, Eq. (2.54) simply reduces to

$$\frac{D}{d_o}\left(v + \frac{y}{d_o} w\right) = n\lambda \tag{2.55}$$

The technique was later extended to measure simultaneously the in-plane components u and v by placing a mask containing two pairs of apertures, one each along the x and y axes, in front of the lens. The corresponding fringe expressions are described by Eqs. (2.51) and (2.55).

Indeed, the fringe equations are independent of the form and direction of the illumination wave. These two factors nevertheless affect the fringe quality by their influence on the movement of the speckle fields in the pupil plane.

2.4 SPECKLE SHEAR INTERFEROMETRY

The potential of speckle techniques for the measurement of the derivatives of surface displacements [93–105] was first put forward independently by Leendertz and Butters [96] and by Hung and Taylor [97]. The principle of this class of methods is the same as that of speckle interferometry, except that the two interfering speckles now originate from the same object and are sheared laterally with respect to each other.

2.4.1 Michelson Type of Shear Interferometer

The schematic of the Michelson type of shear interferometer [96, 98] is shown in Fig. 2.19. The object, illuminated by a collimated beam lying in the xz plane and making an angle θ with the surface normal, is imaged by means of a lens through a Michelson interferometer. A small tilt of one of the two mirrors introduces a lateral shift δx between the two speckle patterns. When the object is deformed, an arbitrary point (x, y) on the object surface is displaced to $(x + u, y + v, w)$, and a neighboring point $(x + \delta x, y)$ is displaced to $(x + \delta x + u + \delta u, y + v + \delta v, w + \delta w)$. Then the relative path length change due to the displacement between the two contributing points is given by

$$\Delta\tau = \delta x \left[\left(\sin\theta + \frac{x}{d_o} - \frac{x^2 + y^2}{2d_o^2}\right)\frac{\partial u}{\partial x} + \frac{y}{d_o}\left(1 - \frac{x^2 + y^2}{2d_o^2}\right)\frac{\partial v}{\partial x} + \left(1 + \cos\theta - \frac{x^2 + y^2}{2d_o^2}\right)\frac{\partial w}{\partial x}\right] \tag{2.56}$$

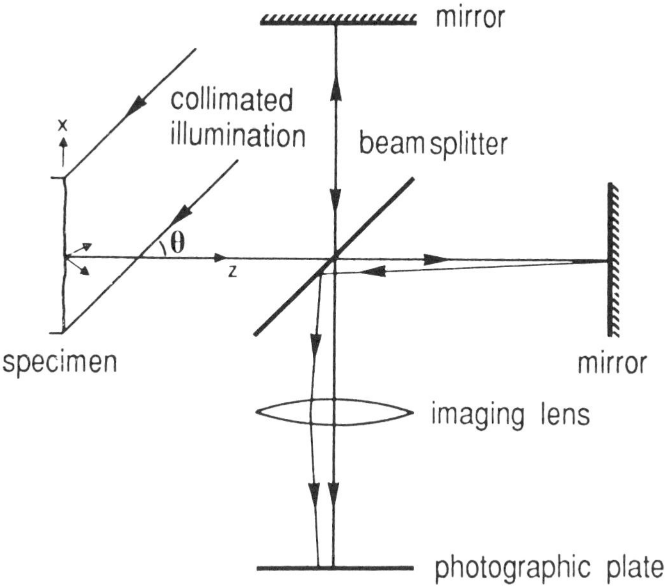

Figure 2.19 An optical setup analogous to the Michelson interferometer for measuring slope change components of a deformed object.

The terms containing the products $u \cdot \delta x$, $v \cdot \delta x$, and $w \cdot \delta x$ have been neglected in writing Eq. (2.56).

In an optical setup designed to make the angle subtended by the imaging lens at the object surface small, the equation for the fringe contour reduces to

$$\delta x \left[\sin\theta \frac{\partial u}{\partial x} + (1 + \cos\theta) \frac{\partial w}{\partial x} \right] = n\lambda \tag{2.57}$$

Incidentally, this equation has the same form as Eq. (2.17). This interferometer has been used to observe fringes depicting slope change contours in the configuration of $\theta = 0°$.

2.4.2 Hung's Biprism Shearing Technique

Due to Hung et al. [101], this interferometer introduces lateral shear by means of a biprism placed in front of the imaging lens focused on the object surface, whose optical axis is parallel to the surface normal. In another variant of the optical system, Hung [102] proposed to cover half of the lens by a small-angle glass wedge and let the light pass unhindered through the uncovered half of the lens. Whatever the optical system employed, the aim of the two systems is to bring the ray scattered from one point on the object surface to interfere with that scattered from a neighboring point. In the same illumination conditions as those employed for the interferometer shown in Section 2.4.1, the fringe formation is described by Eq. (2.56) or in the paraxial approximation of small field angles by Eq. (2.57).

2.4.3 Hung's Double-Aperture Shear Interferometer

Another interferometer is due to Hung and Taylor [97]. The optical system shown in Fig. 2.20 is quite similar to that used in Duffy's technique. A shearing mechanism, in the form of two symmetrically oriented glass wedges, is inserted in front of the two apertures in the mask covering the imaging lens. The lens is focused on the object surface. This wave-front-splitting arrangement brings to focus at two different points on the image plane the two split diffused wave fronts emerging from a point on the object. In other words, the interferometer causes two laterally shifted speckle fields to interfere in the image plane. The magnitude and orientation of the shear can be varied by appropriate control of the wedge plates mounted in front of the two apertures.

The object as illuminated by a plane wave making an angle θ with the surface normal. The plane wave is assumed to lie in the *xz* plane. When the object is deformed, the two shifted points on the object surface undergo different deformations. The mechanism of fringe formation is exactly sim-

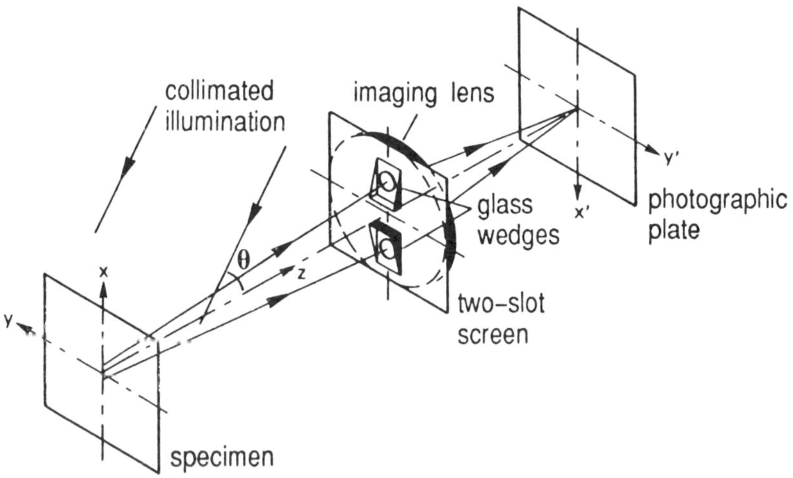

Figure 2.20 Schematic of a two-aperture speckle shear interferometer.

ilar to that described for the Duffy technique with the sole exception that it is the relative difference in optical phase between the lights scattered by the two neighboring points that now causes a shift of the speckle grid. The relative phase change between the two points is given by

$$\Delta\varphi = \frac{2\pi}{\lambda}\left\{\frac{D}{d_o}\left[\left(1 - \frac{3x^2 + y^2}{2d_o^2}\right) u - \frac{xy}{d_o^2} v + \frac{x}{d_o} w\right]\right.$$
$$+ \delta x \frac{\partial u}{\partial x}\left[\sin\theta + \frac{x + D/2}{d_o} - \frac{x^2 + y^2}{2d_o^2}\right]$$
$$+ \delta x \frac{\partial v}{\partial x}\left(\frac{y}{d_o}\right)\left[1 - \frac{x^2 + y^2}{2d_o^2}\right]$$
$$\left. + \delta x \frac{\partial w}{\partial x}\left[1 + \cos\theta - \frac{x^2 + y^2}{2d_o^2}\right]\right\} \tag{2.58}$$

where $x, y, D \ll d_o$, and the terms containing the products $u\,\delta x$, $v\,\delta x$, and $w\,\delta x$ have been neglected. Dropping the second-order terms, expression (2.58) reduces to

$$\Delta\varphi = \frac{2\pi}{\lambda}\left\{\frac{D}{d_o}\left(u + \frac{x}{d_o} w\right) + \delta x\left[\left(\sin\theta + \frac{x + D/2}{d_o}\right)\frac{\partial u}{\partial x}\right.\right.$$
$$\left.\left. + \frac{y}{d_o}\frac{\partial v}{\partial x} + (1 + \cos\theta)\frac{\partial w}{\partial x}\right]\right\} \tag{2.59}$$

Equation (2.59) brings out clearly that its use for the measurement of derivatives should be made with extreme caution. The equation as such cannot be exploited directly. The influence of u and w in Eq. (2.59) can be reduced by lowering the sensitivity of the method to these displacement terms. This is achieved by choosing the distance between the two slots to be small. This modification, when applied in conjunction with small-angle approximation, gives rise to the well-known fringe equation employed to express the response of speckle shear interferometers:

$$\sin\theta \frac{\partial u}{\partial x} + (1 + \cos\theta)\frac{\partial w}{\partial x} = \frac{n\lambda}{\delta x} \tag{2.60}$$

Equation (2.60) shows that in contrast to the Duffy technique, the sensitivity of the method is independent of the separation between the slots. If the object is illuminated at normal incidence, the fringe equation corresponding to slope contours simply becomes

$$\frac{\partial w}{\partial x} = \frac{n\lambda}{2\delta x} \tag{2.61}$$

These fringes represent contours of constant slope change $\partial w/\partial x$. A variant of the method [99] consists of eliminating the glass plates from the optical system, and instead the shear is produced by introducing a slight focusing defect in the imaging system. The system has been analyzed by Brdicko et al. [94]. The advantages of this variant are that the sensitivity of the method can be varied during image analysis and one can display fringes depicting the contours of the derivatives of out-of-plane displacements along any desired direction in the xy plane. The limitations of these methods [97, 99] have been investigated by Keinan and Politch [105].

2.4.4 Three-Aperture Shear Interferometer

This method of determining the first- and second-order derivatives of out-of-plane displacements was reported by Sharma et al. [106]. The optical setup is schematized in Fig. 2.21. The technique consists of imaging the object surface through a lens fitted with a three-slot aperture in front of the camera lens. The center of the central slot is aligned with the optical axis, and the other two are located symmetrically about the lens center. Two small-angle wedges are inserted in front of the two outermost apertures, and a glass plate is mounted in front of the central aperture to compensate for the optical path. In this arrangement the cone of rays

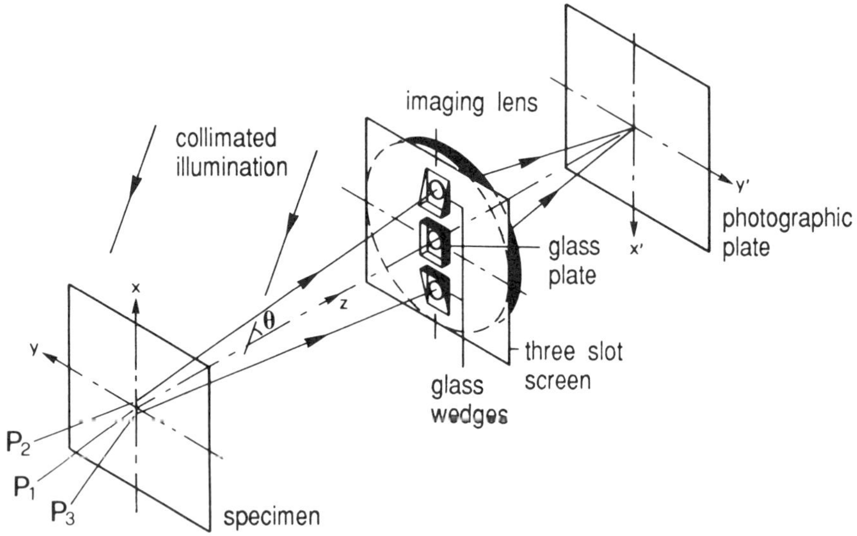

Figure 2.21 Schematic of a three-aperture speckle shear interferometer for measuring slope change and curvature.

diffused by a point on the object is split into three parts. Speckles dependent on the aperture diameter are formed in the image plane. A novel feature of this shearing device is that it brings three points on the object surface to contribute to a single point in the image plane. Three different processes of coherent speckle addition take place simultaneously in the interferometer. As a result of these additions, the speckle pattern is modulated by three systems of grid structures running perpendicular to the lines joining the apertures. Of these three systems of grid structures, one owes its presence to the two outermost slots and the other two, of half the frequency, owe their presence to the outermost slots taken separately in pairs with the central slot.

Stated briefly, the optical system generates a lateral shift $\delta x'$ between two pairs of laterally sheared coherent images in the image plane. Each pair contains two laterally shifted coherent images that interfere with one another. The principles underlying fringe formation are basically similar to those formulated in Section 2.3.4. When the object is deformed, three sets of phase variations occur in the interferometer. These phase variations result in the optical shift of the speckle grids. The relative phase difference between the waves scattered from the object points P_1 and P_2 or P_1 and P_3 are given by expressions similar to Eq. (2.25). The relative phase change between the two pairs of points (P_1, P_2) and (P_1, P_3) can then be written as

$$\Delta(\Delta\varphi) = \Delta\varphi_{12} - \Delta\varphi_{13} \tag{2.62}$$

Neglecting the second-order terms, the computation of the difference of the two phase shifts yields

$$\begin{aligned}\Delta(\Delta\varphi) = \frac{2\pi}{\lambda}\Bigg\{&\frac{\delta x' \cdot D}{d_o}\left(\frac{\partial u}{\partial x} + \frac{x}{d_o}\frac{\partial w}{\partial x}\right) \\ &+ \delta x'\,\delta x\left[\left(\sin\theta + \frac{x + D/2}{d_o}\right)\frac{\partial^2 u}{\partial x^2}\right. \\ &\left.+ \frac{y}{d_o}\frac{\partial^2 v}{\partial x^2} + (1 + \cos\theta)\frac{\partial^2 w}{\partial x^2}\right]\Bigg\}\end{aligned} \tag{2.63}$$

Assuming normal incidence and the small-angle viewing approximation, the equation corresponding to curvature contours follows as

$$\frac{\partial^2 w}{\partial x^2} = \frac{n\lambda}{2\delta x\,\delta x'} \tag{2.64}$$

The fringes depicting curvature contours appear as a beat between two sets of slope contour maps, each arising from the contribution of one of

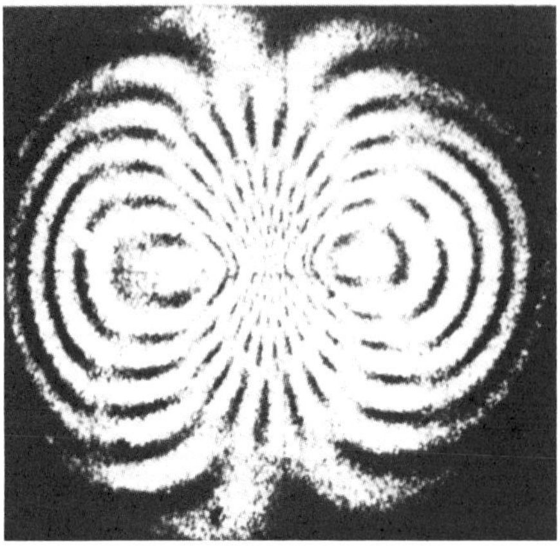

Figure 2.22 An example of a fringe pattern corresponding to $\partial^2 w/\partial x^2$ obtained in the case of a centrally loaded diaphragm clamped along its boundary. (Courtesy of R. S. Sirohi.)

the pairs of points (P_1, P_2) and (P_1, P_3). The curvature contours are observed in the first-order halo of the diffraction pattern obtained by fourier filtering of the double-exposed speckle interferogram. One such fringe pattern is shown in Fig. 2.22. The fringe pattern corresponds to $\partial^2 w/\partial x^2$ of a diaphragm clamped along the edges and subjected to a centrally concentrated load.

Similarly, the phase change between the pair of object points P_2, P_3 leads to a fringe equation identical to Eq. (2.61). The sensitivity of the slope contours is determined by δx ($\delta x = 2\ \delta x'$). The use of the method is, however, subject to the same caution as that of Section 2.4.3. The method is distinguished by its innovative way of combining speckle fields. This feature provides the method with the ability to acquire simultaneously information on both slope change and curvature distributions on the object surface.

In another variant of the method, the shear is produced by introducing a slight defocus in the optical system. Although the method works in principle, the fringes obtained by it are of less satisfactory quality and are more difficult to interpret.

2.4.5 Measurement of Twist

A method of measuring cross second-order derivatives of out-of-plane displacements has been described by Sharma et al. [107]. A four-aperture mask is mounted in front of the imaging lens. These apertures are covered by two similar glass plates in tandem but orthogonally oriented to each other. Each glass plate comprises two halves, one half composed of a plane parallel glass plate and the other composed of a small-angle wedge plate. This ingenious combination of plates gives rise to interesting shearing combinations for different aperture pairs. The spatial filtering of the recorded speckle interferogram permits isolation of the fringe contours. Two types of fringe contours are isolated, (1) the slope contours in two orthogonal directions inclined at 45° to the x or y directions, whose interpretation is governed by Eq. (2.61), and (2) the twist contours whose fringe equation is given by

$$\frac{\partial^2 w}{\partial x\ \partial y} = \frac{n\lambda}{2\ \delta x\ \delta y} \tag{2.65}$$

In writing Eq. (2.65) we have assumed that the imaging lens subtends a small viewing angle at the object surface. The fringes described by Eq. (2.65) represent contours of constant twist.

2.5 COMBINED SPECKLE AND SPECKLE SHEAR INTERFEROMETERS

2.5.1 Simultaneous Measurement of Out-of-Plane Displacements and Slope Change

An elegant method due to Mohanty et al. [108] combines the potentials of speckle and speckle shear interferometry in a single interferometer. The schematic of the optical arrangement is shown in Fig. 2.23. A three-aperture screen is mounted in front of the imaging lens. Of the three apertures, one is covered with a ground glass, one with a small-angle wedge, and the third with a plane glass plate. As a result of the coherent superposition of the beams emerging from the three apertures, each speckle is modulated by grid structures running perpendicular to the lines joining each aperture pair. The mechanism of fringe formation is similar to that of the Leendertz system (Section 2.3.3), Duffy's two-aperture method (Section 2.3.4), and Hung's shear interferometer (Section 2.4.3). The purpose of putting a ground glass in front of aperture A_0 is to generate a diffuse reference wave independent of the object deformation. The aperture pairs (A_0, A_1) and (A_0, A_2) separately give rise to coherent addition of two speckle fields arising from the ground glass and the object. On the other hand, the aperture pair (A_1, A_2) gives rise to a coherent addition of speckle fields arising from two neighboring points on the object surface. The strong point of the arrangement is that it permits one to separate out the phase changes corresponding to the out-of-plane displacement and slope change

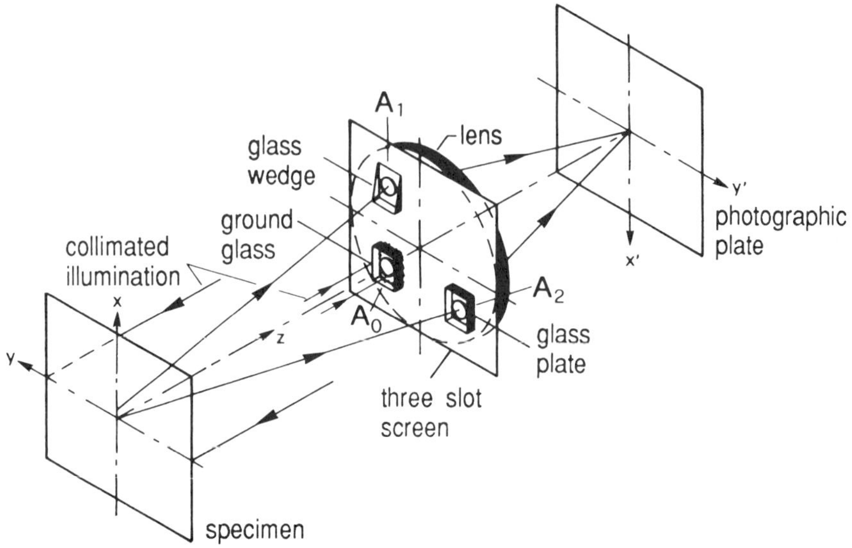

Figure 2.23 An optical configuration combining the subtleties of speckle and speckle shear interferometers. The optical elements leading the collimated light to the object surface and ground glass are not shown in the figure.

during Fourier filtering of the recorded speckle interferogram. The schematic of the diffraction halo at the Fourier plane is shown in Fig. 2.24a. The out-of-plane displacement fringes corresponding to Eq. (2.44) are obtained on filtering through either the A_{01} or A_{02} halo. The fringe contours corresponding to slope distribution, described by Eq. (2.61), are obtained on filtering through the A_{12} halo.

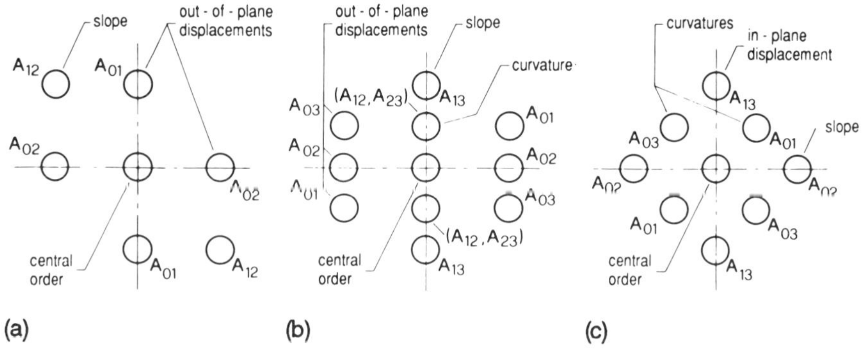

Figure 2.24 Schematic of the diffraction halo at the Fourier plane for the optical configurations of (a) Fig. 2.23, (b) Fig. 2.25, and (c) Fig. 2.28.

2.5.2 Simultaneous Measurement of Out-of-Plane Displacement, Slope Change, and Curvature Distributions

A method of simultaneously recording the information corresponding to the out-of-plane displacements, slope change, and curvature was proposed by Mohanty et al. [108]. The four-aperture optical arrangement employed for recording the speckle interferogram is shown in Fig. 2.25. The properties of both speckle and speckle shear interferometry are combined in the system by a judicious choice of optical and diffuser elements placed in front of the apertures. The speckle fields ensuing from each aperture combine coherently in the image plane to give rise to a five-grid system modulated speckle pattern. The disposition of the halos in the Fourier plane during image analysis is shown in Fig. 2.24b. The type of information that can be retrieved from each halo is also indicated. The interpretation of the out-of-plane, slope change, and curvature contours is governed by Eqs. (2.44), (2.61), and (2.64), respectively. Figure 2.26 shows the out-of-plane displacement fringes obtained in the case of a rectangular diaphragm

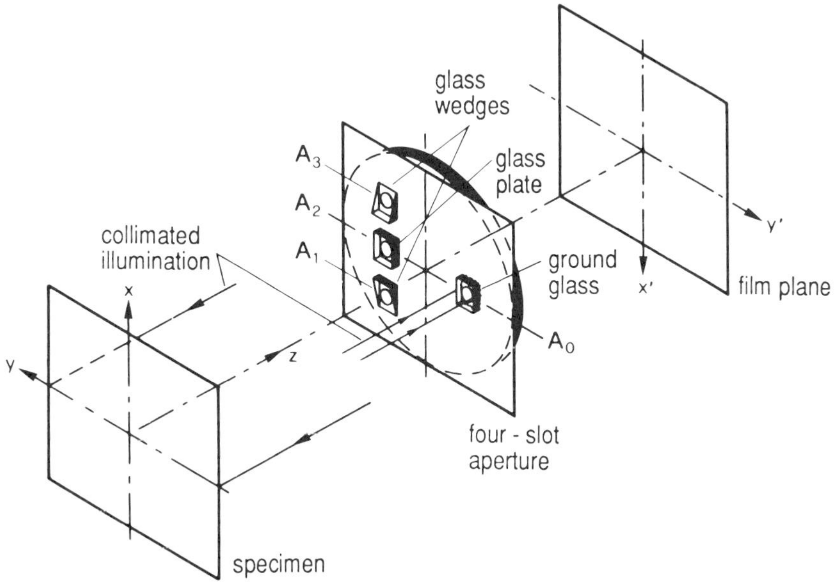

Figure 2.25 An optical configuration for obtaining simultaneous measurement of out-of-plane displacement w and its first and second derivative components $\partial w/\partial x$ and $\partial^2 w/\partial x^2$, respectively. The optical elements leading the collimated light on the object surface and ground glass are not shown in the figure.

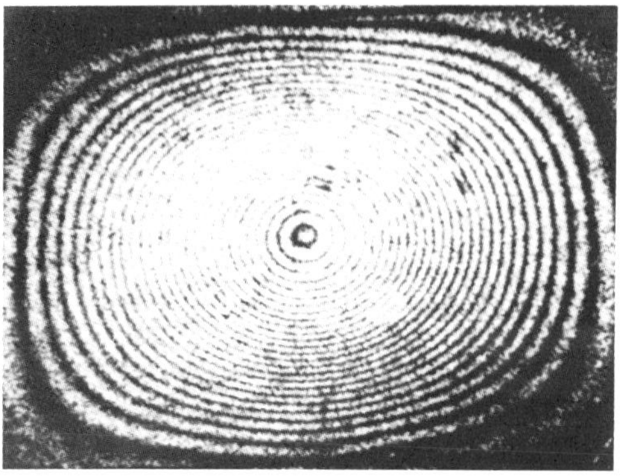

Figure 2.26 An example of a fringe pattern depicting the out-of-plane displacement of a centrally loaded diaphragm clamped along its boundary. (Courtesy of R. S. Sirohi.)

clamped along the edges and subjected to a centrally concentrated load. Figure 2.27 illustrates the slope fringes filtered from the A_{13} halo shown in Fig. 2.24b.

2.5.3 Simultaneous Measurement of In-Plane Displacement, Slope Change, and Curvature Distributions

Another interesting variant of multiaperture speckle and speckle shear interferometers is the method proposed by Sharma et al. [107]. The ap-

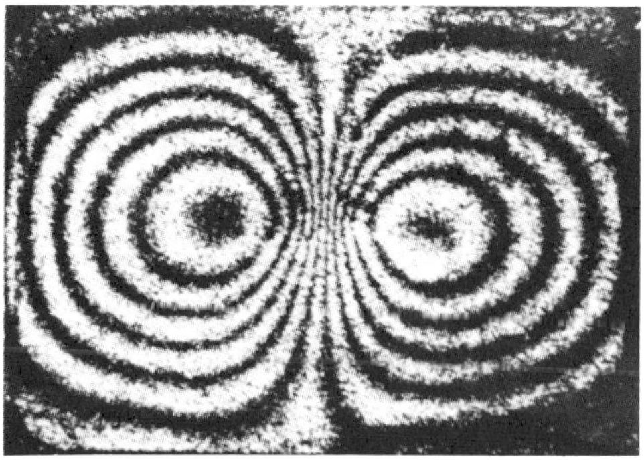

Figure 2.27 Fringe pattern corresponding to the object slope change distribution along the x direction of a centrally loaded diaphragm clamped along its boundary. (Courtesy of R. S. Sirohi.)

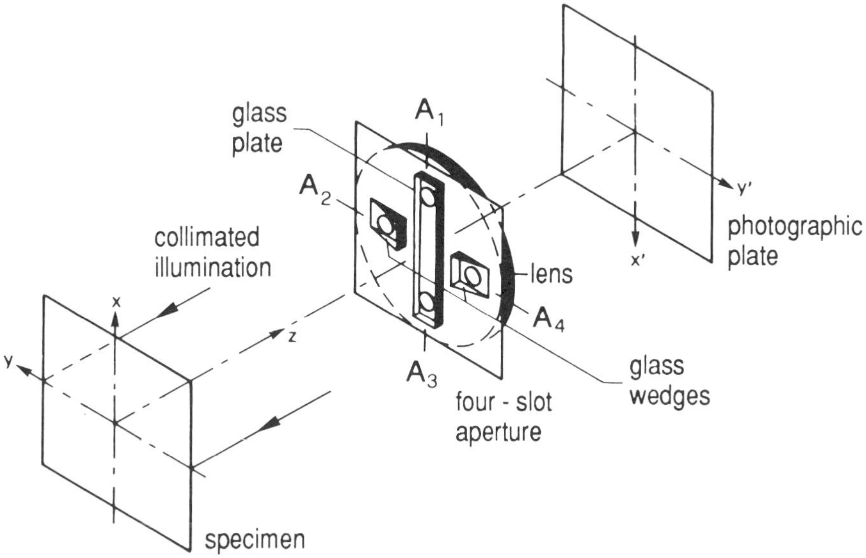

Figure 2.28 An optical configuration for obtaining simultaneous measurement of in-plane displacement, slope change, and curvature distributions.

erture arrangement consists of four slots, of which two are covered by glass plates and the remaining two by small-angle wedges. The arrangement schematized in Fig. 2.28 is symmetrically positioned with respect to the lens center. The interference of the speckle fields from each aperture pair generates a grating pattern aligned orthogonal to the line joining the centers of that particular aperture pair. Deformation of the object gives rise to an optical shift of the multigrid structure embedded in the speckle image. The disposition of the halos in the Fourier plane during fringe visualization is shown in Fig. 2.24c. Filtering through the A_{13} halo yields Duffy-type in-plane displacement fringe contours, whose interpretation is governed by Eq. (2.51). Filtering through the A_{02} halo yields slope change contours whose interpretation is governed by Eq. (2.61). Finally, filtering through the A_{01} halo yields curvature contours whose interpretation is governed by Eq. (2.64).

2.6 SHAPE MEASUREMENT

2.6.1 Immersion/Wavelength Change Methods

Speckle interferometry has been used to determine the shape of a three-dimensional object surface. In the Michelson type of optical arrangement

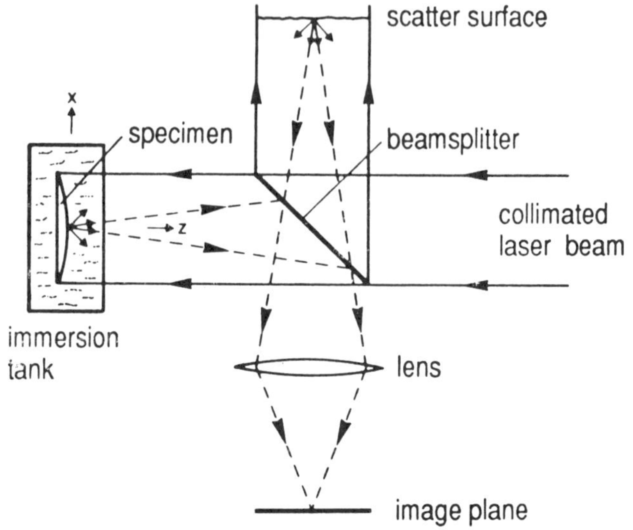

Figure 2.29 A Michelson type of speckle interferometer for determining the shape of a three-dimensional object.

described in Section 2.3.3, the plane object is replaced by a three-dimensional surface whose relief is to be investigated. The optical setup is shown in Fig. 2.29 The resultant speckle pattern is formed by the coherent superposition of the speckle fields issuing from the three-dimensional and scatter surfaces, respectively. The three-dimensional specimen is placed inside an immersion tank filled with a transparent liquid of refractive index n'. A change in the refractive index of the liquid, between two exposures, induces a relative phase change in the interferometer related to the topographic variations on the specimen's surface. Under small-field-angle approximation, the expression for the relative phase change is given by the relation

$$\Delta\varphi = \frac{2\pi}{\lambda}(n'' - n')2z \tag{2.66}$$

where n'' is the new refractive index of the liquid. The set of contour planes are separated by

$$\Delta z = \lambda/2(n'' - n') \tag{2.67}$$

Similarly, surface relief can also be examined by changing the wavelength of the illuminating beam between the two exposures. The contour sensitivity per correlation fringe is given by

$$\Delta z = \lambda_1\lambda_2/2(\lambda_1 - \lambda_2) \tag{2.68}$$

2.6.2 Two-Beam Illumination Method

An alternative approach of contouring a three-dimensional object was recently proposed by Ganesan and Sirohi [109, 110]. A schematic diagram of the optical system is shown in Fig. 2.30. The specimen is illuminated by two beams incident at equal angles θ on both sides of the optical axis. The combined speckle field in the image plane results from the superposition of the individual fields generated by the two illumination beams. The object is rotated by an amount $\Delta\theta_0$ about an axis perpendicular to the plane containing the two illumination beams. The relative phase change introduced in the interferometer as a result of this rotation is given by

$$\Delta\varphi = \frac{4\pi}{\lambda} \sin\theta \sin\Delta\theta_0 \tag{2.69}$$

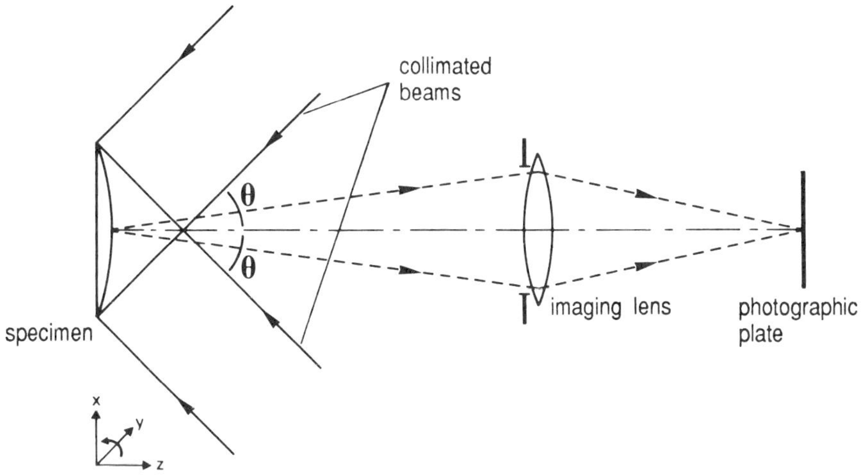

Figure 2.30 Schematic diagram of a speckle interferometer for determining shapes of three-dimensional objects.

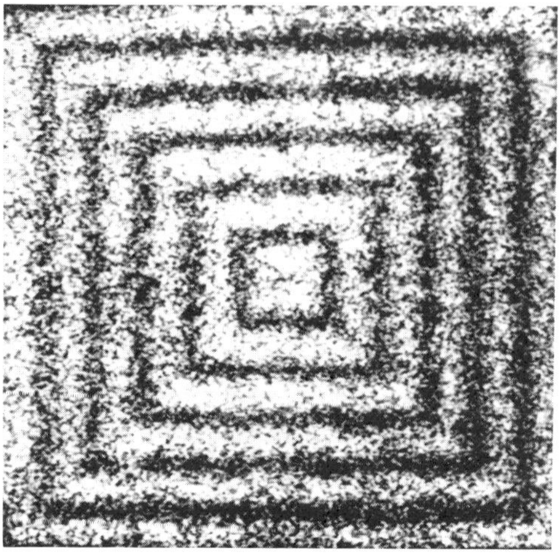

Figure 2.31 Contouring pattern obtained on a pyramidical surface. $\Delta z = 2.23$ mm. (Courtesy of C. Joenathan.)

The increment of height between two contour planes is given by

$$\Delta z = \frac{\lambda}{2 \sin \theta \sin \Delta\theta_0} \tag{2.70}$$

This equation is independent of the direction of observation. Figure 2.31 shows a contouring pattern obtained on a pyramidal surface. The contour planes lie parallel to the xy plane.

2.7 MEASUREMENT OF CONTOUR SLOPES OF AN OBJECT

2.7.1 Measurement of Lateral Slope Contours

The first method to measure contour slopes of an object surface is due to Hung et al. [111]. The schematic of the optical setup is shown in Fig. 2.32. A three-dimensional object is kept immersed in a transparent liquid. The object is imaged on a film plane by means of a lens mounted with a two-slot aperture. One slot is covered by a small-angle wedge and the other by a glass plate. The wave fronts emitted by two neighboring points on the three-dimensional surface superpose coherently in the image plane. As a result, each speckle is modulated by a system of Young's fringes running normal to the line joining the two apertures. The liquid in the tank is now changed to have a refractive index of n''. The change in the refractive index from n' to n'' induces a phase change between the waves scattered from

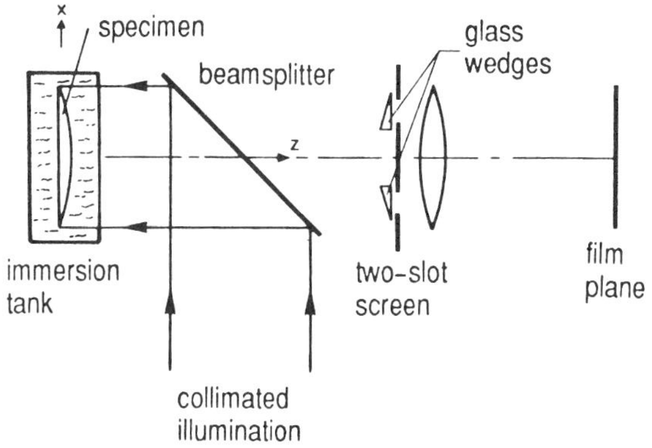

Figure 2.32 Schematic diagram of a speckle shear interferometer for obtaining lateral slope contours of a three-dimensional object.

the two neighboring points separated by a distance δx. This phase change forms the basis of fringe formation relative to slope variations of the three-dimensional object surface. The phase shift is given by

$$\Delta\varphi = \frac{2\pi}{\lambda}(1 + \cos\theta)(n'' - n')\,\delta x \cdot \frac{\partial z}{\partial x} \tag{2.71}$$

If the object is illuminated in a direction parallel to the optical axis, the fringe equation corresponding to slope contours follows as

$$\frac{\partial z}{\partial x} = \frac{n\lambda}{2(n'' - n')\,\delta x} \tag{2.72}$$

A four-slot arrangement enables simultaneous recording of the fringe contours corresponding to $\partial z/\partial x$ and $\partial z/\partial y$.

2.7.2 Measurement of Radial Slope Contours

Joenathan et al. [112] describe an interferometer to obtain radial slope contours of a three-dimensional object. The schematic of the optical arrangement is shown in Fig. 2.33a. A two-aperture screen is placed in front of the split lens. The two segments of the lens are displaced longitudinally with respect to each other along the optical axis. Two differently magnified images are thus brought into interference on the image plane. The refractive index of the liquid is now changed from n' to n''. This change in the refractive index between the two exposures causes the speckle grid to shift.

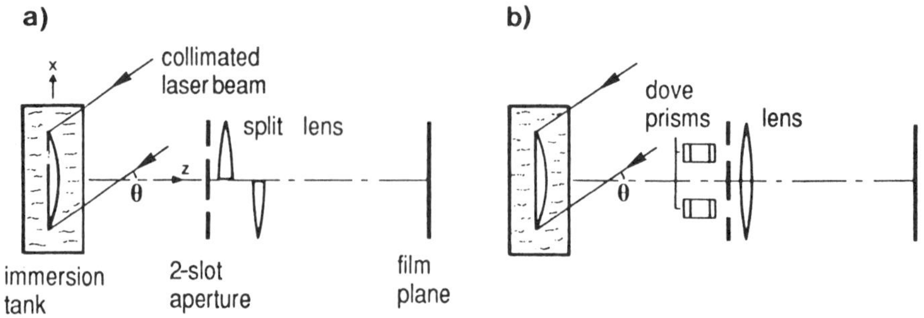

Figure 2.33 Schematic diagrams of speckle shear interferometers for obtaining (a) radial and (b) rotational slope contours of a three-dimensional object.

The relative phase change between the two neighboring radially shifted points is given by

$$\Delta\varphi = \frac{2\pi}{\lambda}(1 + \cos\theta)(n'' - n')\,\delta r \cdot r\,\frac{\partial z}{\partial r} \tag{2.73}$$

where δr is the variable radial shear on the object plane.

2.7.3 Measurement of Rotational Slope Contours

A speckle shear interferometer to obtain rotational slope contours was reported by Joenathan et al. [112]. The optical configuration, schematized in Fig. 2.33b consists of mounting a pair of identical Dove prisms in front of a two-slot-aperture lens. Each prism can be rotated about its axis and translated along its longitudinal axis. Supposing the images formed through the two prisms to be in a state of coincidence, a rotational shear of magnitude $\delta\theta$ is introduced between the two images by rotating one of the Dove prisms by an amount $\delta\theta/2$. The change in the refractive index of the liquid from n' to n'' gives rise to a relative phase change between the two rotationally shifted points. The resultant phase change is given by

$$\Delta\varphi = \frac{2\pi}{\lambda}(1 + \cos\theta)(n'' - n')\,\delta\theta\,\frac{\partial z}{\partial\theta} \tag{2.74}$$

Refraction effects have been ignored in writing Eqs. (2.71), (2.73), and (2.74). An example of a fringe pattern depicting rotational slope contours $\partial z/\partial\theta$ of a cylindrical surface is shown in Fig. 2.34.

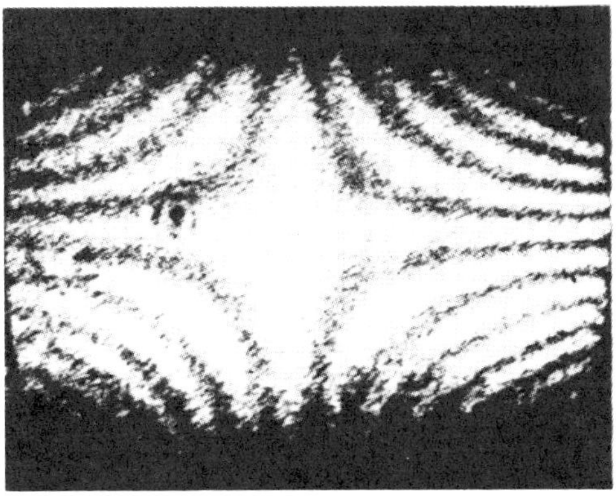

Figure 2.34 Fringe pattern depicting rotational slope contours $\partial z / \partial \theta$ of a cylindrical surface. (Courtesy of R. S. Sirohi.)

2.8 SUMMARY AND CONCLUSIONS

Leading speckle procedures in the deformation analysis and shape measurement of rough object surfaces have been discussed, with special emphasis placed on the analysis of the observed fringes and on the mode of operation of the procedures. These methods comprise three large groups: speckle photography, speckle interferometry, and speckle shear interferometry. Together, these techniques cover a wide band of sensitivities and a large measurement range.

Of the three techniques, speckle photography is the simplest to implement but suffers from drawbacks that include nonconcomitant mode of operation, time-consuming analysis, and relatively poor quality of whole-field fringe display. Important advances are being reported toward making the technique more user-friendly. Quasi-real-time speckle photography has been made possible by the use of photorefractive BSO crystals to record the specklegrams. Progress is being made in overcoming problems associated with fringe analysis. Manual analysis is giving way to automated optoelectronic, heterodyne, and phase-shift readout systems. Image processing techniques are used to improve fringe display.

The remaining two techniques, speckle and speckle shear interferometry, share common basic principles but cater to different measurement needs. While speckle interferometry offers a wide range of variants to measure displacement components, speckle shear interferometry provides a variety of procedures to measure derivatives of displacement components. Important efforts have led to the integration of the subtleties of the speckle and speckle shear interferometers in single optical systems. On the other hand, electronic recording and processing of interferograms or ESPI, with more than 15 years of development behind it, is in an advanced state of development. Its display of information for instant viewing at the video rate has greatly enhanced the appeal of speckle and speckle shear interferometry in nondestructive testing. Potentials for combining these techniques with those of phase shifting are regularly demonstrated. This is the natural outcome of the flurry of activity that surrounded the development of phase-shift techniques in the 1980s. This surge of interest paved the way to intensive use of phase shift techniques in holographic interferometry and moiré and speckle metrology.

The future is bright for speckle techniques. The development of novel ideas of measurement will continue, but at a slower pace. The integration of image processing and phase-shifting methods make speckle-based techniques effective in a wide range of applications in deformation analysis.

REFERENCES

1. A. E. Ennos, Speckle interferometry, in *Laser Speckle and Related Phenomena*, J. C. Dainty, Ed., Springer Verlag, Berlin, 1975, pp. 203–253.
2. K. A. Stetson, A review of speckle photography and speckle interferometry, *Opt. Eng.*, *14*, 482–489 (1975).
3. A. E. Ennos, Speckle interferometry, *Prog. Optics*, *16*, 233–288 (1978).
4. R. K. Erf, *Speckle Metrology*, Academic, New York, 1978.
5. K. A. Stetson, Miscellaneous topics in speckle metrology, in *Speckle Metrology*, R. K. Erf, Ed., Academic, New York, 1978, pp. 295–320.
6. R. Jones and C. Wykes, *Holograpic and Speckle Interferometry*, Cambridge Univ. Press, London, 1983.
7. R. S. Sirohi, *Selected Papers on Speckle Metrology*, SPIE Milestone Ser. MS 35, SPIE Optical Engineering Press, Washington, D.C., 1991.
8. J. A. Leendertz, Interferometric displacement measurement on scattering surfaces utilizing speckle effect, *J. Phys. E: Sci. Instrum.*, *3*, 214–218 (1970).
9. E. Archbold, J. M. Burch, and A. E. Ennos, Recording of in-plane surface displacement by double-exposure speckle photography, *Opt. Acta*, *17*, 883–898 (1970).
10. E. Archbold and A. E. Ennos, Displacement measurement from double exposure laser photographs, *Opt. Acta*, *19*, 253–271 (1972).
11. F. P. Chiang, A new family of 2D and 3D experimental stress analysis techniques using laser speckles, *SM Arch.*, *3*, 1–32 (1978).
12. R. P. Khetan and F. P. Chiang, Strain analysis by one-beam laser speckle interferometry. 1. Single aperture method, *Appl. Opt.*, *15*, 2205–2215 (1976).
13. J. Ebbeni, Application in civil engineering of a sandwich speckle method, *Proc. SPIE*, *136*, 291–293 (1978).

14. W. J. Beranek and A. J. A. Bruinsma, Sandwich speckle interferometry, *Proc. 6th Int. Conference on Experimental Stress Analysis*, Düsseldorf, 1978, pp. 417–425.
15. M. G. Pedretti and F. P. Chiang, Effect of magnification change in laser speckle photography, *J. Opt. Soc. Am.*, *68*, 1742–1748 (1978).
16. M. G. Pedretti and F. P. Chiang, On the lower limit of one beam laser speckle interferometry, *Opt. Laser Technol.*, *11*, 143–147 (1979).
17. V. J. Parks, The range of speckle metrology, *Exp. Mech.*, *20*, 181–191 (1980).
18. H. J. Tiziani, K. Leonhardt, and J. Klenk, Real time displacement and tilt analysis by a speckle technique using $Bi_{12}SiO_{20}$ crystals, *Opt. Commun.*, *34*, 327–331 (1980).
19. K. Nakagawa, T. Takatsuji, and T. Minemoto, Measurement of the displacement distribution by speckle photography using BSO crystal, *Opt. Commun.*, *76*, 206–212 (1990).
20. K. Nakagawa and T. Minemoto, Improvement of Young's fringes visibility in speckle photography by use of BSO-PROM device, *Opt. Commun.*, *70*, 288–292 (1989).
21. G. B. Smith and K. A. Stetson, Heterodyne read-out of specklegram halo interference fringes, *Appl. Opt.*, *19*, 3031–3033 (1980).
22. J. S. Kim, S. Musazzi, V. Perini, and M. Giglio, Heterodyne readout system for dual plate speckle photography: analysis of error sources and performance evaluation, *Appl. Opt.*, *28*, 1862–1868 (1989).
23. G. H. Kaufmann and P. Jacquot, Phase shifting of whole field speckle photography fringes, *Appl. Opt.*, *29*, 3570–3572 (1990).
24. G. H. Kaufmann, A. E. Ennos, B. Gale, and D. J. Pugh, An electro-optical readout system for analysis of speckle photographs, *J. Phys. E: Sci. Instrum.*, *13*, 579–584 (1980).
25. D. W. Robinson, Automatic fringe analysis with a computer image-processing system, *Appl. Opt.*, *22*, 2169–2176 (1983).
26. R. Erbeck, Fast image processing with a microcomputer applied to speckle photography, *Appl. Opt.*, *24*, 3838–3841 (1985).
27. J. M. Huntley, An image processing system for the analysis of speckle photographs, *J. Phys. E: Sci. Instrum.*, *19*, 43–49 (1986).
28. F. Ansari and G. Ciurpita, Automated fringe measurement in speckle photography, *Appl. Opt.*, *26*, 1688–1692 (1987).
29. C. S. Vikram and K. Vedam, Speckle photography of lateral sinusoidal vibrations: error due to varying halo intensity, *Appl. Opt.*, *20*, 3388–3391 (1981).
30. C. S. Vikram, Simple approach to process speckle-photography data, *Opt. Lett.*, *7*, 374–375 (1982).
31. C. S. Vikram and K. Vedam, Processing speckle photography data: circular imaging aperture, *Appl. Opt.*, *22*, 653–654 (1983).
32. C. S. Vikram and K. Vedam, Analysis of Young's fringes in speckle photography: generalized square imaging aperture, *Appl. Phys. B*, *31*, 221–224 (1983).

33. C. S. Vikram, Removing the diffraction halo effect in speckle photography of sinusoidal vibration, *Appl. Opt.*, *29*, 3572–3573 (1990).
34. G. H. Kaufmann, On the numerical processing of speckle photograph fringes, *Opt. Laser Technol.*, *12*, 207–209 (1980).
35. G. H. Kaufmann, Numerical processing of speckle photography data by Fourier transform method, *Appl. Opt.*, *20*, 4277–4280 (1981).
36. J. M. Huntley, Speckle photography fringe analysis by Walsh transform, *Appl. Opt.*, *25*, 382–386 (1986).
37. H. D. Navone and G. H. Kaufmann, Two dimensional digital processing of speckle photography fringes. 3. Accuracy of angular determination, *Appl. Opt.*, *26*, 154–156 (1987).
38. C. S. Vikram and K. Vedam, Selective counting path of Young's fringes in speckle photography for eliminating diffraction halo effects, *Appl. Opt.*, *22*, 2242–2243 (1983).
39. R. Meynart, Diffraction halo in speckle photography, *Appl. Opt.*, *23*, 2235–2236 (1984).
40. C. Joenathan and R. S. Sirohi, Elimination of error in speckle photography, *Appl. Opt.*, *25*, 1791–1794 (1986).
41. J. Georgieva, Diffraction halo effect in speckle photography, *Appl. Opt.*, *25*, 3970–3971 (1986).
42. K. Hinsch, Fringe position in double-exposure speckle photography, *Appl. Opt.*, *28*, 5298–5304 (1989).
43. C. Joenathan, Error in speckle photography: effect of coherent and incoherent speckles, *Optik*, *87*, 179–183 (1991).
44. F. P. Chiang and D. W. Li, Diffraction halo functions of coherent and incoherent random speckle patterns, *Appl. Opt.*, *24*, 2166–2171 (1985).
45. I. Yamaguchi, Fringe formation in speckle photography, *J. Opt. Soc. Am. A*, *1*, 81–86 (1984).
46. J. B. Chen and F. P. Chiang, Statistical analysis of whole field filtering of specklegram and its upper limit of measurement, *J. Opt. Soc. Am. A*, *1*, 845–849 (1984).
47. D. W. Li, J. B. Chen, and F. P. Chiang, Statistical analysis of one-beam subjective laser-speckle interferometry, *J. Opt. Soc. Am. A*, *2*, 657–666 (1985).
48. F. P. Chiang and D. W. Li, Random (speckle) patterns for displacement and strain measurement: some recent advances, *Opt. Eng.*, *24*, 936–943 (1985).
49. J. S. Kim, Range and accuracy of speckle displacement measurement in double exposure photography, *J. Opt. Soc. Am. A*, *6*, 675–681 (1989).
50. H. Barnils, S. A. Ledesma, and J. M. Simon, Fringe visibility in lateral displacement measurement from double-exposure laser photographs, *Appl. Opt.*, *26*, 383–389 (1987).
51. E. Archbold, A. E. Ennos, and M. S. Virdee, Speckle photography for strain measurement—a critical assessment, *Proc. SPIE*, *136*, 258–264 (1977).
52. P. Jacquot and P. K. Rastogi, Influence of out-of-plane deformation and its elimination in white light speckle photography, *Opt. Lasers Eng.*, *2*, 33–55 (1981).

53. K. A. Stetson, The vulnerability of speckle photography to lens aberrations, *J. Opt. Soc. Am.*, *67*, 1587–1590 (1977).
54. H. Tiziani, A study of the use of laser speckle to measure small tilts of optically rough surfaces accurately, *Opt. Commun.*, *5*, 271–274 (1972).
55. H. J. Tiziani, Vibration analysis and deformation measurement, in *Speckle Metrology*, R. K. Erf, Ed., Academic, New York, 1978, pp. 73–110.
56. D. A. Gregory, Basic physical principles of defocused speckle photography: a tilt topology inspection technique, *Opt. Laser Technol.*, *8*, 201–213 (1976).
57. D. A. Gregory, Speckle scatter, affine geometry and tilt topology, *Opt. Commun.*, *20*, 1–5 (1977).
58. D. A. Gregory, Topological speckle and structural inspection, in *Speckle Metrology*, R. K. Erf, Ed., Academic, New York, 1978, pp. 183–223.
59. F. P. Chiang and R. M. Juang, Laser speckle interferometry for plate bending problems, *Appl. Opt.*, *15*, 2199–2204 (1976).
60. K. A. Stetson, Problem of defocusing in speckle photography, its connection to hologram interferometry and its solution, *J. Opt. Soc. Am.*, *66*, 1267–1270 (1976).
61. P. Jacquot and P. K. Rastogi, Speckle motions by rigid-body movements in free-space geometry: an explicit investigation and extension to new cases, *Appl. Opt.*, *18*, 2022–2032 (1979).
62. P. Jacquot, Speckle motions in three dimensional image fields, in *Technical Digest on Hologram and Speckle Metrology*, edited by Optical Society of America, Washington, 1980, pp. MA 4.1–4.4.
63. I. Yamaguchi, Speckle displacement and decorrelation in the diffraction and image fields for small object deformation, *Opt. Acta*, *28*, 1359–1376 (1981).
64. P. K. Rastogi and P. Jacquot, Speckle metrology techniques: a parametric examination of the observed fringes, *Opt. Eng.*, *21*, 411–426 (1982).
65. P. Jacquot, Photographie de speckles: exemples tirés de l'analyse de déformation de corps solides, lasers et applications industrielles, *cour AVCP*, C. Bonjour and M. Matthey, Eds., Presses Polytechniques Romandes, Lausanne, 1982, pp. 149–204.
66. K. A. Stetson and I. R. Harrison, Determination of the principal surface strains on arbitrary deformed objects via tandem speckle photography, *Proc. 6th Int. Conference on Experimental Stress Analysis*, Düsseldorf, 1978, pp. 149–154.
67. R. Dändliker, Heterodyne holographic interferometry, *Prog. Opt.*, *17*, 3–84 (1980).
68. P. Hariharan, Quasi-heterodyne hologram interferometry, *Opt. Eng.*, *24*, 632–638 (1985).
69. K. Creath, Phase-measurement interferometry techniques, *Prog. Opt.*, *26*, 349–393 (1988).
70. J. N. Butters and J. A. Leendertz, A double exposure technique for speckle pattern interferometry, *J. Phys. E: Sci. Instrum.*, *4*, 277–279 (1971).
71. J. N. Butters and J. A. Leendertz, Holographic and video techniques applied to engineering measurement, *Meas. Control*, *4*, 349–354 (1971).
72. Y. Y. Hung and J. D. Hovanesian, Full-field surface strain and displacement

analysis of three-dimensional objects by speckle interferometry, *Exp. Mech.*, *12*, 454–460 (1972).
73. K. A. Stetson, Analysis of double exposure photography with two beam illumination, *J. Opt. Soc. Am.*, *64*, 857–861 (1974).
74. R. Jones, Design and application of a speckle pattern interferometer for measurement of total plane strain fields, *Opt. Laser Technol.*, *8*, 215–219 (1976).
75. D. E. Duffy, Moiré gauging of in-plane displacement using double aperture imaging, *Appl. Opt.*, *11*, 1778–1781 (1972).
76. D. E. Duffy, Measurement of surface displacement normal to the line of sight, *Exp. Mech.*, *14*, 378–384 (1974).
77. F. P. Chiang and R. P. Khetan, Strain analysis by one-beam laser speckle interferometry. 2. Multiaperture method, *Appl. Opt.*, *18*, 2175–2186 (1979).
78. P. K. Rastogi and P. Jacquot, Measurement of difference deformation using speckle interferometry, *Opt. Lett.*, *12*, 596–598 (1987).
79. R. Jones and C. Wykes, Decorrelation effects in speckle pattern interferometry. II, *Opt. Acta*, *24*, 533–550 (1977).
80. J. W. Goodman, Statistical properties of laser speckle patterns, in *Laser Speckle and Related Phenomena*, J. C. Dainty, Ed., Springer Verlag, Berlin, 1975, pp. 9–75.
81. P. K. Rastogi, Visualisation et mesure des déformations tangentielles, et des déformations associées par moiré holographique et interférométrie speckle, Docteur-Ingénieur thesis, Univ. Franche-Comté, France, 1979.
82. M. Spajer, P. K. Rastogi, and J. Monneret, In-plane displacement and strain measurement by speckle interferometry and moiré derivation, *Appl. Opt.*, *20*, 3392–3402 (1981).
83. J. N. Butters, Electronic speckle pattern interferometry: a general review background to subsequent papers, *Engineering Uses of Coherent Optics*, E. R. Robertson, Ed., Cambridge Univ. Press, London, 1976, pp. 155–169.
84. S. Nakadate, T. Yatagai, and H. Saito, Electronic speckle pattern interferometry using digital image processing techniques, *Appl. Opt.*, *19*, 1879–1883 (1980).
85. C. Wykes, Use of electronic speckle pattern interferometry (ESPI) in the measurement of static and dynamic surface displacements, *Opt. Eng.*, *21*, 400–406 (1982).
86. A. R. Ganesan, C. Joenathan, and R. S. Sirohi, Real-time comparative digital speckle pattern interferometry, *Opt. Commun.*, *64*, 501–506 (1987).
87. A. R. Ganesan, C. Joenathan, and R. S. Sirohi, Sharpening of fringes in digital speckle pattern interferometry, *Appl. Opt.*, *27*, 2099–2100 (1988)
88. K. Creath, Phase-shifting speckle interferometry, *Appl. Opt.*, *24*, 3053–3058 (1985).
89. D. W. Robinson and D. C. Williams, Digital phase-stepping speckle interferometry, *Opt. Commun.*, *57*, 26–30 (1986).
90. F. M. Santoyo, M. C. Shellabear, and J. R. Tyrer, Whole field in-plane vibration analysis using pulsed phase-stepped ESPI, *App. Opt.*, *30*, 717–721 (1991).

91. K. Creath, Averaging double-exposure speckle interferograms, *Opt. Lett.*, *10*, 582–584 (1985).
92. R. Höfling and W. Osten, Displacement measurement by image-processed speckle patterns, *J. Mod. Opt.*, *34*, 607–617 (1987).
93. Y. Y. Hung, Displacement and strain measurement, in *Speckle Metrology*, R. K. Erf, Ed., Academic, New York, 1978, pp. 51–71.
94. J. Brdicko, M. S. Olson, and C. R. Hazell, Theory for surface displacement and strain measurements by laser speckle interferometry, *Opt. Acta*, *25*, 963–989 (1978).
95. O. M. Peterson, Digital speckle pattern shearing interferometry: limitations and prospects, *Appl. Opt.*, *30*, 2730–2738 (1991).
96. J. A. Leendertz and J. N. Butters, An image-shearing speckle pattern interferometer for measuring bending moments, *J. Phys. E: Sci Instrum.*, *6*, 1107–1110 (1973).
97. Y. Y. Hung and C. E. Taylor, Speckle-shearing interferometric camera—a tool for measurement of derivatives of surface displacements, *Proc. SPIE*, *41*, 169–175 (1973).
98. Y. Y. Hung, A speckle-shearing interferometer: a tool for measuring derivatives of surface displacements, *Opt. Commun.*, *11*, 132–135 (1974).
99. Y. Y. Hung, I. M. Daniel, and R. E. Rowlands, A new speckle-shearing interferometer—a full-field strain gage, *Appl. Opt.*, *14*, 618–622 (1975).
100. P. Hariharan, Speckle shearing interferometry: a simple optical system, *Appl. Opt.*, *14*, 2563 (1975).
101. Y. Y. Hung, I. M. Daniel, and R. E. Rowlands, Full-field optical strain measurement having post-recording sensitivity and direction selectivity, *Exp. Mech.*, *18*, 56–60 (1978).
102. Y. Y. Hung and C. Y. Liang, Image-shearing camera for direct measurement of surface strains, *Appl. Opt.*, *18*, 1046–1051 (1979).
103. A. Rosenberg and J. Politch, Fringe parameters in speckle shearing interferometry, *Opt. Commun.*, *26*, 301–304 (1978).
104. Y. Y. Hung, Shearography: a new optical method for strain measurement and nondestructive testing, *Opt. Eng.*, *21*, 391–395 (1982).
105. M. Keinan and J. Politch, Second order phenomena in strain measurement by speckle techniques, *Opt. Lasers Eng.*, *15*, 149–182 (1991).
106. D. K. Sharma, R. S. Sirohi, and M. P. Kothiyal, Simultaneous measurement of slope and curvature with a three-aperture speckle shearing interferometer, *Appl. Opt.*, *23*, 1542–1546 (1984).
107. D. K. Sharma, R. S. Sirohi, and M. P. Kothiyal, Multiaperture speckle shearing arrangements for stress analysis, *Opt. Commun.*, *49*, 313–317 (1984).
108. R. K. Mohanty, C. Joenathan, and R. S. Sirohi, Speckle and speckle-shearing interferometers combined for the simultaneous determination of out-of-plane displacement and slope, *Appl. Opt.*, *24*, 3106–3109 (1985).
109. A. R. Ganesan and R. S. Sirohi, New method of contouring using digital speckle pattern interferometry (DSPI), *Proc. SPIE*, *954*, 327–332 (1988).
110. C. Joenathan, B. Pfister, and H. J. Tiziani, Contouring by electronic speckle

pattern interferometry employing dual beam illumination, *Appl. Opt.*, *29*, 1905–1911 (1990).

111. Y. Y. Hung, J. L. Turner, M. Tafralien, J. D. Hovanesian, and C. E. Taylor, Optical method for measuring contour slopes of an object, *Appl. Opt.*, *17*, 128–131 (1978).
112. C. Joenathan, C. S. Narayanmurthy, and R. S. Sirohi, Radial and rotational slope contours in speckle shear interferometry, *Opt. Commun.*, *56*, 309–312 (1986).

3

Speckle Methods in Experimental Mechanics

Rajpal S. Sirohi

Indian Institute of Technology
Madras, India

3.1 INTRODUCTION

Speckle techniques are a class of whole-field methods that offer sensitivities ranging between those obtainable by moiré methods and those of hologram interferometry. Additional advantages accrue, and hence these techniques have found a number of applications in experimental mechanics. A diffuse object illuminated by a coherent beam of radiation produces a grainy structure in space. This grainy structure is called the speckle pattern. The grains (speckles) in the pattern undergo changes in both position and irradiance when the generating surface undergoes changes. Therefore speckle techniques can be basically grouped under two heads: (1) those based on the positional shift of the speckles, and (2) those based on the irradiance changes of the speckles [1]. Speckle photography includes all those methods where positional changes of speckles are monitored, Whereas speckle interferometry includes methods that are based on the measurement of phase changes and hence irradiance changes. If instead of phase change we measure its gradient, the technique is one of speckle shear interferometry. Both speckle interferometry and speckle shear interferometry can be performed using electronic detection, and the technique is broadly known as electronic speckle pattern interferometry (ESPI) [2].

Changes in speckle pattern can be correlated with changes in the generating surface using either objective (free-space) or subjective (image-plane) speckle patterns [3, 4]. For body tilts and translations the objective speckle pattern is used, whereas for deformation studies the subjective speckle pattern is used because a correspondence exists between the object and image regions. I therefore confine our discussion to the subjective speckle pattern.

3.2 SPECKLE PHOTOGRAPHY

A diffuse surface is illuminated by a beam from a laser source as shown in Fig. 3.1. The image of the object is formed on the photographic plate. The photographic plate is assumed to have sufficient resolution. Two recordings are made with the object deformed between the two exposures. The double-exposed plate is developed and processed. To delineate the displacement information, the processed plate, called the double-exposure specklegram, is subjected to either a pointwise or whole-field filtering operation. For pure longitudinal motion of the object, Fourier filtering is also used.

The amplitude distribution at any point (x_i, y_i) in the image of the object can be expressed as

$$U(x_i, y_i) = K \int\int\int\int O(x_o, y_o)P(\xi, \eta) \times \exp\{-ik[\xi(x_i + Mx_o) + \eta(y_i + My_o)]/q\} \times dx_o\, dy_o\, d\xi\, d\eta \tag{3.1}$$

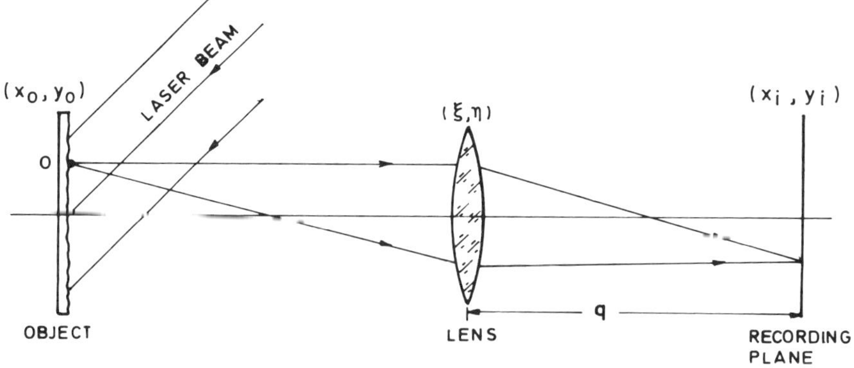

Figure 3.1 Schematic configuration for speckle photography.

where K is a complex constant, $O(x_o, y_o)$ represents the amplitude distribution just in front of the object, $P(\xi, \eta)$ is the pupil function, and M is the magnification of the imaging system. The plate records the intensity distribution, $[(U(x_i, y_i)U^*(x_i, y_i)]$. Deformation of the object causes a positional shift, and hence the amplitude distribution now, for the paired speckle, is $U(x_i + u, y_i + v)$; u and v are the components of positional shift. Therefore the total intensity recorded can be expressed as

$$\begin{aligned} I &= U(x_i, y_i)U^*(x_i, y_i) + U(x_i + u, y_i + v)U^*(x_i + u, y_i + v) \\ &= I(x_i, y_i) + I(x_i + u, y_i + v) \end{aligned} \tag{3.2}$$

Assuming linear recording, the amplitude transmittance of the specklegram can be expressed as

$$t = t_o - \beta T I$$

where T is the exposure time. Now we can use either pointwise or whole-field filtering to obtain the value of the speckle displacement at various locations on the specklegram.

The amplitude distribution, in the far field, when the specklegram is illuminated by a uniform plane wave is

$$\begin{aligned} &\int\int \{t_o - \beta T[I(x_i, y_i) + I(x_i + u, y_i + v)]\} \\ &\quad \times \exp\left[-\frac{2\pi i(x_i x_f + y_i y_f)}{\lambda L}\right] dx_i\, dy_i \end{aligned} \tag{3.3}$$

where L is the distance between the specklegram and the observation plane and λ is the wavelength of light used for illuminating the specklegram. Away from the center (optic axis) the intensity distribution in the far-field diffraction pattern is

$$\left| \int\int I(x_i, y_i) \exp\left[-\frac{2\pi i(x_i x_f + y_i y_f)}{\lambda L}\right] dx_i\, dy_i \times \left\{1 + \exp\left[\frac{2\pi i(x_f u + y_f v)}{\lambda L}\right]\right\}\right|^2$$

This can be rewritten as

$$I(\mu, \nu) = 2\mathscr{I}(\mu, \nu)[1 + \cos 2\pi(\mu u + \nu v)] \tag{3.4}$$

where μ and ν are the spatial frequencies.

The function $\mathcal{G}(\mu, \nu)$ represents the well-known halo distribution. For a circular aperture of diameter D, this is given by [5]

$$\mathcal{G}(\rho) = \begin{cases} K' \left\{ \cos^{-1}\left(\dfrac{\lambda v \rho}{D}\right) - \dfrac{\lambda v \rho}{D}\left[1 - \left(\dfrac{\lambda v \rho}{D}\right)^2\right]^{1/2} \right\}, & \rho \le \dfrac{D}{\lambda v} \\ 0, & \text{otherwise} \end{cases} \tag{3.5}$$

where $\rho = (\mu^2 + \nu^2)^{1/2}$.

The intensity in the diffraction pattern of a single-exposure specklegram decreases monotonically. For a square aperture of the imaging system, the decrease in intensity is linear. It is easy to understand this decrease. The minimum speckle size during recording is governed by the largest subtense, i.e., D/z. The smallest speckle size is therefore $\lambda z/D$. During interrogation each speckle will diffract in the angular range equal to $\pm D/z$. The point on axis receives contributions from all speckles, and hence averaged intensity will be maximum. As one considers points on a plane away from the axis, contributions from less diffracting regions are received, and the intensity therefore decreases. In fact, the intensity distribution is governed by the autocorrelation function of the aperture function [6].

The diffraction halo is modulated by a cosine squared distribution; the modulation term depends on the deformation. If the deformation of the object is uniform, the fringes will be visible in the diffraction halo. On the other hand, no fringes are formed in the diffraction halo when the displacement varies from point to point. The deformation information is then extracted by pointwise or whole-field filtering.

3.2.1 Pointwise Filtering

When a narrow beam is incident on a double-exposure specklegram (Fig. 3.2a), and if there are a sufficiently large number of paired speckles within the beam, the halo distribution is modulated by Young's fringes. The maxima of these fringes occur wherever

$$\cos 2\pi(\mu u + \nu v) = 1, \qquad \mu u + \nu v = m \tag{3.6}$$

where m is an integer. This represents a family of straight-line fringes in the $\mu\nu$ plane; the slope of the fringes is $-u/v$. The fringe spacing s is given by

$$s = \lambda L/(u^2 + v^2)^{1/2} \tag{3.7}$$

It may be noted that $(u^2 + v^2)^{1/2}$ is the magnitude of the displacement vector. The displacement in the interrogation region is thus obtained by the measurement of fringe width. Further, the fringes always run perpen-

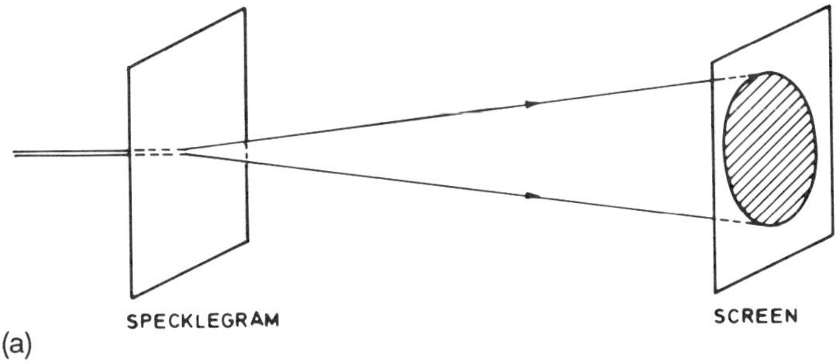

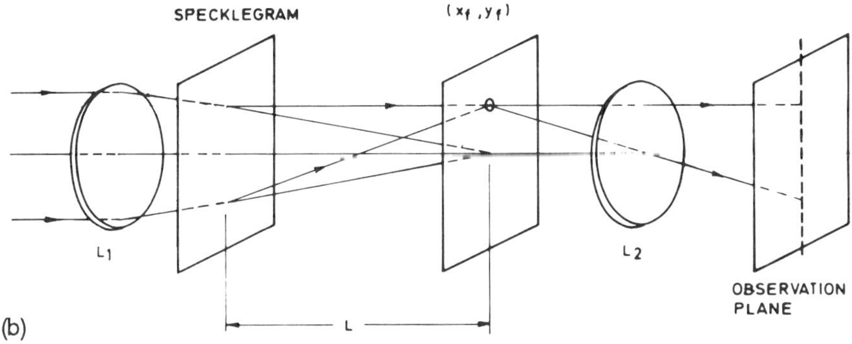

Figure 3.2 (a) Pointwise filtering arrangement. (b) Whole-field filtering arrangement.

dicular to the direction of displacement. The sign of displacement can be obtained by intentionally shifting the recording plate between exposures. Better insight into the fringe formation can be had if we consider paired speckles in the interrogation region. Each speckle pattern produces its own halo distributions, which are phase-shifted, resulting in the formation of interference fringes.

3.2.2 Whole-Field Filtering

The schematic arrangement for whole-field filtering is shown in Fig. 3.2b. The Fourier transform of the transmittance of the specklegram is displayed at the back focal plane of lens L_1. A mask with an aperture placed at this plane allows only certain spatial frequencies for image formation by lens L_2. Let us assume that the aperture is at a distance x_f from the optical axis and that the distance between the specklegram and the Fourier plane is

L. The aperture will allow the spatial frequency in a narrow band around $\mu = x_f/\lambda L$. Now all those paired speckles that diffract in such a way that the diffracted waves meet in phase at the aperture will lead to image formation, that is,

$$u \sin \theta = m\lambda, \qquad m \text{ an integer}$$

or

$$u = m\lambda L/x_f$$

Since m is not known it is not possible to obtain the value of u. On the other hand, an incremental change Δu can be measured, where Δu is given by

$$\Delta u = \lambda L/x_f \tag{3.8}$$

Therefore the fringe pattern, in whole-field filtering, appears on the whole surface. The adjacent fringes correspond to the incremental change of $\lambda L/x_f$. The fringe spacing can be varied by changing x_f. The maximum number of fringes occur when x_f is at the periphery of the halo, thereby giving the maximum sensitivity. However, the contrast of fringes is very poor, as the intensity at or near the periphery of the diffraction halo is low. Therefore the maximum sensitivity may not be utilized in practice. A modification to whole-field filtering is suggested that produces high spatial frequency fringes [7]. It is to be noted that in speckle photography the fringes obtained by filtering correspond to the positional shift of the speckles between exposures. This positional shift can be introduced by in-plane displacement or deformation, tilt, out-of-plane displacement, etc. Filtering can also be performed by locating the aperture on-axis. Circular fringes are obtained from a specklegram containing radially displaced speckle patterns usually obtained for pure out-of-plane displacement. One could also choose a circularly symmetric or other form of filtering aperture.

3.2.3 Sensitivity, Range, and Accuracy of Measurement

It is obvious that the smallest speckle displacement that can be measured will correspond to a single fringe in the diffraction halo when pointwise filtering is used [8, 9]. For quantitative measurement there should be at least two fringes in the halo. In other words, the diameter of the halo should be equal to or smaller than twice the fringe width. We now express fringe width $\bar{x}_f$ and halo size X as

$$\bar{x}_f = \frac{\lambda L}{u} \qquad \text{and} \qquad X = 2D\,\frac{L}{q}$$

where D is the diameter of the imaging lens and q is the image distance. Therefore the smallest displacement that can be measured by speckle photography is given by

$$u|_{\min} = \frac{\lambda q}{D} = \lambda(1 + M)\frac{f}{D} = \lambda(1 + M)F \tag{3.9}$$

where $u|_{\min}$ is the speckle displacement at the specklegram. It may be noted that $\lambda q/D$ is the smallest detail that can be resolved by the imaging lens. In fact, much smaller displacements can be measured by intentionally introducing either in-plane displacement or rotation to the recording plate between exposures [9].

The smallest measurable displacement $u_o|_{\min}$ at the object plane is given by

$$u_o|_{\min} = \lambda F \frac{1 + M}{M}$$

The minimum measurable displacement is proportional to $(1 + M)/M$. The quantity $M/(1 + M)$ can be defined as the sensitivity; the sensitivity increases with increasing magnification. However, for magnifications greater than 4, the increase in sensitivity is marginal. Further demagnification is not advised for speckle photography.

It is seen that as speckle displacement increases, the Young's fringes become narrower. For large displacements the Young's fringes may become nondiscernible owing to speckle noise, and the corresponding displacement may be the largest that speckle photography can measure. Ironically, the fringe modulation begins to decrease at much smaller values of displacement. There are three factors that mainly contribute to the decrease in fringe modulation [10]. The decorrelation due to structural changes in the speckles is one of the contributory factors. It is also likely that some of the paired speckles fall outside the interrogating region. Third, the speckle displacement may be nonuniform within the area of the interrogating beam. The superposition of a large number of fringe patterns of different fringe widths leads to a decrease in the fringe contrast. The modulation decreases as sinc $\{\pi x_i(u_2 - u_1)/\lambda L\}$, where $(u_2 - u_1)$ gives the variation of speckle shift. In spite of these detrimental factors, the range of applicability of speckle photography is very large.

Similar considerations apply for whole-field filtering. For example, if the displacement of the speckles is such that the path difference between the diffracted waves from any paired speckles, even at the periphery of the halo, is less than one wavelength, the whole object will appear uniformly bright for any position of the filtering aperture. The sensitivity can, however, be increased by intentionally introducing a shift of the plate between exposures. The contrast of the fringes depends on the size of the filtering aperture; too large a diameter allows a large band of spatial frequencies, resulting in a decrease in the fringe modulation, while too small a diameter results in highly speckled fringes.

The displacement at any point is calculated using the measured fringe width. Owing to the nonuniform halo distribution, the locations of fringe maxima get shifted. Furthermore, there is always a nonuniform background in practical situations, and hence the positions of the minima of the fringes are also shifted. This leads to an error in the measurement of fringe width and consequently of the speckle displacement. Both analytical and numerical methods have been suggested to take into account the influence of nonuniform halo distribution [11–14].

In order to create the deformation/displacement field over the object, pointwise filtering is done at a large number of points. Consequently there is an enormous amount of data to be handled. Manual calculations are time-consuming and susceptible to errors. Therefore automatic fringe analysis procedures have been developed. Owing to the highly speckled nature of the fringes, the procedures are different from those used for evaluating interferometric fringes. Chapter 9 contains details on the various procedures adopted for speckle fringe pattern analysis.

3.3 SPECKLE INTERFEROMETRY

A specular or speckled reference beam is added to the speckled image of the object at the recording plane. The reference wave codes the phase of the object wave. Figure 3.3 shows a schematic diagram of a speckle interferometer.

The intensity distribution at any point on the image plane is expressed as

$$I_1(x_i, y_i) = a_1^2 + a_2^2 + 2a_1a_2 \cos \theta \tag{3.10}$$

where $\theta = (\theta_1 - \theta_2)$ is the random phase. a_1, a_2 and θ_1, θ_2 are the amplitudes and phases, respectively, of the object and reference waves at the image plane. a_1 and θ_1 are random variables. The intensity distribution after the object is deformed is given by

$$I_2(x_i, y_i) = a_1^2 + a_2^2 + 2a_1a_2 \cos (\theta + \delta) \tag{3.11}$$

where δ is the phase difference introduced by deformation. In writing the above equation it is assumed that the deformation is very small compared to the other distances involved and that the microstructure of the surface is not altered.

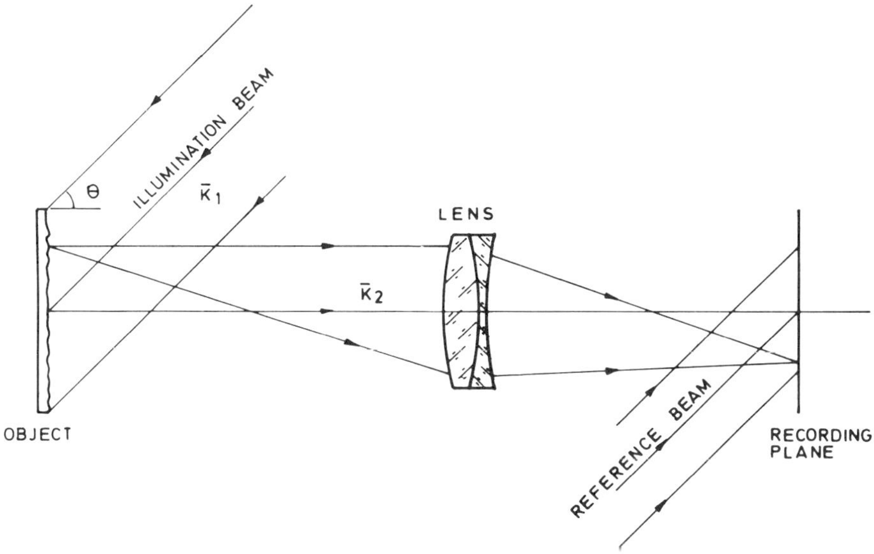

Figure 3.3 A schematic diagram of a speckle interferometer.

The amplitude transmittance of this double-exposed specklegram is expressed as

$$t = t_o - \beta T(I_1 + I_2) \tag{3.12}$$

When the specklegram is whole-field-filtered, the correlation fringes are obtained. The bright fringes occur wherever $\delta = 2m\pi$. The correlation coefficient $\rho(\delta)$ between two intensity distributions is obtained from [6]

$$\rho(\delta) = \frac{\langle I_1 I_2 \rangle - \langle I_1 \rangle \langle I_2 \rangle}{(\langle I_1^2 \rangle - \langle I_1 \rangle^2)^{1/2} (\langle I_2^2 \rangle - \langle I_2 \rangle^2)^{1/2}} \tag{3.13}$$

where the brackets $\langle \cdot \rangle$ signify time average.

Equation (3.13) can be evaluated by noting that a_1, a_2, and θ are independent random variables and can be averaged separately. Therefore,

$$\langle \cos \theta \rangle = \langle \cos (\theta + \delta) \rangle = 0$$
$$\langle i^2 \rangle = 2\langle i \rangle^2$$

where i $(= a^2)$ is the intensity of a beam. If we assume $i_1 = ri_2$, then

$$\rho(\delta) = \frac{1 + r^2 + 2r \cos \delta}{(1 + r)^2} = \frac{1 + \cos \delta}{2} \qquad \text{when } i_1 = i_2 \tag{3.14}$$

For $\delta = 2m\pi$, $\rho(\delta) = 1$, the two patterns are fully correlated, and for $\delta = (2m + 1)\pi$, they are uncorrelated. Therefore, for $\delta = 2m\pi$, bright fringes are formed. The fringes may be termed correlation fringes; correlation fringes are obtained by the addition of two or more speckle patterns. However, we often use subtraction correlation fringes in ESPI (Section 3.7). It may be worth mentioning that the correlation coefficient $\rho(\delta)$ can be obtained from the visibility of the Young's fringes obtained by pointwise filtering.

The phase difference δ can be expressed as

$$\delta = (\bar{k}_2 - \bar{k}_1)\cdot\bar{L} \tag{3.15}$$

where $\bar{k}_1$ and $\bar{k}_2$ are the propagation vectors in the directions of illumination and observation and $\bar{L}(u, v, w)$ is the deformation vector. Since the fringes are loci of constant phase difference, the deformation vector components can be measured using appropriate configurations. For example, in the configuration of Fig. 3.4, the arrangement senses only the out-of-plane component w, since $\delta = 4\pi w/\lambda$. We can modify the configuration so that only the in-plane component is sensed. Consider the experimental arrangement shown in Fig. 3.5. The object is illuminated by two collimated beams making angles of θ_1 and $-\theta_2$ with respect to the local normal. Essentially a fringe pattern (a grating structure) is formed on the object whose spacing is $\lambda/(\sin\theta_1 + \sin\theta_2)$. The object is imaged on the recording material. The fringe structure on the object is too fine to be resolved by the imaging lens. Therefore a speckle pattern rather than the image of the grating is

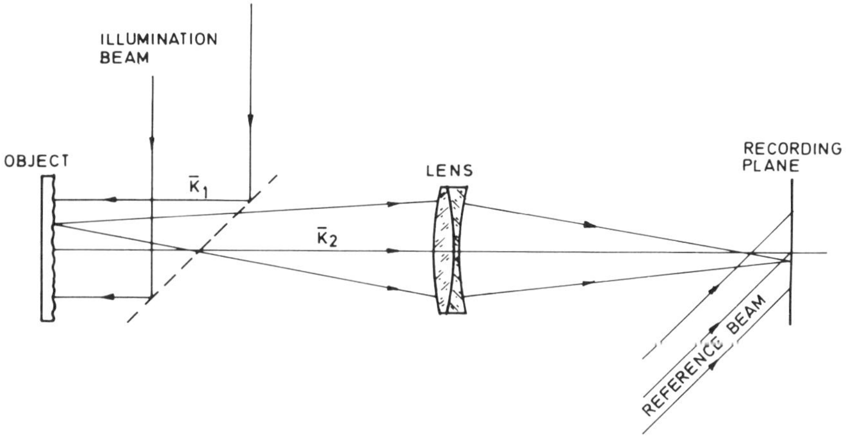

Figure 3.4 An arrangement for measuring the out-of-plane component of displacement.

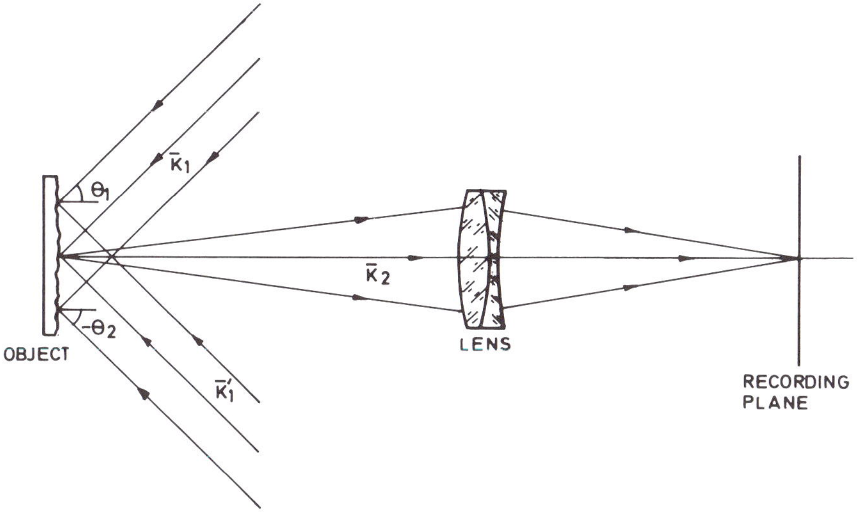

Figure 3.5 An arrangement for measuring the in-plane component of displacement.

formed at the image plane. When the object is deformed, the phase difference between the illuminating waves at any point will change. Consequently, the intensity of the speckle at that point and hence of the speckle pattern over the whole object will vary. If the phase change δ is a multiple of 2π, the two patterns, before and after deformation, will be correlated, and the bright fringe will result along regions of correlated speckles.

The phase change δ is

$$\delta = \delta_2 - \delta_1$$

where $\delta_1 = (\bar{k}_2 - \bar{k}_1)\cdot\bar{L}$ and $\delta_2 = (\bar{k}_2 - \bar{k}_1')\cdot\bar{L}$ are the phase changes due to deformation considering each illumination wave. Therefore, the phase change δ can be expressed as

$$\delta = \frac{2\pi}{\lambda}\left[u(\sin\theta_1 + \sin\theta_2) + w(\cos\theta_1 - \cos\theta_2)\right] \tag{3.16}$$

where u, v, w are the components of vector $\bar{L}$ along the x, y, and z dircctions, respectively. The bright fringes are formed when

$$u(\sin\theta_1 + \sin\theta_2) + w(\cos\theta_1 - \cos\theta_2) = m\lambda \tag{3.17}$$

We can analyze this expression for various geometrical arrangements as discussed in the following subsections.

3.3.1 In-Plane Displacement Measurement

When the object is illuminated symmetrically with respect to the local normal, the condition for fringe formation becomes

$$2u \sin \theta = m\lambda$$

or

$$u = m\lambda/(2 \sin \theta) \tag{3.18}$$

where $\theta_1 = -\theta_2 = \theta$. The arrangement is sensitive only to the in-plane component. The effect of the out-of-plane component is fully compensated for in the symmetric illumination arrangement [15]. The orthogonal component of in-plane displacement can be measured by rotating either the object or the illumination beams by 90° about the normal to the object. Two pairs of symmetrical beams orthogonal to each other can be used to obtain the u and v components simultaneously.

3.3.2 Desensitized Out-of-Plane Measurement

The illumination beams are incident on the object from the same side of the normal. In such a situation the bright fringes are formed wherever [16]

$$u(\sin \theta_1 - \sin \theta_2) + w(\cos \theta_1 - \cos \theta_2) = m\lambda \tag{3.19}$$

Usually $w \gg u$. Further, under the condition that the illumination is nearly grazing ($\theta_1 \doteq \theta_2 \rightarrow \pi/2$), the condition $(\cos \theta_1 - \cos \theta_2) > (\sin \theta_1 - \sin \theta_2)$ is satisfied. Hence $u(\sin \theta_1 - \sin \theta_2)$ can be ignored in comparison with $w(\cos \theta_1 - \cos \theta_2)$. Therefore the bright fringes are formed wherever

$$w(\cos \theta_1 - \cos \theta_2) = m\lambda$$

or

$$w = \frac{m\lambda}{\cos \theta_1 - \cos \theta_2} \tag{3.20}$$

In a conventional arrangement (Fig. 3.1) the fringes are formed for every $\lambda/2$ change in w, whereas in this case the fringes are formed whenever w changes by $\lambda/(\cos \theta_1 - \cos \theta_2)$. Therefore a desensitization by a factor $2/(\cos \theta_1 - \cos \theta_2)$ takes place. As an example we taken $\theta_1 = 85°$ and $\theta_2 = 86°$. This gives desensitization by a factor of 115. Therefore a large out-of-plane displacement can be measured with this arrangement.

3.3.3 Multiple Apertures

Instead of using the full aperture of the lens, a mask containing two or more apertures is placed at the entrance pupil of the lens. This modifies the optical transfer function of the lens. From the speckle metrology point of view, the speckle size increases at the image plane. Also the depth of focus or depth of field increases. Furthermore, each speckle pattern is modulated by Young's-type fringes. During filtering these Young's fringes diffract light in well-defined directions. Therefore there is no need to place an aperture at the filtering plane while carrying out whole-field filtering. Because there is considerably more light available in these so-called diffraction halos, there is an enormous gain in the signal-to-noise ratio, resulting in almost unit contrast fringes [17]. There are other advantages as well. For example, both u and v in-plane components can be measured simultaneously. Multiplexing using theta modulation or/and frequency modulation is possible [18].

Let us consider the experimental arrangement shown in Fig. 3.6. A collimated beam illuminates the object. The lens has a mask with two circular apertures of diameter b and an intercenter distance of c. An image of the object is formed by each aperture, and the two images are super-

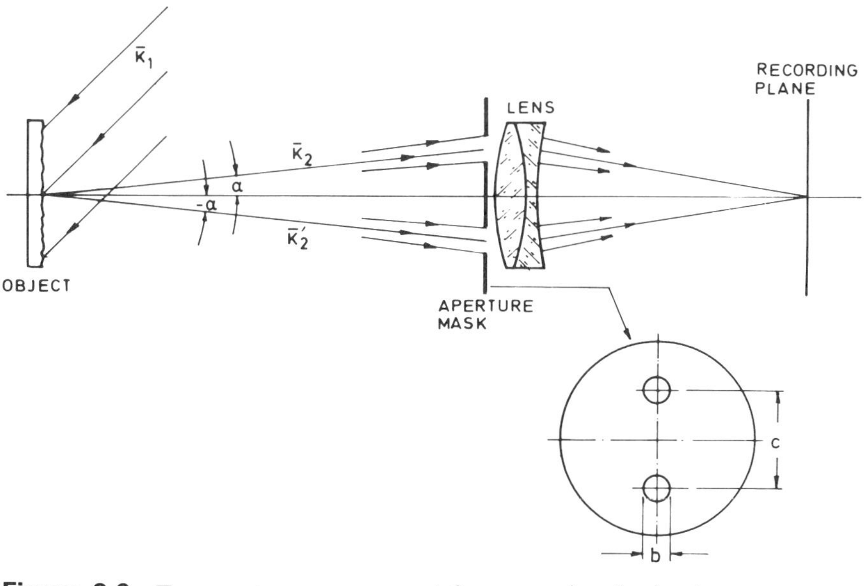

Figure 3.6 Two-aperture arrangement for measuring the in-plane component of displacement.

posed. The average speckle size in the image is $2.44\lambda q/b$. The fringe spacing due to the superposition of the two images is $\lambda q/c$. Therefore each speckle has fringes numbering approximately c/b. We can also look at the situation in which a wave from one aperture acts as a reference for the wave from the other aperture. When the object is deformed, the phases of both waves change, and hence the net phase change δ is

$$\delta = \delta_2 - \delta_1$$

where $\delta_1 = (\bar{k}_2 - \bar{k}_1)\cdot\bar{L}$ and $\delta_2 = (\bar{k}_2' - \bar{k}_1)\cdot\bar{L}$. The net phase change δ is then expressed as

$$\delta = (\bar{k}_2' - \bar{k}_2)\cdot\bar{L} \tag{3.21}$$

For the case when the apertures are placed symmetrically with respect to the optical axis, Eq. (3.21) gives

$$2u \sin \alpha = m\lambda$$

or

$$u = m\lambda/(2 \sin \alpha) \tag{3.22}$$

This equation is similar to Eq. (3.18) except that θ is replaced by α. The angle α is governed by the aperture of the lens, and hence the sensitivity of this arrangement is inherently poor. However, it can be increased by using holographic lenses for imaging [19, 20]. The out-of-plane component is fully compensated for while there is sensitivity for the in-plane component along the direction joining the centers of the two apertures. In-plane components can be measured simultaneously using four apertures at the corners of a square placed symmetrically on the optical axis. Even different sensitivities for the two components can be obtained by arranging the apertures at the corners of a rectangle.

Equation (3.22) can also be expressed as

$$u = m\lambda/(2M \sin \alpha') \tag{3.23}$$

where M is the magnification of the imaging system and α' is the half-angular subtense in the image space. For $M = 1$, $\alpha = \alpha'$, and hence Eq. (3.22) can be used for further analysis. It is seen that the spacing of Young's fringes in each speckle is $\lambda/(2 \sin \alpha')$. Therefore in regions where the in-plane component is an integral multiple of fringe spacing, the grating-like structure in each speckle will be retained with high modulation after the second exposure. On the other hand, in regions where the in-plane displacement is an odd half-integer multiple of the fringe width, the modulation of the grating structure in the speckle will be nearly zero. The double-exposure specklegram therefore contains regions of high modulation and regions of very low modulation. The regions of high modulation will diffract strongly in the whole-field filtering setup and appear bright in the filtered image, while the regions of low modulation will diffract poorly and thus appear dark in the image. Therefore the image of the object appears covered by dark and bright regions, the fringes.

It may be noticed that the two techniques thus far described can be classified as using (1) two directions of illumination and a single direction of viewing [Eq. (3.18)], and (2) a single direction of illumination and two directions of viewing [Eq. (3.22)]. In fact, both these techniques can be combined into one that offers the advantage of higher sensitivity. The experimental arrangement is shown in Fig. 3.7. The object is illuminated symmetrically by two illumination beams, and the observation is also made along two directions by using two apertures in front of the imaging lens. We can examine this arrangement by considering one illumination beam at a time and two observation directions. The phase change δ due to deformation can be expressed as

$$\begin{aligned}\delta &= (\bar{k}_1 - \bar{k}_1')\cdot\bar{L} - (\bar{k}_1' - \bar{k}_1)\cdot\bar{L} \\ &= 2(\bar{k}_1 - \bar{k}_1')\cdot\bar{L} \\ &= 8\pi u \sin\theta/\lambda\end{aligned}$$

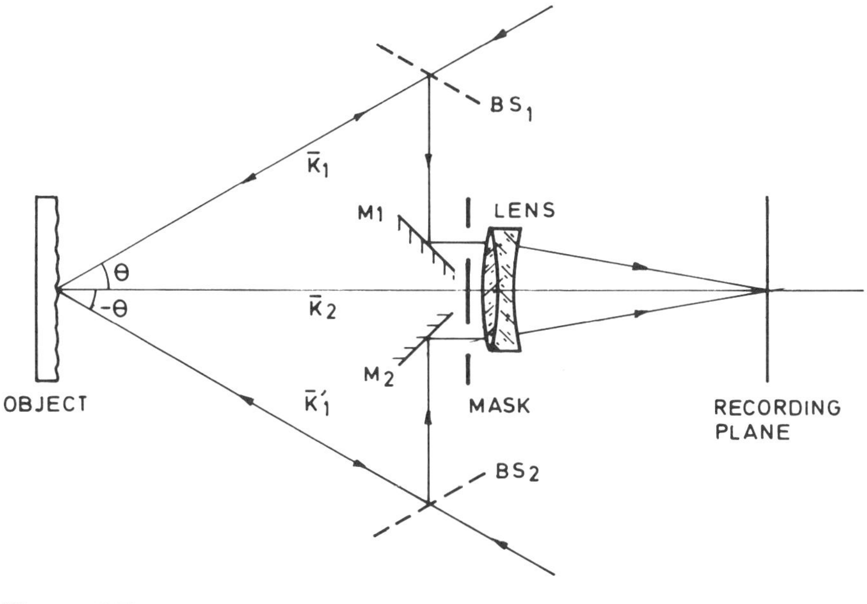

Figure 3.7 A schematic diagram of an arrangement for obtaining increased sensitivity for the in-plane component, BS, beamsplitter.

The sensitivity is twice that obtainable with dual illumination and a single observation direction. The increase in sensitivity is, however, obtained at the expense of loss of light and greater system complexity.

The configuration of Fig. 3.6 can be modified to sense only the out-of-plane component. Figure 3.8 shows a schematic diagram of the experimental arrangement [21]. A ground glass plate is placed at one of the apertures and illuminated by a direct beam. This generates a diffuse reference beam. The other aperture receives the light from the object that is used for image formation. A negligibly small fraction of light from the object may be transmitted by the ground glass, but this can be neglected in the analysis. To make more light available for image formation, the aperture mask has been dispensed with and a ground glass plate is used over half of the lens aperture [22].

The condition for bright fringe formation is

$$\frac{2\pi}{\lambda}[u(\sin\theta - \sin\alpha) + w(\cos\theta + \cos\alpha)] = 2m\pi \tag{3.24}$$

When the object is illuminated normally and placed sufficiently far away from the imaging lens, Eq. (3.24) can be expressed as

$$w = m\lambda/2 \tag{3.25}$$

Therefore the arrangement senses only out-of-plane component and offers a sensitivity of half a wavelength. Indeed it is possible to modify the aperture arrangement on the mask so as to make the configuration sensitive for all the components of the deformation vector. Figure 3.9a shows such an aperture arrangement on the mask. It contains five apertures; the central

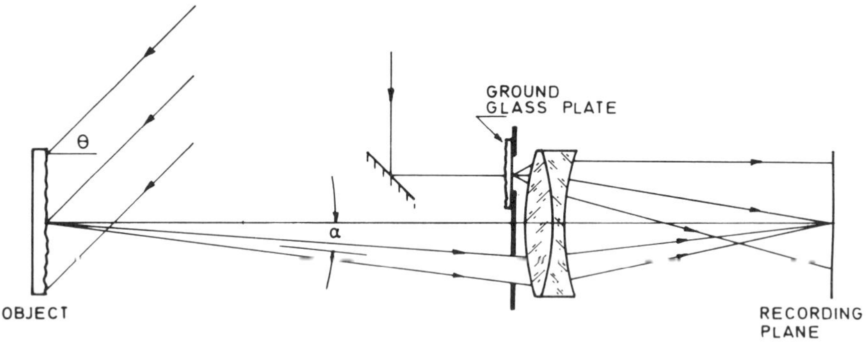

Figure 3.8 An arrangement for measuring the out-of-plane displacement component.

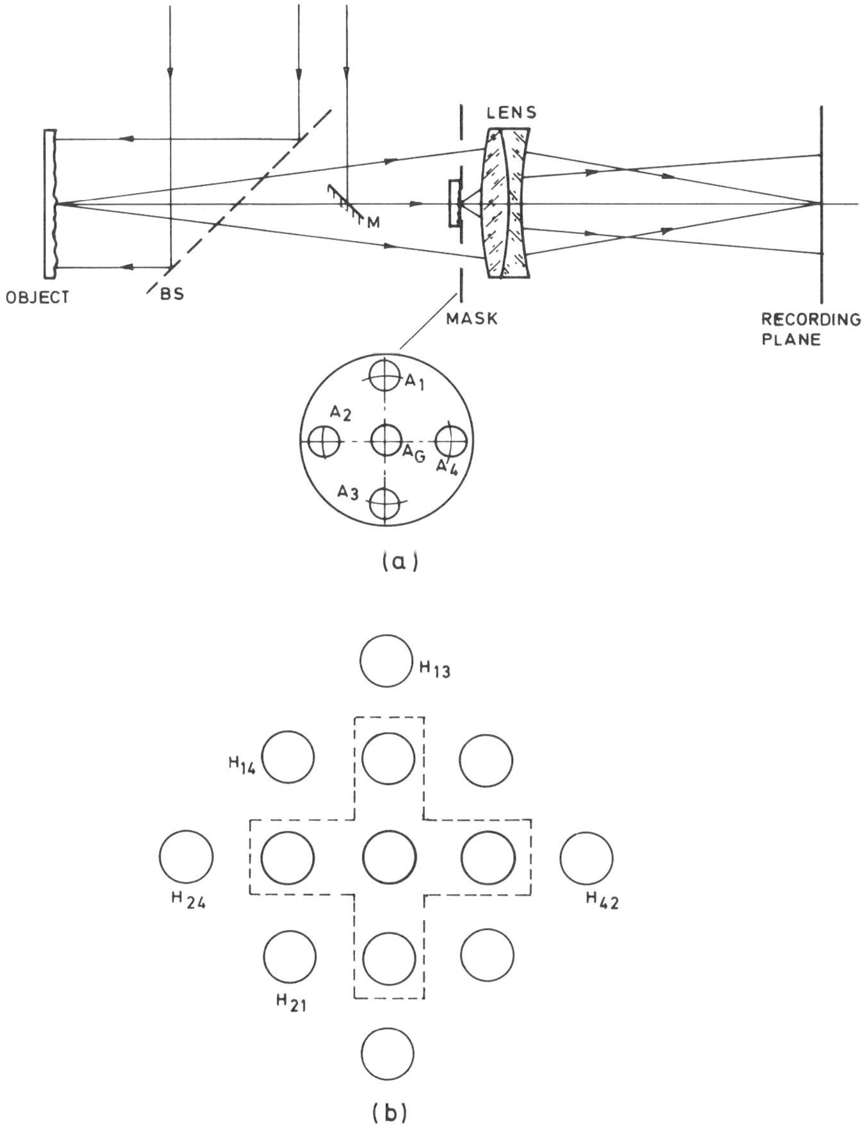

Figure 3.9 (a) Arrangement of apertures in a mask for measuring both in-plane and out-of-plane components. (b) Halo distribution at the Fourier plane (filtering plane), BS, beamsplitter.

aperture carries the ground glass plate that provides the reference beam. The image formation takes place via the remaining four apertures; the four images are superposed. Each speckle in the image is now crisscrossed by a number of Young's fringes of different orientations and fringe widths. The specklegram therefore diffracts in well-defined directions, and 13 halos are formed in the Fourier plane as shown in Fig. 3.9b. These halos are spatially isolated; each one can be filtered for image formation in the whole-field filtering setup. The fringes depicting the out-of-plane displacement are formed if any of the halos within the dashed-line region except the zeroth-order is used for filtering. The *u* family of fringes are formed when halo H_{42} is used for filtering, and the *v* family of fringes are formed when halo H_{13} is used for filtering. Halos H_{14} and H_{21} give the in-plane displacements along the diagonal directions. The sensitivities are different for in-plane and out-of-plane components, i.e., $\lambda/(2 \sin \alpha)$ and $\lambda/2$, respectively. Figure 3.10a–c show the fringe patterns obtained when a cantilever is subjected to in-plane and out-of-plane deformations using two micrometers in orthogonal directions. The fringe pattern in Fig. 3.10c indicates that the loading is not symmetrical. It may be noted that some halos are due to

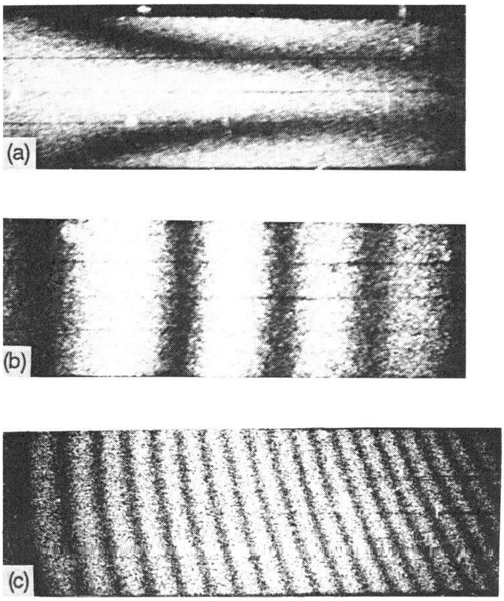

Figure 3.10 Interferograms showing the fringes due to (a) and (b) in-plane components, and (c) out-of-plane component of deformation of a cantilever.

the superposition of two or more halos. This results in a decrease in the fringe contrast when image formation is done via such halos. By proper arrangement of apertures on the mask, halo superposition can be avoided. For example, three apertures at 120° instead of four as discussed here may be sufficient to yield in-plane components without halo superposition.

3.4 HOLOSPECKLE INTERFEROMETRY

Out-of-plane displacement can be measured with $\lambda/2$ or better sensitivity by hologram interferometry. On the other hand, speckle photography offers acceptable sensitivity for in-plane displacement components. Therefore, attempts were made to combine these two methods for obtaining components of the three-dimensional deformation vector [23, 24]. Recently Chiang and Li [25] proposed a versatile approach combining these two methods, and the new technique is called holospeckle interferometry. The information from a double-exposure recording can be extracted holographically or via speckle photography. The technique possesses the attributes of both hologram interferometry and speckle photography. The underlying working principle of holospeckle interferometry can be easily followed from the analysis of a double-exposure hologram. When a double-exposure hologram is interrogated, two object waves belonging to initial and deformed states of the object are released, which generate the interference pattern [26]. We can also record a double-exposure hologram with two reference waves, using one reference wave for each recording. The double-exposed hologram is interrogated by both reference waves simultaneously, releasing the two object waves, which interfere to generate the interference pattern. This two-reference-beam arrangement was proposed for holospeckle interferometry by Chiang and Li [27]. Figure 3.11 shows a schematic diagram of the experimental arrangement, which incorporates an optical device to rotate the polarization of one of the reference waves. A double-exposure record is made with each reference wave, and the object is deformed between exposures. The polarization states of the reference and object illumination waves are the same during each recording. The double-exposed hologram is then interrogated with both reference waves simultaneously. Assuming linear recording, the intensity distribution in the reconstructed image is given by

$$I(p) = 2i_1(p)(1 + \mu_{12} \cos \delta) \tag{3.26}$$

where δ is the phase difference due to deformation.

It is assumed that the amplitudes of reconstructed waves are equal. The visibility of the fringes (μ_{12}) depends on the transverse displacement of the

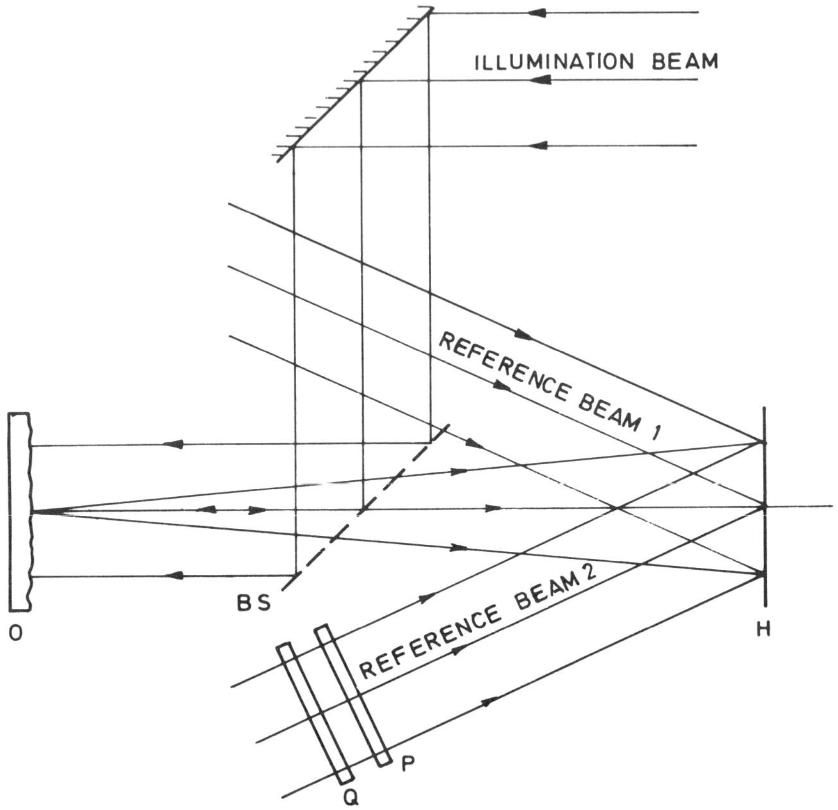

Figure 3.11 A schematic diagram of an arrangement for performing holospeckle interferometry. (Q) $\lambda/4$ plate, (P) polarizer, (O) object, (BS) beam splitter, (H) hologram.

speckles between exposures. The visibility μ_{12} can be controlled by changing the polarization state of the reference beams. If the angle between the linear polarization states between the two reference beams is θ, then the visibility is given by $\mu_{12} \cos \theta$. Obviously, when $\theta = 0°$, the fringes of maximum contrast, governed by speckle displacement and other factors, are formed. For $\theta = 90°$, there is no interference term, and hence only two speckled object fields are generated. Whole-field filtering can then be performed to obtain an isothetic fringe pattern. It is assumed, however, that the two speckled fields generated with orthogonal polarizations remain correlated. This has been confirmed by experiments. Furthermore holospeckle interferometry can also be performed in real time [27].

3.5 SPECKLE SHEAR INTERFEROMETRY

In speckle shear interferometry, a point in the image plane receives contributions from two or more points on the object. Alternatively, a point on the object may be imaged as two or more points in the image plane. A speckle shear interferometer consists of an imaging element and a shearing element. The imaging lens is used as either a single aperture or multiaperture system. Let us now consider a situation when the contributions from two object points (x, y) and $(x + \Delta x_o, y)$ are received at the same image point. Here Δx_o is called the object plane shear and is related to the image plane shear Δx_i through the magnification of the imaging lens. The waves from the two points at any image point can be represented by $a_1 e^{i\theta_1}$ and $a_2 e^{i\theta_2}$; one wave acts as a reference for the other. Therefore the intensity distribution I_1 recorded at the image plane is

$$I_1 = a_1^2 + a_2^2 + 2a_1 a_2 \cos \theta, \qquad \theta = \theta_1 - \theta_2$$

When the object is deformed, these waves arrive at the image point with additional phases δ_1 and δ_2. These waves are expressed as $a_1 e^{i(\theta_1 + \delta_1)}$ and $a_2 e^{i(\theta_2 + \delta_2)}$. Thus the intensity distribution I_2 recorded in the second exposure is

$$I_2 = a_1^2 + a_2^2 + 2a_1 a_2 \cos (\theta + \delta), \qquad \delta = \delta_1 - \delta_2$$

The phases δ_1 and δ_2 are different because the deformation vectors at the two sheared points are different. The phases δ_1 and δ_2 are expressed as

$$\delta_1 = (\overline{k}_2 - \overline{k}_1) \cdot \overline{L}(x, y)$$

and (3.27)

$$\delta_2 = (\overline{k}_2 - \overline{k}_1) \cdot \overline{L}(x + \Delta x_o, y)$$

where $\overline{L}(x, y)$ and $\overline{L}(x + \Delta x_o, y)$ are the deformation vectors at the two points (x, y) and $(x + \Delta x_o, y)$, respectively. Let the components of the deformation vectors at these points be (u, v, w) and $(u + \Delta u, v + \Delta v, w + \Delta w)$, respectively. The phase difference δ can then be expressed as

$$\delta = \frac{2\pi}{\lambda} \left[\frac{\partial u}{\partial x} \sin \theta + \frac{\partial w}{\partial x} (1 + \cos \theta) \right] \Delta x_o \tag{3.28}$$

where the illumination beam lies in the xz plane and makes an angle θ with the z axis.

The bright fringes are formed wherever

$$\frac{\partial u}{\partial x} \sin \theta + \frac{\partial w}{\partial x} (1 + \cos \theta) = \frac{m\lambda}{\Delta x_o} = \frac{m\lambda M}{\Delta x_i} \tag{3.29}$$

Table 3.1 Shear Types and Corresponding Fringe Pattern

Shear types	Superposition of fields	What fringes represent (for normal illumination)
1. Linear/lateral	$U(x, y)$ and $U(x + \Delta x, y + \Delta y)$	$\partial w/\partial x$, $\partial w/\partial y$
2. Radial shear	$U(r, \theta)$ and $U(r + \Delta r, \theta)$	$r(\partial w/\partial r)$
3. Rotational shear	$U(r, \theta)$ and $U(r, \theta + \Delta\theta)$	$\partial w/\partial\theta$
4. Inversion shear	$U(r, \theta)$ and $U(r, \theta + \pi)$ or $U(x, y)$ and $U(-x, -y)$	$w(r) - w(-r)$
5. Folding shear	$U(x, y)$ and $U(-x, y)$ or $U(x, y)$ and $U(x, -y)$	$w(x, y) - w(-x, y)$ or $w(x, y) - w(x, -y)$

For normal illumination, $\theta = 0$, and hence

$$\frac{\partial w}{\partial x} = \frac{m\lambda}{2\,\Delta x_o} = \frac{m\lambda M}{2\,\Delta x_i} \tag{3.30}$$

The fringe pattern represents the partial x slope. The fringes representing the partial y slope are obtained by introducing shear along the y direction. Both slope and strain information can be obtained by recording specklegrams for two different directions of illumination of the object [16].

We have tacitly assumed that the shear is linear and/or lateral. There are other kinds of shear as well. They can be easily implemented. Table 3.1 gives the types of shear, the superposition of fields, and the derivatives that the fringes represent. A review of speckle shear interferometry is available elsewhere [28].

3.5.1 Shear Interferometers

Single-Aperture

Figure 3.12 shows an experimental arrangement. This is a Michelson interferometer. The object is imaged by the lens L via the mirrors of the interferometer. When the mirrors are equidistant from the beamsplitter and aligned perpendicular to the optical axes and hence to each other, both images of the object are fully overlapped. One of the mirrors can now be tilted to introduce shear along any desired direction. The object can be illuminated obliquely or normally. The double-exposure specklegram is whole-field-filtered to obtain fringes.

The shear has been introduced with the help of some other optical elements such as plane parallel plates, wedges, beamsplitters, and gratings.

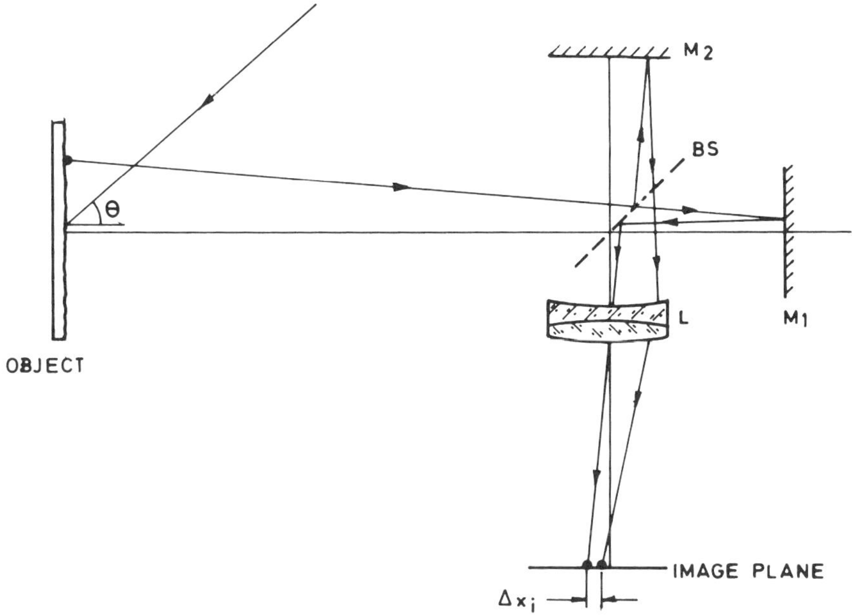

Figure 3.12 Michelson interferometer configuration used for speckle shear interferometry.

The shear elements are placed in front of the imaging lens. Shear has also been introduced by defocusing, that is, by shifting the recording plane slightly from the image plane.

Multiaperture

A mask containing a number of apertures is placed in front of the imaging lens. One of the apertures carries a shearing element. As an example, we consider a two-aperture mask symmetrically placed on the optical axis. One aperture carries a wedge plate, the thicker end toward the periphery of the lens, and the x axis is taken along the direction of maximum slope of the wedge. The shear Δx_o is then $(\mu - 1)\alpha p$, where μ is the refractive index, α is the refracting angle of the wedge, and p is the object distance. Greater shear can be realized by placing an oppositely sloping wedge in the other aperture. Shears along the x and y directions can be obtained by employing a mask with four apertures with wedges appropriately aligned. Billet's split lens performs the dual function of imaging and shearing. The shear can be continuously varied [29].

The fringes are formed by filtering via any of the halos. It may be noted that the fringes will depict either the x slope or y slope depending on whether the shear is along the x or y direction irrespective of the arrangement of apertures. As shown in Section 3.3.3, the two-aperture arrangement is intrinsically sensitive to the in-plane component, and hence the fringes will have in-plane contributions as well [30]. To demonstrate this we write the phase difference δ as

$$\begin{aligned} \delta &= (\overline{k}_2 - \overline{k}_1)\cdot\overline{L}' - (\overline{k}_2' - \overline{k}_1)\cdot\overline{L} \\ &= (\overline{k}_2 - \overline{k}_1)\cdot(\overline{L}' - \overline{L}) - (\overline{k}_2' - \overline{k}_1)\cdot\overline{L} + (\overline{k}_2 - \overline{k}_1)\cdot\overline{L} \\ &= (\overline{k}_2 - \overline{k}_1)\cdot\frac{\partial \overline{L}}{\partial x}\,\Delta x_o + (\overline{k}_2 - \overline{k}_2')\cdot\overline{L} \end{aligned} \tag{3.31}$$

Assuming normal illumination, and with $\overline{k}_2$, $\overline{k}_2'$, and $\overline{k}_1$ lying in the xz plane, we obtain

$$\delta = \frac{2\pi}{\lambda}\, 2\, \frac{\partial w}{\partial x}\, \Delta x_o + \frac{2\pi}{\lambda}\, 2u \sin\alpha \tag{3.32}$$

In the absence of shear, $\Delta x_o = 0$, and hence we obtain the phase difference due to in-plane displacement, as expected. Therefore the derivative fringes will always carry the in-plane contribution. This problem does not arise when the full aperture of the lens is used. Figure 3.13 shows a fringe pattern when a diaphragm is given both in-plane and out-of-plane displacements. The contribution of the in-plane component is clearly seen

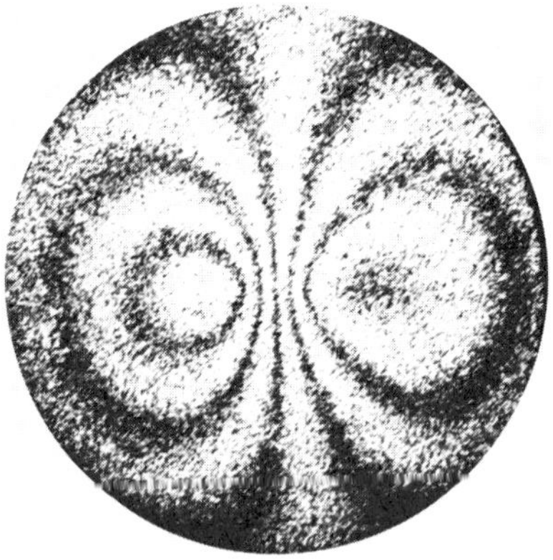

Figure 3.13 Asymmetry in fringe pattern showing sensitivity to in-plane displacement when a two-aperture arrangement is used in speckle shear interferometry.

as a lack of symmetry in the fringe pattern. It is therefore desirable to have an experimental setup where the contribution of the in-plane component can be independently sensed. This can be done by proper arrangement of apertures in a mask. Let us consider a mask with three apertures as shown in Fig. 3.14a. One aperture carries a wedge plate for shearing. The aperture separations are so chosen that the halos are not superposed; seven halos are formed instead of the five formed with three equispaced apertures; the halo distribution is shown in Fig. 3.14b. The fringe pattern

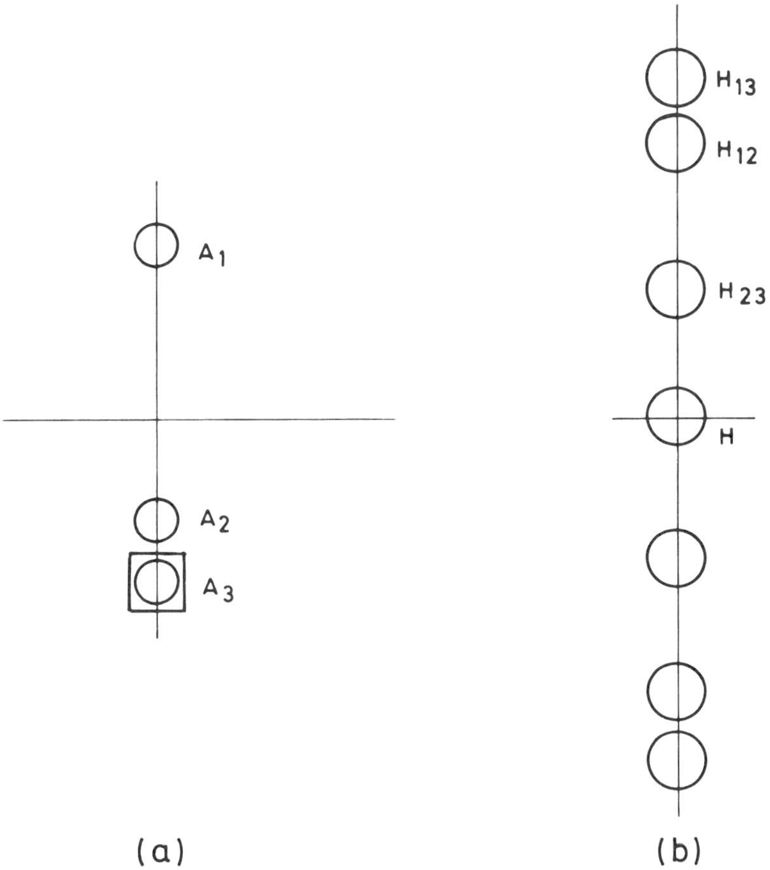

Figure 3.14 (a) Aperture arrangement to sense in-plane component as well as combined in-plane and slope components. (b) Halo distribution obtained with the aperture mask of (a).

corresponding to the in-plane component is obtained by filtering via halo H_{12} while the pattern corresponding to combined slope and in-plane components is obtained by filtering via halos H_{13} or H_{23}. The in-plane sensitivities are different but can be scaled to subtract the effect of the in-plane contribution from that of the combined pattern.

Sandwich speckle shear interferometry is yet another modification that provides a technique for the cancellation of rigid-body displacements [31].

The curvature fringes are obtained as a moiré pattern of the slope fringes. Instead of mechanical superposition of two slope patterns, curvature fringes are obtained by an optical method. An arrangement in which contributions from three points on the object are received at a single point on the image plane is used to obtain both slope and curvature fringes [32].

3.6 MULTIPLEXING

It is shown in Section 3.3.3 that in-plane and out-of-plane components can be separately obtained from a double-exposure specklegram if the lens aperture is properly masked. In another situation it may be required to record successive loading stages on the same specklegram and retrieve the corresponding information. This can be achieved by multiplexing that uses either frequency modulation or theta modulation. These modulation methods can be combined to record still more information content provided the dynamical range of the recording medium permits. In frequency modulation the spacing between the two apertures in the mask is varied. Let us consider N equidistant apertures in a mask placed in front of the lens. The initial exposure is recorded with all the apertures open. The object is loaded and the second exposure is made with apertures 1 and 2 open and the remaining apertures closed. The object is further loaded, and another exposure is made with apertures 1 and 3 open. In this manner a large number of exposures corresponding to different states of the object are made. In whole-field filtering a large number of halos are formed. Filtering via appropriate halos generates an in-plane displacement fringe pattern in this case. Care must be taken in masking the lens aperture so that during filtering various halos do not overlap. If N apertures are placed equidistantly on a line, $2N - 1$ halos are formed on filtering; some halos are superposed. If N apertures are placed in line such that no halo superposition takes place, then $(N - 1)^2 + N$ halos are formed.

Similarly, by angularly shifting the apertures for various exposures, theta multiplexing is performed.

Multiplexing can be incorporated for configurations sensitive to the in-plane or out-of-plane component. It can also be incorporated in speckle shear interferometry [18].

3.7 ELECTRONIC SPECKLE PATTERN INTERFEROMETRY

Speckle size is governed by the *f* number of the imaging lens. Furthermore, by adding a reference beam coaxially, the speckle size is doubled. It is therefore possible to record the speckle pattern using a standard television camera. In other words, speckle interferometry can be performed with electronic detection. Video processing can be used to generate correlation fringes eqivalent to those obtained photographically. This method is known as electronic speckle pattern interferometry (ESPI) [2, 33]. The main feature of ESPI is its ability to present correlation fringes in real time. Intensity correlation in ESPI is observed by the process of either video signal subtraction or addition.

In digital speckle pattern interferometry (DSPI) the speckle pattern data are processed digitally on a host computer; this is in contrast to ESPI where processing is done using analog electronic circuits. The analog video signal from the TV camera is sent to a high-speed analog-to-digital converter (ADC). The ADC samples the incoming video signal to yield a digital picture made up of an array of picture cells or pixels, and the intensity at each pixel is quantized to a set of gray levels. The number of pixels and the number of gray levels depend on the frame memory of the system used. Two frames of this digital picture corresponding to the object in its initial and deformed states are stored in the host computer, which is interfaced to the frame store. These two frames are subtracted pixel by pixel and sent to a digital-to-analog converter (DAC). The processed data are then displayed on the monitor. Real-time differencing can be achieved by means of a recursive processor incorporated in the digital image processing system. The difference signal is subjected to nonlinear processing such as level slicing or level windowing to enhance fringe contrast.

Earlier ESPI systems used a He-Ne laser as a source. The semiconductor lasers used now possess good thermal stability and give out radiation of long coherence length [34]. They are therefore used in ESPI systems, making such systems portable; commercial systems using semiconductor lasers are available. It may be of interest to know that ESPI has been performed in the far-infrared at CO_2 wavelength [35]. Fiber optics is used, in some configurations, for illuminating the object [36]. Further phase shifting has been incorporated in DSPI to obtain the phase map and hence the deformation map from the interference pattern [37]. The theory and applications of ESPI are discussed in detail in Chapter 4.

3.8 APPLICATIONS IN FRACTURE MECHANICS

Experimental fracture mechanics is an area of study in determining the strength or lifetime of modern-day structures. The theory of fracture mechanics is based on the concept that a crack in an elastic medium will propagate if the strain energy stored in the medium is sufficient to supply the energy to form new surfaces. This energy is related to the stress and displacement field of the body. When the load is less than that required for the crack to propagate, the stress and displacement field near the tip are defined in terms of the stress-intensity factor K_I through a system of equations. These are exact for two-dimensional theory. But numerous approximate solutions to three-dimensional problems have been obtained. An accurate determination of the displacement field in the neighborhood of the crack tip would provide a check on the validity of these approximate solutions [38].

Photoelasticity and the method of caustics have been frequently used in fracture mechanics. Photoelasticity requires a birefringent specimen. However, an optically smooth surface is needed to apply the method of caustics. Laser speckle photography has no such restrictions; it can be applied to all materials with no surface preparation. The measurement of displacements around crack tips in sheets of Araldite CT200 has been carried out by Luxmoore [39] using speckle photography at unit magnification.

Speckle photography is applied to the study of the displacement field around a central crack in an aluminum plate subjected to tensile load [9]. The double-exposure specklegram is whole-field filtered, and Fig. 3.15a shows the vertical displacement field (v-family). Each fringe represents a constant displacement of 0.005 mm. The displacement values along the axis obtained from the fringes are plotted in Fig. 3.15b along with the theoretical results. A good deal of important fracture mechanics information can be obtained from such a fringe pattern. The crack opening displacement (COD) at the middle of the crack can be quickly obtained by counting the number of fringes missing as one traces the fringe order from one side of the crack to the other. The number of fringes missing in the case plotted in Fig. 3.15 was estimated to be 13.2, which gave the COD value of 0.0625 mm as against the theoretical value of 0.06 mm. Furthermore, there is an asymmetry in the fringe pattern, suggesting that the specimen did not deform symmetrically.

In another example speckle photography was used to study the displacement field near the tip of a sharp side crack in a PMMA plate [40]. The displacement field was obtained by pointwise filtering. The stress intensity factor K_I was calculated from the experimental data using two different methods, least squares fitting and evaluation of the J integral. The results from the two methods of data analysis are in reasonable agreement, although the J method gives results that are consistently lower than those obtained with the method of least squares.

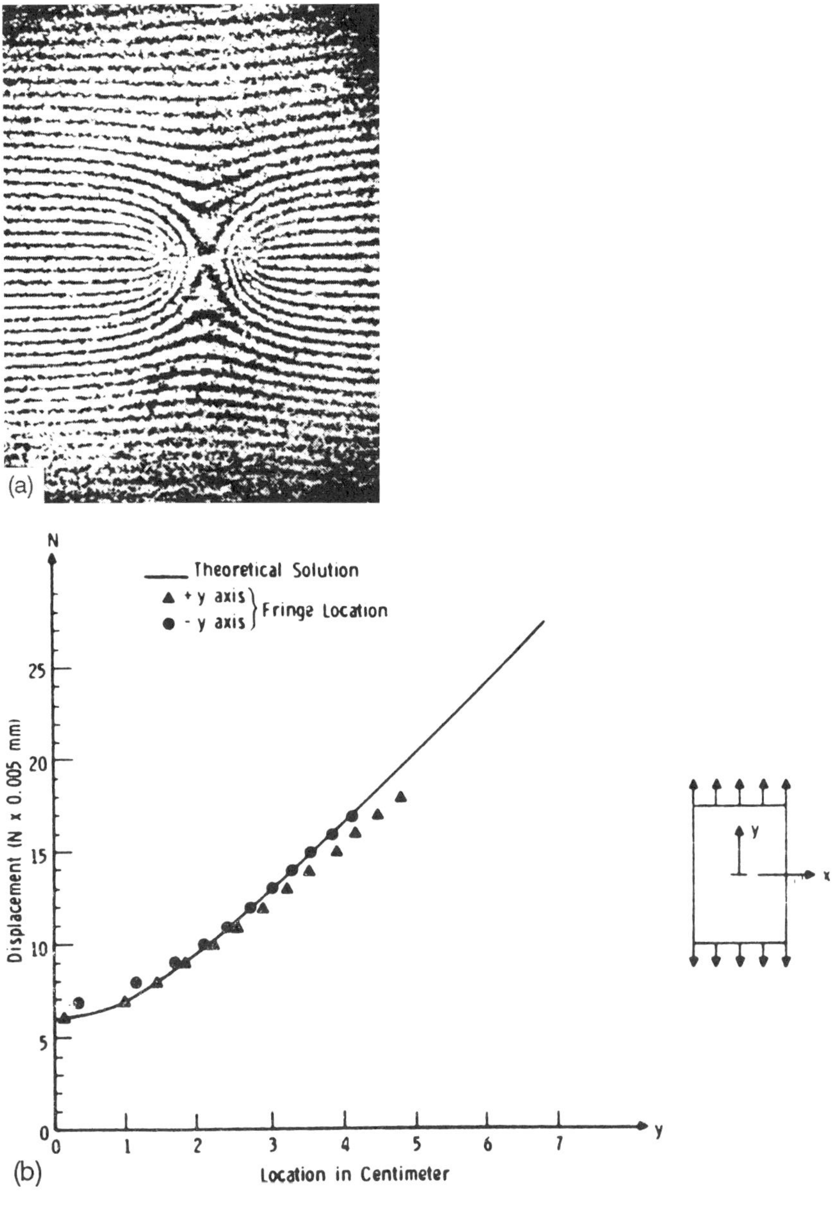

Figure 3.15 (a) v-Isothetic fringe pattern around a central crack in an aluminum plate (from Ref. 9). (b) Measured and theoretical displacement components along y direction (from Ref. 9).

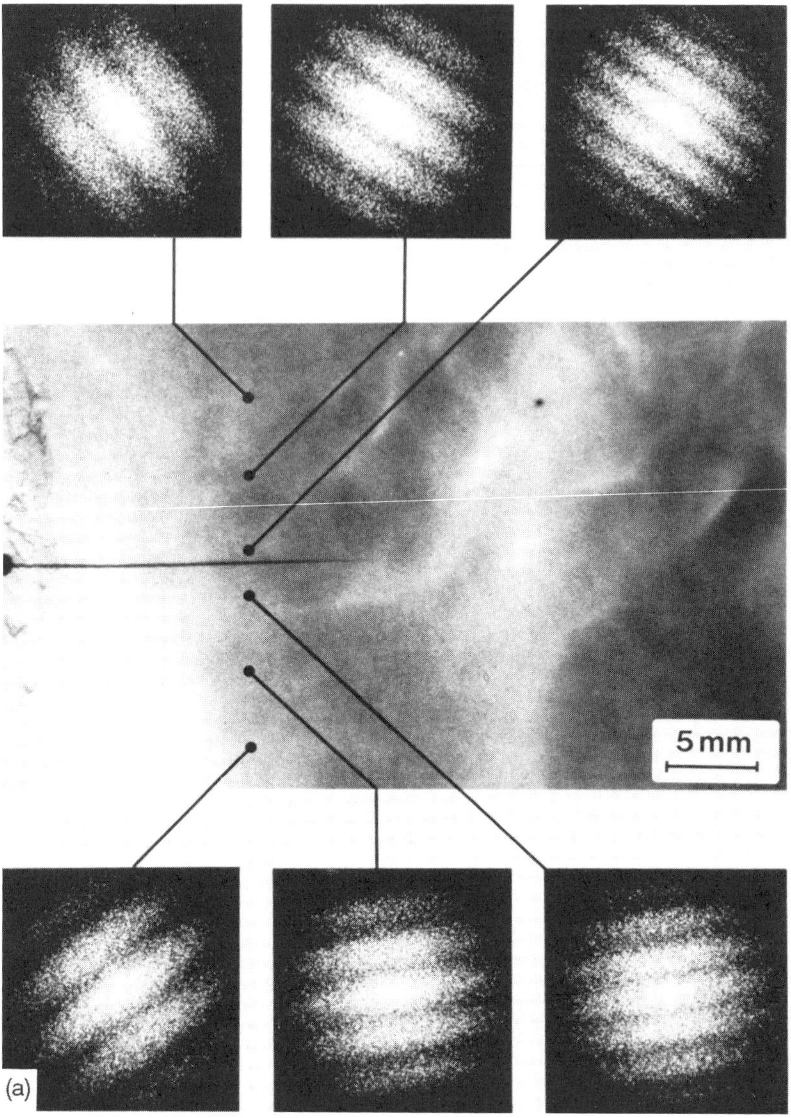

Figure 3.16 (a) Photograph of a specklegram along with pointwise filtered fringes (from Ref. 41). (b) Displacement field obtained from the analysis of specklegram of (a).

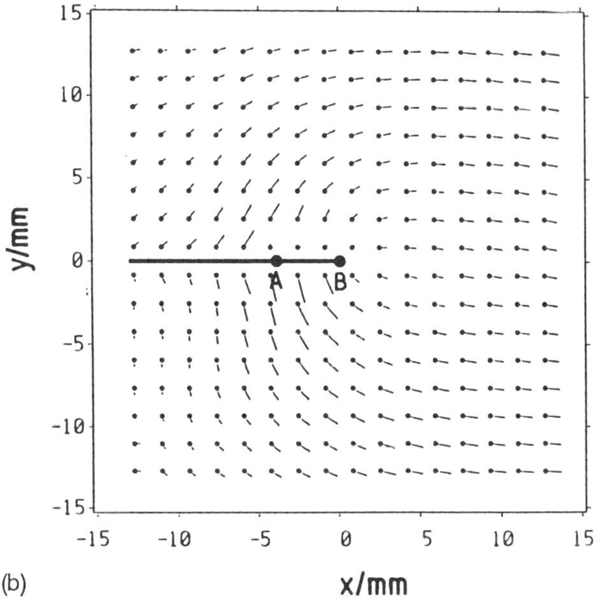

(b)

Figure 3.16 Continued.

These examples illustrate the use of speckle photography for the calculation of the stress intensity factor K_I when the crack remains effectively stationary. However, in a dynamic situation when the crack tip moves with a high velocity, the stress intensity factor K_{Id} would be expected to be quite different from that of the stationary crack. In fact, there is considerable interest in the relationship between K_{Ic} and K_{Id}, where K_{Ic} is the critical stress intensity factor just prior to the onset of the fast crack. Huntley et al. [40] studied the propagation of a fast crack in a plate. The crack was initiated by firing a small detonator in the shaft of a chisel positioned in the mouth of a blunt starter notch. The propagation of the crack was monitored by an independent method. To record specklegrams in the case of a dynamic crack, a pulsed ruby laser is used. It is not necessary to mode-lock and Q-switch the laser. The pulses from a free-running laser can be used for speckle photography, unlike for holography where Q-switched pulses with intercavity etalons are required. Figure 3.16a shows a photograph of the specklegram of the plate along with the fringe patterns obtained by pointwise filtering at various locations. The displacement field obtained from the specklegram is shown in Fig. 3.16b. The dynamic stress intensity factor K_{Id} is calculated using the least squares method and is found to be much higher than the static stress intensity factor K_{Is}.

Other interesting applications of speckle photography include the measurement of crack-opening displacement in the interior of a transparent medium embedded with scattering centers [38] and the study of cracks in the thickness of the coating on a ceramic [41]. Displacements in a part-circular external crack in a PMMA cylinder were measured using speckle photography and also calculated using the finite-element method [42]. The cylinder was loaded by applying an internal pressure. There is reasonably good agreement between experimental and numerical results.

Determination of all the components of the deformation vector near the crack tip requires the combined use of moiré and speckle interferometry

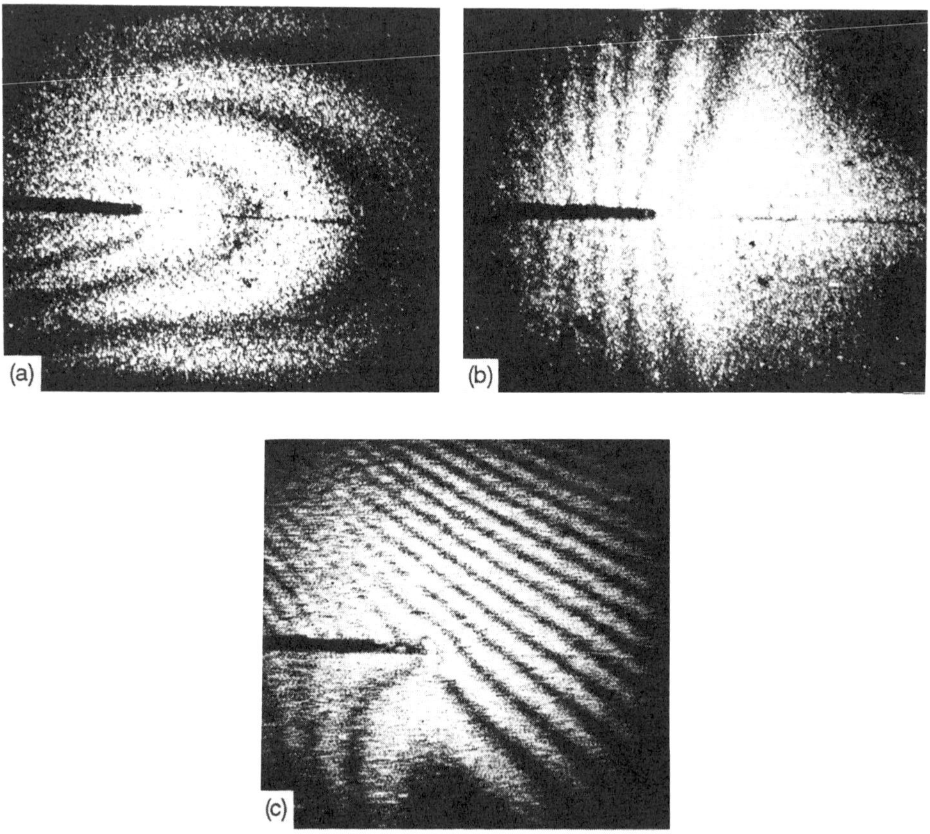

Figure 3.17 u-Component, v-component, and w-component of displacement around the crack. The fringe patterns are filtered from a single specklegram recorded using experimental arrangement of Fig. 3.9a.

[43]. Three-dimensional deformation measurement has also been attempted by speckle interferometry alone. The configuration used is described in Section 3.3.3. Fringe patterns corresponding to all three components are filtered out via appropriate halos. It may be pointed out that the sensitivity for the out-of-plane component is much higher than for in-plane components compared to the earlier method [43]. Figures 3.17a–c show the fringe patterns corresponding to the three components of displacement around a crack.

Crack detection in pressure vessels is a practical application of speckle methods. The thinning (thinness) of a pressure vessel wall was investigated by Gregory [44] using a ruby pulsed laser in defocused speckle photography. Whole-field filtering of a double-exposure specklegram yields fringes that are due to bulging of thin regions of the object when pressure is changed between exposures. Speckle photography methods require a development step, necessitating a longer time or time delay in obtaining useful data. ESPI has therefore been used for detecting the cracks. A number of loading methods have been studied, and it is found that the loading method should open out the crack for ESPI to unambiguously detect crack location. Therefore pressurization of vessels is an ideal method for crack detection.

Electronic speckle pattern interferometry has been applied to the study of crack propagation in epoxy resins [45]. The ease with which the interferogram is displayed is an added advantage of ESPI. A number of experiments were conducted to detect simulated cracks using DSPI. The specimens were Araldite plates 300 $\times$ 100 $\times$ 5.5 mm with programmed cracks at the middle of the plate. A small region around the crack was

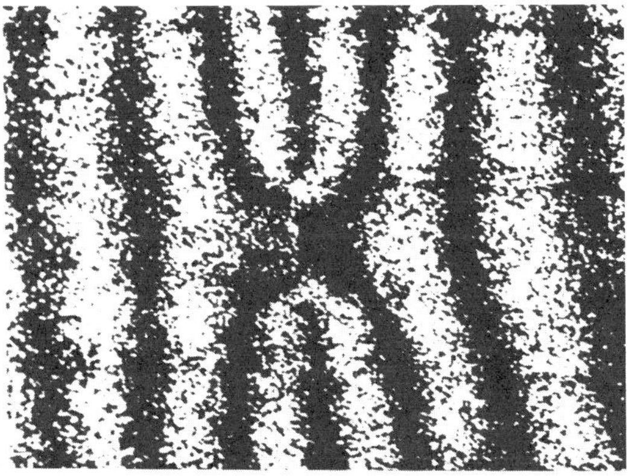

Figure 3.18 Interferogram on a monitor when DSPI was performed on a specimen having a simulated crack. An in-plane-sensitive configuration was used, and the specimen was mechanically loaded near the site of crack.

imaged. It was observed that the selection of a loading mechanism was crucial to the success of the experiment. Figure 3.18 shows an interferogram of a simulated crack; an in-plane-sensitive configuration was used, and the specimen was given an out-of-plane deflection of 400 μm near the site of the crack. For higher deflections, fringe contrast decreases owing to decorrelation. Loss of contrast was greater for tensile loading.

3.9 VIBRATION STUDIES

Speckle photography is used for the measurement of vibration amplitudes in the case of both in-plane and out-of-plane vibrations. In the case of in-plane vibration, the image of the speckle is drawn in a line, rather it getting doubled when double-exposure speckle photography is performed. Indeed a pulsed laser can be used to record double-exposure specklegrams covering various phases of vibration. When analyzed, these specklegrams yield vibration amplitude changes during the time between two exposures. Instead of recording a double-exposure specklegram, a time-average specklegram over a period that is much longer than the time of vibration can be made; the image of the speckle in such a specklegram will be drawn into a line. Let us consider an object that is vibrating in its own plane along the y direction. Let the instantaneous amplitude of vibration be given by $y = y_o \sin \omega t$, where ω is the frequency of vibration. The speckle draws a line of length $2My_o$ in the image plane, where M is the magnification of the imaging lens. The position probability density function $P(y)$ of a speckle is given by [46]

$$P(y) = \begin{cases} \dfrac{1}{\pi M y_o}\left[1 - \left(\dfrac{y}{My_o}\right)^2\right]^{-1/2}, & -My_o < y < My_o \\ 0, & \text{otherwise} \end{cases} \tag{3.33}$$

The probability is highest when the speckle is at the ends of travel. The local transmittance of a time-average specklegram will be governed by the position probability density function. The diffraction pattern of such a specklegram will have the diffraction pattern modulated by the fringe pattern. The fringe pattern is obtained from the Fourier transform of the position probability density function,

$$\begin{aligned} U(\nu) &= \int_{-My_o}^{My_o} \frac{1}{\pi M y_o} \frac{\exp(2\pi i \nu y)}{[1 - (y/My_o)^2]^{1/2}}\, dy \\ &= \frac{1}{\pi}\int_{-\pi/2}^{\pi/2} \exp(2\pi i \nu M y_o \sin\theta)\, d\theta \end{aligned} \tag{3.34}$$

where ν is the spatial frequency in the observation plane. The amplitude distribution in the fringe pattern is then given by

$$U(\nu) = J_o(2\pi M y_o \nu) \tag{3.35}$$

The intensity distribution in the diffraction pattern of the specklegram is thus that of the autocorrelation of the aperture function (diffraction halo) modulated by the function $J_o^2(2\pi M y_o \nu)$. The argument of the modulating function depends on the amplitude of vibration, and hence its value can be extracted from the fringe pattern.

In principle it is possible to record the in-plane vibration of more complex motions. For example, a surface may execute a circular motion in its own plane; the speckle in the image plane will also trace a circular path unlike the linear path in the case of simple harmonic motion. More complex speckle trajectories can be recorded by vibrating the objects with two independent shakers whose amplitudes and phases can be controlled [47]. The fringes in the diffraction pattern are characteristic of the trajectory of the speckle. However, it is not always possible to relate such a diffraction pattern back to the speckle trajectory that gave rise to it.

The out-of-plane vibration can also be studied using speckle photography. The camera is focused not on the surface of the object rather on a plane in front of it. The out-of-plane motion therefore results in the in-plane shift of the speckles at this plane. Defocused speckle photography, however, yields fringes that represent partial slopes [48].

Let us consider the experimental setup shown in Fig. 3.19. At some state of vibration a speckle formed at S_1 moves to another position S_2 at another state of vibration due to a change in slope. The displacement d in the plane P can be expressed as

$$\begin{aligned} d = |\bar{d}| &= p[\tan(2\phi + \alpha) - \tan\alpha] \\ &= p\,\frac{\tan 2\phi(1 + \tan^2\alpha)}{1 - \tan 2\phi \tan\alpha} \\ &= 2p\phi \qquad \text{when both } \phi \text{ and } \alpha \text{ are small} \end{aligned} \tag{3.36}$$

The amplitude of vibration $A(x, y; t)$ can be expressed as

$$A(x, y; t) = A_o(x, y) \sin \omega t \tag{3.37}$$

where $A_o(x, y)$ is the amplitude of vibration. Due to the vibration, the slope will also vary with time, and hence we could write the slope as

$$\phi(x, y; t) = \phi_o(x, y) \sin \omega t \tag{3.38}$$

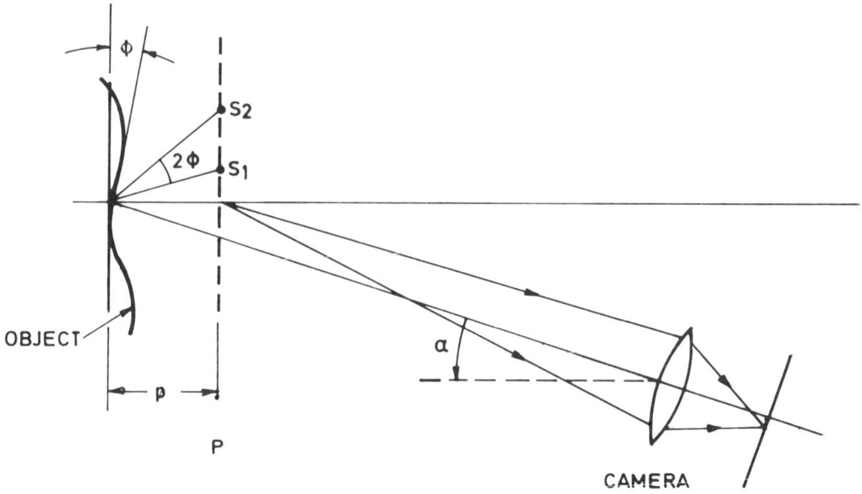

Figure 3.19 Experimental arrangement to measure vibration amplitude of an object.

where $\phi_o(x, y)$ is the extremum value of the slope. Therefore the displacement of speckle at the plane P is

$$d = 2p\phi_o(x, y) \sin \omega t \tag{3.39}$$

A specklegram is recorded for a time period T that is much longer than the period of vibration. The total exposure $E(x', y')$ recorded will be given by

$$E(x', y') = c \int_0^T |U(x' + u, y' + v)|^2 \, dt \tag{3.40}$$

where $U(x', y')$ is the amplitude at any point (x', y') on the recording plane, and u and v are the components of the displacement vector d/M at the recording plane.

When such a specklegram is pointwise filtered, the intensity distribution in the (x_f, y_f) plane will be

$$I(x_f, y_f) = cI_s(x_f, y_f)J_o^2 \left[\frac{2pkM}{L}(x_f\phi_{ox} + y_f\phi_{oy})\right] \tag{3.41}$$

where $k = 2\pi/\lambda$. The minima of the fringes occur at the zeros of the Bessel function J_o. This is the same result as we obtained for in-plane motion.

On the other hand, if the specklegram is whole-field filtered by placing an aperture along either the x_f or y_f directions, then the object image is covered with fringes that represent partial slopes. The fringe formation is governed by the equations

$$\frac{\partial w}{\partial x} = \phi_x = \frac{N\lambda L}{4\pi p x_f M} \tag{3.42a}$$

and

$$\frac{\partial w}{\partial y} = \phi_y = \frac{N\lambda L}{4\pi p y_f M} \tag{3.42b}$$

Here N is the fringe order for which $J_o^2(N)$ is maximum, that is, $N = 0$, 3.84, 7.02, ... The centerlines of zeroth-order ($N = 0$) fringes are the partial antinodal lines where the slope along the filtering direction is zero. Figure 3.20 shows the partial slope fringes along seven different directions of a vibrating square plate of plexiglass; the filtering is done via the aperture locations on the transform plane (T.P.) shown alongside. The technique is applied to the study of vibration modes of plates of composite materials

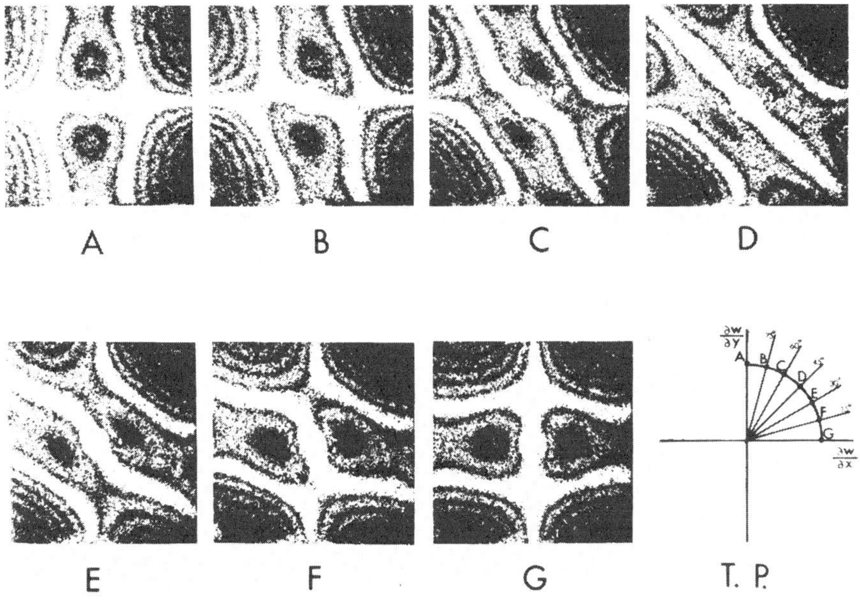

Figure 3.20 Fringe patterns showing partial slopes of a vibrating square plate (from Ref. 48).

and plates with mirrorlike surfaces [49]. The transient vibrations of a cantilever impacted at the tip have also been studied experimentally and theoretically [50].

Vibration analysis can also be performed by speckle interferometry. Consider the image of an object illuminated with a coherent beam. If the object as a whole is moving toward or away from the camera lens there will be no change in the speckle pattern in the image provided it is within the depth of focus, since the brightness of the speckle is independent of the phase of the illuminating wave. However, the situation is completely different when a reference beam is added; the speckle pattern undergoes cyclic variation in brightness as the object moves. A visual speckle interferometer first developed by Archbold et al. [51] and later modified and improved by Stetson [52] is suitable for studying the vibration modes of an object. In Stetson's design a uniform reference wave is superposed on the object field; the reference wave is brighter than the average brightness of the speckle field.

Though speckle interferometry with photographic emulsions has not been reported for vibration studies, probably because of the intermediate photoprocessing step, electronic speckle pattern interferometry has been successfully applied. In many situations it is preferred to hologram interferometry. Apart from its use in the study of the modes of simple systems such as plates [9, 53], ESPI has been applied in vibration analysis of an engine block under test [54, 55] and a car body [56]. The quality of ESPI fringes has been considerably improved by a technique of speckle averaging [57]. Interpretation of the vibration pattern is simplified by modulating the reference beam phase at a frequency near the object frequency [58, 59]. The fringes oscillate at the beat frequency, enabling the relative phases of different vibrational modes to be clearly observed.

A comparison of the resonance structure with its master can also be performed. The object and the master are driven at the same frequency; the fringes will indicate the difference in amplitudes in subtractive ESPI. A fringe-free field would indicate that the two systems are oscillating with identical amplitudes. Some other interesting applications of ESPI include vibration analysis at high temperatures [60], vibration mode analysis [61, 62], high-frequency vibration analysis [63], and three-dimensional vibration measurements [64]. The pulsed laser electronic pattern interferometer is an instrument capable of being used in an industrial environment without the need for vibration isolation and photographic processing. Optoelectronic addition of speckle patterns followed by high-pass filtering is employed. Earlier systems used a ruby laser [53], but Nd:Yag lasers [59, 65] are now employed for pulsed ESPI.

Phase shifting has been incorporated in ESPI to generate deformation maps of the object [59, 66, 67].

3.10 IN-PLANE DISPLACEMENT AND STRAIN MEASUREMENT

One of the very impressive applications of speckle photography for the measurement of in-plane displacement is presented by Tiziani [46] wherein the displacement at a number of grid points on a simply supported centrally loaded beam was obtained by pointwise filtering. One arrangement to extract information from a large number of grid points in parallel uses composite hololenses [68]. The displacement thus obtained is divided by the magnification of the imaging lens to obtain the displacement field at the object. It may be noted that the image plane is sensitive to in-plane displacement whereas its sensitivity to tilt and out-of-plane displacement is exceedingly poor. However, there is a plane, called the tilt plane, on which speckle displacement is due to the tilt of the object only. In between the image and tilt planes, the speckle displacement is due to both in-plane displacement and tilt. Unfortunately, the location of the tilt plane depends on the curvature of the illuminating wave, and hence the experimental arrangement should be properly chosen.

Speckle photography is applied to the study of the behavior of adhesive joints under shear stress [69]. The technique permits determination of the shear modulus of the thin adhesive layer. The behavior of concrete subjected to uniaxial tension is studied using the speckle method [70] with emphasis on detecting cracks and measuring the width of a microcrack and its growth. Double-exposure photography is limited to static in-plane displacements. It has been shown that a multiple-exposure specklegram of a specimen undergoing deformation can be analyzed unambiguously to give the in-plane displacement at the time of each exposure [71]. The multiple-exposure specklegram is analyzed by pointwise filtering as for double-exposure specklegrams. To study the time evolution of transient strain fields, high repetition rate lasers must be employed. The speckle technique has been extended to the analysis of displacements on rotating parts by the use of a sandwich procedure and a pulsed ruby laser [72].

The deformation of an automobile engine connecting rod was studied by Happe [73]. The connecting rod transmits the forces from the piston to the crankshaft and therefore is subjected to traction forces due to the inertia of the piston. Figure 3.21a shows a photograph of a connecting rod. The big end bearing, by means of which the rod is connected to the crankshaft, consists of two halves that are clamped together by bolts. The experimental setup to measure (1) oval deformation of the bearing and (2) the relative movement of the halves (circled in the photograph) when the rod is subjected to a tensile force is shown in Fig. 3.21b. The big end bearing of the connecting rod has been fitted to a journal on a heavy steel box. The front of the journal is in plane with the side surface of the connecting rod's big

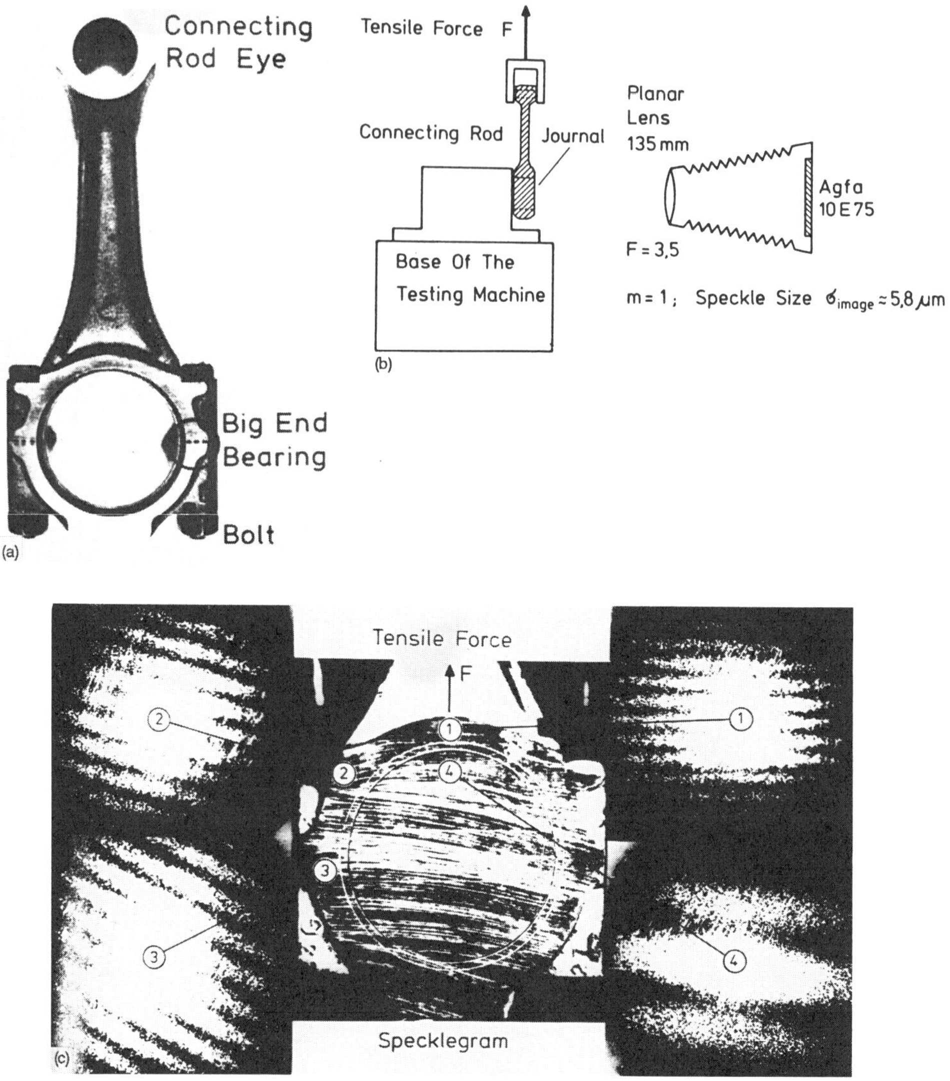

Figure 3.21 (a) Photograph of the connecting rod. (b) Schematic of the experimental setup. (c) Photograph of the specklegram along with pointwise-filtered fringes (from Ref. 73).

end. The steel block is clamped to the base of a testing machine, which subjects the connecting rod to various tensile loads. The surfaces of the big end bearing and journal are imaged on the photographic plate at unit magnification. Double-exposure specklegrams were recorded with incremental loads. These are pointwise-filtered. Figure 3.21c shows the photograph of the specklegram along with the Young's fringes obtained by filtering at various locations. The displacement field is then generated. Considerable error may arise due to lens aberration when speckle photography is applied at its highest sensitivity to measure strain fields [74].

A linear encoder using speckle correlation techniques has been described by Yamaguchi [75]. Figure 3.22 shows an experimental arrangement. A diode laser (780 nm) is used as a source, and the speckle field is detected by a linear diode array detector. The speckle displacement detected by the arrangement of Fig. 3.22 is

$$A_x = a_x \left[\frac{L_o \cos^2 \theta_s}{L_s \cos^2 \theta_o} + \cos \theta_o \right] \tag{3.43}$$

where a_x is the displacement of the object. Speckle displacement is calculated at the peak positions of the binary cross-correlation of the detector outputs.

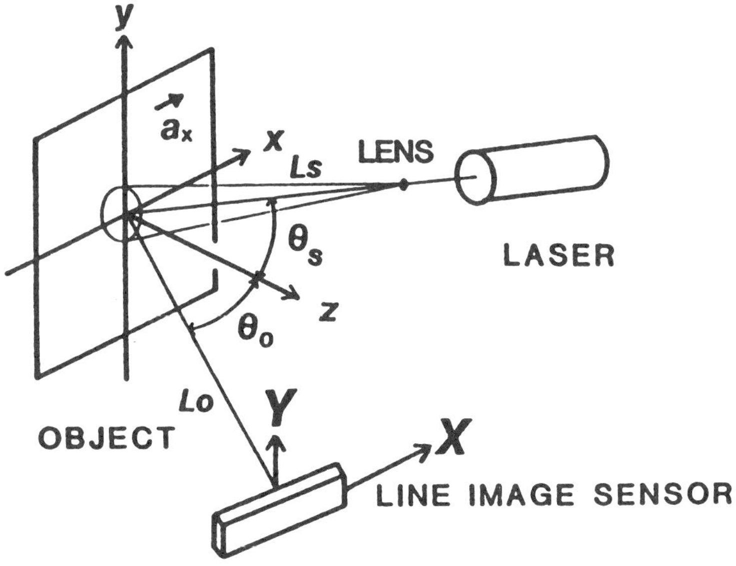

Figure 3.22 A laser encoder.

Symmetrical illumination of the object is used to measure the in-plane component of displacement with interferometric precision. One of the applications of this is to measure the linear strain on the surface of a polythene sleeve joint connecting two pipes made of similar materials [76]. The experimental arrangement was modified to generate symmetrical illumination beams. A double-exposure specklegram was recorded with a deliberate vertical displacement of the photographic plate so that the filtering could be done via the minima of Young's fringes to generate correlation fringes. The component of strain along the axial direction of the sleeve is calculated from the fringes. The sign of the strain could not be obtained from the interferogram, but live fringe is used to establish it. It is to be noted that strain changes sign from tensile to compressive.

Duffy's two-aperture method [77] has also been used for measuring in-plane displacement, but its sensitivity is extremely poor. It measures the displacement component along the direction joining the centers of the two apertures. This drawback was removed by the use of a mask having four apertures in front of the imaging lens so that orthogonal components of in-plane displacement could be measured [17]. Further, the sensitivity of this method has been enhanced by the use of hololenses [19, 20].

A fully automatic laser speckle strain gauge has been reported by Yamaguchi [78, 79]. In one configuration, the specimen is symmetrically illuminated by collimated beams, and one linear diode array detector is used [78], while in the second configuration the specimen is illuminated nearly normally, and two linear array detectors are symmetrically placed to receive the speckle field [79]. Excellent correlation is found to exist between the strain values measured by the laser strain gauge and the resistance strain gauge.

Use of symmetric illumination with ESPI for the measurement of in-plane displacement is very convenient, as we can follow the time history without much difficulty [80, 81]. Preater [82, 83] used the symmetrical illumination arrangement of Leendertz [15] to analyze in-plane strain in high-speed rotating components. A pulse laser is used to freeze the rotation of test components. Engineering problems over a wide range of speeds up to 12,500 rpm can be analyzed. The use of fiber optics extends the application of the technique to components with limited optical access.

For the direct measurement of strain and slope, various shear interferometers have been proposed [28]. These methods have been applied to the study of objects such as plates and cantilevers [84].

3.11 MEASUREMENT OF OUT-OF-PLANE DISPLACEMENT

Speckle photography is absurdly less sensitive for out-of-plane displacement measurement [85]. Further, the decorrelation effect restricts the range of its applicability. It is, however, well suited for studying deformations of large structures illuminated by white light. This topic is covered in Chapter 9. The double-exposure laser specklegram, on Fourier filtering, yields a fringe pattern that may be localized at various planes depending on the experimental conditions [86].

It has been shown that defocused speckle photography offers sensitivity comparable to that of hologram interferometry for the measurement of out-of-plane deformation [87]. The out-of-plane displacement is obtained in terms of the local slope measurement. High sensitivity is achievable if an aberration-free lens is used and experimental parameters are optimized. Experiments carried out on the hydraulic cylinder with hologram interferometry and defocused photography demonstrate their comparable sensitivities. Out-of-plane deformation can also be monitored with speckle shear interferometry; large deformations for which too many fringes are obtained with speckle interferometry can be measured with speckle shear interferometry. Indeed, demagnification of speckle interferometry by a fairly large factor is achievable using two illumination beams near grazing incidence [16]. Very little desensitization is possible by changing the direction of illumination.

The speckle interferometry, on the other hand, offers interferometric sensitivity of $\lambda/2$ per fringe. A convenient method to record out-of-plane displacement is shown in Fig. 3.9. The object is a cantilever that has been loaded to provide out-of-plane and in-plane displacements between exposures. The specklegram is whole-field filtered via appropriate halos to yield out-of-plane and in-plane displacement fringes shown in Fig. 3.10.

Similar results on cantilevers and other structures are obtained using ESPI [88, 89]. ESPI offers the advantage of a phase-shifting algorithm to yield more accurate values of the displacement. The comparative DSPI is a technique for the comparison of deformation profiles of two objects and hence their responses to the external loading agency. No reference wave is used as the wave field from one object acts as a reference to the other [90]. A number of researchers have considered measurement of three-dimensional deformation of an object because it provides significantly more information [91–95]. It is, however, necessary to measure the contour of the object along with three-dimensional deformation.

3.12 PLATE-BENDING PROBLEM

The deflection of a plate can be measured by speckle interferometry and ESPI. However, sometimes the experimentalist requires the slope values and cross-slope values in addition to strains. Therefore defocused speckle photography [87, 96, 97] and speckle shear interferometry [28, 98, 99] are used. In order to apply defocused speckle photography to obtain slope fringes, the camera is focused at a plane in front of the object, wherein the in-plane displacement is due to a change of slope. If the change of slope is accompanied by in-plane displacement of the object, recording should be made at the tilt plane. Under the assumption that there is no in-plane displacement, tilts are small, and other angles in the experimental geometry are small, the displacement at the defocused plane is

$$d = 2p\phi$$

where p is the distance between the object plane and the plane on which the camera is focused.

The displacement at the image plane will be Md. A double-exposure specklegram that is whole-field filtered yields bright fringes wherever

$$\frac{ux_f + vy_f}{L} = N\lambda, \qquad N = 0, \pm 1, \pm 2, \ldots \tag{3.44}$$

where L is the distance between the specklegram and the Fourier plane, u, v are the components of d along the x and y axes, respectively, and (x_f, y_f) are the coordinates of the filtering aperture. If the filtering is done by placing an aperture at (x_f, 0) and (0, y_f), respectively, the fringes corresponding to

$$\phi_x = \frac{\partial w}{\partial x} = \frac{n\lambda L}{2px_f} \tag{3.45a}$$

and

$$\phi_y = \frac{\partial w}{\partial y} = \frac{n\lambda L}{2py_f} \tag{3.45b}$$

are obtained.

In fact, defocusing is one of the methods of obtaining shear. There are a large number of other methods reported in the literature [28]. One of the common methods is to use a multiaperture mask in front of the imaging lens and appropriate shearing elements on the apertures [18, 100]. A four-aperture arrangement with two wedges properly positioned gives slope fringes along the x, y, and diagonal directions. Since the recording is done at the image plane and a two- or four-aperture arrangement is inherently sensitive to in-plane displacement, it should be applied with caution. Figure 3.23a gives x-slope fringes for a rectangular diaphragm clamped at the edges and centrally loaded.

Radial slope fringes can be obtained by using a Billet's split lens whose elements are axially separated such that the object is imaged by both with the inverse magnification condition satisfied. A photograph of the radial

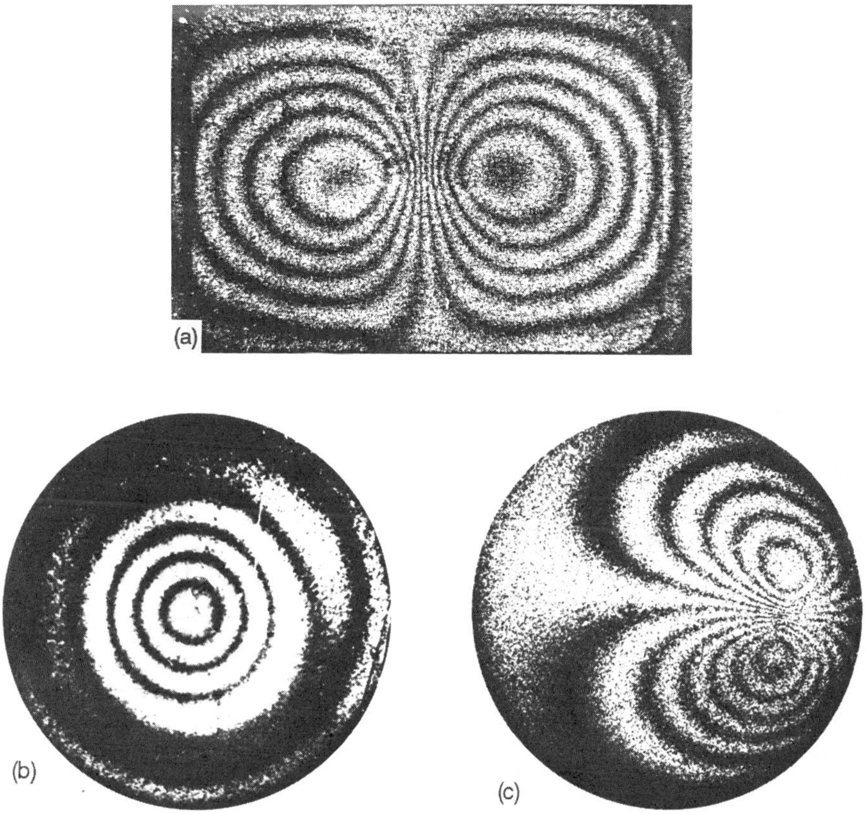

Figure 3.23 (a) x-Slope fringe pattern of a rectangular diaphragm. (b) Radial slope fringes represented by $r(\partial w/\partial r)$ of a circular diaphragm centrally loaded. (c) Theta-slope fringes of a circular diaphragm loaded away from the center.

slope fringes represented by $r(\partial w/\partial r)$ is shown in Fig. 3.23b. Since the radial shear increases linearly from the center toward the periphery, the radial shear is most sensitive to the defects at or near the periphery, for example, loose clamping of the diaphragm.

Rotational shear can be introduced again in a variety of ways; the common arrangement uses a pair of Dove prisms. The magnitude of the shear can be varied by rotating the Dove prism about the ray direction. The interesting feature of rotational shear is that no fringes are formed for any value of the deflection if the object is circularly symmetric and the rotational shear is introduced about the center of symmetry. On the other hand, fringes are formed when the deflection profile is asymmetrical, as is evident from Fig. 3.23c. For this reason theta shear is used in the nondestructive testing of circularly symmetric objects so that they can be deflected to a large extent to bring out changes in and around the defective region.

Apart from obtaining derivatives of the displacement components along different directions, it is found that rotational, radial, inversion, and folding shears can be employed to measure the tilt of diffuse objects with a variable sensitivity [101]. Also, the axis of tilt can be found in some situations.

The fringe pattern corresponding to curvature can be realized by superposing two displaced slope patterns. This requires three equidistant apertures in the mask, the outer apertures carrying oppositely sloping wedges. Filtering via an appropriate halo yields curvature fringes, $\partial^2 w/\partial x^2$ or $\partial^2 w/\partial y^2$. If the curvature fringes in orthogonal directions are simultaneously required, a five-aperture mask is used. The four apertures are located at the corners of a square and the fifth aperture at the center. The mask is so aligned that the lines having three apertures are along the x and y directions. The wedge plates with their refracting edges near the center are placed on the four apertures. A double-exposure specklegram is recorded with the plate deformed between exposures. The specklegram is whole-field filtered. At the Fourier plane the intensity distribution consists of 17 diffraction halos. The filtering via different halos gives fringe patterns corresponding to $\partial w/\partial x$, $\partial w/\partial y$, $\partial w/\partial x|_{45°}$, $\partial^2 w/\partial x^2$, $\partial^2 w/\partial y^2$, and $\partial^2 w/(\partial x \partial y)$. The sensitivity of slope fringes is twice as large as that obtainable with a single-wedge arrangement.

Shear interferometry has also been implemented on ESPI and applied to the study of centrally loaded plates and vibrating plates [102].

3.13 THERMAL STRESS ANALYSIS

Thermal strains in structures are sometimes difficult to measure because the very instrument that is used for strain measurement is itself often influenced by the temperature. Transient thermal strains such as those due to thermal shock are even more difficult to monitor because compensation techniques are usually not applicable. Furthermore, the contact method is ruled out for high-temperature measurement as the transducer itself may disintegrate. Laser speckle techniques, on the other hand, seem to be ideal for this purpose; these are noncontact techniques and hence can be applied to a specimen at any temperature. Radiation from the body can be blocked by a suitable laser filter. However,the decorrelation effect due to the turbulence in the intervening medium and changes in the microstructure of the surface of the specimen degrades the fringe contrast and limits the range of measurement. The influence of the intervening medium can be avoided by using a pulsed laser. In one experiment, thermal strains on an aluminum plate were studied using speckle photography [9]. The plate was heated with a propane torch. A number of double-exposure specklegrams were recorded with a ruby laser. The time difference between exposures was governed by the charging of capacitor banks and was about 30 sec. The plate cooled owing to natural convection during this period. These specklegrams were whole-field filtered to obtain isothetic fringe patterns corresponding to u and v displacements. The thermal strains were then calculated and compared with the theoretical values. Excellent agreement between experimental and theoretical results was obtained, thereby justifying the use of speckle photography for thermal strain measurements.

The speckle method is also used to measure temperature profiles normal to the cylinder head in a motored two-cycle engine [103]. This kind of study provides information regarding surface heat transfer and other fluid characteristics such as boundary layer thickness.

There is interest in the study of strain developed in the material of frescoes, i.e., plaster or other wall material. Variation of ambient temperature and humidity may introduce strains that lead to crack development and hence damage the frescoes. The speckle hologram method is used for the study of frescoes [104]. The thermal strain is developed by blowing hot air for a certain time. An argon ion laser is used for the recording.

Studies on the deterioration of stones and mural paintings were carried out by ESPI [105]. A plastered stone with various mural colors was chosen as a specimen and heated by irradiation from a sun-simulating lamp. Due to the different absorption coefficients of the various colored parts of the plaster, these regions acquired different temperatures. Figure 3.24 shows out-of-plane displacement fringes. It is seen that owing to the differences in temperature, the deformations of different colored regions are different and irregular. Further, the fringes show pronounced discontinuities exactly along the cracks. In the real world all mural paintings are covered by cracks and are regularly influenced by temperature variations. Over a period of time there is movement of the mural carrier material due to the decay process, and this has also been monitored using DSPI. The movement of walls is also a subject of considerable interest due to increased emphasis on preservation [106].

Electronic speckle pattern interferometry has also been used at much higher temperatures of the specimen almost approaching 3000°C [60, 107]. It has been used for the study of turbine blades heated to temperatures above 800°C [108] and vibrations of panels at temperatures up to 900°C [60]. It is observed that there is a decrease in the resonance frequencies of panels and also a decrease in the Q-factor for the different modes with temperature [60].

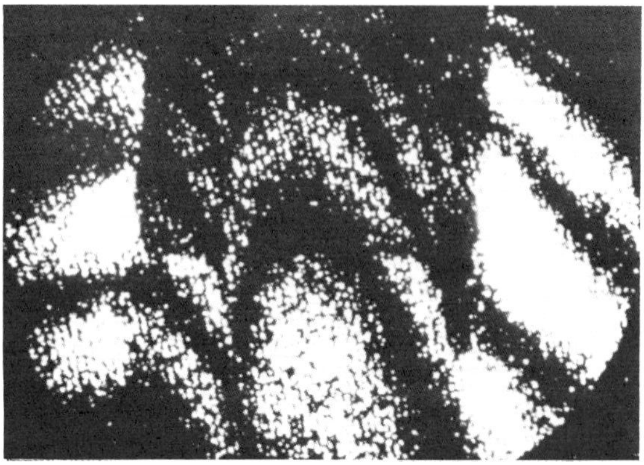

Figure 3.24 Fringes due to thermal loading of frescoes (from Ref. 105).

Symmetric illumination about the normal for the study of thermal strain in the vicinity of a hydride blister on the surface of a heated sample of a zirconium pressure tube used in a nuclear reactor has been used by Mader and Gnoyke [109]. A piston for a marine diesel engine cast from aluminum-base alloy and fitted with a burning chamber of heat-resistant ceramic was studied for thermal strain using a symmetrical illumination configuration in DSPI [110]. The mismatch between thermal expansion coefficients causes a complicated deformation. The test object is heated to temperatures up to 600 K using an 800-W heating element.

3.14 NONDESTRUCTIVE TESTING

Speckle nondestructive testing (SNDT) is akin to holographic nondestructive testing (HNDT). It requires some kind of loading of the object between exposures. The flaws and/or defects in the object are detected by comparison of two states of the object; the comparison is performed by any of the techniques—pointwise filtering, whole-field filtering, speckle correlation, etc. Defocused speckle photography has been applied to detect debonding in layer composites [9]. The deterioration of stone surfaces has been studied by speckle photography [111]. A record of the speckled image of the stone surface is made, the stone is subjected to corrosive environment, and a second exposure is made. The photographic plate is displaced by a small amount in its own plane between exposures. On pointwise filtering, Young's fringes of varying visibility are obtained from different regions of the specklegram. The decrease in visibility is due to changes in the microstructure of the surface due to corrosion. An edge-mounted diaphragm, though a simple system, presents considerable difficulties in nondestructive testing; these difficulties arise mainly because a small deformation due to flaws is superposed over a large deformation due to loading. The theta shear is found to be insensitive to circularly symmetric deformation, and hence only the differential change due to flaws that violates circular symmetry is seen. A local thinness of the diaphragm material may be treated as another diaphragm with an elastic boundary. Both laser speckle photography [112] and shear speckle interferometry [113] have been employed for the NDT of diaphragms. Delamination in composite laminates has been studied by speckle shear interferometry [114]. Time-average shearography is applied to detect flaws in glass-fiber-reinforced plastic beams [115]. The beams are excited at the resonance frequencies. The sandwich method has been combined with holospeckle interferometry for studying damage in concrete [116]. Because of the real-time capabilities of ESPI, it has been adopted for automatic inspection and measurement [117–119]. A status report on ESPI for inspection was prepared by Bergquist et al. [120]. ESPI has also been applied to the study of the decay of natural stones in buildings [121] and to the study of composites [122, 123].

Recently DSPI was applied to the study of the debonding of zinc layers on galvanized steel and the nondestructive testing of weldments. Steel plates 100 mm × 100 mm × 3 mm were used as objects. The galvanization was done by both hot dipping and electroplating. Adhesion of zinc onto the steel is due to the formation of several layers of zinc-iron alloys of various concentrations. Debonding of layers is due to either inadequate cleaning of the surface or the evolution of hydrogen during galvanization leading to blisters on the surface. An out-of-plane-sensitive configuration was used and thermal loading was applied. Figure 3.25a shows the fringe pattern corresponding to a well-galvanized plate, while Fig. 3.25b shows the fringe pattern for a plate for which debonding was clearly visible as a kink in the fringe pattern. The weldment between two steel plates was examined by DSPI and a radiographic method. Figure 3.26 shows a fringe

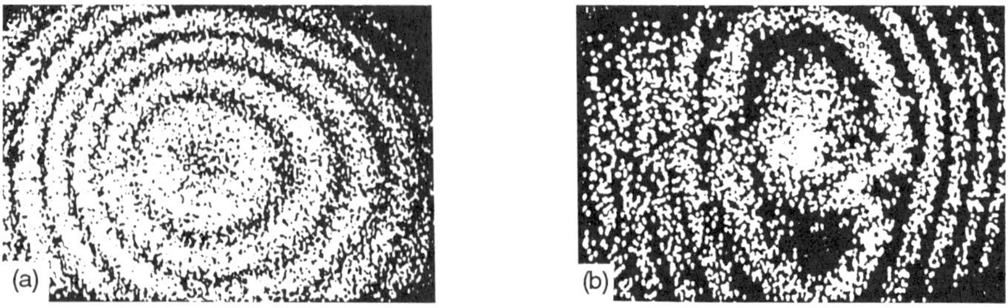

Figure 3.25 (a) Interferogram of a well-galvanized iron plate. (b) Interferogram of a poorly galvanized iron plate; the kink shows the debonding region. (Courtesy J. Sivaganthan, University of Malaya.)

pattern obtained with DSPI; the fringes show a discontinuity. This is due to the presence of a crack, which was independently verified by radiography. It was also possible to detect incomplete penetration and porosity of the weldment.

The comparison of the responses of two objects subjected to the same loading simultaneously is offered by ESPI. The comparative ESPI can be used to evaluate responses of systems like loudspeakers subjected to vibratory loading. If the responses are identical, no fringes are formed for identical loading; the compensation of deformation profiles has been demonstrated [124].

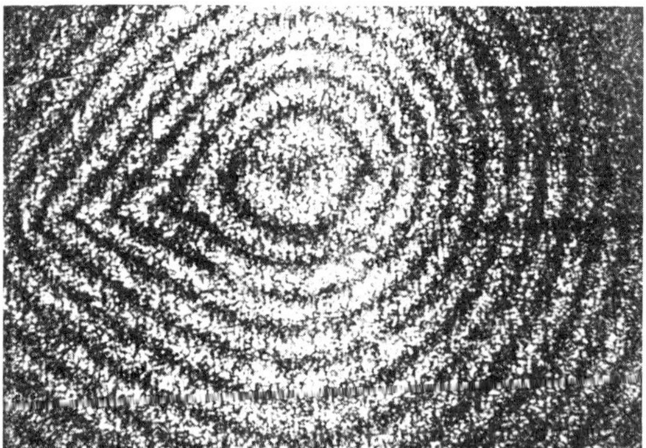

Figure 3.26 Interferogram of a weldment with the crack. The presence of the crack was independently verified by radiography. (Courtesy of J. Sivaganthan, University of Malaya.)

3.15 CONCLUSION

This chapter presents a number of speckle techniques that have been applied in experimental mechanics and describes selected examples.

REFERENCES

1. A. E. Ennos, Speckle interferometry, *Prog. Opt.*, *16*, 233–288 (1978).
2. R. Jones and C. Wykes, *Holographic and Speckle Interferometry*, Cambridge Univ. Press, Cambridge, 1983, 2nd Ed. (1988).
3. I. Yamaguchi, Fringe formation in deformation and vibration measurements using laser light, *Prog. Opt.*, *22*, 271–340 (1985).
4. P. Jacquot and P. K. Rastogi, Speckle motions induced by rigid body movements in free space geometry: an explicit investigation and extension to new cases, *Appl. Opt.*, *18*, 2022–2032 (1979).
5. J. W. Goodman, *Introduction to Fourier Optics*, McGraw-Hill, New York, 1968.
6. J. W. Goodman, Statistical properties of laser speckle patterns, in *Laser Speckle and Related Phenomena*, J. C. Dainty, Ed., Springer-Verlag, New York, 1975, pp. 9–75.
7. P. J. Gielisse and M. Tu, High frequency optical Fourier transform analyser for whole-field specklegram reconstruction, *Opt. Eng.*, *30*, 830–837 (1991).
8. A. E. Ennos, Comparative accuracy of holographic interferometry and speckle photography for out-of-plane deformation measurement, *Opt. Commun.*, *33*, 9–12 (1980).
9. F. P. Chiang, J. Adachi, R. Anastasi, and J. Beatty, Subjective laser speckle method and its application to solid mechanics problems, *Opt. Eng.*, *21*, 379–390 (1982).
10. F. P. Chiang and D. W. Li, Random (speckle) patterns for displacement and strain measurement: some recent advances, *Opt. Eng.*, *24*, 936–943 (1985).
11. G. H. Kaufmann, On the numerical processing of speckle photograph fringes, *Opt. Laser Technol.*, *12*, 207–209 (1980).
12. C. S. Vikram and K. Vedam, Speckle photography of lateral sinusoidal vibrations: error due to varying halo intensity, *Appl. Opt.*, *20*, 3388–3390 (1981).
13. C. Joenathan and R. S. Sirohi, Elimination of errors in speckle photography, *Appl. Opt.*, *25*, 1791–1794 (1986).
14. K. Hinsch, Fringe positions in double-exposure speckle photography, *Appl. Opt.*, *28*, 5298–5304 (1989).
15. J. A. Leendertz, Interferometric displacement measurement on scattering surfaces utilising speckle effect, *J. Phys. E: Sci. Instrum.*, *3*, 214–218 (1970).
16. Y. Y. Hung, Displacement and strain measurements, in *Speckle Metrology*, R. K. Erf, Ed., Academic, New York, 1978, pp. 51–71.
17. F. P. Chiang and R. P. Khetan, Strain analysis by one beam laser speckle interferometry. 2. Multiaperture method, *App. Opt.*, *18*, 2175–2186 (1979).

18. C. Joenathan, R. K. Mohanty, and R. S. Sirohi, Multiplexing in speckle shear interferometry, *Opt. Acta*, *31*, 681–692 (1984).
19. C. Joenathan and R. S. Sirohi, Enhancement of sensitivity in in-plane measurement in speckle photography, *Appl. Opt.*, *25*, 1380–1380 (1986).
20. C. Shakher, Industrial applications of holographic lenses: present and future, in *Holography and Speckle Phenomena and Their Industrial Applications*, R. S. Sirohi, Ed., World Scientific, Singapore, 1990, pp. 51–71.
21. R. K. Mohanty, C. Joenathan, and R. S. Sirohi, Speckle interferometric methods of measuring small out-of-plane displacements, *Opt. Lett.*, *99, 475–477 (1984).*
22. P. A. Klumpp and E. Schnack, Speckle interferometric camera for displacement measurements, *Exp. Mech.*, *30*, 411–415 (1990).
23. F. D. Adams and G. E. Maddux, Synthesis of holography ad speckle photography to measure 3-D displacements, *Appl. Opt.*, *13, 219–220 (1974).*
24. C. Shakher and R. S. Sirohi, Hologram interferometry and speckle photography combined for stress analysis, *Optik*, *51*, 141–146 (1978).
25. F. P. Chiang and Q. B. Li, The theory of holospeckle interferometry, presented at SEM's 5th International Congress, Montreal, Canada, June 1984, SEM Brookfield, Conn., 1984.
26. C. M. Vest, *Holograpic Interferometry*, Wiley, New York, 1979.
27. F. P. Chiang and Q. B. Li, The theory of holospeckle interferometry, *SPIE*, *814*(Part I), 90–97 (1987).
28. R. S. Sirohi, speckle shear interferometry—a review, *J. Opt.* (*India*), *13*, 95–113 (1984).
29. R. Krishnamurthy, R. S. Sirohi, and M. P. Kothiyal, Speckle shearing interferometry: a new method, *Appl. Opt.*, *21*, 2065–2067 (1982).
30. Y. Iwahashi, K. Iwata, and R. Nagata, Influence of in-plane displacement in single-aperture and double-aperture speckle shearing interferometry, *Appl. Opt.*, *24*, 2189–2192 (1985).
31. H. A. Lamprecht and E. Von Rooyen, Sandwich shearography: a technique for the cancellation of unknown rigid body displacements in shearography, *SPIE*, *1162*, 201–212 (1990).
32. D. K. Sharma, N. Krishna Mohan, and R. S. Sirohi, A holographic speckle shearing technique for the measurement of out-of-plane displacement, slope and curvature, *Opt. Commun.*, *51*, 230–235 (1986).
33. O. J. Løkberg and G. A. Slettemoen, Basic electronic speckle pattern interferometry, in *Applied Optics and Optical Engineering*, R. R. Shanon and J. C. Wyant, Eds., Academic, New York, 1987, pp. 455–504.
34. C. Wykes and M. Flanagcn, The use of diode laser in an ESPI system, *Opt. Laser Technol.*, *19*, 37–39 (1987).
35. O. J. Løkberg and O. Kwon, Electronic speckle pattern interferometry using a CO_2 laser, *Opt. Laser Technol.*, *16*, 187–192 (1984).
36. J. L. Santos, T. P. Newson, and D. A. Jackson, Electronic speckle pattern interferometry using single mode fibers and active fringe stabilization, *Opt. Lett.*, *15*, 573–575 (1990).

37. K. Creath, Phase-shifting speckle interferometry, *SPIE*, *556*, 337–346 (1985).
38. M. E. Fourney, Scattered light speckle interferometry, in *Speckle Metrology*, R. K. Erf, Ed., Academic, New York, 1978, pp. 281–293.
39. A. Luxmoore, Measurement of displacements around crack tips, in *Speckle Metrology*, R. K. Erf, Ed., Academic, New York, 1978, pp. 257–266.
40. J. M. Huntley, S. J. P. Palmer, and J. E. Field, Automatic speckle photography fringe analysis—application to crack propagation and strength measurement, *SPIE*, *814*, 153–160 (1987).
41. E. J. Barbero, G. H. Kaufmann, and S. R. Idelsohn, Fracture analysis of a surface coated ceramic by speckle photography and finite-elements, *Opt. Laser Technol.*, *22*, 17–22 (1990).
42. G. H. Kaufmann, A. M. Lopergolo, S. R. Idelsohn, and E. J. Barbero, Evaluation of finite-element calculations in a part-circular crack by coherent optics techniques, *Exp. Mech.*, *29*, 154–157 (1987).
43. H. V. Tippur and F. P. Chiang, Analysis of combined moiré and laser speckle grating method used in 3D crack tip deformation measurement, *Appl. Opt.*, *30*, 2748–2756 (1991).
44. D. A. Gregory, Topological speckle and structures inspection, in *Speckle Metrology*, R. K. Erf, Ed., Academic, New York, 1978, pp. 183–223.
45. D. P. Herbert, A. H. M. Al-Hassani, and M. O. W. Richardson, The use of electronic speckle pattern interferometry (ESPI) in the crack propagation analysis of epoxy resins, *Opt. Lasers Eng.*, *5*, 249–262 (1984).
46. H. J. Tiziani, Vibration analysis and deformation measurement, in *Speckle Metrology*, R. K. Erf, Ed., Academic, New York, 1978, pp. 73–110.
47. E. Archbold and A. E. Ennos, Two dimensional vibrations analysed by speckle photography, *Opt. Laser Technol.*, *7*, 17–21 (1975).
48. F. P. Chiang and R. M. Juang, Vibration analysis of plate and shell by laser speckle interferometry, *Opt. Acta*, *23*, 997–1009 (1976).
49. C. J. Lin and F. P. Chiang, Laser speckle method for the analysis of steady state of vibration of plates, *J. Acoust. Soc. Am.*, *69*, 456–459 (1981).
50. P. K. Gupta and F. P. Chiang, Laser speckle interferometry applied to studying transient vibrations of a cantilever beam, *J. Sound Vibration*, *133*, 251–259 (1989).
51. E. Archbold, A. E. Ennos, and P. A. Taylor, A laser speckle interferometer for the detection of surface movements and vibration, in *Optical Instruments and Techniques*, J. Home Dickson, Ed., Oriel Press, New Castle-Upon-Tyne, England, 1970, pp. 265–275.
52. K. A. Stetson, New design of laser image-speckle interferometer, *Opt. Laser Technol.*, *2*, 179–181 (1970).
53. T. J. Cookson, J. N. Butters, and H. C. Pallard, Pulsed lasers in electronic speckle pattern interferometry, *Opt. Laser Technol.*, *10*, 119–124 (1978).
54. D. W. Robinson, Holographic and speckle interferometry in the UK, *SPIE*, *814*, 330–337 (1987).
55. D. Kerr and J. R. Tyrer, The application of phase stepping to the analysis of ESPI fringe patterns, *SPIE*, *814*, 379–389 (1987).

56. J. T. Malmo and E. Vikhagen, Vibration analysis of a car body by means of TV holography, *Exp. Tech.*, *12*, 28–30 (April 1988).
57. P. Montgomery and B. D. Bergquist, Contrast enhancement of ESPI patterns by speckle averaging in a video frame store. *SPIE*, *599*, 201–206 (1985).
58. O. J. Løkberg and K. Høgmoen, Holographic methods made useful by phase modulated ESPI, *SPIE*, *136*, 222–225 (1977).
59. J. R. Tyrer, Critical review of recent developments in electronic speckle pattern interferometry, *SPIE*, *604*, 95–111 (1986).
60. J. T. Malmo and O. J. Løkberg, Vibration analysis and deformation measurements by TV holography, *SPIE*, *1162*, 270–278 (1988).
61. O. J. Løkberg, Mapping of in-plane vibration modes by electronic speckle pattern interferometry, *Opt. Eng.*, *24*, 356–359 (1985).
62. A. P. M. Hurden, Vibration mode analysis using electronic speckle pattern interferometry, *Opt. Laser Technol.*, *14*, 21–25 (1982).
63. S. Ellingsrud and O. J. Løkberg, Analysis of high frequency vibrations using TV holography, *J. Phys. D*, *21*, 11–13 (1988).
64. M. C. Shellabear and J. R. Tyrer, Application of ESPI in three-dimensional vibration measurements, *Opt. Lasers Eng.*, *15*, 43–56 (1991).
65. J. R. Tyrer, Application of high repetition pulsed lasers to electronic speckle pattern interferometry (ESPI), *SPIE*, *746*, 98–103 (1987).
66. F. M. Santoyo, M. C. Shellabear, and J. R. Tyrer, Whole field in-plane vibration analysis using pulsed phase-stepped ESPI, *Appl. Opt.*, *30*, 717–721 (1991).
67. S. Nakadate, Vibration measurement using phase-shifting speckle pattern interferometry, *Appl. Opt.*, *25*, 4162–4168 (1986).
68. K. Hinsch and W. Arnold, Spatially multiplexed optical correlation for displacement mapping in speckle metrology, *SPIE*, *1136*, 327–334 (1989).
69. M. F. Vallat, P. Martz, J. Fontaine, and J. Schultz, Application of coherent optics to the study of adhesive joints. I. Speckle photography, *J. Appl. Polym. Sci.*, *31*, 309–321 (1986).
70. F. Ansari, Stress-strain response of micro-cracked concrete in direct tension, *ACI Mater. J.*, *8*, 481–490 (1987).
71. J. M. Huntley and J. E. Field, Measurement of time varying displacement field by multiple exposure speckle photography, *Appl. Opt.*, *25*, 1662–1667 (1986).
72. C. H. Chien, J. L. Turner, W. F. Swinson, and W. F. Ranson, Extending the laser specklegram technique to strain analysis of rotating components, *Exp. Mech.*, *22*, 434–440 (1982).
73. A. Happe, Deformation measurement on connecting rods by speckle photography, *SPIE*, *136*, 270–276 (1977).
74. E. Archbold, A. E. Ennos, and M. S. Virdee, Speckle photography for strain measurement—a critical review, *SPIE*, *136*, 258–264 (1977).
75. I. Yamaguchi, Encoder and strain gauge using laser speckle, in *Holography and Speckle Phenomena and Their Industrial Applications*, R. S. Sirohi, Ed., World Scientific, Singapore, 1990, pp. 299–312.

76. E. Archbold and A. E. Ennos, Applications of holography and speckle photography to the measurement of displacement and strain, *J. Strain Anal.*, *9*, 10–16 (1974).
77. D. E. Duffy, Moiré gauging of in-plane displacement using double aperture imaging, *Appl. Opt.*, *11*, 1178–1181 (1972).
78. I. Yamaguchi, A laser speckle strain gauge, *J. Phys. E: Sci. Instrum.*, *14*, 1270–1273 (1981).
79. I. Yamaguchi, Advances in the laser speckle strain gauge, *Opt. Eng.*, *27*, 214–218 (1988).
80. A. J. Moore and J. R. Tyrer, An electronic speckle pattern interferometer for complete in-plane displacement measurement, *Meas. Sci. Technol.*, *1*, 1024–1030 (1990).
81. A. A. M. Maas and H. A. Vrooman, In-plane strain measurement by digital phase-shifting speckle interferometry, *SPIE*, *1162*, 248–256 (1990).
82. R. W. T. Preater, Pulsed and fiber optic ESPI for the analysis of in-plane strain on rotating components, *SPIE*, *1136*, 359–363 (1989).
83. R. W. T. Preater, Pulsed ESPI rotating component displacement measurements, *SPIE*, *1319*, 270 (1990).
84. Y. Y. Hung, Shearography: a new optical method of strain measurement and non-destructive testing, *Opt. Eng.*, *21*, 391–395 (1982).
85. E. Archbold and A. E. Ennos, Displacement measurement from double exposure laser photographs, *Opt. Acta*, *19*, 253–271 (1972).
86. C. Joenathan, C. S. Narayanamurthy, and R. S. Sirohi, Localisation of fringes in speckle photography that are due to axial motion of the object, *J. Opt. Soc. Am.*, *A5*, 1035–1040 (1988).
87. A. E. Ennos and M. S. Virdee, Laser speckle photography as a practical alternative to holographic interferometry for measuring plate deformation, *Opt. Eng.*, *21*, 478–482 (1982).
88. C. Wykes, Use of electronic speckle pattern interferometry (ESPI) in the measurement of static and dynamic surface displacements, *Opt. Eng.*, *21*, 400–406 (1982).
89. D. P. Herbert, Inspection of out-of-plane surface movements over small areas using electronic speckle pattern interferometry, *Opt. Lasers Eng.*, *4*, 229–239 (1983).
90. R. S. Sirohi, Speckle interferometry, *Phys. Educ.*, *6*, 95–103 (1989).
91. C. C. Kin and F. P. Chiang, Objective laser speckle method for 3D displacement measurement on curved surfaces/objects, *Opt. Eng.*, *22*, 3532–3536 (1983).
92. X. P. Wu and F. P. Chiang, Sandwich holospeckle interferometry for three dimensional displacement determination, *Opt. Eng.*, *25*, 736–737 (1986).
93. J. R. Tyrer, Structural analysis using phase-shifted, double pulse ESPI, *SPIE*, *1162*, 144–155 (1990).
94. S. Winther, 3D displacement measurement using ESPI, *Opt. Lasers Eng.*, *8*, 45–57 (1988).

95. G. Gulker, O. Haack, K. Hinsch, C. Holscher, J. Kuls, and W. Platen, Investigations on three dimensional deformation of corroding natural stones using ESPI (electronic speckle pattern interferometry), *SPIE*, *1319*, 412–413 (1990).
96. F. P. Chiang, A new family of 2D and 3D experimental stress analysis techniques using laser speckles, *SM Arch.*, *3*, 1–32 (1978).
97. H. Schwieger and J. Banken, Speckle photography for deformation analysis of bent plates, *Mater. Prüf.*, *27*, 153–156 (1985).
98. K. V. S. G. Rao and D. V. K. Rao, Investigation of the strain distribution in a warped square plate using speckle shear interferometry, *Opt. Acta*, *28*, 1009–1012 (1981).
99. K. V. S. G. Rao and D. V. K. Rao, Investigation of moment distribution in a bent circular plate by shearing speckle pattern interferometry, *Opt. Lasers Eng.*, *4*, 187 (1983).
100. F. P. Chiang, K. C. Chin, and W. B. Chang, Time average specklegram of plate vibration using multi-aperture recording, *Appl. Opt.*, *20*, 1123–1129 (1981).
101. R. K. Mohanty, C. Joenathan, and R. S. Sirohi, High sensitivity tilt measurement by speckle shear interferometry, *Appl. Opt.*, *25*, 1661–1664 (1986).
102. S. Nakadate, T. Yatagai, and H. Saito, Digital speckle pattern shearing interferometry, *Appl. Opt.*, *19*, 4241–4246 (1980).
103. P. V. Farrell and D. D. Verhoeven, Heat transfer measurements in a motored engine using speckle interferometry, SAE Tech. Paper Ser., paper 870456, SAE, Warrendale, Penna., 1987.
104. D. Paoletti, A. D'Altario, and S. Amadesi, Double exposure speckle-hologram for strain measurements in frescoes diagnostics, *Opt. Commun.*, *48*, 309–312 (1984).
105. G. Gulker, K. Hinsch, C. Holscher, A. Kramer, and H. Neunaber, In-situ application of electronic speckle pattern interferometry (ESPI) in the investigation of stone decay, *SPIE*, *1162*, 156–167 (1989).
106. G. Gulker, K. Hinsch, C. Holscher, A. Kramer, and H. Neunaber, Electronic speckle pattern interferometry system for in-situ deformation monitoring on buildings, *Opt. Eng.*, *29*, 816–820 (1990).
107. J. T. Malmo, O. J. Løkberg, and G. A. Slettemoen, Interferometric testing at very high temperatures by TV-holography (ESPI), *Exp. Mech.*, *28*, 315–321 (1988).
108. O. Slettemoen, Vibration analysis of turbine wheels by TV-holography, Master's thesis, Norwegian Institute of Technology, 1977.
109. D. L. Mader and R. E. Gnoyke, Vacuum speckle interferometry for long term monitoring of in plane strains at direct sites in nuclear pressure tubes, *SPIE*, *746*, 113–120 (1987).
110. P. Aswendt, R. Hofling, and W. Totzauer, Digital speckle pattern interferometry applied to thermal strain measurements of metal-ceramic compounds, *Opt. Laser Technol.*, *22*, 278–282 (1990).
111. K. Hinsch, Coherent optics in environmental monitoring, *SPIE*, *673*, 502–507 (1986).

112. C. Shakher, Laser speckle photography for NDT of diaphragm, *Appl. Opt.*, *21*, 367–369 (1982).
113. R. Krishna Murthy, R. S. Sirohi, and M. P. Kothiyal, Detection of defects in plates and diaphragms using split lens speckle-shearing interferometer, *NDT Int.* (*UK*), *15*, 329–333 (1983).
114. R. F. Anastasi, S. M. Serabian, R. J. Shuford, and D. K. Das-Gupta, Non-destructive detection of simulated delaminations in composite laminates by laser speckle shearography, *Exp. Tech.*, *11*, 28–31 (June 1987).
115. S. L. Toh, H. M. Shang, F. S. Chau, and C. J. Tay, Flaw detection in composites using time-average shearography, *Opt. Laser Technol.*, *23*, 25–30 (1991).
116. D. Paoletti and G. S. Spagnolo, Sandwich holospecklegram: practical tool for studying damage in concrete, *Meas. Sci. Technol.*, *1*, 1067–1071 (1990).
117. B. P. Holownia, Non-destructive testing of overlap joints using electronic speckle pattern interferometry, *Opt. Lasers Eng.*, *6*, 79–90 (1985).
118. O. D. D. Soares, Electronic speckle pattern interferometry (ESPI) for dynamic studies, *SPIE*, *699*, 10–21 (1987).
119. K. Paler, Automatic electronic speckle interferometry—a tool for inspection, *SPIE*, *654*, 90–94 (1986).
120. B. D. Bergquist, P. C. Montgomery, F. Mendoza-Santoyo, P. Henry, and R. J. Tyrer, The present status of electronic speckle pattern interferometry (ESPI) with respect to automatic inspection and measurement, *SPIE*, *654*, 95–100 (1986).
121. H. Neunaber, G. Gulker, K. Hinsch, C. Holscher, and A. Kramer, Deformation measurements of natural stones in-situ by electronic speckle pattern interferometry (ESPI), *SPIE*, *1010*, 68–74 (1989).
122. J. T. Malmo and O. J. Løkberg, Detection of defects in composite materials by TV holography, *NDT Int.* (*UK*), *21*, 223–228 (1988).
123. E. Vikhagen and O. J. Løkberg, Detection of defects in composite materials by television holography and image processing, *Mater. Eval.*, *48*, 244–248 (1990).
124. R. S. Sirohi and A. R. Ganesan, Some applications of digital speckle pattern interferometry (DSPI), *Proc. of Intl. Conf. on Holography*, University of Bath, England, 227–232 (1989).

4

Recent Developments in Video Speckle Interferometry

Ole J. Løkberg

The Norwegian Institute of Technology
Trondheim, Norway

4.1 INTRODUCTION

The main purpose of this chapter is to describe recent developments in video speckle interferometry, but it seems appropriate to honor the pioneers by starting with a short background to the emergence and development of video speckle interferometry.

The use of video systems to record holograms and speckle patterns for interferometric purposes was proposed independently and almost simultaneously by three groups, [1–3], who approached the problem from different grounds of reasoning. Butters and Leendertz [1] in Loughborough considered it a speckle technique development—an assumption that quite naturally followed from Leendertz's pioneering work and contributions in this field. Macovski et al. [2] at Stanford were working with video transmission of ordinary holograms, and it was obvious that their system could also be used for measuring tasks. Schwomma [3] at the Austrian company Eumig considered his technique a development of hologram interferometry.

After the triple announcement of the technique, most holographers agreed that the solution to the industrial applications of hologram interferometry was found, but the big breakthrough never came—at least not through the original inventors. Macovski did not pursue the research any further, and, strangely enough, research work in the United States died out for more than a decade. The Eumig company made a commercial instrument, but general economic problems in the company stopped the work after some years. The Loughborough group had a very fruitful period with new ideas and applications constantly put forward (see, e.g., Ref. 4). However, their commercial instrument was as hard to sell as the Eumig instrument.

A general apathy regarding the technique came after the initial flurry, probably because of the limited success of the commercial systems. In addition, key patents discouraged new groups with distant commercial goals from entering the field. The research was carried on by a few university groups around the world at a limited level of ambition and funding. Still, there was steady progress in the field. Elaborate schemes for in-plane measurement [5] and contouring [6] were developed by the Loughborough group. Vibration analysis was developed to a high degree of perfection by the author's group in Trondheim, culminating with speckle averaging [7, 8] and measurement of low vibration amplitudes of very unstable objects [9, 10]. The group of Saito in Japan demonstrated the potential for computer integration in the system [11]. Most basic ideas were, in fact, put forward and discussed in this period, but the lack of powerful, fast, and cheap computer power prevented any real breakthrough in performance and acceptance. The change came around 1985. At that time the determination of the optical phase by digital computers and image processing had been making gradual progress from classical interferometry [12] to hologram interferometry [13] and finally came to video speckle interferometry [14–16].

With the introduction of computerized fringe analysis, fringe interpretation became easy, which removed one of the biggest obstacles to the general acceptance of video speckle interferometry (as well as to conventional hologram interferometry) as a diagnostic measuring tool. In addition, the emergence of computer analysis was coupled with the introduction of exciting new hardware such as sophisticated CCD cameras, fiber optics, and lasers, which gave an extra boost to the field of video speckle interferometry.

In this chapter, I try to give an overview of the main developments in the field, stressing the basic features and limitations of video speckle interferometry. For a more detailed presentation of the development up to 1991, the reader should consult papers from the proceedings of the Society of Experimental Mechanics, 25th Anniversary for Hologram Interferometry [17]. The proceedings of the annual SPIE meeting in San Diego also contain numerous papers on video speckle interferometry, (see, e.g., Ref. 18).

I wish to comment on the choice of names used to describe the various techniques. More than a dozen different names have been used in the literature for essentially the same techniques, which must be a source of deep confusion for librarians and newcomers in the field. There are essentially three subtechniques based on the arrangement of the interfering waves. I use the name *video speckle interferometry* to represent all these subtechniques, which then are introduced by their most common name as they are described. Whether this approach increases the confusion further remains to be seen.

4.2 THEORY AND SYSTEM REQUIREMENTS

The basic principles of video interferometry using speckle waves have been thoroughly discussed elsewhere [19, 20]. Here I give only a brief overview of the terms and physical principles involved and comment on how the resulting constraints influence the requirements of recording media and hardware.

4.2.1 The Interference Process

The speckle wave from the illuminated test object can be written as $A_o(x, y) \exp [i(\phi_o(x, y)]$, where the (x, y) coordinates refer to the image plane. Similarly, the reference wave is denoted by $A_r(x, y) \exp [i(\phi_r(x, y)]$, where the spatial variation indicates that the reference wave might be a uniform or speckle wave. The intensity $I(x, y)$ due to interference between the two waves, which are supposed to be perfectly coherent, is then given by the standard equation

$$I(x, y) = |A_o(x, y)|^2 + |A_r(x, y)|^2 + 2A_o(x, y)A_r(x, y) \cos [\phi_o(x, y) - \phi_r(x, y)] \tag{4.1}$$

As in hologram interferometry, the first two terms—the self-interference terms—represent noise terms that we have to suppress or discard as their presence in the final interferometric image leads to signal-to-noise ratios often on the border of no detection. I have not expanded the third term into two exponential terms as is normally done in holographic theory, where diffraction in the reconstruction process spatially separates the two terms as side-order waves. However, in video speckle interferometry we work on the real part of the signal where the cross-interference between the reference and object waves represents the information. For most interferometric purposes we are interested in the global position of the test object, which is encoded in the phase argument of the cosine term.

4.2.2 The Spatial Frequency Spectra

We will now discuss the spatial frequency content of the basic waves—the speckle wave, the uniform wave, and their interferometric combination [21].

The speckle pattern from a laser-illuminated object image is given by the mutual interference of all the wavelets leaving the exit aperture of the imaging lens. As, in video speckle interferometry, we operate with small angles, we use the paraxial approximation $\mu = \phi/\lambda$ for the spatial frequency μ in the interference pattern between two waves interfering at an angle ϕ and wavelength λ. This simple equation gives us the maximum spatial frequency of the image speckle pattern as $\mu_o = D/d_i\lambda$, where D is the diameter of the lens and d_i is the lens–image distance.

The distribution of spatial frequencies is given by the autocorrelation of the lens aperture as shown (in one direction) in Fig. 4.1 assuming a circular aperture. The average size of the speckle is equal to that of the Airy's disk of the lens.

In Fig. 4.2 are shown the resulting frequency spectra due to interference between a speckle (image) wave and a uniform spherical wave introduced at the center b of the exit aperture. The uniform wave is normally much stronger than the speckle wave. From considerations of the waves involved in the interference, we find the cross-interference term (bold lines) to have a constant value. Its maximum frequency is half the maximum frequency in the object self-interference term or the speckle pattern (thin lines). The self-interference spectrum of the reference wave will ideally be a delta function, but some broadband frequency noise is always present owing to diffraction on dust particles, interference due to reflections, and so on. A

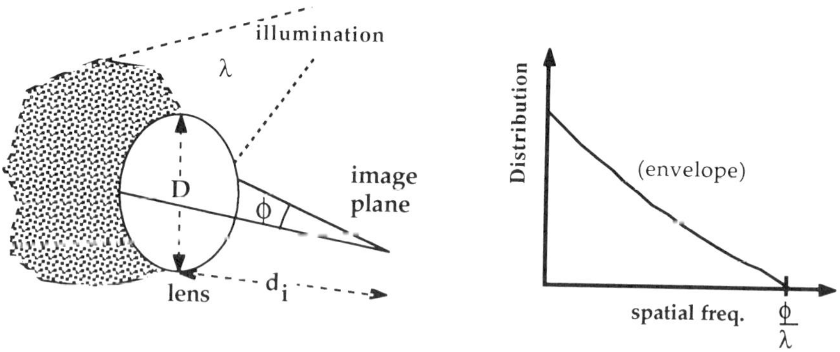

Figure 4.1 The spatial frequency content of a speckle pattern.

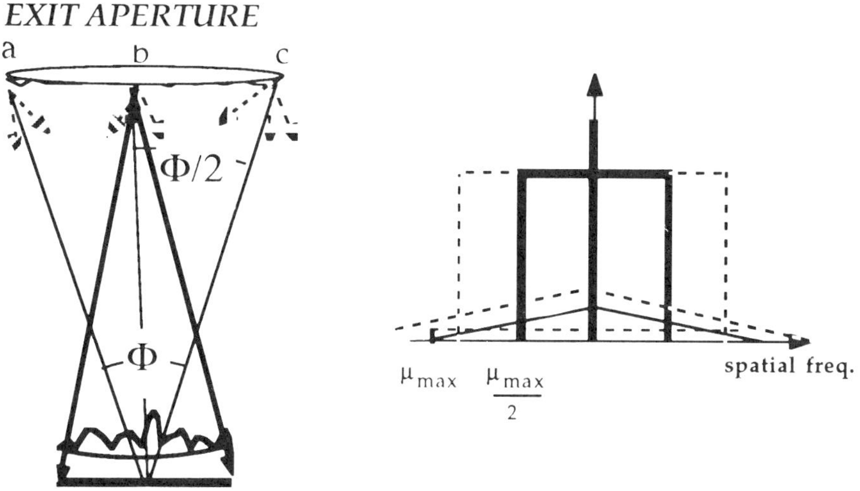

Figure 4.2 Spatial frequency spectra due to interference between a uniform spherical wave and a speckle wave.

strong reference wave increases the cross-interference term relative to object self-interference terms. However, as the self-interference noise in the reference wave will also increase, we are not free to increase the reference wave indefinitely. The optimum is found where the noise from the two self-interference terms is about equal [22].

The optimum size of the imaging aperture has been the source of much discussion, and we still lack experimental evidence to back up the theory. According to the theoretical consideration of Slettemoen [23], the optimum aperture should be when the maximum spatial frequency of the intermodulation band is just being resolved by the recording medium. We may, however, deviate considerably from this criterion as the signal-to-noise ratio increases slowly as the aperture is opened from its optimum setting.

The effect of opening the aperture is indicated by dashed lines in Fig. 4.2. The self-interference terms are increased in both total size and frequency range (only the object speckle term is indicated). The cross-interference term (or the information) is also increased, but the new information is located outside the resolution band of the recording medium. Therefore, when sufficient laser power is available, we should be gaining nothing by opening the lens aperture. On the contrary, we are losing depth of focus. At low light levels there might be some gain in opening the lens as we are able to bring more signal into the linear region of the video camera; however, no stringent analysis or experiment supports this assumption. If we have "satellite" reference waves due to reflections in the path, a gain might be possible if the satellite cross-interference terms are not symmetrically located in the frequency domain.

The situation is quite different when we analyze the interference between two speckle waves as shown on Fig. 4.3. For simplicity the illumination of the two rough surfaces generating the speckle waves is not shown. We are here discussing the case where two separate surfaces are involved in the interference, but the discussion is valid also for a single surface where the two waves originate either from double illumination or from splitting and displacing the image (shear). The self-interference and cross-interference terms of the resulting intensity distribution are shown on the right-hand side of the figure. In the figure we assume that the two waves are equal in intensity, which gives the optimum ratio. The dotted spectra in the figure show the effect of increasing the lens aperture. The spectra grow broader and contain more energy as expected. However, in contrast to the uniform reference system, we here also increase the energy within the resolution band. Therefore, whenever we are dealing with cross-interfer-

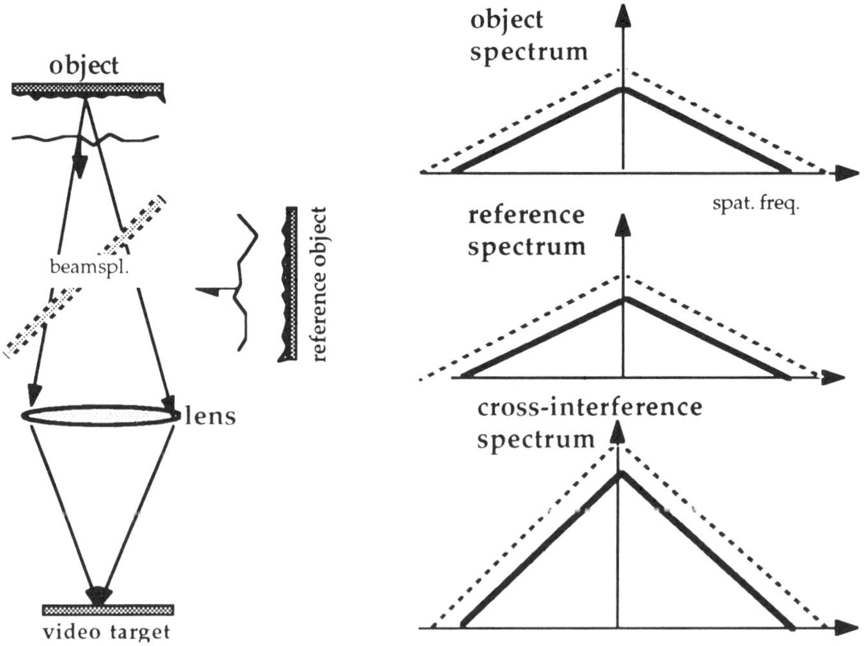

Figure 4.3 Spatial frequency spectra due to interference between speckle waves.

ence between speckle waves, the lens aperture should be opened until the saturation level of the camera is reached if the reduced depth of focus is acceptable. This difference between uniform and speckle reference systems is understood when we consider the speckle reference wave as numerous elementary reference waves emanating from the aperture. This continuum of reference sources contributes to cross-interference at low spatial frequencies regardless of the size of the aperture.

We are still faced, however, with the problem that the self-interference (noise term) and the cross-interference (information term) frequency spectra completely overlap. We cannot use interferometric gain as in uniform reference systems to overpower the noise because the self-interference spectra also grow in proportion. Fortunately, modern video techniques allow frame subtraction, a feature not easily duplicated in pure optical techniques. We assume that the object and reference self-interference spectra (or speckle patterns) are unchanged in the two frames. As this assumption is fulfilled in normal measuring situations, the subtraction process removes the two noise terms. The phase in the cross-interference spectrum is, however, changing cyclically with the optical path length in either branch, resulting in fringe patterns in the final processed image.

The subtraction process is also used extensively to remove noise in the uniform reference system, but here subtraction is not absolutely necessary to get readable fringes.

4.3 THE BASIC SYSTEMS

In this section the optical part of the three main systems in video speckle interferometry is briefly described. For a more detailed description, the reader should consult Refs. 19 and 20.

4.3.1 Uniform Reference

The uniform reference system has so far been the most popular system. Consequently, the technique has also been blessed with the most names, the most common being ESPI (electronic speckle pattern interferometry), DSPI (digital speckle pattern interferometry), TV holography, and electronic holography.

Figure 4.4 shows only some of the elements of a uniform reference system, where for simplicity I have omitted the laser and some of the standard components used in every two-branch interferometer. The illuminated test object is imaged through the beam combiner (BC) onto the video target. The reference wave is added in-line to the image wave by the beam combiner and adjusted to be centered in the exit aperture of the lens. To keep the drawing simple I have used a plane beamsplitter as beam

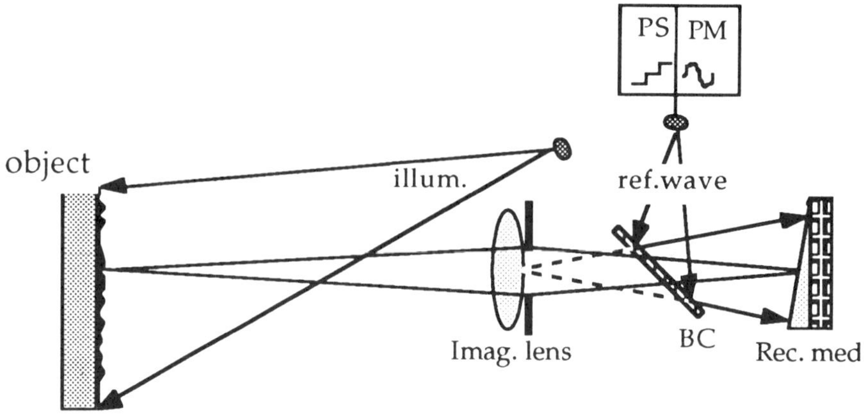

Figure 4.4 Uniform reference system.

combiner. This is not an optimum solution because of secondary reflections of the reference wave. Numerous solutions to the beam combiner problem have been tried, for example, using a wedge or a cube beamsplitter, introducing the reference by using a monomodal fiber, focusing and/or reflecting it from a small sphere, or focusing the wave through a pinhole in a mirror. With the advent of image processing, cosmetic variations can be smoothed, and most techniques are acceptable provided the reference wave intensity at the target is reasonably uniform. The choice of beam combiner is therefore more a question of ease of adjustment. Reflections in the video target result in a similar cosmetic problem. A glass wedge coupled to the front surface of the video target reduces the effect of Haidinger fringes caused by secondary reflections of the reference wave. The wedge translates these fringes to spatial frequencies above the cutoff frequency of the camera. This technique worked well with most vidicon-type cameras, but in the CCD the effect is only marginal due to added reflections from the underlying structure. A CCD target without cover glass solves the problem but is very fragile. I should perhaps warn readers that the previous cheerful note that image processing may solve all cosmetic problems should be accepted with caution. We have experienced serious errors in computed values when the reference wave contained noise spikes due to large dust particles.

Returning to Fig. 4.4, the box labeled PS/PM in the reference branch indicates the introduction of controlled phase changes in the interferometer. The phase change is effected either as calibrated phase steps (PS) or as continuous, periodic changes—phase modulation (PM). The phase change may be implemented as displacement of mirrors, elongation of optical fibers, or double refraction in electro-optical modulators. As phase-stepping routines have become integrated in all modern video speckle interferometers, the phase box will be omitted from discussions of the remaining systems.

Before describing a system where the reference wave is more fictive, I should point out that the uniform reference wave can be replaced by a separate speckle field as reference wave as indicated in Fig. 4.2. As previously discussed, we normally need subtraction to separate the cross-interference term from the self-interference. This is not the case in the elegant system invented by Slettemoen [7], where a beam combiner consisting of transmitting, reflecting, and opaque slits separates the various terms without the need for subtraction.

4.3.2 Double-Illumination System

The double-illumination system was originally suggested and extensively used by the Loughborough group [24]. The system has a clear parallel to laser-Doppler velocimetry (LDV).

Figure 4.5 shows the principal construction details of such a system. The system has no separate reference wave, but the test surface is illuminated by two waves. (We could call either of the reflected waves the object/reference wave.)

To simplify the fringe interpretation, the illumination waves should preferably be plane and symmetrical to the normal of the (plane) test surface.

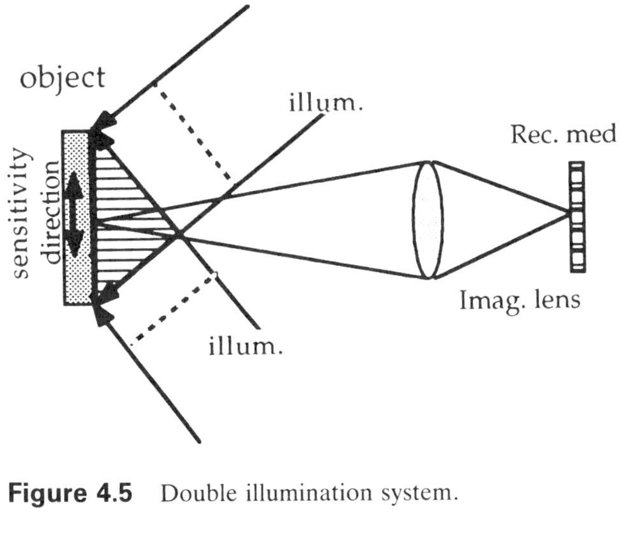

Figure 4.5 Double illumination system.

From LDV we know that the sensitivity vector will be normal to the interference planes as indicated in Fig. 4.5, which gives in-plane sensitivity to the displacement. As the in-plane component of the displacement is of importance to engineers computing stress and strain levels, this system is technically very important. To determine the *x* and *y* components alike, we have to rotate the component or use a "double" double illumination.

The fringe quality is inferior to that of a uniform reference system because of the speckle statistics. The measuring range of the system is also lower, owing to speckle decorrelation.

4.3.3 SHEAR SYSTEM

Leendertz and Butters [25] first suggested and demonstrated the feasibility of speckle shear interferometry using film and optical filtering; the possibility of using a video system in shear is also specifically mentioned in their paper. Nakadate et al. [11] were the first to demonstrate the actual use of video-speckle, and Slettemoen and Løkberg [26] described its use for vibration analysis with no subtraction needed. Hung [27] has developed speckle shear interferometry into a highly accepted tool in interferometric nondestructive testing.

There are numerous ways to construct a speckle shear system; in fact, there are probably as many system constructions as there are workers in the field. Some systems are very simple, which is one of the attractions of shear interferometry. In Fig. 4.6 we use a Michelson interferometer to

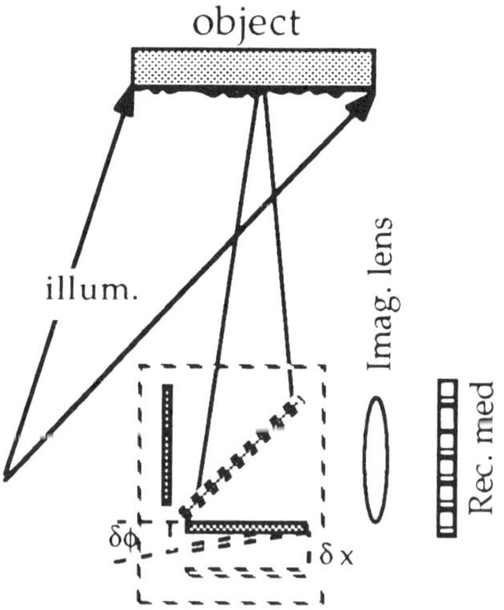

Figure 4.6 Shear system.

provide the shearing of the wave front. By tilting one of the mirrors a small angle $\delta\phi$, we move the wave fronts relative to each other. Linear shear is given in any desired direction by the tilt angle of the mirror. Similarly, if we move the mirror a distance δx, we obtain interference between images of different magnifications, or radial shear. Rotational shear can be obtained by replacing the mirrors with right angle prisms in reflection, which is rotated to give the desired shear. In practice, linear shear is most often used as it provides constant shear across the image with fringe patterns fairly easy to interpret. When the shear is small compared to the displacement, the resulting fringe pattern is indicative of the derivative of the displacement in the direction of the shear. As a result, uniform movements such as tilt are canceled, giving only a constant fringe order, which the resulting fringe pattern "rides" on. The shear interferogram displays only abrupt changes in the displacement, which are of key interest in nondestructive testing.

As this is a common-path interferometer, the effect of turbulence is also negligible for small shear. It is not strictly correct to claim that the interferometer is completely insensitive to uniform movements and turbulence as speckle decorrelation could reduce the fringe contrast to an unacceptable level. We want to keep the shear as small as possible to ensure stability and a pure derivative fringe pattern that is reasonably easy to interpret. However, as the shear decreases we enter the regime of speckle photography where the contrast is very poor. In addition, interferometric sensitivity is reduced to a level where defect detection is difficult even with image processing. In practice, most shear interferometers are operated with a large shear compared to the smallest deformation details. The interferometer acts then more like a speckle reference interferometer with a very spatially phase-modulated reference. However, even at this larger shear the interferometer is still far superior to the uniform reference interferometer in terms of stability because of the common path and the compensation for tilts in the shear direction.

4.4 THE FRINGE FUNCTIONS

Like classical interferometers, video speckle interferometry measures changes in the optical path length defined as distance as measured in wavelength units multiplied by the refractive index of the surrounding medium. When changes are caused by displacements only,the dynamic or static deformation of the test object can be studied and measured. Changes in the refractive index lead us into the regime of classical interferometry where the study of temperature and pressure distribution in gases is of extreme importance. Changes in refractive index and in wavelength may, as in hologram interferometry, be used to contour surfaces. Contouring can also be achieved by tilting the illumination beams in various ways (see, e.g., Ref. 28). For a thorough discussion of contouring by video speckle interferometry, the reader should consult the book by Jones and Wykes [19].

As in hologram interferometry, video speckle interferometry obtains interferometric information during the exposure or by combining several exposures. Basically, we have two different modes of operation depending on whether data acquisition takes place during a single frame or two or more frames are combined.

4.4.1 Single-Frame Exposure

This mode can be further subdivided depending on whether the exposure is continuous (time averaged) or broken up into subexposures.

The time-average methods are similar to their holographic counterparts and are used mainly for measurements of objects vibrating at a single frequency. Assuming that the vibration amplitude and phase are given by $a_o(x, y)$ and $\varphi_o(x, y)$, respectively, we get the well-known fringe function for the reconstructed image intensity:

$$I_{\mathrm{TA}}(x, y) \approx J_o^2 \left[\frac{2\pi}{\lambda} \cdot g(\phi) \cdot a_o(x, y) \right] \tag{4.2}$$

where J_o is the Bessel function of zero order and of the first kind, and $g(\phi)$ is the common interferometric sensitivity factor containing the illumination and observation directions.

The time-average vibration method can be further improved by introducing phase modulation at equivalent amplitude and phase $a_r(x, y)$ and $\varphi_r(x, y)$. This leads to a slightly modified reconstructed intensity distribution:

$$I_{\mathrm{TA/PM}}(x, y) \approx J_o^2 \left[\frac{2\pi}{\lambda} \cdot \sqrt{[g(\phi)a_o(x, y)]^2 + a_r^2 - 2g(\phi)a_o(x, y)a_r \cos\,[\varphi_o(x, y) - \varphi_r]} \right] \tag{4.3}$$

We see that the fringe function is unchanged from Eq. (4.2), but its argument is now given by the vectorial difference between the object and reference vibrations. By either observing the zero-order fringe or working on the linear part of the zero order, we open the way for determination of phase distribution and extended measurement range as described in early publications [9, 29, 30].

Stroboscopic operation of the interferometer can be used in the study of harmonically vibrating objects [22]. In this case short exposures effectively freeze the displacement in two positions to give the intensity distribution

$$I_{add} \approx \cos^2\left\{\frac{2\pi}{\lambda}\cdot g(\phi)\cdot a_d(x, y)\right\} \tag{4.4}$$

where $a_d(x, y)$ is the displacement between exposures. If the exposures are placed at the extrema of the vibration cycle, $a_d(x, y)$ represents the peak-to-peak displacement.

If the movement is nonharmonic—for example, caused by running machinery—we have to sample the motion with only two pulses per frame, giving the same fringe function as in Eq. (4.3). As this kind of object is normally moving at a large velocity, the only practical alternative is a pulsed laser, preferably a repetitively pulsed laser that can be run at video-frame rates [31].

4.4.2 Frame Combination

In most measuring situations we combine the information from two or more frames (note that we are not yet concerned with phase-shifts techniques), for example as in the analysis of slow deformations or in experiments such as contouring. We have to record a reference frame, which is then subtracted from the frames coming from the video camera. The subtraction process is preferential for noise reduction and essential for some techniques such as shearing and in-plane.

We get essentially the same fringe function for addition processes like the stroboscopic operation in Eq. (4.3) except that the zero-order fringe is dark, which is an advantage in the visual detection of small displacements (see Fig. 4.8).

The resulting intensity distribution I_{sub} for a fringe pattern generated by subtraction will be given by

$$I_{sub} \approx \sin^2\left\{\frac{2\pi}{\lambda}\cdot g(\phi)\cdot a_d(x, y)\right\} \tag{4.5}$$

For digital image processing of all the fringe functions mentioned here, several frames are combined for alternative displays and numerical analysis as discussed in a later section.

4.4.3 The Fringe Carrier

The fringe functions can be considered envelopes modulating the carrier wave, the speckle pattern. The contrast and appearance of the speckle pattern are important for the resolution of the secondary fringes. The basic calculations of the speckle contrast in video speckle interferometry have been discussed by Slettemoen [23, 32], and the reader should consult these references for details.

As speckle patterns are not regular interference patterns, the classical definition of contrast based on two-wave interference does not hold. Instead we have to use a definition based on the statistical properties of the pattern. We define the speckle contrast as $C_s = \sigma_I/\langle I \rangle$, where σ_I and $\langle I \rangle$ are the standard deviation and mean intensity, respectively, of the speckle pattern.

For a fully developed polarized speckle pattern, the intensity distribution of the speckles is given by negative exponential statistics [21] with a speckle contrast $C_s = 1$. The same speckle contrast is obtained in a hologram reconstruction due to the spatial separation of the side-order waves by the diffraction process. Consequently the fringes in hologram interferometry ride on a carrier wave of unity speckle contrast.

The video speckle interferometry system developed by Stetson and coworkers [33, 34] combines three phase-shifted frames that simulate a holographic reconstruction where the speckles follow the normal size distribution and have unity contrast. The self-interference term due to the object field is also removed, so the fringe minima go completely dark.

In other cases of video speckle interferometry we do not simulate the spatial separation of the twin waves. We only reduce the devastating effect of the self-interference terms (or the zero orders) by temporal filtering or eliminate it by subtraction. In both cases the contrast of the processed speckle pattern will be different from that of holography.

For a setup where a uniform reference wave is used, the statistical distribution of speckle intensity changes into the more unfavorable χ^2 distribution. The speckle contrast in this case is $\sqrt{2}$, which means that the "dark" speckles are more dominant than in the normal speckle pattern. In addition, the contribution from the self-interference term will be noticeable in the fringe minima if it is not removed by subtraction as suggested by Creath and Slettemoen [35]. Use of the double-slit aperture suggested by Biedermann and Ek [36] also completely removes the self-interference, but at the expense of image resolution.

For setups where two speckle waves contribute to the primary interference (speckle reference, double illumination, shear), the contrast of the speckles in the processed image is still greater. In this case the contrast increases from $\sqrt{2}$ to $\sqrt{5}$ as we increase the size of the imaging aperture. This surprisingly great contrast can be explained as spatial field variations in both waves contributing to the resulting variations of the cross-interference term.

The speckle contrast will be effectively reduced if images with the same fringe function, but with uncorrelated speckle patterns, are added after

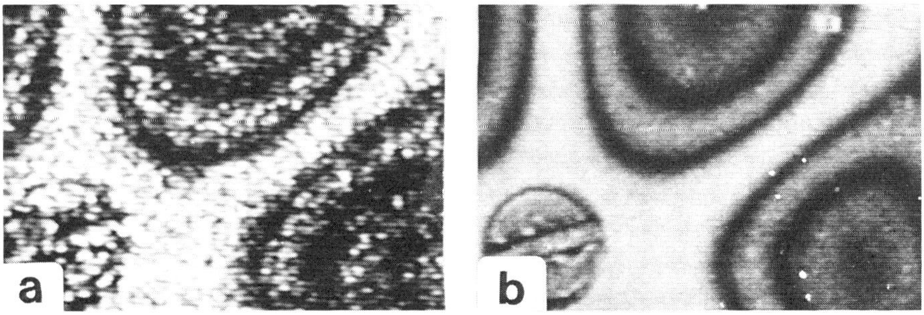

Figure 4.7 Speckle averaging. Area shown about 1/40 of monitor screen. (a) Single frame; (b) 50 averaged frames.

the rectifier [7, 8]. If N uncorrelated patterns are added, the average speckle contrast will be

$$C_{av} = C_s/\sqrt{N} \tag{4.6}$$

We see that already after a single averaging steps the speckle contrast in a standard uniform system will be similar to the holographic speckle contrast. The smoothing of speckle contrast is gradually diminished as N increases, and the visual improvement is hardly noticeable after about N = 50. The speckles are decorrelated by changing the direction of illumination, rotating a diffuser in the illumination, or rotating a subaperture in the imaging lens. Note that as we must ensure that uncorrelated speckle patterns are added,the effect of speckle averaging will depend on the magnification of the system. At large magnification, the imaging aperture in the object space will be large. Therefore, we have to tilt the illumination by such large angles to achieve decorrelation that the fringe pattern might also be smoothed owing to variations in the sensitivity vector.

The image from a large number of speckle-averaged frames is nearly free of speckles, and the increase in quality is convincing. In Fig. 4.7 a small portion of the monitor image has been magnified to clearly show the improvement in speckle contrast and image resolution.

4.5 SYSTEM CONSTRUCTION

4.5.1 Layout

No principal new systems have appeared, which is to be expected as there is not much to be done within the one degree of angular freedom offered by the resolution of the video system.

As already mentioned, Stetson and coworkers [33, 34] have developed an impressive uniform reference system where powerful computing support provides real-time displays of high quality. The operation of this system is described later. Davies et al. [37] have used fiber optics very extensively to create a compact system linked to a strong central laser.

The Loughborough group has refined the work on in-plane systems with a pulsed laser. Using polarizing properties of metallic surfaces, they have even been able to record simultaneously the *x* and *y* components of the displacement [38]. Similar work on in-plane recording using two continuous-wave lasers has been reported by the Oldenburg group [39].

In shear, Hung [27] has transferred his pioneering work with film recording to video systems, and commercial shear systems are available for detection of defects, for example, in industrial environments. Shear systems based on holographic optical elements seem to be a promising solution as they can be very compact and versatile.

An interesting high-speed system has been developed by Brown and Smith [40]. Using a concept originally described by Ottonnello and Pontiggia [41] for analyzing vibrations, the system is based on using cylindrical lenses for one-dimensional illumination and imaging. Using an array-type detector, the system is able to sample dynamic displacements in the microsecond range instead of in milliseconds with an ordinary video system. Although the system spatially samples in only one direction, the increased time resolution will be of vital interest in industrial research.

4.5.2 Components

Apart from the increasing use of fiber optics, exemplified by the work of Davies and coworkers [37] mentioned earlier, not much has happened on the optical component side. Monomodal fibers are used as carriers of strong central lasers, for object illumination, and to introduce and filter the reference wave, while coherent fiber bundles are used to transfer images from, for example, cavities to the system. Phase shift and/or phase modulation of either wave is achieved by winding the fiber on a piezocylinder, which stretches the fiber in accordance with the voltage input to the piezo crystal. However, the same effect can be troublesome if the fiber is unintentionally moved; a change of polarization might result in a large intensity fluctuation in the interference whereas extra phase shifts due to fiber loading give errors in numerical analysis. An extensive overview of the problems in this respect and their possible solution are given in Ref. 42.

As for continuous laser sources for the interferometer, there has been a gradual shift from the standard He-Ne laser to the diode laser although the change is taking place at a slower pace than was predicted some years ago. The Oldenburg group has made extensive use of laser diodes in their instrumentation [39]. The main problem is to keep the wavelength of the diode laser stable. As diode lasers become more powerful and their useful wavelength creeps toward the center of the visible spectrum, we can expect them to completely banish the He-Ne and Ar lasers from the field. For example, the fact that they are easily modulated at a peak effect substantially higher than the average power allows interesting applications within stroboscopic work.

Pulsed-laser work is mainly done with YAG lasers, which are preferred because they can be repetitively double-pulsed at the video frame rate. Their main problem is unstable pulse-to-pulse intensity, which results in a great number of frames at low or even zero fringe visibility as witnessed by Spooren's investigations [43, 44]. The double-cavity YAG laser is a better (but more expensive) solution to the pulse-to-pulse stability problem as reported by the Loughborough group [31]. We can expect great progress in solid-state lasers in both efficiency and stability; if the prices are reduced also, these lasers will be the natural light source for industrial and medical purposes. Ideally, we would also like to be able to run these lasers at longer pulse lengths or even in a continuous wave mode to avoid the safety problem as a pulsed laser like the YAG is a potential eye-killer of the first class!

In terms of cameras, CCDs have completely taken over the field from vidicons. Although even the best standard CCDs do not compete with a high-resolution vidicon in terms of image resolution, their linearity, small size, and robustness make them the natural choice in this field. Frame transfer allows us to separate the two fields in a video frame, which allows us to subtract the exposures in a double-pulse recording and thereby improve the fringe quality [45, 46] as shown in Fig. 4.8. This figure also demonstrates how the visual detection of low vibration amplitudes is im-

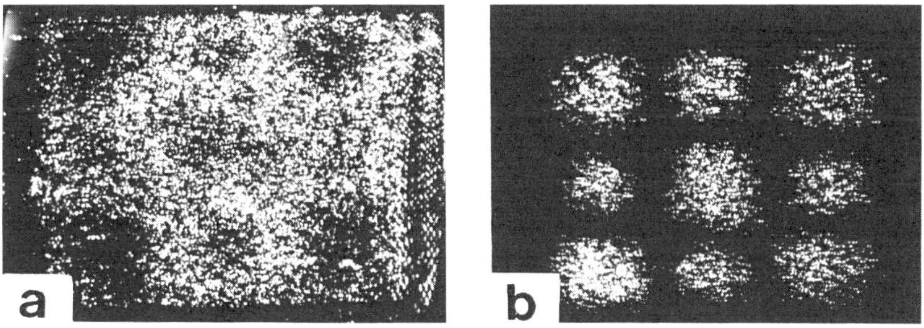

Figure 4.8 (a) Addition and (b) subtraction fringes in pulsed-laser TV holography.

proved by going from addition to subtraction. The ability to easily shorten and lengthen the exposure time can be an added bonus. We must expect the development of CCDs to continue. For example, standard resolution will soon increase to 2048 × 2048 or higher compared to 512 × 512 today (it is to be hoped that the price will not also be quadrupled), and the cameras will become digital, allowing most signal manipulation to take place on the chip at video rates.

4.5.3 Computer Integration

The most important development has been the digital storage and manipulation of the recorded video frames. Increased data capacity combined with greatly reduced prices have made image processing available to academic institutions on low budgets. The most commonly used framegrabber board is able to record 16 separate frames, which can be manipulated directly on the board. This saves time compared to transferring the data and processing in the computer [47]. Subtraction, addition, and even Fourier transforms can be performed very fast.

The incorporation of the computer into video speckle interferometry has allowed manipulation of the interferometric speckle pattern to improve the visual display and provide numeric information. For example, documentary text on the video image, which is highly desired but was almost unattainable not many years ago, is now a standard feature in most computer programs. And standard subtraction procedures can be easily incorporated to clean up noise from the reference wave.

A single-frame video interferogram recording is noisy, and averaging several frames is very beneficiary both visually and for data acquisition. The averaging is, of course, most efficient and most easily performed in vibration analysis where the same pattern can be recorded and displayed over an extended period of time. I have already described the effect of speckle averaging whereby the speckle carrier is changed (decorrelated) from frame to frame to further improve the final, averaged image. Speckle averaging is easily performed automatically, and the result can be displayed in real time using the computer and a frame store.

Using a fringe-improvement technique specially developed for the study of unstable objects, Vikhagen [48] resolved up to 100 time-average fringe orders across the image (see Fig. 4.9), which is quite remarkable considering the low intensity of the J_o^2 fringes at such high orders. (This technique gives essentially the same fringe function and quality as the Stetson system to be explained next.)

An example of the extensive computerization of modern video speckle interferometry systems is the system developed by Stetson et al. [33, 34].

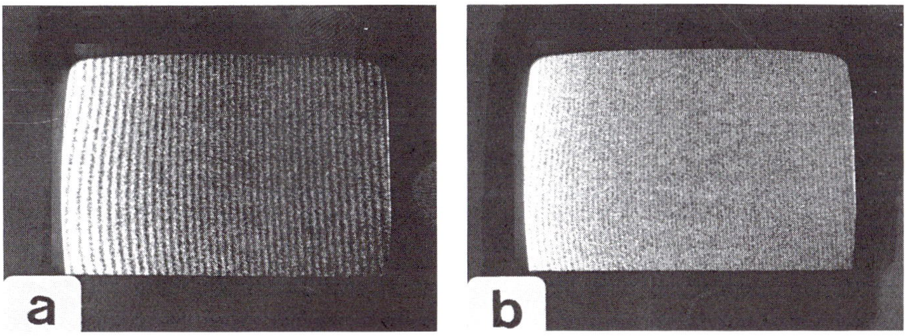

Figure 4.9 Recording of high amplitude by fringe scanning (a) 35 and (b) 75 fringe orders.

The basic concept was originally developed for video transmission of true holograms [49]; however, its favorable speckle statistics and high contrast in the final image makes it most interesting for high quality speckle interferometry. The system is based on the continuous acquisition and computing of frames. The phase is stepped 90° in four steps between frames.

The procedure is somewhat different for vibration and deformation testing, and readers should consult the original papers for a full account. For vibration testing we can write the intensity of each frame N as

$$I_N(x, y) = I_r + I_o(x, y) + 2\sqrt{I_o(x, y)\cdot I_r}\cos[\alpha(x, y) + \delta_N]\cdot J_o(\Omega) \tag{4.7}$$

where δ_N is the phase step value (0°, 90°, 180°, and 270°) and Ω is the standard argument of the vibration function.

The computer combines the four frames in the algorithm

$$I_{\text{out}} = \sqrt{(I_1 - I_3)^2 + (I_2 - I_4)^2} \tag{4.8}$$

which is solved to give the resulting output frame,

$$I_{\text{out}}(x, y) = 4\sqrt{I_o(x, y)I_r}\,|J_o(\Omega)| \tag{4.9}$$

We see that the frame combination reconstructs only the cross-interference term, eliminating the self-interference term from the object and reference waves. The speckle contrast is also reduced as discussed previously. This results in the dark fringes being completely free of background speckles, which reduces the resolution of higher order fringes in standard systems. In addition, the reconstruction of the absolute value of the J_o function instead of the squared function in theory increases the detection of higher order fringes. This advantage is, however, only marginal. As the phase-stepped new frames are being continuously sent into the processor, the system works in real time. The penalty for the excellent fringe quality is a more complex system totally dependent on the computer and to a stronger degree stability in the system.

4.6 FRINGE ANALYSIS

As I have stressed several times, the surge of interest in video speckle interferometry is based mainly on the use of computers in the system. In that context, most important is the computer's ability to analyze the fringe pattern and assign numerical values of displacement to each pixel (see, e.g., Ref. 50).

It is relevant to divide numerical analysis of fringe patterns in video speckle interferometry into three major groups: the analysis of "static" displacement, repetitive displacements, and dynamic, single-event displacements. Each group demands a different type of approach to fringe analysis. We now discuss the specific problems for these groups and some solutions, stressing solutions that are not commonly used in classical interferometry. At the end of this section, I comment on the general problems encountered in fringe analysis of speckle patterns recorded by a low-resolution medium such as a video system.

In the discussion to follow we assume that the direction of the displacement is known and is parallel to the sensitivity vector. That is, in Eqs. (4.2)–(4.5) we set $g(\phi) = 2$. In practice, the direction of the displacement vector is often unknown and varies across the test surface. In such cases we only know that we are recording a fringe pattern where the optical path length change is the projection of the displacement onto the sensitivity vector.

In video speckle interferometry we have at our disposal the in-plane technique, which provides sensitivity vectors not available in hologram interferometry. If we combine the information from a uniform reference system with two orthogonal recordings from an in-plane system, we obtain directly the x, y, and z displacements from which the total displacement vector can be found. This approach with multiple cameras and light sources has been used by the Loughborough [51] and Oldenburg [39] groups. Another approach is to change the direction of illumination in a uniform reference system; this approach was used by Winther [52] to map the displacement field where the essential surface contours were also mapped in the same setup. Which technique is the best has not been investigated. Intuitively, the separate determination of the components seems most promising, but it may be that the less favorable speckle statistics of the in-plane system makes the uniform reference system competitive.

4.6.1 Static Displacement

Static displacement fringe analysis consists of measurements where we have a no load–load situation or the displacement is slow or controlled as in an experiment using heat or pressure increments. We want to measure the displacement between separate frames 1 and 2 where the displacement during the recording of one frame is negligible or zero. The primary results after such an experiment are intensity distributions $I(x, y)$,

$$I_{1,2}(x, y) \approx I_o(x, y) + I_r(x, y) + 2\sqrt{I_o(x, y)I_r(x, y)} \cos \alpha_{1,2}(x, y) \qquad (4.10)$$

where $I_o(x, y)$ is the intensity of the object speckle wave, $I_r(x, y)$ is the intensity of the (speckle) reference wave, $\alpha_{1,2}(x, y)$ are the speckle phase terms indicating the relative phase between object and reference waves before and after displacement.

The fringe analysis problem is to separate and quantify $\alpha_{1,2}(x, y)$ so we can determine the optical phase change $\theta(x, y) = \alpha_1(x, y) - \alpha_2(x, y)$ and quantify the displacement. The most common solution is to use phase stepping, where the phase in one of the interferometric branches is stepped in $N + 1$ equal and controlled steps [50]. The corresponding $N + 1$ intensity distributions or frames are recorded in a frame store. By combining the resulting $N + 1$ equations we eliminate the trivial unknowns in the interference equations to solve for $\alpha_{1,2}(x, y)$.

Most commonly used is $N = 3$, where four frames $I_N(x, y, N\cdot\pi/2)$ are acquired. Combining the four frames gives the solution for the phase angle $\alpha(x, y)$:

$$\alpha(x, y) = \arctan \left[\frac{I_3(x, y) - I_1(x, y)}{I_0(x, y) - I_2(x, y)} \right] \qquad (4.11)$$

The great advantage of the use of this phase-stepping routine is the elimination of uneven illumination, contrast variations, etc. This is not the case with a one-step routine [53] where two images $I_{1,0}(x, y)$ are combined with a phase step of $\pi/2$ to give the simple equation

$$\alpha(x, y) = \arctan \frac{I_1(x, y)}{I_0(x, y)} \qquad (4.12)$$

This algorithm demands uniform intensities and constant fringe contrast to yield correct answers, which rarely is the case in practical situations. However, by using preprocessing such as Fourier transform filtering, the images can be normalized to a sufficient degree.

As the basic phase-stepping routines are well known from the optics literature [50], I will describe the max-min scanning routine that has been developed for defect detection under unstable conditions [54, 55]. This routine is based on the constant production of interferograms in the video system and is difficult to envision being developed in hologram interferometry.

Defects in the test object give rise to a nonhomogeneous movement of the surface. The image processing procedure is aimed at accentuating these (small) inhomogeneities, which are hidden in the fringe pattern. In addition, the procedure is well suited for use on test objects that are moving slowly, owning to thermal loading, for example.

The algorithm is outlined with reference to the flowchart of Fig. 4.10. The first step is to determine the phase Φ of the wave front reflected from the test object before deformation (the object phase as defined relative to the phase of the reference wave front). This is done by a maximum-minimum scanning whereby the object phase drifts through values from 0 to 2π by natural drift or controlled by a phase-shifting device. During the cycle the extreme values of the phase are recorded. The absolute sign of the object phase is determined by adding a known phase shift α and analyzing the resulting intensity shift.

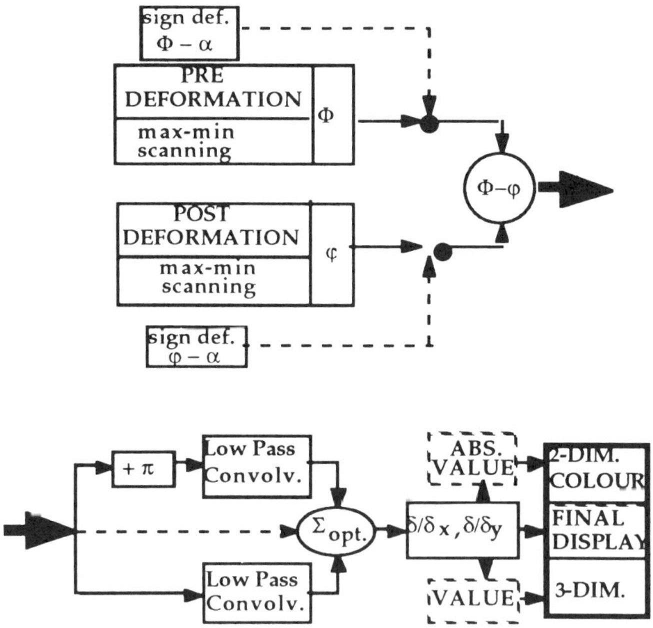

Figure 4.10 Max-min scanning.

The object is thereafter deformed—for example, by heating—and the new "deformed" phase φ is determined by the same procedure as above. The phase values Φ and φ representing the undeformed and deformed states are thereafter subtracted. The subtracted phase distribution is sent through a low-pass filter. However, a straightforward low-pass filtering introduces excessive noise in the regions where there is a 2π phase shift. This is avoided by shifting the image by π and low-pass filtering (convolving) both images. The two images are subsequently combined into one optimum image by the Σ_{opt} operator. This operator combines the best parts of the two images, a process controlled by the pixel values in the original phase distribution.

Finally, the gradient of the resulting image is calculated, and the results are displayed on the video screen in a two-dimensional contour map, false colors, or mesh plots as shown on the examples in Fig. 4.11. Here the technique has been used to visualize damage in a composite material caused by the standard drop test.

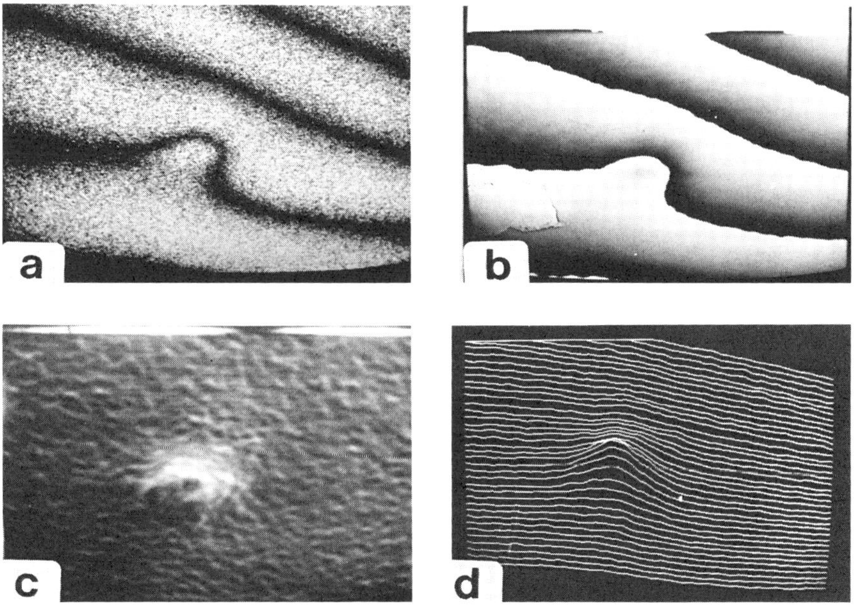

Figure 4.11 Subsurface damage in a laminate revealed by slight heating of the sample. (a) "Raw" fringe pattern; (b) phase map; (c) two-dimensional display of gradient; (d) mesh plot.

4.6.2 Vibration Analysis

In deformation analysis we are mainly concerned with the absolute value and direction of the displacement across the deformed surface. In vibration analysis, we find the projected displacement vector or the amplitude readily by time-average recordings. However, we need to know during what part of the vibration cycle the maximum displacement is reached; that is, we need to know the distribution of the vibration phase across the surface. In classical modes, the phase within each antinode is constant and the antinodes vibrate in phase or out of phase. As the values thus shift between 0° and 180°, it is fairly straightforward to analyze the vibration from the displacement map only. In real life, however, the classical mode is most often replaced by a superposition of modes, which makes the phase vary continuously across the surface. A typical example is that of vibrations in loudspeakers where uniform movements can be observed only at the lowest frequencies. In conclusion, to obtain proper information about a general vibration, we need to measure amplitude and phase distribution alike.

Here we consider a technique [56] based on time-average recordings using sinusoidal phase modulation [9] [see Eq. (4.4)]. The numerical amplitude and phase values of the test object are found by subtracting the unknown vibration vector from a reference vibration vector that is given calibrated changes in amplitudes and phases.

The technique is interesting because it can be used to measure vibration amplitudes and phase values at very low excitations. Using low excitation levels ensures that the object vibrates within the linear region, which makes the comparison with finite-element calculations more reliable. In addition, there is a definite need to measure low vibration amplitudes as we go to high-frequency analysis where the displacements are in the submicrometer region. There are a few cases where the method cannot be used because the excitation has to be at a certain level to induce the vibration, the most notable example being pressure-induced vibrations in pipe structures [57].

The fringe function for a time-average recording of an object vibration at low amplitude levels is shown in Fig. 4.12. The middle part of the function is linear around the working point a_w, and we adjust the excitation so the amplitude of the object is within this linear part. In the measuring procedure, we start with the object at rest. The reference mirror is vibrated at amplitudes corresponding to the extremal parts $a_{r,\max}$ and $a_{r,\min}$ of the linear part of the fringe function. The corresponding $I_{\max}$ and $I_{\min}$ are recorded as the first two frames. These frames are used to calibrate the sensitivity of the system in terms of intensity versus amplitude factor k. Alternatively, we may record the intensity values I_0 and I_{dark}, where I_0

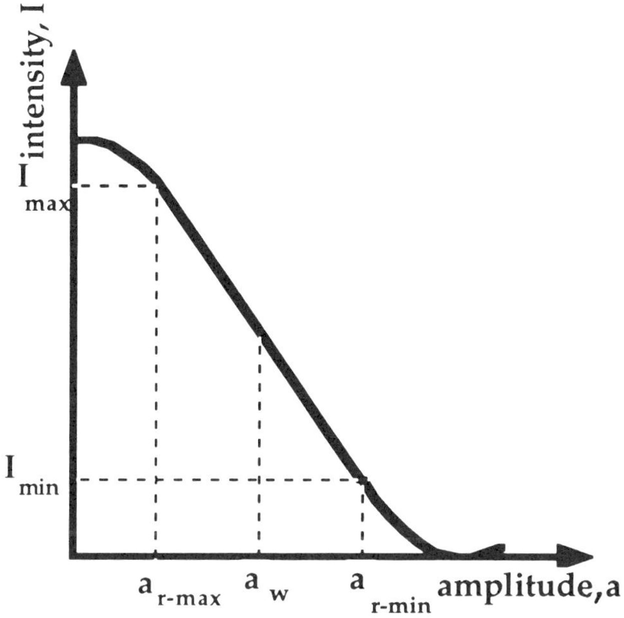

Figure 4.12 J_{o^2} function at low arguments.

corresponds to no modulation and I_{dark} corresponds the first minimum of the fringe function. A simple calibration factor gives the correct k value. This procedure has the advantage that it gives a more accurate k value because of the larger dynamic range in the initial measurements.

The reference mirror is vibrated at an amplitude a_w while the object is vibrated at an excitation level that keeps its maximum amplitude within the linear part of the fringe function, that is, about 30 nm. We step the acoustical phase of the reference wave by 0°, 90°, 180°, and 270° and record the corresponding intensities I_0, I_{90}, I_{180}, and I_{270}. These four recordings are related by the following equations, where the use of (x, y) coordinates refers to pixel values in the digitized image:

$$\begin{aligned} I_0(x, y) &= I_b(x, y) - k(x, y)\cdot a_o(x, y) \cos \varphi_o(x, y) \\ I_{90}(x, y) &= I_b(x, y) + k(x, y)\cdot a_o(x, y) \sin \varphi_o(x, y) \\ I_{180}(x, y) &= I_b(x, y) + k(x, y)\cdot a_o(x, y) \cos \varphi_o(x, y) \\ I_{270}(x, y) &= I_b(x, y) - k(x, y)\cdot a_o(x, y) \sin \varphi_o(x, y) \end{aligned} \tag{4.13}$$

where $I_b(x, y)$ is an (unknown) bias intensity.

From Eqs. (4.13) we solve for the unknown of interest, the amplitude $a_o(x, y)$ and the phase $\varphi_o(x, y)$. The vibration phase is found by the usual combination of Eqs. (4.13), which gives

$$\varphi_o(x, y) = \arctan\left[\frac{I_{90}(x, y) - I_{270}(x, y)}{I_{180}(x, y) - I_0(x, y)}\right] \tag{4.14}$$

As we have determined $\varphi_o(x, y)$ from Eq. (4.14) and we know the sensitivity factor $k(x, y)$ from the calibration procedure, we find the unknown amplitude from the expression

$$a_o(x, y) = \frac{\{[I_{180}(x, y) - I_0(x, y)]^2 + [I_{90}(x, y) - I_{270}(x, y)]^2\}^{1/2}}{2k(x, y)} \tag{4.15}$$

Figure 4.13 shows an example of the calculated amplitude and phase distribution of a crystal vibrating at 7.680 MHz with a maximum amplitude of about 20 nm. The vibration pattern consists of a central peak surrounded by a broken ring in antiphase. In this case the information is presented as a mesh plot; this gives a good visual display, but exact interpretation is more difficult. A two-dimensional contour plot is better for that purpose; for a discussion, see Ref. 58.

There are other ways to analyze time-average vibrations that provide amplitude and phase distributions. Pryputniewicz and Stetson [59] use a somewhat similar algorithm based on phase modulation. This algorithm also works at higher excitation levels, but as it relies on measurements of J_o^2 fringes of higher orders and corresponding lower intensities, the signal-to-noise ratio is probably lower than for the previous technique.

Vibration analysis can also be done with stroboscopic techniques where the displacement is measured in several parts of the vibration cycle using normal phase-shifting techniques. From these measurements, both the amplitude and vibration phase can be calculated. The stroboscopic technique should also be of interest when we are dealing with nonsinusoidal vibrations, as sufficient sampling gives information about the higher harmonics as well. In general, whenever the presence of nonsinusoidal vibrations is suspected, a stroboscopic or double-pulse technique should be used for analysis and measurement of the vibration instead of a time-average method.

The most challenging problem in vibration analysis by speckle interferometry is vibration induced by white noise excitation. Quite extensive work by my group has failed to come up with a good solution. A technique reported in hologram interferometry, referred to as holographic vibration

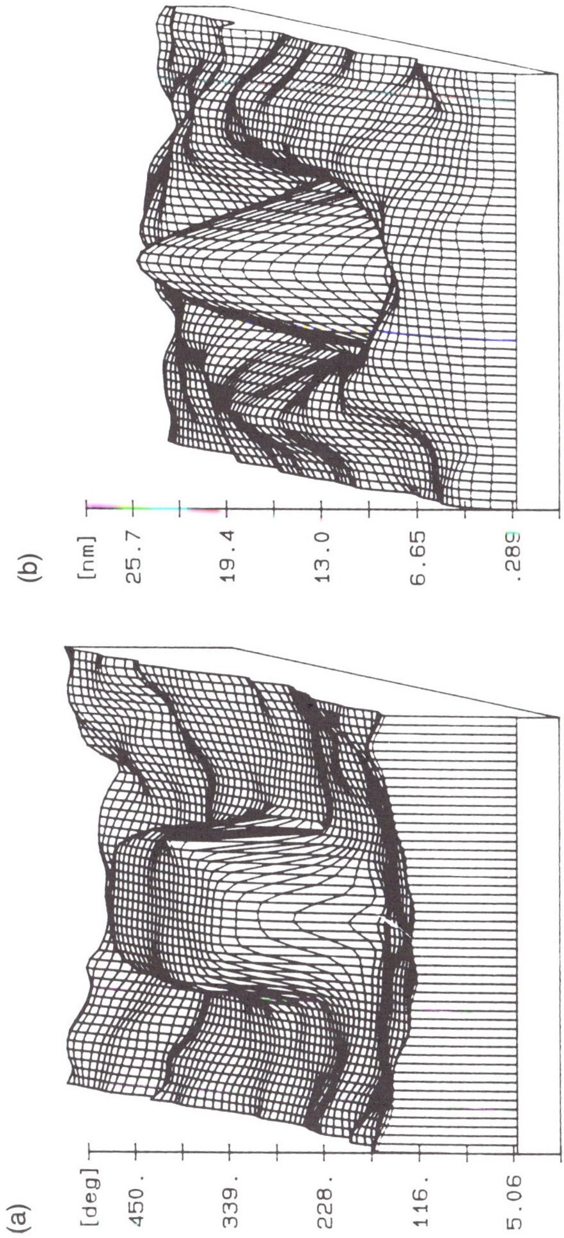

Figure 4.13 (a) Phase and (b) amplitude distribution of a vibrating crystal at f = 7.680 MHz.

spectroscopy [60], turned out to be of extremely marginal value. Numerous experiments with digital analysis of time average techniques, with and without phase modulation, in addition to pulsed-laser recordings have given interesting information about the time history of white and pink noise vibration. However, we are still looking for a good solution that provides information about the statistical amplitude and phase (the vibration phase is in this context somewhat vague; the problem is more to determine which areas contribute positively (in phase) to the average sound pressure). Straight time-average recordings give part of the solution as they give information about the areas where excessive vibrations exist.

A different approach to the analysis of vibration due to general excitation—for example, due to a motor running—is to measure the resonant frequencies and their relative sound levels by acoustic means under real working conditions [61]. Keeping the setup and surroundings unchanged, the resonance vibration patterns are thereafter excited separately and analyzed by the digital methods just described. In this way, we gain enough information to synthesize the vibration due to random excitation. The entire procedure is rather time-consuming.

4.6.3 Single Dynamic Events

In the experiments described so far, we have been able to control the displacement or the displacement has been stable and periodic. In real life we are more often faced with situations where the information has to be gathered within the time of a single video frame. Typical examples are explosive events in industry or studies of live biological samples where there are rapid speckle decorrelations due to multiple scattered light. Apart from the fact that pulsed lasers with associated stability problems have to be used in these experiments, the main problem is that all the aforementioned phase-shift techniques break down. We have no possibility to combine several consecutive frames, and we have to analyze single frames, which obviously reduces the measuring sensitivity and range.

There are several ways to attack the single-frame problem. The most obvious solution is to use the trick from visual fringe analysis in classical interferometry, where reference tilt fringes are added to the deformation to be analyzed. I should point out that as we are here talking about double exposure with a pulse separation in the microsecond range, it is not a trivial problem to add calibrated, controlled fringes. To automate and improve the analysis of the resulting fringe pattern, the Fourier spectrum of the fringe pattern should be analyzed; see, for example, Ref. 62. Because of the speckle noise in the straight-line fringe carrier, we must expect lower accuracy than in classical interferometry. Using 10–15 reference fringes across the field, we should be able to measure with an accuracy on the order of $\lambda/25$.

The Steinbichler group [63] uses an ingenious technique based on a uniform reference setup where the reference wave impinges on the target at a small angle. This angle is chosen so that there is a 120° phase shift between consecutive pixels. Three neighboring pixels are assumed to be within the same speckle, something that is achieved by reducing the aperture of the imaging system or using a camera of high resolution. By treating identical phase pixels as individual frames in the readout we obtain three phase-shifted interferograms due to a deformation and can perform the corresponding algorithm and unwrap the phase. The technique has been proven to work well in experiments with double-pulsed lasers. The drawback is that there is a loss of resolution and light, which has been partly compensated for by using a high-resolution CCD camera. The adjustment of the reference wave is rather critical.

The same three-phase concept can be implemented by using three cameras and a complex maze of polarizing elements to get the necessary phase shifts [64]. The problem here is to get a pixel-by-pixel alignment of the three cameras.

Finally, we can use the method suggested and used by Kwon [65] for analyzing Smartt interferograms produced by pulsed lasers. Here a Ronchi-type grating provides three phase-shifted image interferograms on the video target. Each image can be read out separately and solved for the optical phase distribution.

In some phase-shift techniques the phase information is spread over different pixels or even totally different locations on the target. In these cases, we have to be very careful to calibrate the sensitivity of the individual pixels and also to correct for image distortion due to the imaging lens.

4.6.4 Problems Using Digital Fringe Analysis in Video Speckle Interferometry

When we use digital techniques in video speckle interferometry to analyze the displacement, we face several problems and sources of errors [15]. Most of these, such as drift, unwanted vibrations, uneven phase settings, electronic noise, nonlinear recordings, and quantization errors, are common to classical and holographic interferometry and will not be commented on here.

The speckle pattern has areas where the intensity is low or even zero. These areas do not contribute to the interferometric signal as their modulation is too low for reliable detection. In addition, there are dust particles on the target, diffraction and interference patterns on the reference wave, etc. All this adds up to a substantial number of "bad" pixels in a single video frame. These pixels leave holes in the information mapping and give rise to noise spikes in the final results. To correct for these problems, it is often necessary to combine several processing techniques. We have to use algorithms that search for the pixels having modulation too low for reliable calculations and replace their values with the mean of the neighboring pixels. If possible, several frames should be added together. In analysis of repetitive displacement, the contrast of the speckle pattern can be reduced by speckle averaging as already described. The speckle-averaging technique has proved very efficient in reducing the number of rejected pixels. In addition, the computed values can be presmoothed by reducing the initial 512 × 512 pixels to, for example, a 100 × 100 matrix, while a spatial $N \times N$ filter smooths noise in the final image. However, excessive spatial filtering reduces the spatial resolution and should be used with care, especially if fine details in the displacement patterns are expected.

Another fundamental problem is the low resolution of the video camera. In video analysis of ordinary hologram interferometry the video camera is used to analyze interferograms recorded on, and reconstructed from, a high-resolution medium. In this case the video sensor is recording secondary fringes with low speckle noise and hardly any decorrelation effects, whereas in video speckle interferometry the video sensor records the primary interferogram from which the secondary fringes are produced. Because of the low resolution we have to work with small apertures, which results in speckle decorrelation as tilting of the object's surface brings new speckles into the aperture. Opening the aperture does not help, as the resolution bandwidth acts as an effective aperture. Together with the basic low resolution of the camera, the speckle decorrelation reduces the contrast of the fringes and thus the number of fringes across the object that can be allowed. Speckle decorrelation is a greater problem in techniques based on speckle reference waves as the speckle statistics are more unfavorable. Therefore a uniform reference setup is preferable to avoid decorrelation. Most experimenters avoid the speckle decorrelation problem by keeping the number of fringes small, which makes accurate (sub)fringe measurement essential. Whenever we are faced with a large deformation that can be controlled, the best solution is to subdivide it into many deformation steps, which are then integrated into a total deformation.

4.8 APPLICATIONS

There has been strong growth in the general application of video speckle interferometry, and within the scope of this chapter it is possible to mention only a few examples. (All documented examples stem from my own group, solely for the convenience in acquiring documentary pictures.) For a more extensive update, readers should consult conference proceedings such as Ref. 18.

It is appropriate to suggest that video speckle interferometry has been gradually taking over the routine testing formerly done with hologram interferometry. The speed, ease of handling, and ability to work outside the laboratory are most interesting for industrial research, and we must expect this market to grow strongly in the immediate future.

Most commonly accepted within the industry is definitely video shear interferometry, which is recognized as an official testing method for defects and delamination in the airplane industry. It is obvious that the shear technique will be the inspection method [27] of choice whenever it is desired to pinpoint abnormal displacements indicating failure, weakness, or other sources of trouble and there is no concern about the exact analysis of the movement. As the shear system can be very simple and inexpensive in construction, it is tempting to predict that such systems will be used in great numbers for continuous surveillance of critical structures.

The standard uniform reference system has also proved to be capable of applications outside the laboratory. In vibration testing especially, this system has taken over most of the work formerly done by hologram interferometry. A typical example is the routine testing of vibration in turbine blades. The system has been used extensively for testing of various vibrations inside and outside airplanes of all sizes. As most of the work has been done with a low-power He-Ne laser, the surfaces have to be covered with retroreflective coatings. Davies and Buckberry [66] have used their fiber-optics system on a routine basis for the analysis of vibrations and deformations in automobile bodies and motor parts, working on the factory floor. As their system uses a powerful Ar laser as the central light source, no retroreflective coating is needed in their case. A typical example of vibration testing of automobiles by my own group [67] is shown in Fig. 4.14. The side of the car (a Mazda 818) is vibrating at 131 Hz. The recorded area was covered with retroreflective coating, enabling us to use a 5-mW laser.

It has also been proved that digital analysis can be used to analyze vibrations under rather rough conditions even when the system is based on continuous-wave lasers. A typical example is the analysis of vibrating objects [68, 69] heated to 1000°C where the recordings at high temperature were less noisy than at room temperature. This somewhat unexpected result is probably due to more effective speckle averaging taking place at the high temperatures. In standard work we also like to mention measurements of very high frequency vibrations [70], where the amplitude and phase distribution have been mapped at frequencies in the 10 MHz range. The

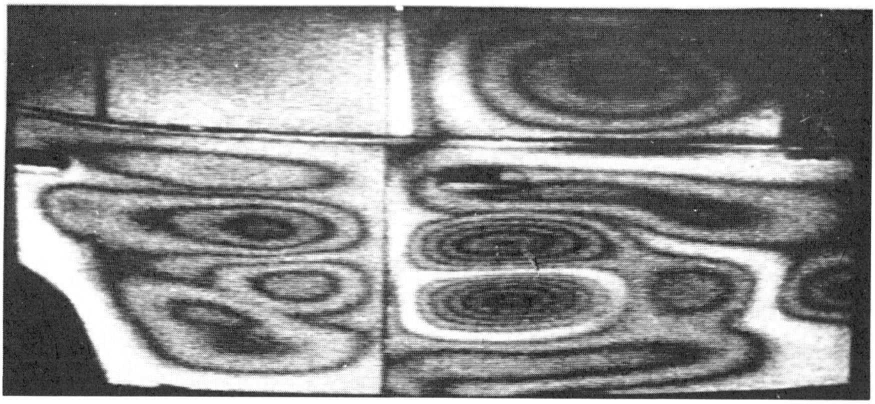

Figure 4.14 Car body (Mazda 818) vibrating at 131 Hz.

same digital procedure has also been used to measure vibration of passive sonar transducers in water tanks [71].

Measurements under very unstable, rough conditions can also be done by using a pulsed-laser system as demonstrated by the Loughborough group [38], who developed the pulsed system for in-plane measurements in both directions using the polarizing properties of metal surfaces. Pulsed lasers have also been used for measurement of in-plane stresses of large rotating structures [72].

In pure deformation testing, the stability properties of video speckle interferometry are not comparable to vibration testing because different frames, which might be several seconds apart, are being compared, whereas information about vibration is acquired within single frames. Still, the system has proven usable in practice in this mode. The best example is probably the work of the Oldenburg group [39] who followed and measured cracks in a church wall due to heating and cooling over a period of weeks. The instruments were firmly attached to the wall, so gross movement of the building was effectively canceled. In fact, the behavior of concrete and stones in loading situations has been a favorite subject for several groups. The development of microcracks in concrete and larger cracks in sandstone has been studied by my group [73]. An example of this work is shown in Fig. 4.15, where a small area of concrete is observed at increasing pressures. To aid interpretation of the images, the interferometric gradient images have been overlaid with a regular white light image. The images show microcracks (white lines) in a matrix of irregular stones in the concrete. Note how the main crack (centered) creeps toward and reflects from the

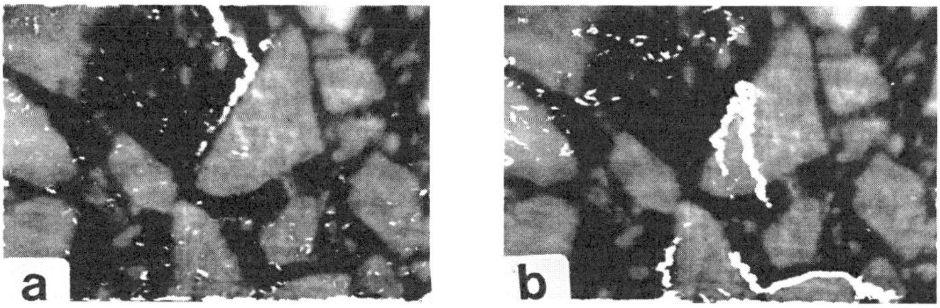

Figure 4.15 Crack propagation in concrete mix. Area about 25 mm × 40 mm. The compression pressure is increasing from (a) to (b).

stone's edge at the lower pressure. As the pressure is increased, the stone is broken by the crack. The upper part of the crack has disappeared, probably because of the tension being released. Observations of the progression of cracks in real time and analysis of still recordings like those of Fig. 4.15 have given new and important information about the mechanism of concrete failure [74].

Very interesting work is reported by Takemoto and Tsubo [75], who used TV holography to monitor irregular displacements in the wall of a mine. The information is used to detect earthquakes at an early stage. Similar work has been reported in progress in California [76].

Surprisingly enough, there are rather few papers to be found within the biomedical field. In my opinion, interferometric studies of live organisms should offer the most challenging, fruitful, and difficult area of research. The most exciting report has been presented by Leith et al. [77], who report that imaging through thick layers of tissue is possible by picosecond interferometric imaging using a standard form of video speckle interferometry. The paper also shows very convincingly how frame averaging improves the signal-to-noise ratio in the image. The continuous production of new reconstructed image holograms made possible by video speckle techniques will be of utmost importance in this field of research and is a point often overlooked in comparing video speckle with conventional holography. In general, it is obvious that for any of the potential uses proposed for hologram interferometry to find their way into routine clinical diagnosis, the recording has to be based on a real-time medium like a video system.

ACKNOWLEDGMENT

I wish to thank Svein Ellingsrud, Jan T. Malmo, Rudie Spooren, and Eiolf Vikhagen for their kind permission to use documentary pictures in this chapter.

REFERENCES

1. J. N. Butters and J. A. Leendertz, Holographic and videotechniques applied to engineering measurements, *J. Meas. Control*, *4*, 349–354 (1971).
2. A. Macovski, D. Ramsey, and L. F. Schaefer, Time lapse interferometry and contouring using television systems, *Appl. Opt.*, *10*, 2722–2727 (1971).
3. O. Schwomma, German Pat. 298830 (1972); see also *Neuer Züricher* Beilage "Forschung und Technik" (in German), p. 257 (1975).
4. J. N. Butters, R. Jones, and C. Wykes, Electronic speckle pattern interferometry, in *Speckle Metrology*, R. K. Erf, Ed., Academic, New York, 1978, pp. 111–158.
5. R. Jones, The design and application of a speckle pattern interferometer for total strain field measurement, *Opt. Laser. Techn.*, *8*, 215–219 (1976).
6. R. Jones and C. Wykes, The comparison of complex object geometries using a combination of electronic speckle pattern interferometric difference contouring and holographic illumination elements, *Opt. Acta*, *25*, 449–472 (1978).
7. G. Å. Slettemoen, Electronic speckle pattern interferometric system based on a speckle reference beam, *Appl. Opt.*, *19*, 616–623 (1980).
8. O. J. Løkberg and G. Å. Slettemoen, Improved fringe definition by speckle averaging in ESPI, ICO-13 Proceedings (Sapporo, Japan), 1984, pp. 116–117.
9. K. Høgmoen and O. J. Løkberg, Detection and measurement of small vibrations using electronic speckle pattern interferometry, *Appl. Opt.*, *16:*, 1869–1875 (1976).
10. O. J. Løkberg, Use of chopped laser light in electronic speckle pattern interferometry, *Appl. Opt.*, *18*, 2377–2384 (1979).
11. S. Nakadate, T. Yatagai, and H. Saito, Electronic speckle pattern interferometry using digital image processing technique, *Appl. Opt.*, *19*, 1879–1885 (1980).
12. C. Koliopoulos, Interferometric optical phase measurement techniques, Ph.D. Dissertation, Optical Science Center, Univ. Arizona, 1981.
13. P. Hariharan, B. F. Oreb, and N. Brown, Real-time holographic interferometry: a microcomputer system for measurement of vector displacements, *Appl. Opt.*, *22*, 876–880 (1983).
14. S. Nakadate and H. Saito, Fringe scanning speckle pattern interferometry, *Appl. Opt.*, *24*, 2172–2180 (1985).
15. K. Creath, Phase-shifting speckle interferometry, *Appl. Opt.*, *24*, 3053–3058 (1985).

16. D. Robinson and D. W. Williams, Digital phase speckle pattern interferometry, *Opt. Commun.*, *57*, 26–30 (1986).
17. K. A. Stetson and R. Pryputniewicz, eds., *Hologram Interferometry and Speckle Metrology*, Proceedings of the Society of Experimental Mechanics, Baltimore, 1991.
18. *Laser Interferometry IV; Computer-Aided Interferometry*, Proceedings of the Society of Photo-Instrumentation Engineers (SPIE), Vol. 1553, San Diego, 1991.
19. R. Jones and C. Wykes, *Holographic and Speckle Interferometry*, Cambridge Univ. Press, London, 1983.
20. O. J. Løkberg and G. Å. Slettemoen, Basic electronic speckle pattern interferometry, *Appl. Opt. Opt. Eng.*, *10*, 455–504 (1987).
21. J. W. Goodman, Statistical properties of laser speckle patterns, in *Laser Speckle and Related Phenomena*, J. C. Dainty, Ed., Springer-Verlag, Berlin, 1975, pp. 9–75.
22. H. M. Pedersen, O. J. Løkberg, and B. M. Førre, Holographic vibration measurement using a TV-speckle interferometer with silicon target vidicon, *Opt. Commun.*, *12*, 421–426 (1974).
23. G. Slettemoen, General analysis of fringe contrast in electronic speckle pattern interferometry, *Opt. Acta*, *26*, 313–327 (1979).
24. J. N. Butters and J. A. Leendertz, A double exposure technique for speckle pattern interferometry, *J. Phys. E: Sci. Instrum.*, *4*, 277–279 (1971).
25. J. A. Leendertz and J. N. Butters, An image shearing speckle pattern interferometer for measuring bending moments, *J. Phys. E: Sci. Instrum.*, *6*, 1107–1110 (1973).
26. G. Å. Slettemoen and O. J. Løkberg, Speckle reference ESPI: in practice, *Appl. Opt.*, *20*, 3467–3469 (1981).
27. Y. Y. Hung, Electronic shearography versus ESPI for nondestructive testing, *SPIE*, *1554B*, 692–700 (1991).
28. S. Winther and G. Å. Slettemoen, An ESPI contouring technique in strain analysis, *SPIE Symp. Opt.*, *473*, 44–47 (1984).
29. O. J. Løkberg and K. Høgmoen, Use of modulated reference wave in electronic speckle pattern interferometry, *J. Phys. E*, *9*, 847–851 (1976).
30. O. J. Løkberg and K. Høgmoen, Vibration phase mapping using electronic speckle pattern interferometry, *Appl. Opt.*, *15*, 2701–2704 (1976).
31. J. Tyrer, Continuous pulsed electronic speckle pattern interferometry, *SPIE*, *732*, 110–115 (1987).
32. G. Å. Slettemoen, First-order statistics of displayed speckles in electronic speckle pattern interferometry, *J. Opt. Soc. Am.*, *71*, 474–482 (1981).
33. K. A. Stetson and W. R. Brohinsky, Electro-optic holography system for vibration analysis and nondestructive testing, *Opt. Eng.*, *26*, 1234–1239 (1987).
34. K. A. Stetson, W. R. Brohinsky, J. Wahid, and T. Bushman, An electro-optic holography system with real-time arithmetic processing, *J. Nondestruct. Eval.*, *8*(2), 69–76 (1989).

35. K. Creath and G. Å. Slettemoen, Vibration-observation techniques for digital speckle-pattern, *J. Opt. Soc. Am. A*, *2*, 1629–1636 (1985).
36. K. Biedermann and L. Ek, A recording and display system for hologram interferometry with low resolution imaging devices, *J. Phys. E: Sci. Instrum.*, *8*, 571–576 (1975).
37. J. C. Davies, C. H. Buckberry, J. D. C. Jones, and C. N. Pannel, Developments in electronic speckle pattern interferometry for automotive vibration analysis, *Proc. SPIE*, *952*, 260–274 (1988).
38. A. J. Moore and J. Tyrer, Surface strain measurement with ESPI applied to fracture mechanics, *Hologram Interferometry and Speckle Metrology*, Proc. SEM, 1990, pp. 192–198.
39. G. Gülker, K. Hinsch, C. Hölscher, A. Kramer, and H. Neunaber, An ESPI system for in situ monitoring on buildings, *Opt. Eng.*, *29*, 816–820 (1990).
40. G. M. Brown and G. H. Smith, Dynamic computer-aided video holometry, *SPIE*, *1162*, 36–45 (1989).
41. P. Ottonello and C. Pontiggia, An ESPI apparatus with multichannel data acquisition, *Opt. Commun.*, *30*, 20–23 (1979).
42. C. Joenathan and B. M. Khorana, Phase measuring fiber optics electronic speckle pattern interferometry: phase step calibration and phase drift minimization, *Opt. Eng.*, *31*, 315–321 (1992).
43. R. Spooren, TV-holography using a double pulsed single oscillator Nd:YAG laser; restrictions and possibilities, *Hologram Interferometry and Speckle Metrology*, Proc. SEM, 1990, pp. 260–267.
44. R. Spooren, Double-pulse characteristics of a single-oscillator Nd:YAG laser affecting its performance in TV-holography, *Appl. Opt.*, *31*, 208–216 (1992).
45. R. Spooren, Fringe quality in pulsed TV-holography, *SPIE*, *1508*, 118–127 (1991).
46. R. Spooren, Double pulse subtraction TV-holography, *Opt. Eng.*, *31*, 1000–1007 (1992).
47. G. O. Rosvold, Fast measurements of phase using a PC-based frame grabber and phase stepping technique, *Appl. Opt.*, *62*, 237–241 (1990).
48. E. Vikhagen, Vibration measurement using phase shifting TV-holography and digital image processing, *Opt. Commun.*, *69*, 214–218 (1989).
49. C. B. Burkhardt and L. H. Enlow, Television transmission of holograms with reduced resolution requirements on the camera tubes, *Bell. Syst. Tech. J.*, *48*, 1529–1535 (1969).
50. K. Creath, Phase-measurement interferometry techniques, *Prog. Opt.*, *26*, 349–393 (1988).
51. M. C. Shellabear and J. Tyrer, Application of ESPI to three-dimensional vibration measurement, *Opt. Laser Eng.*, *15*, 43–56 (1991).
52. S. Winther, 3D strain measurements using ESPI, *Opt. Lasers Eng.*, *8*, 45–57 (1988).
53. D. Kerr, F. Mendoza-Santoyo, and J. R. Tyrer, Extraction of phase data from electronic speckle pattern interferometric fringes using a single phase step method: a novel approach, *J. Opt. Soc. Am. A*, *7*, 820–826 (1990).
54. E. Vikhagen, Nondestructive testing by use of TV-holography and deformation phase gradient calculation, *Appl. Opt.*, *29*, 137–144 (1990).

55. E. Vikhagen, TV-holography: spatial resolution and signal resolution in deformation analysis, *Appl. Opt.*, *30*, 420–425 (1991).
56. S. Ellingsrud and G. O. Rosvold, Analysis of a data-based TV-holography system used to measure small vibration amplitudes, *J. Opt. Soc. Am. A*, *9*, 237–251 (1992).
57. O. J. Løkberg and O. K. Ledang, Vibration of flutes studied by electronic speckle pattern interferometry, *Appl. Opt.*, *23*, 3052–3056 (1984).
58. S. Ellingsrud and O. J. Løkberg, Full field amplitude and phase measurement of loudspeakers using TV-holography and digital imaging processing, *J. Sound Vibr.*, in press.
59. R. J. Pryputniewicz and K. A. Stetson, Measurement of vibration patterns using electro-optic holography, *Proc. SPIE*, *1162*, 456–467 (1989).
60. R. Politsch, Spectroscopic holography—a new method for mode analysis of randomly excited structures, *J. Opt. Soc. Am. A*, *7*, 1355–1361 (1990).
61. B. Brevig, Vibration measurement using ESPI and acoustical frequency analysis, M.Sc. thesis at Norwegian Institute of Technology, 1984.
62. R. Józwicki, M. Kujawinska, and L. Salbut, New contra old wavefront measurement concepts for interferometric optical testing, *Opt. Eng.*, *31*, 422–433 (1992).
63. H. Steinbichler, S. Leidenbach, J. Engelsberger, E.-H. Nösekabel, and J. Sun, Optical measuring systems help to find constructional weak spots, *SPIE*, *1553*, 240–249 (1991). (Note: the algorithm was mentioned in the presentation but hardly described in the proceedings.)
64. R. Smyth and R. Moore, Instantaneous phase measuring interferometry, *Opt. Eng.*, *23*, 361–364 (1984).
65. O. Y. Kwon, Multichannel phase-shifted interferometer, *Opt. Lett.*, *9*, 59–61 (1984).
66. J. Davies and C. Buckberry, Applications of a fibre optic TV holography system to the study of large automotive structures, *SPIE*, *1162*, 279–292 (1989).
67. J. T. Malmo and E. Vikhagen, Vibration analysis of a car body by means of TV-holography, *Exp. Techn.*, *12*, 28–30 (1988).
68. J. T. Malmo, O. J. Løkberg, and G. Å. Slettemoen, Interferometric testing at very high temperatures by TV-holography (ESPI), *Exp. Mech.*, *28*, 315–321 (1988).
69. O. J. Løkberg, G. O. Rosvold, J. T. Malmo, and S. Ellingsrud, Computerized vibration analysis of hot objects, *SPIE*, *1508*, 153–160 (1991).
70. S. Ellingsrud and O. J. Løkberg, Analysis of high frequency vibrations using TV-holography and digital image processing, *SPIE*, *1162*, 402–407 (1989).
71. J. T. Malmo, O. J. Løkberg, and S. Ellingsrud, Interferometric measurements of objects in water by TV-holography, *Opt. Lasers Eng.*, *15*, 25–41 (1991).
72. R. W. T. Preater and R. C. Swain, In-plane measurements in hostile environments, *SPIE*, *1554A*, 727–738 (1991).
73. O. J. Løkberg, Electronic speckle pattern interferometry and its applications in rock mechanics, in *Laser Holography in Geophysics*, S. Takemoto, Ed., Ellis Horwood/John Wiley, New York, 1989, pp. 168–196.

74. E. Aasved-Hansen, Real time TV-holography observation of the fracture process zone, in *Micromechanics of Failure of Quasi-Brittle Materials*, S. P. Shah, S. E. Swartz, and M. L. Wang, Eds., Elsevier, London, 1990, pp. 504–513.
75. S. Takemoto and T. Tsubo, Application of holography and electronic speckle pattern interferometry techniques to earthquake prediction, *SPIE*, *952*, 230–234 (1989).
76. D. M. Rosenthal, J. D. Trolinger, and D. C. Weber, The use of double pulsed ESPI for earthquake mitigation of large structures, *SPIE*, *1553*, 175–184 (1991).
77. E. Leith, H. Chen, Y. Chen, D. Dilworth, J. Lopez, R. Masri, J. Rudd, and J. Valdmanis, Electronical holography and speckle methods for imaging through tissue using femtosecond gated pulses, *Appl. Opt.*, *30*, 4204–4210 (1991).

5

Novel Applications of Speckle Metrology

Chandra S. Vikram

The University of Alabama in Huntsville
Huntsville, Alabama

5.1 NOVEL APPLICATIONS

Around 1970, it was firmly established that speckle patterns, otherwise considered a source of noise in imaging, were very useful for metrological applications. This usefulness resulted in an enormous amount of systematic activity, as is evident from chapters of this book. Because of the universal applicability of these patterns, there are unique applications and techniques in very diverse fields. In this chapter, I make an attempt to cover these novel situations.

5.2 SPECKLE-BASED PROBES

In unique situations, a probe object can be subjected to a deformation influenced by the quantity to be measured. The deformation of the probe can then be studied by the speckle method. An example is measuring the change in diameter of circular objects by double-exposure laser speckle photography of the jaws of a lens holder [1]. Holding objects of different diameters, the motions (in-plane) of the jaws and the support ring of the holder will be different. The motion can be measured by speckle photography and related to the diameter change. In this section I describe some speckle-based probes.

5.2.1 Supersensitive Dial Gauge

Commercial gauges with micrometer-order resolution are common. The resolution corresponds to the detectable change in the needle position on the dial scale. The visually measurable change in the needle tip is of the order of 1 mm, whereas the common speckle method is capable of measuring lateral changes in the micrometer range. Thus, if the displacement of the needle tip is measured by double-exposure speckle photography, the sensitivity can be increased by orders of magnitude, limited only by the backlash error and other mechanical factors. This capability has been experimentally demonstrated [2]. On the other hand, the photography, chemical processing, analysis of the Young's fringes, etc. are time-consuming and thus unsuitable for many applications. However, a remedy is available based on the rotation of objective laser speckles [3], as shown in Fig. 5.1. In this method, out-of-plane rotation or tilt of the side of the needle is observed. Aluminum foil can be glued to the needle if its side is very thin. A laser beam is incident on the side of the needle or the foil, and the objective laser speckles are formed in space. Rotation of the needle will rotate these speckles. For near-normal illumination and observation

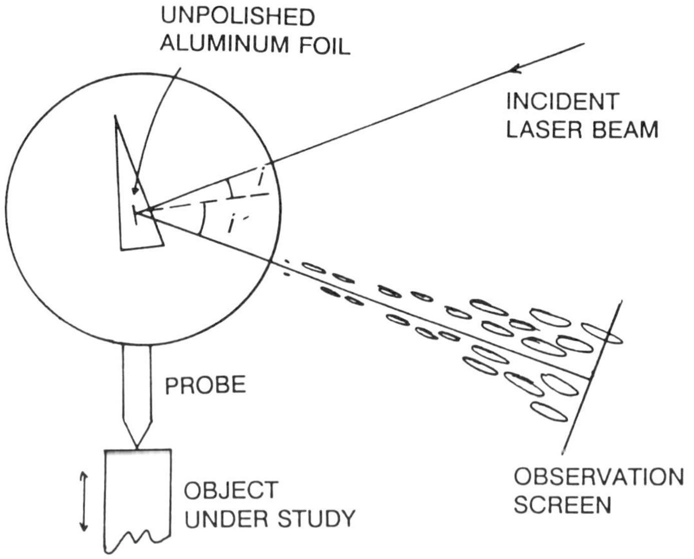

Figure 5.1 Diagram representing the experimental setup. The observation screen can be replaced by a closed-circuit TV system with or without a lens.

($i = i' = 0$), the speckle rotation is twice the needle rotation. The rotation can be observed at a distance on a ground glass. Lenses and/or a vidicon-monitor system can also be used to save space and process the data. Again, the sensitivity of the conventional dial gauge can be increased by orders of magnitude [3]. An interesting application of this approach is to enhance the sensitivity of push-rod dilatometers for ultralow thermal expansion measurements [4]. In this application, the usual LVDT (linear variable differential transformer) for recording the movement of the push rod is replaced by the dial gauge. Again, the sensitivity of the dilatometer can be enhanced by a few orders of magnitude [4].

5.2.2 Fiber-Optic Current Sensor

Ramprasad and Radha Bai [5] presented this speckle-based current-sensing technique. The concept is schematically represented in Fig. 5.2. The light from a 1-mW He-Ne laser is focused on the fixed end of a glass fiber of 0.5 mm diameter. The fiber is fixed as a cantilever with a free length of 6.3 cm. A lens collimates the output light beam of the fiber. A nickel jacket is mounted on the fiber to act as the probe. A solenoid with a soft iron core is energized by a suitable dc power supply with current variation capability. The attraction of the nickel jacket toward the solenoid varies with the current. Therefore, the Young's fringes produced from double-exposure photographs are a measure of the current or magnetic field. Experiments were performed [5] with varying distances of the jacket from

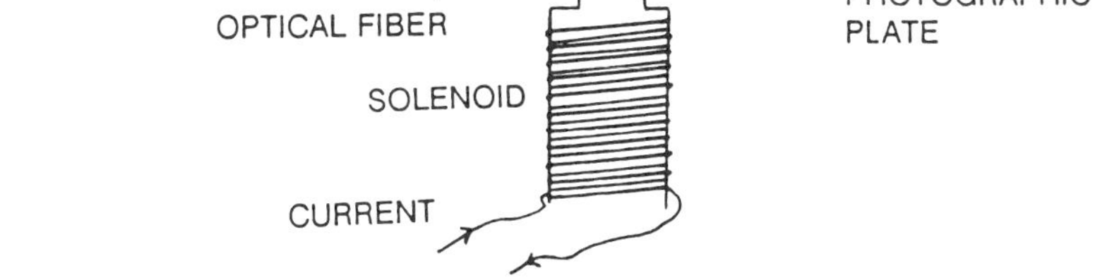

Figure 5.2 Schematic diagram of the speckle-based fiber-optic current or magnetic field sensor. (Modified from Ref. 5, with permission.)

the fixed end. It was found that the fiber behaved like an optical lever rather than a loaded cantilever.

Thus, it is established that the current can be measured by the speckle method. The current, the movement of the nickel jacket, and the frequency of the resulting Young's fringes are interrelated. As such, the measurement is a two-step procedure. However, the movement of the nickel jacket can be measured by other speckle approaches discussed throughout this book. An example is the dial-gauge procedure of Section 5.2.1.

5.2.3 Ultralow Thermal Expansion Measurements

There have been applications of laser speckles to thermal expansion measurement problems using conventional double-exposure laser speckle photography [6, 7] as well as Michelson-type [8] arrangements. Another unique approach [9], represented in Fig. 5.3, has been proposed and tested. This method achieves ultralow thermal expansion measurement of small samples in real time. The sample and reference rods (the latter of known thermal expansion properties) are kept on a platform directly on a vibration isolation table. The furnace is mechanically isolated from the table. A probe bar is kept on the sample and reference rods. To define the contact points, the tops of the rods are edged or a thin wire is glued on the probe bar to

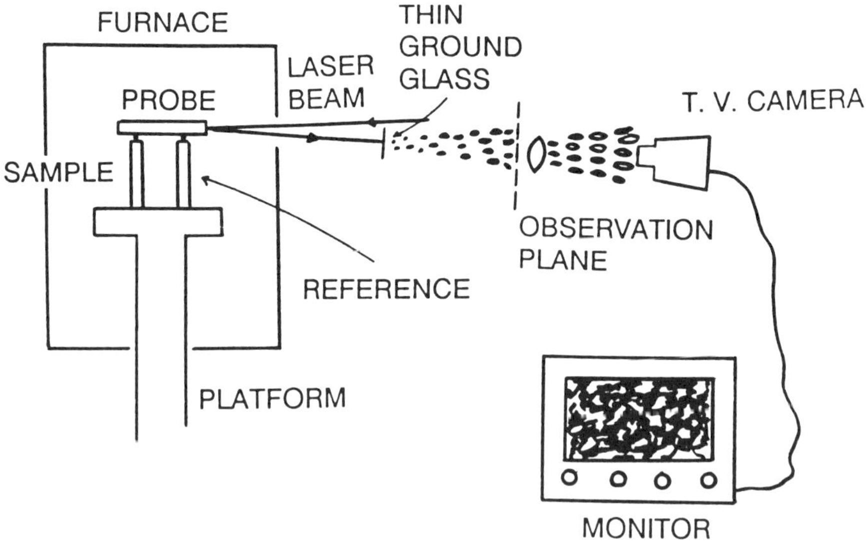

Figure 5.3 Experimental setup for measuring ultralow thermal expansion by observing rotation of laser speckles.

form a spacer. The side of the probe bar is polished and coated to make it reflective. Against unequal thermal expansion of the sample and reference rods, the probe bar is rotated by [9]

$$\Delta\theta = L(\alpha_1 - \alpha_2)\Delta T/D \tag{5.1}$$

where L is the length of the rods, α_1 and α_2 are the thermal expansion coefficients of the reference and the sample material, respectively, D is the distance between the sample and reference rods' contact points with the probe bar, and ΔT is the temperature change of the system.

The incident laser beam reflected by the probe bar side illuminates a thin ground glass and creates laser speckles in space. When the probe bar rotates by $\Delta\theta$, the speckles are rotated by $2\,\Delta\theta$. The observation plane can be magnified using a lens and ultimately viewed on a TV monitor. The sensitivity of measuring $\alpha = 10^{-8}\,°C^{-1}$ with 2.5-cm-long samples at 25°C temperature changes is easily obtained [9]. One main advantage of the method is that it has very poor sensitivity for several unwanted linear changes such as the thermal expansion of platform and probe materials. The closed-circuit TV system is very helpful but not necessary. The speckles can be observed on a white paper or a wall at a far distance and the method will still have a good sensitivity [10].

5.3 REMOTE APPLICATIONS AND LARGE STRUCTURES

There have been several applications of speckle metrology to the study of remote objects. Television transmission of the picture containing the speckles is one of them. Fiber optics can be used to illuminate a remote object as well as to transmit the picture to a suitable location. Then there are large geophysical structures such as glaciers that have to be studied remotely because of limited physical access as well as the need to cover a large area simultaneously.

5.3.1 Television Transmission of the Speckle Field

This application is described by Da Costa [11] in dealing with the exploitation of heavy hydrocarbons. Samples from a depth of 1 km in a well will change in mechanical structure if they are examined at the surface. However, an experimental module can be kept at the well's bottom and the speckle picture can be transmitted to a laboratory on the surface.

Figure 5.4 illustrates the experimental setup. Using a beamsplitter, the laser beam in the experimental module illuminates an auxiliary fixed diffuser and the object under study. The TV camera transmits two side-by-side speckle fields due to the object and auxiliary surfaces. The object is

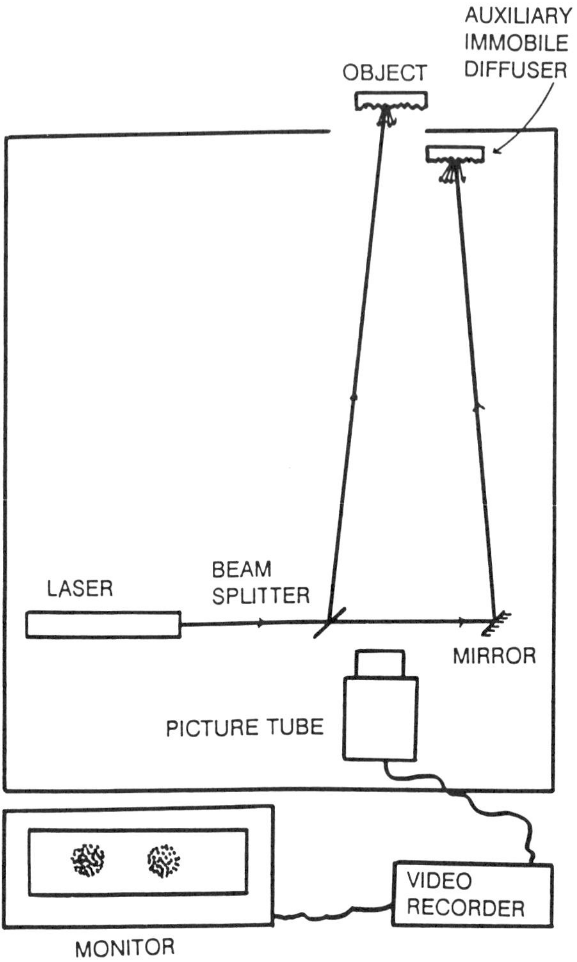

Figure 5.4 Arrangement showing the TV transmission of object and reference laser speckle patterns from a remote location.

subjected to the desired mechanical or thermal loads to study the response. Thus, different frames will contain moving and/or changing object speckles. The monitor screen is photographed and processed. Two film frames can be superimposed, and pointwise analysis by the Young's fringes method can be used to study the deformation of the object's surface. The role of the auxiliary (immobile) diffuser is to properly superimpose the photographs from different frames. When two reference fields exactly superimpose each other, the Young's fringes disappear in the reference region. The Young's fringes in the object region will then yield the true object surface movement between the frames.

Although developed for petroleum wells, the technique is universal and can be used in other physical situations. The arrangement of Fig. 5.4 is different from that due to Da Costa [11] because the useful light does not pass twice through the beamsplitter, which gives better light economy. Other arrangements such as those employing optical fibers (see Section 5.3.2) are also possible.

5.3.2 Applications of Fiber Optics

Besides simplifying the experimental system, optical fibers can be used for remote or hostile situations. The role of optical fibers and bundles in speckle metrology is twofold: to illuminate the test surface and to relay the speckle field to a desired location. Gilbert et al. [12] demonstrated the role of flexible multimode fiber optics in white light speckle metrology. As shown in Fig. 5.5, a multimode fiber bundle illuminates the test surface. The test surface has black paint splattered on a white background.

A microscope objective is used to image the test surface onto the entrance of a coherent multimode fiber bundle. Here, *coherent* means that

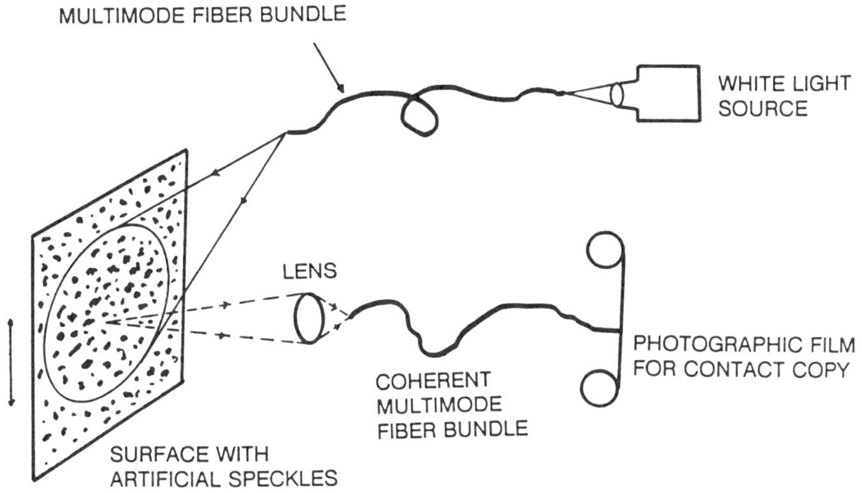

Figure 5.5 Schematic diagram of the experimental setup of white light speckle photography using fiber bundles. (Modified from Ref. 12, with permission.)

individual fibers in the bundle are in a close-packed array so that a pattern at one end is relayed as its mirror image at the other end. The transmitted white light speckle pattern at the exit of the coherent fiber bundle can then be contact-copied onto a suitable photographic film. Double-exposure speckle patterns can be recorded with a displaced object (or speckle patterns), and the transparency can be analyzed using the Young's fringes approach [12].

Rather than optical processing, electronic processing via digitization is also possible [12, 13]. The photographic film of Fig. 5.5 can be replaced by a video camera, a digitizer, and a computer system. The digital fields corresponding to shifted speckle patterns can be correlated. Practical demonstrations of the optical as well as electronic processes are available [12, 13]. Thus, it is established that subject illumination as well as transmission of white light speckle fields can be performed via flexible optical fiber bundles. The conventional photographic approach to Young's fringes and numerical correlation of the digital images are also available. Matthys et al. [14] compared the results with those obtained by finite-element analysis as well as moiré techniques to establish the fiber-assisted white light speckle technique.

The role of optical fibers in laser speckle metrology has been considered in detail by Dudderar et al. [15]. The applications are pointwise in nature because then the speckles can generally be larger than the individual fibers in the bundle. For illuminating the object, a single monomode optical fiber is very useful.

Three types of arrangements described by Dudderar et al. [15] are shown in Fig. 5.6. In Fig. 5.6a, the objective speckles are created in free space by back-illuminating a frosted glass. At a distance, a lensless input end of the coherent fiber bundle picks up the pattern, which is transmitted directly to a vidicon camera. The camera output is connected to a digitizer and processor for numerical correlation. Thus, in-plane surface movements can be studied. Figure 5.6b corresponds to the case when the illuminating light is transmitted via a single monomode optical fiber. Finally, as shown in Fig. 5.6c, a photographic plate can be placed directly in contact with the output of the coherent fiber bundle. Double-exposure speckle photography can thus be performed, and the Young's fringes approach can be used for the analysis.

The experiments prove the usefulness of fiber optics for remote speckle metrology. The sensitivity of this remote-objective speckle metrology has also been considered in detail [16]. Hayashi and Kitagawa [17] used single-mode optical fiber for illumination in speckle metrology of rotation-angle measurement of a cylinder.

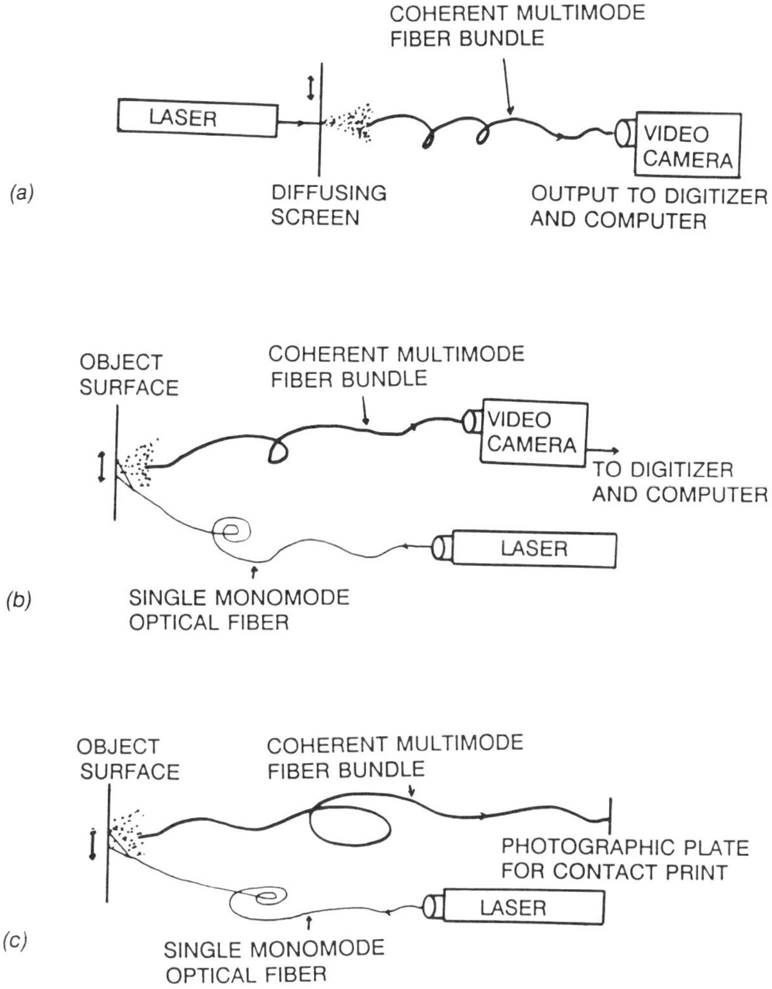

Figure 5.6 Some experimental arrangements for remote laser speckle metrology using fiber optics. (a) Objective laser speckle intensity distribution transmitted to a video camera using a coherent multimode fiber bundle. (b) Single monomode optical fiber for the subject illumination and coherent multimode fiber bundle for transmitting the speckle intensity distribution to the video camera. (c) Same as (b) but with the video camera and electronics replaced by a photographic plate. (Modified from Ref. 15, with permission.)

5.3.3 Large Structures

In engineering, there is always a need to study minute object deformations of real-life structures located remotely. Early efforts to use laser speckle photography for such structures are described in detail by Gregory [18]. Basically, these efforts employed the usual concepts of laser speckle photography adapted to nonlaboratory objects and situations. Divergent rather than collimated beams are generally used to practically cover large objects without the use of heavy and large optical components. These practical situations as well as details of analytical aspects are considered by Gregory [18]. One approach is to use multiple emulsions arrayed along the normal to the camera axis to measure different types of motion such as surface tilt, rotation, and in-plane displacement. Several practical applications are possible, including the study of antenna dish distortions, of aircraft wing integrity, and of distortions in solar array boom joints and the inspection of pressure vessels [18]. Thus, with suitable power lasers and appropriate system mountings and analytical techniques, laser speckle photography is a powerful tool for use with practical engineering structures.

Another interesting development in this connection is the role of white light speckle photography (see Chapter 7). For very large objects such as bridges, buildings, and glaciers, it is very difficult if not impossible to provide laser light illumination. Safety aspects are also of serious concern in those outdoor situations. On the other hand, even natural sunlight and already existing patterns on these objects provide an excellent situation for white light speckle photography. Cloud and Conley [19–24] provided several practical applications of time-lapse, double-exposure white light speckle photography of geological surfaces, large buildings, etc. There are endless possible applications including the study of tectonic plate motion, nuclear waste repositories [25], and the flow of glaciers, sea ice, and earthen surfaces. Sensitivity and limitations due to atmospheric turbidity are discussed by Conley and Cloud [22–24]. Results of glacier field experiments [19–23] at Nisqually Glacier in the state of Washington are very encouraging. In fact, the technique has a potential use in the study of several geophysical mechanisms from space [21]. The study of sea waves from an aircraft is reported by He and He [26].

5.4 TRANSPARENT OBJECTS

Several common applications of speckle metrology dealing with transparent objects are detailed in this volume. One example is particle image velocimetry, where seed particles or bubbles carry information about fluid motion. Seeding or embedding is also well known in white light speckle photography of transparent solid objects. In this section we discuss certain unique situations dealing with transparent objects when the deflections are viewed through rough screens or surfaces. Owing to refractive index changes or variations in the transparent media, the speckles shift in the recording plane. The desired parameters such as temperature or density can then be studied by speckle photography or interferometry. Köpf [27] and Debrus et al. [28] describe the application of speckle photography to the measurement of light deflection by refractive index changes. The following subsections give more specific applications. These are mainly speckle photography, with one example of speckle interferometry.

5.4.1 White Light Speckle Schlieren Technique

In the conventional schlieren approach, the cross-sectional area to be examined is dictated by the size of the optics. The white light speckle schlieren technique introduced by Giglio et al. [29] is easily applicable to areas of up to a few square meters without using large optical components. A good quality large-format camera, a flash lamp, and a flat retroreflective screen are enough for the information storage. Figure 5.7 depicts the recording arrangement. *L* is the thickness of the transparent medium, *a* is the distance

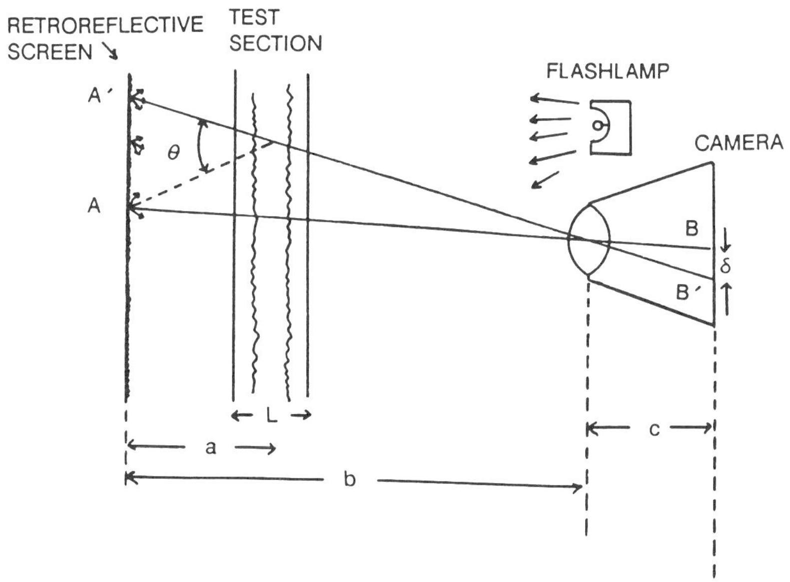

Figure 5.7 Speckle displacement δ caused by the refractive index gradient. (Modified from Ref. 29, with permission.)

of its central position from the screen, b is the screen–camera lens distance, and c is the lens–film distance. Without any optical perturbation, a point at A will be imaged at point B on the film. Now suppose a vertical refractive index gradient $dn/d\mathbf{r}$ is applied. $\mathbf{r}$ is in the direction of the normal to the line of sight. The apparent position of point A will shift to A′, and the corresponding image position to B′. The angle θ in Fig. 5.7 will then be $L\,dn/d\mathbf{r}$. The image or white light speckle displacement on the film plane is

$$\delta = \frac{acL}{b}\frac{dn}{d\mathbf{r}} \tag{5.2}$$

A double-exposure photograph can be recorded on the same film. The first exposure is without and the other with the optical gradients. Observation of the film with proper frequency plane filtering can yield the usual semi-quantitative information of the classical schlieren technique [29]. However, a large cross-sectional test area is possible now without the need for large curved mirrors.

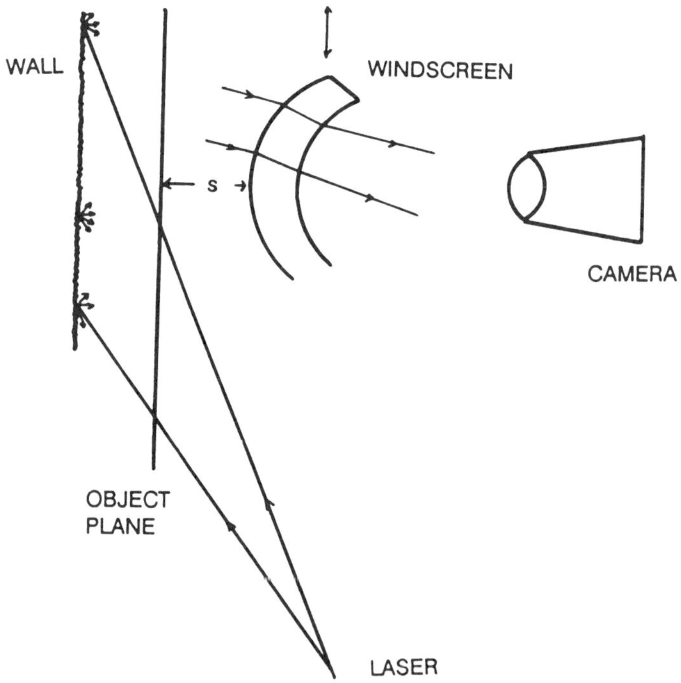

Figure 5.8 Laser speckle photography for the study of aircraft windscreen binocular disparity.

5.4.2 Binocular Disparity of Aircraft Windscreen

A pilot looking through a windscreen may face a variation in the angular deviation for the two eyes. This double vision or binocular disparity can be studied by laser speckle photography as described by Harris et al. [30] and shown in Fig. 5.8. Moving the windscreen to simulate the binocular effect, a double-exposure laser speckle photograph is taken. The object plane is behind a wall illuminated by a divergent laser beam. When point-wise analysis is performed by the Young's fringes approach, the apparent displacement of the speckle in the object plane is

$$d = \lambda Z / mL \tag{5.3}$$

where L is the fringe spacing, m is the photographic magnification, Z is the distance between the observation screen and the negative, and λ is the wavelength of the readout laser beam. The desired change ϕ in the direction of the line of sight between the two positions is [30]

$$\phi = \arctan\,(d/s) \tag{5.4}$$

5.4.3 Refractive Field Studies

The resulting in-plane speckle movement in the object plane as shown in Sections 5.4.1 and 5.4.2 can be due to the refractive index change as well. The refractive index itself can change because of temperature or density changes. Consequently, there have been applications to study these quantities by laser speckle photography. Ramprasad et al. [31] showed that refractive indices of liquid mixtures as small as 0.0005 can be measured by double-exposure laser speckle photography. A speckle field created by a laser beam incident on a ground glass is passed through the test cell. Two exposures, one before and the other after the change in the fluid mixture, are recorded in the same frame. Analysis by the Young's fringes method yields the refractive index change. The approach has also been used to test incandescent lamps and was found to be a useful tool for distinguishing between good and defective lamps [31]. Sivasubramanian et al. [32] demonstrated the use of the technique to measure temperature gradients in liquids. As shown in Fig. 5.9, the objective speckle patterns are created by the laser beam (a 15-mW He-Ne laser) incident on a ground glass. Upon passing through the test cell, the speckle patterns can be recorded on a photographic plate. A double-exposure photograph is taken, one exposure of the unheated test section and the other made on the same plate after heating from the bottom to achieve the temperature gradient. Since no lens is used, the transparency can be analyzed using Eq. (5.3) but with $m = 1$. The temperature gradient can then be determined from the speckle

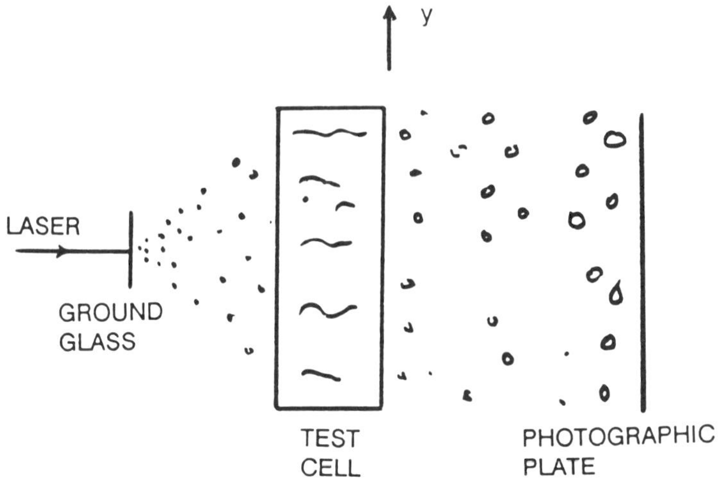

Figure 5.9 Schematics of the procedure for the temperature gradient measurement using double-exposure objective laser speckle photography.

shift obtained from the Young's fringes and the known refractive properties and geometrical parameters. From the one-dimensional temperature gradient experiment, it is established that the speckle technique is simple and reliable. Also, it can measure higher temperature gradients than holographic interferometry [32].

Farrell and Hofeldt [33] concentrated on quantitative aspects of the technique to examine the sensitivity, accuracy, and repeatability for gas-phase temperature measurement. The object was a cylindrical propane flame. The arrangement was in principle similar to that shown in Fig. 5.8, but the object was now a gas burner. The burner was on during the first exposure and off during the second exposure. Pointwise analysis was used. Equation (5.3) can be used to find the speckle displacement at the object plane of the camera. The refractive index in the axisymmetric field is then determined from the speckle shift. The radial temperature variation is then obtained by applying the Lorenz–Lorentz equation. A thermocouple was used to compare the results with those obtained by the speckle approach. It was established that for an axisymmetric field the speckle has several advantages. Besides the relaxed stability requirement and the simplicity of the optical system, there is no need for numerical differentiation to calculate the inverse Abel transform.

Thus, double-exposure speckle photography is established as a powerful tool for the study of refractive index changes and related parameters. For example, Fomin et al. [34] found an application in a turbulent density field. The status of the deflection symmetry can be easily found out by the speckle method [34]. Bernardo and Soares [35] used the approach for birefringence measurements. Ordinary and extraordinary images yield shifted speckle patterns. The shift, measured by the frequency of the Young's fringes, can then be related to the birefringence. The study assumes that the crystal is uniaxial and that the *c* axis is oblique to the surface of incidence.

A rather interesting variant of the basic approach deals with the analysis of ordinary as well as gradient index (GRIN) rod lens systems. As described by Komatsu et al. [36], the speckle pattern of a coherently illuminated diffuser is formed in the image plane of the system. Against the lateral translation of the diffuser, the speckle decorrelation can be related to the system MTF. Thus, a double-exposure laser speckle photograph is taken, one exposure before and the other after the translation. The contrast of the Young's fringes obtained from such a photograph yields the system MTF. The desired spatial frequency response can be obtained by different lateral shifts of the diffuser.

5.4.4 Large Deflections

The optical arrangements described in Section 5.4.3 are limited to small deflections. In refractive index fields of large diameters or wide extensions, the deflections may be considerable. The ultimate result is decorrelation of two speckle fields in the image plane leading to the disappearance (zero contrast) of the Young's fringes. To overcome this difficulty, Wernekinck and Merzkirch [37] introduced a variable sensitivity procedure, practically without limits.

The optical system is shown in Fig. 5.10. The collimated laser light passes through the test field, and a lens focuses a plane of the field onto a ground-glass plate. A second lens images a plane at a distance l from the ground glass. The usual two exposures, one with and the other without the test field, are taken. Usual and deflected rays will arrive at the ground glass at the same point. However, for the unit magnification case, they will form the same angle ϵ between them. In the plane at a distance l from the ground glass, the rays (and the speckles) are separated by $\Delta = \epsilon l$. Thus, the speckle displacement can be changed by changing l. The arrangement was applied to a heated plate, and the results were successfully compared [37] with those obtained from a Mach–Zehnder interferometer. Applications in a mildly heated, turbulent airstream in a low-speed wind tunnel [38] and to the density field of a flame [39] have proved valuable.

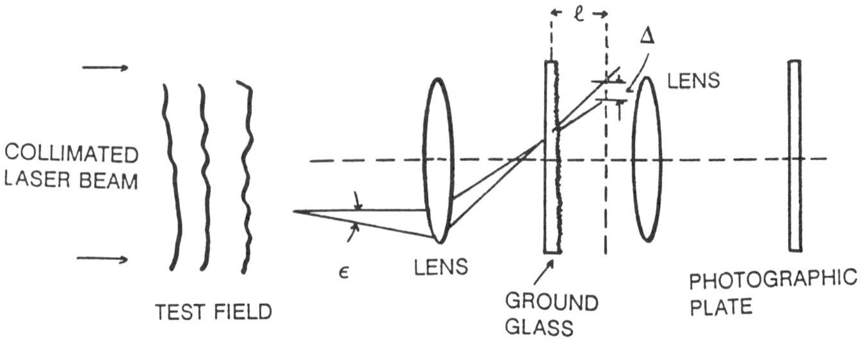

Figure 5.10 Arrangement for light deflection measurement of spatially extended refractive index fields. (From Ref. 37, with permission.)

5.4.5 Speckle Tomography

For asymmetric refractive index fields, information is required from different viewing directions. Three-dimensional information can then be reconstructed from various projections. It has been found that speckle photography is an adequate tool for such tomographic reconstructions [40, 41].

There are two techniques for the optical implementation of speckle tomography. For asymmetric but steady laminar flows, the object can be mounted on a turntable [40]. The table can be rotated for storing the speckle photographs in different directions (projections). Liu et al. [40] considered a nonaxisymmetric helium jet for the determination of helium concentration.

Simultaneous measurements along different directions are necessary in a dynamic or unsteady process. As in the arrangement described by Blinkov et al. [41], beamsplitters can be used to divide a laser beam into many beams. Multidirectional illumination of the test field and subsequent speckle photography can then be performed. The procedure [41] was applied to reconstruct the two-dimensional temperature distribution in a candle flame.

5.4.6 Speckle Interferometry

Our entire discussion on transparent objects so far has involved speckle photography. In common terminology, "speckle photography" means double-exposure photography of the object illuminated by a single beam. The analysis is performed pointwise by the Young's fringes approach or whole-field by the spatial filtering technique. Common speckle interferometry uses two beams during the recording process. The changes in the correlation

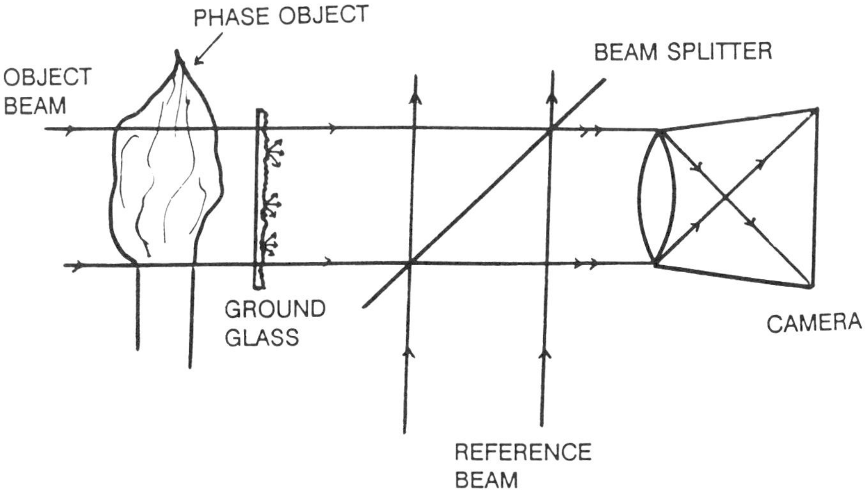

Figure 5.11 Optical system for density measurements using speckle interferometry. (Modified from Ref. 42, with permission.)

between two speckle patterns is used to form the fringes for the analysis. The reference beam or the second beam of speckle interferometry can be introduced using a beamsplitter. The arrangement for transparent media described by Verhoeven and Farrell [42] is shown in Fig. 5.11. The first exposure is taken of the image of the diffuser along with the reference beam. The test field is uniform. The second exposure involves the activated test section. The photograph yields speckle correlation fringes. Temperature variation in a bunsen burner flame (cylindrical) was measured, and the results were compared with those obtained by laser speckle photography [42].

5.5 SMALL MOTIONS

Conventional lower measurement ranges are limited by typical wavelengths of the light used or the fringe spacings. For measuring movements that are very small compared to these limits, several techniques can be found in the literature. One successful approach for which even commercial systems are available is based on laser–Doppler studies or heterodyne detection [43]. For rough objects, the speckles are formed and their parameters are important for system performance. Otherwise, the Doppler technique is a subject beyond the scope of the current discussion. In this section, I emphasize attempts to enhance the sensitivity of speckle photography and interferometry.

5.5.1 Heterodyne Speckle Photogrammetry

It is notable that the spacing in the Young's fringe pattern can be determined to within 1% accuracy [44, 45]. However, the smallest measurable object displacement is also set by the need for at least one fringe in the halo. This and other practical limitations set the lower measurable speckle displacement as 0.2 μm [46]. In applications such as strain measurement, nanometer domain displacement sensitivity is required. In conventional speckle photography, this means the requirement of a low *f* number lens, but this has many associated problems such as aberrations and very limited depth of focus [46, 47]. Incidentally, there has been good progress in solving the aberration problem using holographic optical elements (see Chapter 10 for details). However, these lenses are not common, and the depth-of-focus problem still remains with large-aperture optics.

The concept of heterodyne readout of the halo fringe patterns introduced by Smith and Stetson [46] is a very important development in this connection. The procedure involves two separate single-exposure speckle photographs taken before and after the deformation. For the data analysis, the corresponding diffraction halos are combined but with a frequency shift between the beams. Halo fringes are then scanned continuously. A photodetector in the halo or the observation plane will yield sinusoidal signals whose phase is proportional to the fringe order. With sets of detector pairs and their translations, the complete surface strain tensor can be determined [46]. Stetson [48] provides a review of the status of this technology.

5.5.2 Addition of a Reference Motion

This approach [49, 50] is based on the known addition of the motion to be determined by the speckle method. The mismatch [50] is said to be linear if the two motions are parallel or antiparallel. Otherwise, the mismatch is called rotational because the combined effect will rotate the fringes. If the known additional displacement is large enough to be measured conventionally, then the unknown motion can be determined from the experimental combined data.

For the orthogonal addition of lateral motion, rotational mismatch yields a very sensitive tool by measuring the orientation of the pattern of the combined Young's fringes [49]. As shown in Fig. 5.12, suppose σ is the unknown lateral displacement and Σ is the known lateral but orthogonal displacement. The combined displacement D is

$$D = (\Sigma^2 + \sigma^2)^{1/2} \tag{5.5}$$

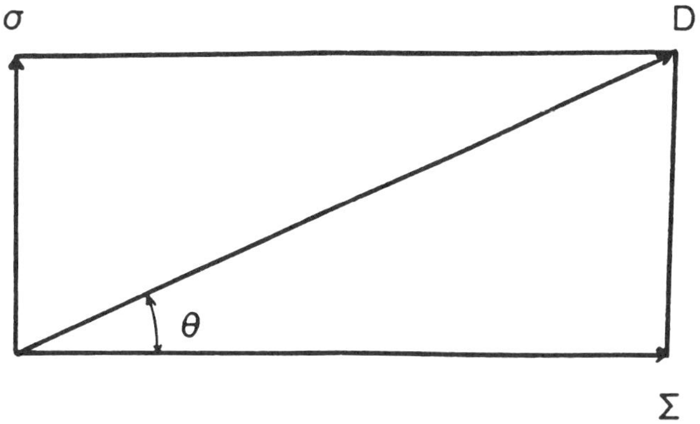

Figure 5.12 Schematic diagram showing the resultant D of the combined orthogonal lateral displacements σ and Σ.

The rotation angle θ (from the state $\sigma = 0$) of the Young's fringe pattern is

$$\theta = \arctan(\sigma/\Sigma) \tag{5.6}$$

Minimum measurable θ sets the lower limit of σ/Σ. Measurable σ/Σ of 0.1 demands measurable θ as large as 5.8°. θ measurement capability of 1° yields σ/Σ measurement capability of 0.0175. Obviously, the approach can easily improve the conventional sensitivity by an order of magnitude; this has been experimentally demonstrated [49]. It is interesting to note that a fringe orientation measurement capability better than 0.2° is available [51].

It is not necessary to know a priori the direction of the unknown motion. If both the directions incidentally become parallel or antiparallel, the technique becomes a linear mismatch rather than rotational. The approach is also valid for whole-field filtering. If sensitivity is not a problem, the mismatch procedure can be applied for to determine the sign of the unknown motion. These aspects and further experimental verifications are provided by Chiang et al. [50].

5.5.3 Speckle Photography with Electron Microscopes

The sensitivity of conventional speckle photography depends upon the size of the smallest speckle that can be resolved by the system. The smallest speckle diameter in the object space is $1.2\lambda F(1 + m)/m$, where m is the photographic magnification, λ is the wavelength of the illumination, and F is the f number of the imaging lens. The minimum value, at high mag-

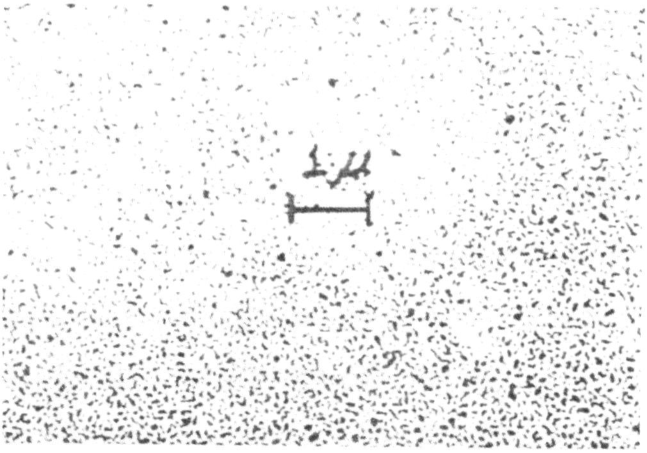

(a)

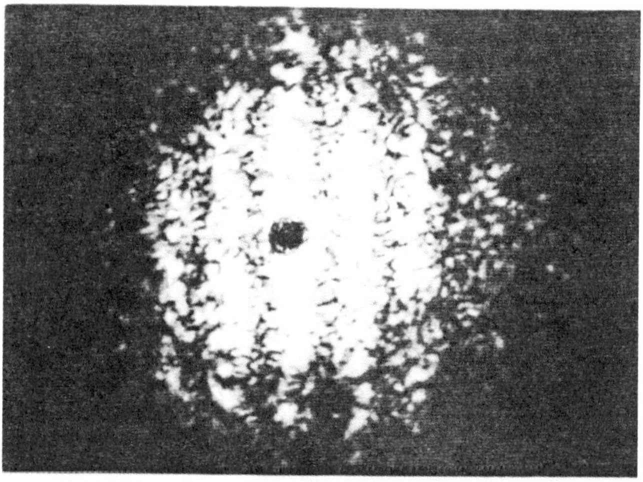

(b)

Figure 5.13 White light speckle photography using an electron microscope. (a) Speckle pattern of gold particles recorded by scanning electron microscope. (b) Corresponding Young's fringes. (From Ref. 52, with permission.)

nifications, approaches $1.2\lambda F$. Thus, the practical sensitivity in the visible region remains in the micrometer domain. Significantly improved sensitivity then requires a very low wavelength.

Laser speckle photography in the very low wavelength region depends on the availability of a suitable source such as an X-ray laser. However, very small incoherent speckles can be generated. These speckles can be viewed through an electron microscope. Subsequent double-exposure white light speckle photography with the electron microscope is then possible. The first results of this pioneering application are described by Chiang and Li [52]. The application used clusters of "islands" that were formed during the thin-film deposition process. Although these patterns were not detectable by visible light, an electron microscope readily resolved them. Double-exposure speckle photography using scanning electron microscopy (SEM) before and after the deformation is possible. Photographs of two different states can also be superimposed later. Figure 5.13a is a speckle pattern of gold particles, and Fig. 5.13b shows the Young's fringes from two such patterns superimposed.

Besides giving higher resolution, SEM has very high magnification (on the order of 10^5) and a depth of field 300 times that of an optical microscope [52]. Thus, the problems associated with an object surface deformed out of focus under loadings are practically eliminated by the SEM technique.

Huntley et al. [53] successfully applied the SEM technique to measure the displacement around crack tips in aluminum. Apart from this application, the SEM speckle technique can be applied to the study of strains in individual grains or grain boundary sliding in metallic materials, microprocessor chips, and other miniature elements [53].

Transmission electron microscopy (TEM) can further enhance the sensitivity [53]. However, TEM is applicable to extremely thin specimens.

5.6 REAL-TIME APPLICATIONS

Use of the photographic emulsion is very common in speckle metrology. The procedure is slow because of the need for wet chemical processing. Therefore, it is always important to seek fast alternatives. Hariharan [54] reported successful speckle interferometry using an "instant" Polaroid film. This enabled quick viewing (in 30 sec) of the interference fringes. Ueha et al. [55] demonstrated the use of thermoplastic photographic materials. They successfully performed time-average laser speckle photography of sinusoidal vibration. Tiziani [56] used thermoplastic material for comparing the results obtained with BSO (bismuth silicon oxide) crystals (see Section 5.6.2 for speckle metrology with BSO crystals). Again, there is no wet processing, and the fringes can be viewed within seconds after the exposure.

The above procedures are logical in the quest to reduce cycle time and also to avoid the wet procedure. However, modern electro-optical crystals and systems yield nearly instantaneous outputs. Of these, the well-known electronic speckle pattern interferometry (ESPI) is covered in Chapter 4 of this book. Some other novel procedures are reported in this section.

5.6.1 Liquid Crystal Light Valves and Television

Takemori et al. [57] demonstrated the use of liquid crystal light valves (LCLVs) for speckle photography of sinusoidal vibration. The time integration property of such a valve helps time averaging for vibration analysis. Real-time incoherent-to-coherent conversion yields instantaneous fringe visualization. In the experiment [57], an argon laser (write beam) light passed through a laterally vibrating ground glass. The resulting vibrating objective laser speckles were incident on the LCLV plane to form the time-average pattern. Another beam from a He-Ne laser (read beam) illuminated the LCLV plane, and a lens was then used to observe the Young's fringes in the frequency plane. The fringe positions yield the vibration amplitude.

The use of liquid crystal television (LCTV) [58] in this connection was an important development. The basic arrangement is shown in Fig. 5.14. The pattern can be displayed directly from a microcomputer or a video

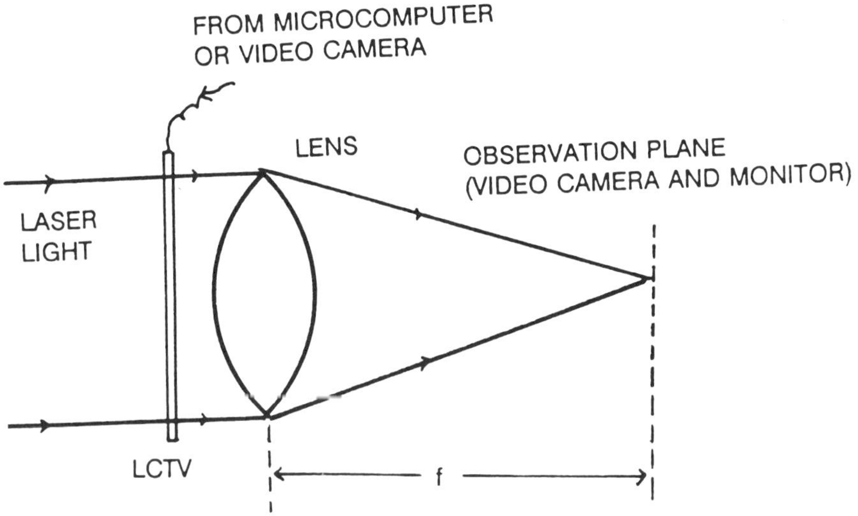

Figure 5.14 Basic diagram showing optical Fourier transform of the pattern displayed on liquid crystal television (LCTV).

camera. A laser beam and a lens can obtain the Fourier transform of the pattern instantly. In the application of Bates and Miller [58], a digitized double-exposure speckle pattern from a microcomputer was displayed on one LCTV. The optical Fourier transform was then obtained, which contained the Young's fringes. At this stage, the fringe frequency was high enough for displacement analysis. However, with the help of a vidicon, they displayed the Young's fringe pattern on another LCTV for a second optical Fourier transform. The resulting separation of the correlation spots can directly yield the original speckle displacement. Bates and Miller [59] described several practical aspects of speckle metrology using LCLVs.

The important aspect of LCTV and other modern spatial light modulator (SLM) technology is the fast and convenient data feeding as well as processing. As in other aspects of optical information processing, these devices are very useful in speckle metrology.

5.6.2 Photorefractive Crystals

Speckle metrology with crystals of bismuth silicon oxide ($Bi_{12}SiO_{20}$; BSO) [56, 60, 61] or bismuth germanium oxide ($Bi_{12}GeO_{20}$; BGO) [62, 63] has been successfully performed. The application with BSO crystals involved displacement and tilt analysis [60] and vibration analysis [61]. The general experimental arrangement is shown in Fig. 5.15. The crystal used had absorption coefficients of 2 cm^{-1} and 0.28 cm^{-1} for wavelength's of 514.5 nm and 632.8 nm, respectively. Thus, the argon laser was used to create the photorefraction effect of the speckle pattern on the crystal. A low-power He-Ne laser simultaneously illuminated the pattern (double-exposure, time-average, etc.) to obtain the Young's fringes for quasi-real-time analysis (cycle time ~1 sec). In the arrangement of Fig. 5.15, the polarizer and analyzer were used to avoid stray light. For in-plane displacement and vibration, the image of the object is formed in the crystal. For tilt analysis and torsional oscillations, the crystal is placed in the Fourier transform plane of the imaging lens. An aperture is used to reduce stray light, and a filter to transmit only the reading light. Successful comparisons of results with those obtained with thermoplastic materials [26] and laser–Doppler vibrometry [62] have been performed. High photographic sensitivity and short times for recording and erasure make BSO a very suitable material. Nakagawa and Minemoto [62] describe more recent applications and the readout properties of the BSO crystal.

The experimentation with BGO crystals [63, 64] involved double-exposure speckle photography for velocimetry applications. A *Q*-switched (double-pulsed) frequency-doubled Nd:YAG laser was used for the recording and a He-Ne laser for the readout. In-plane motion of a bead-blasted acrylic disk was studied.

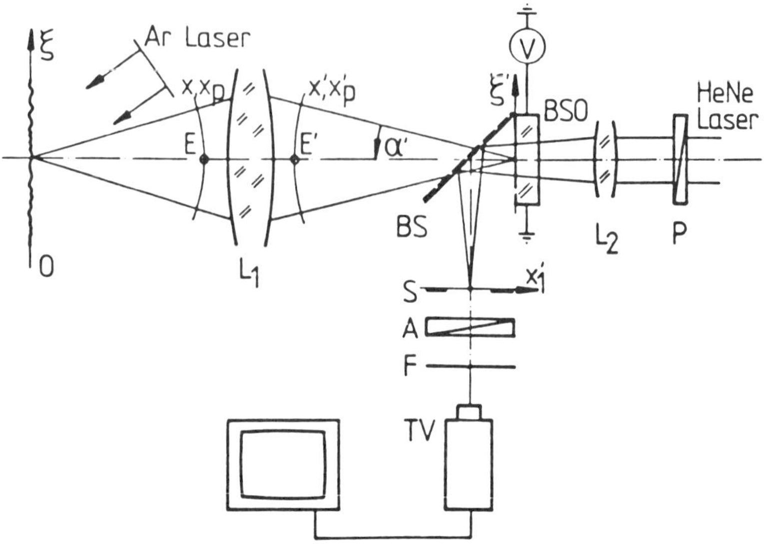

Figure 5.15 In-plane motion analysis in quasi-real time using an electro-optical BSO crystal. (From Ref. 61, with permission.)

These photorefractive materials, combined with modern processing methods, can permit real-time measurements.

5.6.3 Direct Observation

A laser-illuminated object surface produces speckles in the image plane of the camera lens. Typically, the speckles are too small (see Section 5.5.3) to be viewed by the naked eye. However, a cascade of microscope objective lenses can be used to magnify the speckles [65] and project their image onto a paper or ground glass. The final image on a vidicon–monitor system is very helpful. It further magnifies (monitor/vidicon size ratio) the speckles and avoids laser light to the eyes. Thus, the speckles can be made large enough for viewing in real time. Data from the camera can be used directly for further processing. However, the method is convenient for pointwise analysis; there are light intensity and field-of-view problems at large magnifications. Figure 5.16 shows a pair of laterally displaced speckle patterns as seen on a TV monitor [65]. These speckles were produced by a 15-mW He-Ne laser (λ = 632.8 nm) illuminated on an abraded metallic piece attached to an *XYZ* translation stage. Three microscope objectives and a vidicon–monitor system produced a net 10,000× magnification. The result is a 1-cm shift on the monitor screen for 1-μm object surface displacement.

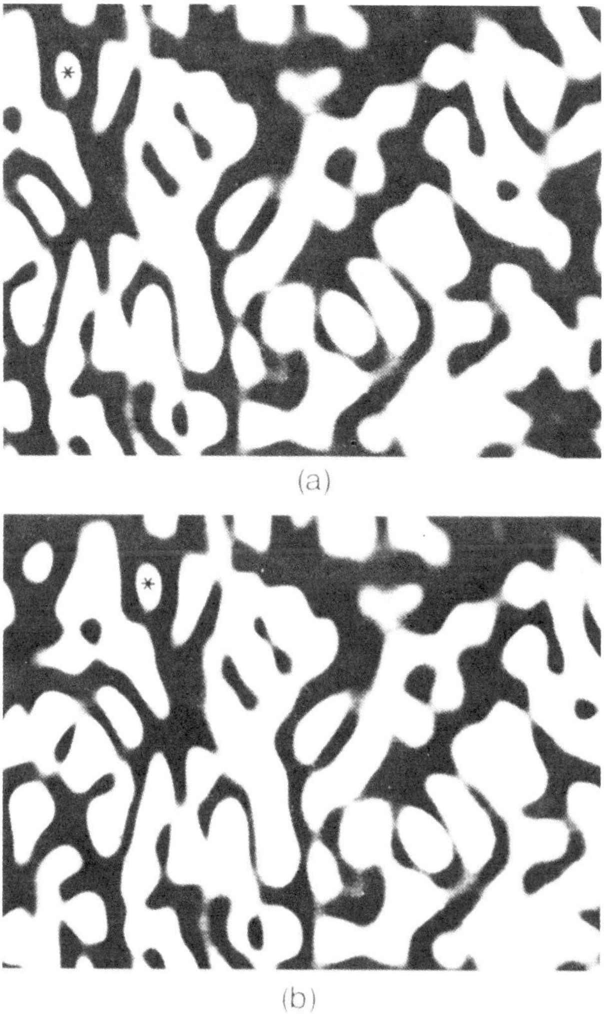

Figure 5.16 Speckle patterns with 10,000× magnification at the screen of the TV monitor. Three microscope objectives were used. A typical speckle is marked with an asterisk. (a) The original pattern; (b) pattern after 1-μm lateral shift in the object plane. (From Ref. 65.)

Obviously, submicrometer resolution can easily be achieved. Available digital speckle correlation techniques [12, 13, 66, 67] can further enhance the power of direct observation approaches. With these techniques, cross-correlation between two speckle patterns can directly give the strain information.

5.6.4 Real-Time Holospeckle Approach

This is a novel extension of the concepts of real-time holographic interferometry. In real-time holographic interferometry, a hologram of the object is taken and the image is reconstructed. If the actual object is also illuminated by the original object beam and viewed along with the reconstructed image, live or real-time interference fringes are formed. These fringes yield the displacement between the two positions—the actual object and the reconstructed image.

If the polarization states of the two fields are mutually orthogonal, real-time holographic fringes will disappear because of the lack of interference. However,the speckle fields are still correlated, making speckle photography and interferometry possible. Chiang and Li [68] applied this principle to speckle photography with the help of a $\lambda/2$ plate in the reference beam and a polarizer in front of the object surface. They obtained two mutually orthogonally polarized but correlated speckle fields. Superimposing speckle fields of a plate deformation resulted in the expected whole-field as well as pointwise Young's fringes.

The approach was successfully applied to two-beam speckle interferometry [69]. In fact, the concept that two speckle fields may remain correlated even in the absence of holographic interference fringes is general. An example is the defocused and delocalized holospeckle techniques reported by Li and Chiang [69]. Darskij and Markov [70] measured very small displacements (compared to the speckle size) by the use of holography with a reference speckle wave. The method is basically speckle field intensity autocorrelation measurement.

5.7 MISCELLANEOUS APPLICATIONS AND TECHNIQUES

There are several other unique methods and applications in speckle metrology. We consider some of them here.

5.7.1 Combining Objective Speckles and Reflection Holography

When extraneous object motion is very large, the photographic plate can be rigidly attached to the object. Interferometry using reflection holography, as described by Neumann and Penn [71], is therefore insensitive to rigid-body translations of the object.

Interferometry with normally illuminated reflection holography is sensitive to longitudinal motions. However, the backscattered light forming the object wave is basically an objective speckle pattern on the recording plane. Thus, in-plane object displacements can be obtained from the speckle pattern. The photographic emulsion can be bleached to obtain speckle fringes in the usual transmission mode. Cao and Chen [72] introduced a reflection display mode to obtain the speckle fringes without bleaching the emulsion. Figure 5.17 shows the schematic diagram of the technique. Figure 5.17a corresponds to the usual recording procedure of reflection holograms where objective laser speckles are also stored. For speckle photography data analysis, Figs. 5.17b and 5.17c correspond to pointwise and whole-field filtering procedures, respectively, in the reflection mode. The approach has been experimentally verified with a couple of object and loading conditions [72].

5.7.2 White Light Source for Coherent Speckles

The well-known technique of white light speckle photography is described in detail in Chapter 7. In those applications, a natural texture on the surface is used or a random pattern is artificially created by painting or other means.

However, Chiang [73] reported an interesting use of a thermal white light source to generate coherent speckles. Most thermal sources have sufficient coherence to create speckles at small distances from the object surface. Thus, coherent speckle photography is possible when a laser source cannot be used because of safety or other requirements.

5.7.3 Mechanical Characteristics of Paper and Wood

Castagnede [74] reported the development of speckle photography applicable to paper. There are serious problems with ordinary contact (including acoustical) techniques owing to their poor accuracy and reliability. Speckle photography can provide a solution [74]. However, due to the translucent character and fibrous structure of paper, speckles exhibit substantial decorrelation. To limit the problem, coatings and paints are very useful [74].

Benckert [75] described the usefulness of white light speckle photography for studying deformation and crack development in wood during drying. The surface was painted with glass beads 50 μm in diameter for retroreflection. Series of 25 single exposures with a time interval of about 24 hr were taken during the drying process. Two negatives were then combined for the analysis. It has been established that the speckle tech-

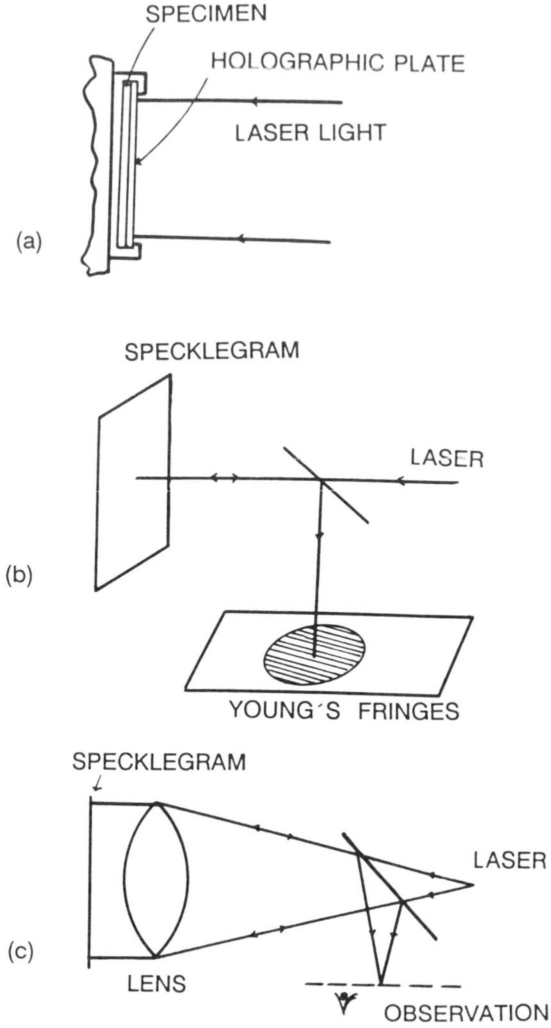

Figure 5.17 Combining reflection holography and objective laser speckle photography. (a) Recording; (b) pointwise analysis in the reflection mode; (c) whole-field analysis in the reflection mode. (From Ref. 72, with permission.)

nique provides a good amount of information scarcely obtainable by classical means such as the use of strain gauges.

5.7.4 Periphery Camera for Curved Surfaces

Many engineering structures are curved. Cylindrical surfaces are common. Interpretation of the fringes obtained from them is difficult because of perspective effects. Chiang et al. [76] used a periphery camera for moiré and speckle techniques. For laser speckle photography, the general experimental arrangement can be represented by Fig. 5.18. A slit near the film plane limits the width of the object or the film used at a given step. Rotation of the camera and translation of the film results in the entire cylindrical surface being photographed as a plane. Ultimately, the analysis can be performed using the simple procedures applicable to flat surfaces.

Experiments [76] with white light and laser speckle methods provided acceptable pointwise results through Young's fringes. Whole-field fringes were of poor quality due to decorrelation. However, the camera proved to be a valuable tool for cylindrical objects.

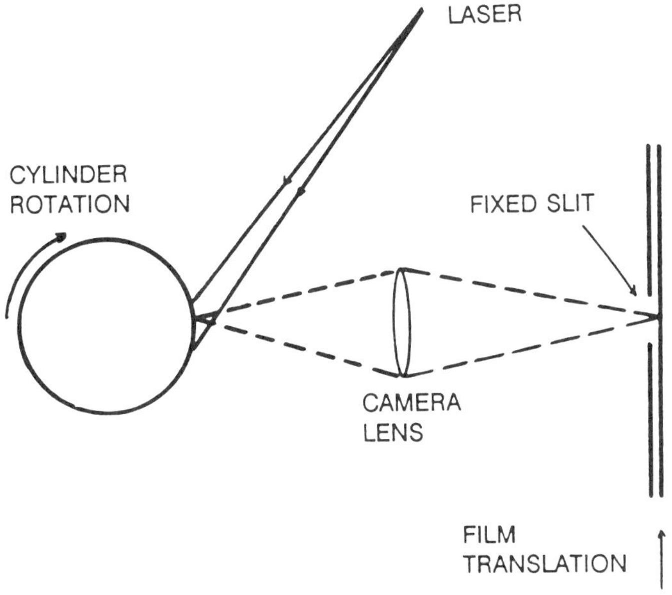

Figure 5.18 Diagram showing laser speckle photography using a periphery camera. (From Ref. 76, with permission.)

5.7.5 Laser Speckle Photography in Daylight

In field conditions or where speckle photography is performed along with other measurement techniques, it is difficult to obtain complete darkness. Georgieva [77] addressed this issue to determine the conditions for laser speckle photography in natural daylight. Several experiments were performed with ORWO MA 8 film and a 5-mW He-Ne laser. The camera was supplied with a red filter. The results show that it is possible to use laser speckle photography in the presence of significant background noise. Similar experiments can be performed in a particular situation and with a particular film type, etc. to determine the acceptable limits.

5.7.6 Partial Laser Beam Coverage in Pointwise Analysis

In the pointwise analysis of speckle photographs, the readout laser beam is small enough to define a point but large enough to cover at least two identical speckles. However, the object itself may be small. An example is a pin specimen. In those situations, an ordinary unexpanded laser beam (typical diameter 1.2 mm) may contain only a fraction of the tiny specklegram. Moslehy et al. [78] addressed this issue by performing a series of experiments in which a double-exposure laser speckle specklegram was recorded as usual but pointwise filtering was performed by changing the area of the cross section of the laser beam covering the specklegram. With as little as one-tenth of the beam diameter covering the specklegram, Young's fringes were observable. Thus, for very small object (or image) sizes compared to the readout beam cross section, the usual unexpanded laser beam may be sufficient.

5.7.7 Detection of Loose Contacts

Waterworth and Reid [79] introduced a unique application of objective laser speckles. A focused laser beam illuminates the surface of the contact, and the reflected objective laser speckle pattern is collected by a photodiode. The active area of the detector matches with a typical speckle size. The system or assembly is excited with a constant frequency, and the resulting movement of the contact is random if the contact is loose. Thus, repetitive or random photodetector output indicates good or loose contact, respectively. The photodetector output can be directed to an oscilloscope to visualize this repetitive or random output. On the other hand, the output connected to an earphone will yield a "pure tone" or a "crackle" depending upon whether the contact is "good" or "loose," respectively. In this way, an operator can check the contact by using a pair of headphones. Waterworth and Reid [79] aimed at developing in situ applications in telephone exchanges. Their initial system could test up to 20 relay contacts in less than 1 min. However, the approach should be useful for other applications such as checking the tightness of rivets, nuts, or bolts [79].

5.7.8 Etch Depth in Metallic Specimens

Radha and Ramprasad [80] presented this novel application. The surface of a metallic specimen is illuminated with an objective laser speckle field at an angle. When material is removed, speckles appear to be shifted on the surface. Double-exposure photography of these speckles and resulting Young's fringes can then yield the thickness change.

5.7.9 Biomedical Applications

There are several applications of laser speckles in biomedical metrology. One well-established application is in optometry. When objective laser speckles on a surface are viewed, the apparent speckle motion in comparison with the head movement indicates far-sighted, near-sighted, or perfect eye. This simple property of the image formation by a lens (an eye in the present situation) resulted in considerable interest in speckle optometry. The early developments are summarized by Hennessy and Leibowitz [81]. The principle deals with the functional aspect of vision rather than just the refractive errors. Sinclair [82] used different wavelengths to address chromatic aspects also. Phillips et al. [83] addressed the question of the cylindrical axis of power. Mathematical derivations and literature reviews are also available [84, 85]. Bahuguna et al. [86] introduced a white light speckle optometer that is competitive with the laser speckle optometer. More recently, Bahuguna et al. [87] gave further insights into the perceived size of laser speckles.

Another interesting application related to the eye is retinal blood flow studies [88, 89]. A single-exposure laser speckle photograph of the retina will yield contrast variations in the speckles depending upon the velocity variations.

Temporal statistics of laser speckles have shown applications in skin blood flow studies. The study can be important for early diagnostics of some diseases and healing processes. In the method due to Fujii et al. [90], a He-Ne laser illuminated the skin tissue and the scattered objective laser speckles were observed. The fast speckle fluctuation was due to the blood flow in the capillary network. The intensity fluctuation was used to obtain the power spectral distribution (PSD). Different regions of the skin surface yielded different PSDs. The PSD can also change substantially as the flow varies due to say, inflated or deflated cuffs in the region [90]. In another approach [91], multichannel recording and digital processing were used to obtain a spatial map of the blood flow. Basically, image speckles were observed on a CCD array sensor, and scanning procedures covered a surface area of the skin. Pixel-by-pixel analysis of the data of a time-varying speckle field can ultimately yield a three-dimensional plot or a color graphic display of a microcirculation map. In one experiment, it was found [91] that the wound-healing process induces a high level of blood circulation around the scar. The injection of a drug also had a pronounced effect on blood circulation. Tanin et al. [92] considered scattering and polarization aspects and also studied the speckle fluctuations from the muscle-developed force under isotonic loads. Tuchin et al. [93] consider several theoretical aspects and relationships regarding the scattered field and the blood flow in vessels. They also describe modified interferometry and sensing approaches. The study of dynamic biospeckles can offer valuable information about the blood flow in the skin surface, internal organs, and ocular fundus [94].

Early detection of periodontal disease is possible by measuring increases in tooth mobility. Rydén et al. [95] introduced a very effective, simple, and noncontact method with the desired sensitivity under clinical conditions. They called the technique the laser reflection method. However, the tooth surface being a complex one, the resulting reflection can best be described as a laser speckle pattern. That is why they observed a very specific pattern corresponding to a given point on the tooth surface so that positioning for repeated measurements was simple. The basic arrangement is shown in Fig. 5.19. The tooth is illuminated by a laser beam, and the

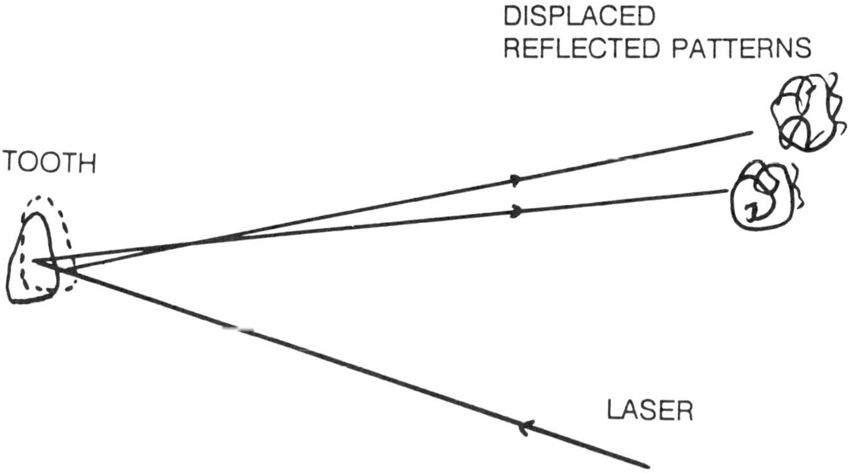

Figure 5.19 Tooth mobility studies using reflected complex pattern.

scattered complex pattern is observed on a screen. A change in the position of the tooth by an angle α results in a change of 2α of the scattered beam. The change in the position on the screen can thus yield data on tooth mobility versus conditions. The method was applied for clinical purposes [96], the apparatus was improved [97], and tooth mobility was studied in a group of patients [98].

In another parallel development by Vogel [99], two speckle methods for determining transverse motions and applications in tooth deformation studies are described. One is a real-time speckle motion determination approach up to an accuracy of 20 nm. The other is speckle photography with an accuracy of better than 100 nm.

Kasprzak et al. [100] present results of double-exposure laser speckle photography of broken leg bones. Different fixing devices under axial loading were considered. Holographic interferometry was used for comparing the results and led to similar conclusions. The experimentation established the usefulness of speckle metrology in orthopedic biomechanics.

Mechanical activity of the heart results in chest wall movements and vibrations. More important, it has been established [101] that the displacement of the left ventricular apex during open-chest experiments can be related to those near the same area with the intact chest. Thus, functional aspects of the heart can be studied from the knowledge of chest wall movements. Ramachandran and Singh [102] report progress in speckle metrology techniques applied to this effect. They performed time-average laser speckle photography of chest wall movements and compared the results with those obtained by electrocardiography (ECG). Comparative studies between healthy subjects and cardiac patients were also made. It was established that any tissue abnormality can be studied by the speckle approach. The abnormality may impair the function of the cardiac chambers. Thus, the noninvasive and noncontact technique may be a useful tool in the diagnosis of heart conditions.

5.8 CONCLUDING REMARKS

It has been shown that the speckle technique has vast applications beyond traditional activities in solid mechanics and general metrology. It is not possible to do justice here to every work. The well-known novel application of speckle interferometry in astronomy is not covered. The use of the Wigner distribution function [103], image derotating for rotating objects [104], polarization characteristics [105], camera testing and other aberration-related aspects [106, 107], application in point diffraction interferometry [108], alignment applications [109, 110], and radial metrology using a panoramic annular lens [67] is only a partial list of contributions not described in detail. Wu et al. [111] introduced and demonstrated the usefulness of a fiber specklegram sensor. Korneev and Stepanov [112] used a non-steady-state photo-EMF effect of the photodetector to study very small vibration amplitudes. With semi-insulating GaAs:Cr, vibration amplitudes as small as 10^{-5} of the speckle diameter can be studied [112]. Irisarri et al. [113] successfully demonstrated that multiple-exposure speckle photography can yield kinematic parameters of linear motion. These recent developments show ongoing enthusiasm for the application of speckles. However, the purpose of this chapter was to highlight the universal usefulness of speckle metrology and to indicate that it is limited only by our imagination. Fortunately, a series of rather complete bibliographies on "speckle" are available [114].

REFERENCES

1. C. S. Vikram and K. Vedam, Testing the trueness of circular surfaces by laser speckle photography, *Appl. Opt.*, *18*, 2351–2353 (1979).
2. C. S. Vikram and K. Vedam, Laser speckle photography for supersensitive dial gauge, *Opt. Lett.*, *5*, 441–442 (1980).
3. C. S. Vikram and K. Vedam, Supersensitive dial-gage measurement by observing rotation of laser speckles, *Exp. Tech.*, *9*, 16–17 (1985).
4. C. S. Vikram, D. K. Agrawal, R. Roy, and H. A. McKinstry, Application of dial-gage and laser speckles for ultra-low thermal expansion measurements using push-rod dilatometers, *Opt. Lasers Eng.*, *7*, 221–226 (1986/87).
5. B. S. Ramprasad and T. S. Radha Bai, Speckle-based fiber-optic current sensor, *Opt. Laser Technol.*, *16*, 156–158 (1984).
6. F. P. Chiang, R. Anastasi, J. Beatty, and J. Adachi, Thermal strain measurements by one-beam laser speckle interferometry, *Appl. Opt.*, *19*, 2701–2704 (1980).
7. P. Koulev, P. Aswendt, V. Balavessoy, R. Höfling, I. Roussev, and V. Grosser, Two fringe evaluation methods applied to thermal expansion coefficient measurements by speckle-photography: a comparison, *Proc. SPIE*, *1183*, 382–388 (1990).
8. G. A. Costa, P. Ottonello, and E. Piano, Thermal expansion measurements by speckle interferometry, *Appl. Opt.*, *19*, 1032–1033 (1980).
9. C. S. Vikram and H. A. McKinstry, A laser speckle method for measuring ultra-low thermal expansion, *Opt. Lasers Eng.*, *6*, 91–100 (1985).
10. C. S. Vikram, D. K. Agrawal, R. Roy, and H. A. McKinstry, A simple laser speckle dilatometer for thermal expansion measurements, *Mat. Lett.*, *3*, 482–484 (1985).
11. G. Da Costa, Transient phenomena analysis, in *Speckle Metrology*, R. K. Erf, Ed., Academic, New York, 1978, Chap. 9.4.

12. J. A. Gilbert, T. D. Dudderar, and J. H. Bennewitz, The application of fibre optics to remote speckle metrology using incoherent light, *Opt. Lasers Eng.*, *3*, 183–195 (1982).
13. T. D. Dudderar and J. A. Gilbert, Fiber-optic measurement of the deformation field on a remote surface using numerically processed white-light speckle, *Appl. Opt.*, *21*, 3520–3527 (1982).
14. D. R. Matthys, T. D. Dudderar, J. A. Gilbert, M. A. Taher, and H. S. Johnson, Speckle metrology combined with finite-element modeling for stress analysis, *Opt. Eng.*, *25*, 749–753 (1986).
15. T. D. Dudderar, J. A. Gilbert, A. J. Boehnlein, and M. E. Schultz, Application of fiber optics to speckle metrology—a feasibility study, *Exp. Mech.*, *23*, 289–297 (1983).
16. T. D. Dudderar, J. A. Gilbert, J. H. Bennewitz, and M. A. Taher, Displacement sensitivity in remote-objective-speckle metrology, *Exp. Mech. 25*, 392–398 (1985).
17. A. Hayashi and Y. Kitagawa, High-resolution rotation-angle measurement of a cylinder using speckle displacement detection, *Appl. Opt.*, *22*, 3520–3525 (1983).
18. D. A. Gregory, Topological speckle inspection, in *Speckle Metrology*, R. K. Erf, Ed., Academic, New York, 1978, Chap. 8.
19. G. Cloud and E. Conley, A whole-field interferometric scheme for measuring strain and flow rates of glacier and other natural surfaces, *J. Glaciol.*, *29*, 492–497 (1983).
20. E. Conley and G. Cloud, Practical applications of double-exposure noncoherent-light speckle photography, *Appl. Opt.*, *25*, 2246–2248 (1986).
21. E. Conley and G. Cloud, Whole-field measurement of ice displacement and strain rates, *J. Offshore Mech. Arctic Eng.*, *110*, 169–171 (1988).
22. E. Conley and G. L. Cloud, Enhancing sensitivity of the white-light speckle method in terrestrial applications, *Exp. Mech.*, *29*, 138–143 (1989).
23. E. Conley and G. Cloud, White-light speckle for measuring geophysical surface motions, in *Hologram Interferometry and Speckle Metrology*, The Society for Experimental Mechanics, Bethel, Conn., 1990, pp. 88–94.
24. E. Conley and G. Cloud, Resolution experiments using the white light speckle method, *Appl. Opt.*, *30*, 795–800 (1991).
25. E. Conley and J. Genin, Application of speckle metrology at a nuclear waste repository, *Proc. SPIE*, *1332*, 798–801 (1990).
26. D. He and M. He, Basic use of acoustic speckle pattern for metrology and sea wave study, *Proc. SPIE*, *1332*, 808–819 (1990).
27. U. Köpf, Application of speckling for measuring the deflection of laser light by phase objects, *Opt. Commun.*, *5*, 347–350 (1972).
28. S. Debrus, M. Françon, C. P. Grover, M. May, and M. L. Robin, Ground glass differential interferometer, *Appl. Opt.*, *11*, 853–857 (1972).
29. M. Giglio, S. Musazzi, and U. Perini, A white light speckle schlieren technique, *Opt. Commun.*, *36*, 117–120 (1981).
30. J. S. Harris, K. G. Harding, and S. H. Mersch, Techniques for evaluation of aircraft windscreen optical distortion, *Opt. Eng.*, *20*, 115–122 (1981).

31. B. S. Ramprasad, T. S. Radha, and E. S. R. Gopal, Laser speckle photography for the measurement of changes in refractive index in phase media, in *Nondestructive Characterization of Materials*, P. Höller et al., Eds., Springer-Verlag, Berlin, 1989, pp. 817–824.
32. M. S. Sivasubramanian, R. Cole, and P. C. Sukanek, Optical temperature gradient measurements using speckle photography, *Int. J. Heat Mass Transfer*, *27*, 773–780 (1984).
33. P. V. Farrell and D. L. Hofeldt, Temperature measurement in gases using speckle photography, *Appl. Opt.*, *23*, 1055–1059 (1984).
34. N. A. Fomin, U. Wernekinck, and W. Merzkirch, Speckle photography of a turbulent density field, in *Optical Methods in Dynamics of Fluids and Solids*, M. Pichal, Ed., Springer-Verlag, Berlin, 1985, pp. 159–165.
35. L. M. Bernardo and O. D. D. Soares, Birefringence measurements by double speckle photography, *Appl. Opt.*, *26*, 769–772 (1987).
36. S. Komatsu, K. Sumi, and H. Ohzu, Double exposure speckle technique for imaging system analysis and use in rod lens arrays, *Appl. Opt.*, *28*, 1869–1873 (1989).
37. U. Wernekinck and W. Merzkirch, Speckle photography of spatially extended refractive-index fields, *Appl. Opt.*, *26*, 31–32 (1987).
38. R. Erbeck and W. Merzkirch, Speckle photographic measurement of turbulence in an air stream with fluctuating temperature, *Exp. Fluids*, *6*, 89–93 (1988).
39. J.-Z. Shu and J.-Y. Li, Speckle photography applied to the density field of a flame, *Exp. Fluids*, *5*, 422–424 (1987).
40. T. C. Liu, W. Merzkirch, and K. Oberste-Lehn, Optical tomography applied to speckle photographic measurement of asymmetric flows with variable density, *Exp. Fluids*, *7*, 157–163 (1989).
41. G. N. Blinkov, N. A. Fomin, M. N. Rolin, R. I. Soloukhin, D. E. Vitkin, and N. L. Yadrevskaya, Speckle tomography of a gas flame, *Exp. Fluids*, *8*, 72–76 (1989).
42. D. D. Verhoeven and P. V. Farrell, Speckle interferometry in transparent media, *Appl. Opt.*, *25*, 903–906 (1986).
43. C. S. Vikram and T. E. McDevitt, Simple spectrum analysis in laser Doppler studies of sinusoidal vibrations, *Opt. Eng.*, *28*, 922–925 (1989).
44. A. Luxmore and F. A. A. Amin, Measurement of crystal length changes, in *Speckle Metrology*, R. K. Erf, Ed., Academic, New York, 1978, Chap. 9.2.
45. B. Ineichen, P. Eglin, and R. Dändliker, Hybrid optical and electronic image processing for strain measurements by speckle photography, *Appl. Opt.*, *19*, 2191–2195 (1980).
46. G. B. Smith and K. A. Stetson, Heterodyne readout of specklegram halo interference fringes, *Appl. Opt.*, *19*, 3031–3033 (1980).
47. K. A. Stetson, The use of heterodyne speckle photogrammetry to measure high-temperature strain distributions, *Proc. SPIE*, *370*, 46–55 (1983).
48. K. A. Stetson, The theory and applications of heterodyne speckle interferometry, in *Holography and Speckle Phenomena and Their Industrial Applications*, R. S. Sirohi, Ed., World Scientific, Singapore, 1990, pp. 313–324.

49. C. S. Vikram and K. Vedam, Measurement of subspeckle-size changes by laser-speckle photography, *Opt. Lett.*, *4*, 406–407 (1979).
50. F. P. Chiang, J. Adachi, R. Anastasi, and J. Beatty, Subjective laser speckle method and its application to solid mechanics problems, *Opt. Eng.*, *21*, 379–390 (1982).
51. D. J. Chen and F. P. Chiang, Digital processing of Young's fringes in speckle photography, *Opt. Eng.*, *29*, 1413–1420 (1990).
52. F. P. Chiang and D. W. Li, Random (speckle) patterns for displacement and strain measurement: some recent advances, *Opt. Eng.*, *24*, 936–943 (1985).
53. J. M. Huntley, M. B. Whitworth, and S. J. P. Palmer, Automatic speckle photography fringe analysis: application to electron microscopy and high speed photography, in *Hologram Interferometry and Speckle Metrology*, The Society for Experimental Mechanics, Bethel, Conn., 1990, pp. 337–342.
54. P. Hariharan, Double-exposure speckle interferometry with "instant" Polaroid film, *Opt. Commun.*, *26*, 325–326 (1978).
55. S. Ueha, N. Kobayashi, T. Saito, and J. Tsujiuchi, Thermoplastic photographic material for vibration measurements by speckle photography, *Opt. Commun.*, *34*, 340–344 (1980).
56. H. J. Tiziani, Real time metrology with BSO crystals, *Opt. Acta*, *29*, 463–470 (1982).
57. T. Takemori, S. Ueha, and J. Tsujiuchi, Real-time in-plane vibration measurements by speckle interferometry with the aid of liquid crystal light valve, *Opt. Commun.*, *32*, 24–26 (1980).
58. B. Bates and P. C. Miller, Liquid crystal television in speckle metrology, *Appl. Opt.*, *27*, 2816–2817 (1988).
59. B. Bates and P. C. Miller, Speckle metrology employing LCTV spatial light modulators, *Opt. Lasers Eng.*, *14*, 341–349 (1991).
60. H. J. Tiziani, K. Leonhardt, and J. Klenk, Real-time displacement and tilt analysis by a speckle technique using $Bi_{12}SiO_{20}$ crystals, *Opt. Commun.*, *34*, 327–331 (1980).
61. H. J. Tiziani and J. Klenk, Vibration analysis by speckle technique in real time, *Appl. Opt.*, *20*, 1467–1470 (1981).
62. K. Nakagawa and T. Minemoto, Readout properties of the specklegram recorded in photorefractive $Bi_{12}SiO_{20}$ crystal, *Appl. Opt.*, *30*, 2386–2392 (1991).
63. S. Collicott and L. Hesselink, Photorefractive recording of speckle for real-time velocimetry and metrology, *Proc. SPIE*, *814*, 700–705 (1988).
64. S. Collicott and L. Hesselink, Real-time photorefractive recording and optical processing for speckle velocimetry, *Opt. Lett.*, *13*, 348–350 (1988).
65. C. S. Vikram and K. Vedam, Direct observation of laser speckles for real-time analysis of lateral motions, *Opt. Lett.*, *6*, 511–513 (1981).
66. W. H. Peters III, M. A. Sutton, W. F. Ranson, W. P. Poplin, and D. M. Walker, Whole-field experimental displacement analysis of composite cylinders, *Exp. Mech.*, *29*, 58–62 (1989).
67. D. R. Matthys, J. A. Gilbert, and P. Greguss, Endoscopic measurement using radial metrology with digital correlation, *Opt. Eng.*, *30*, 1455–1460 (1991).

68. F. P. Chiang and Q. B. Li, Real-time laser speckle photography, *Appl. Opt.*, *23*, 4469–4470 (1984).
69. Q. B. Li and F. P. Chiang, Studies of holospeckle interferometry, *Opt. Eng.*, *27*, 200–206 (1988).
70. A. M. Darskij and V. B. Markov, Measurement of small displacements by speckle holography, *Opt. Laser Technol. 21*, 198–203 (1989).
71. D. B. Neumann and R. C. Penn, Off-table holography, *Exp. Mech.*, *15*, 241–244 (1975).
72. Z. Cao and F. Chen, Information extracting and application for the combining objective speckle and reflection holography, *Proc. SPIE*, *1332*, 358–364 (1990).
73. F. P. Chiang, Some recent advances in speckle techniques for photomechanics and optical metrology, *Proc. SPIE*, *661*, 241–261 (1986).
74. B. Castagnede, Mechanical characterization of paper by laser speckle interferometry, in *Nondestructive Characterization of Materials II*, J. F. Bussière, J.-P. Monchalin, C. O. Ruud, and R. E. Green, Jr., Eds., Plenum, New York, 1987, pp. 745–751.
75. L. Benckert, Wood drying studies using white light speckle photography, *Measurement*, *10*, 24–30 (1992).
76. F. P. Chiang, R. M. Juang, and Y. Z. Dai, Application of moiré and speckle techniques to curved surfaces using a periphery camera, *Proc. SPIE*, *814*, 249–256 (1988).
77. J. Georgieva, Laser speckle photography in the presence of natural day light, *Optik*, *83*, 178–180 (1989).
78. F. A. Moslehy, S. L. Rice, and S. Elmi, Effect of partial laser beam coverage of point-wise filtering of specklegrams, *Exp. Techn.*, *13*, 29–31 (1989).
79. P. Waterworth and D. C. J. Reid, Loose contact detection using laser speckle, *Opt. Laser Technol.*, *7*, 135–137 (1975).
80. T. S. Radha and B. S. Ramprasad, A simple method of measuring the etch depth in metallic specimens using speckle photography, *J. Electrochem. Soc. India*, *34*, 149–150 (1985).
81. R. T. Hennessy and H. W. Leibowitz, Laser optometer incorporating the Badal principle, *Behav. Res. Meth. Instrum.*, *4*, 237–239 (1972).
82. D. C. Sinclair, Demonstration of chromatic aberration in the eye using coherent light, *J. Opt. Soc. Am.*, *55*, 575–578 (1965).
83. D. Phillips, W. Sterling, and W. O. Dwyer, Validity of the laser refraction technique for determining cylindrical error, *Am. J. Optom. Physiol. Opt.*, *52*, 328–331 (1975).
84. W. N. Charman, On the position of the plane of stationarity in laser refraction, *Am. J. Optom. Physiol. Opt.*, *51*, 832–838 (1974)
85. R. D. Bahuguna, Speckle motion in laser refraction, *J. Opt. Soc. Am. A*, *1*, 1085–1086 (1984).
86. R. D. Bahuguna, D. Malacara, and K. Singh, White-light speckle optometer, *J. Opt. Soc. Am. A*, *1*, 132–133 (1984).
87. R. D. Bahuguna, K. K. Gupta, and K. Singh, Some experiments on the perceived size of laser speckles, *Proc. SPIE*, *954*, 444–447 (1989).

88. J. D. Briers, Optical filtering techniques to enhance speckle contrast variations in single-exposure laser speckle photography, *Optik*, *63*, 265–276 (1983).
89. A. F. Fercher, M. Peukert, and E. Roth, Visualization and measurement of retinal blood flow by means of laser speckle photography, *Opt. Eng.*, *25*, 731–735 (1986).
90. H. Fujii, T. Asakura, K. Nohira, Y. Shintomi, and T. Ohura, Blood flow observed by time-varying laser speckle, *Opt. Lett.*, *10*, 104–106 (1985).
91. H. Fujii, K. Nohira, Y. Yamamoto, H. Ikawa, and T. Ohura, Evaluation of blood flow by laser speckle image sensing. Part 1, *Appl. Opt.*, *26*, 5321–5325 (1987).
92. L. V. Tanin, A. S. Rubanov, I. V. Markhvida, S. K. Dick, and L. I. Rachkovsky, The investigation of laser speckle fluctuations depending on the state of muscular tissue, *Proc. SPIE*, *1183*, 425–431 (1990).
93. V. V. Tuchin, A. V. Ampilogov, G. Bogoroditsky, E. M. Rabinovich, V. P. Ryabukho, S. S. Ul'yanov, and M. E. V'yushkin, Laser speckle and optical fiber sensors for micromovement monitoring in biotissues, *Proc. SPIE*, *1429*, 62–73 (1991).
94. Y. Aizu and T. Asakura, Bio-speckle phenomena and their application to the evaluation of blood flow, *Opt. Laser Technol.*, *23*, 205–219 (1991).
95. H. Rydén, H. Bjelkhagen, and P.-O. Soder, The use of laser beams for measuring tooth mobility and tooth movement: an in vitro study, *J. Periodontol.*, *45*, 283–287 (1974).
96. H. Rydén, H. Bjelkhagen, and P.-Ö. Söder, The use of laser beams for measuring tooth mobility and tooth movements, *J. Periodontol.*, *46*, 421–425 (1975).
97. H. Rydén, H. Bjelkhagen, and U. Sandström, A laser instrument for measuring tooth movements, *J. Periodontol.*, *50*, 265–269 (1979).
98. H. Rydén, H. Bjelkhagen, and P.-Ö. Söder, Movements of healthy and periodontally involved teeth measured with laser reflection technique, *J. Periodontol.*, *53*, 439–445 (1982).
99. A. Vogel, Comparison of two sensitive speckle methods applied to the measurement of tooth deformation, *Optik*, *72*, 95–101 (1986).
100. H. Kasprzak, H. Podbielska, and D. Pennig, Speckle photography for investigation of bones supported by different fixing devices, *Proc. SPIE*, *1429*, 55–61 (1991).
101. A. E. Aubert, L. Welkenhuysen, J. Montold, L. de Wolf, M. Geivers, J. Minten, H. Kesteloot, and H. Gees, Laser method for recording displacement of the heart and the chest wall, *J. Biomed. Eng.*, *6*, 134–140 (1984).
102. G. Ramachandran and M. Singh, Three-dimensional reconstruction of cardiac displacement patterns on the chest wall during the P, QRS and T-segments of the ECG by laser speckle interferometry, *Med. Biol. Eng. Comput.*, *27*, 525–530 (1989).
103. J. Widjaja, J. Uozumi, and T. Asakura, Method for evaluating the displacement of objects using the Wigner distribution function, *J. Opt. (Paris)*, *21*, 253–259 (1990).

104. K. A. Stetson, The use of an image derotator in hologram interferometry and speckle photography of rotating objects, *Exp. Mech.*, *18*, 67–73 (1978).
105. Y. Landsberg, J. Shamir, A. Brunfeld, and D. Peri, Complex amplitude correlations of polarized speckle patterns, *Appl. Opt.*, *29*, 1455–1465 (1990).
106. L. H. Tanner, Camera testing by use of speckle patterns, *Appl. Opt.*, *13*, 2026–2034 (1974).
107. A. Kumar and K. Singh, Effect of spherical aberration on elongated laser speckles, *Optik*, *88*, 25–29 (1991).
108. M. Giglio, U. Perini, and E. Paganini, Speckle-tracking point diffraction interferometer for fluid studies, *Appl. Opt.*, *27*, 197–199 (1988).
109. H. X. Bai and G. Indebetouw, Focusing and aligning by laser speckle, *Appl. Opt.*, *22*, 1609–1611 (1983).
110. E. N. Hogert, J. J. Lunazzi, and G. Gaggioli, Visual speckle technique for alignment and centering of a translucent rough surface, *Appl. Opt.*, *28*, 722–725 (1989).
111. S. Wu, S. Yin, and F. T. S. Yu, Sensing with fiber specklegrams, *Appl. Opt.*, *30*, 4468–4470 (1991).
112. N. A. Korneev and S. I. Stepanov, Measurement of small lateral vibrations of speckle patterns using a non-steady-state photo-EMF in GaAs:Cr, *J. Mod. Opt.*, *38*, 2153–2158 (1991).
113. M. A. C. Irisarri, A. Garcia, R. Fernandez, and M. C. Vazquez, Kinematic study by speckle photography, *Opt. Lett.*, *17*, 1–3 (1992).
114. K. Singh, A. Kumar, and S. Baskar, Laser speckles: a bibliography: 1985, 1986, *J. Pure Appl. Phys.*, *1*, 129–149 (1989).

6

Particle Image Velocimetry

K. D. Hinsch

University of Oldenburg
Oldenburg, Germany

6.1 OPTICAL METHODS IN FLOW ANALYSIS

6.1.1 Limits of Flow Visualization

Moving fluids have always been an object of fundamental physical research. An understanding of the dynamic behavior of streaming liquids and gases is the basis for disciplines in engineering ranging from the design of aircraft and ships to harbor development, wavebreaker projects, cavitation erosion, or sound production and attenuation. Modern nonlinear dynamics even tackles the classical problem of turbulence, encouraging untraditional approaches to yet unsolved questions such as weather prediction.

Ever since people first developed an interest in flows, they have attempted to visualize the motion of transparent media. Wind speed has been derived from journeying clouds or drifting smoke plumes; the motion of flowing streams has been judged by observing swirling branches or entrained air bubbles. The motion of the invisible medium has thus been studied mostly by tracing foreign matter, naturally existent in the flow or artificially added, assuming that it is carried along without any lag. Large-area movement is inferred from the drift of big structures like complete cloud fields; pointwise measurement requires the tracking of tiny particles or even molecules as in modern fluorescence techniques. There exist a variety of tools in diagnostic flow visualization [1]. Flow phenomena become most impressive when they are visualized; a good example are the familiar dye or smoke lines depicting von Kármán vortex streets.

Some nonintrusive techniques utilize inhomogeneities inherent in the flow, such as density or temperature structures, for visualization. Interferometric records of shocks in high-speed flows or satellite-based thermographic plots of the Gulf Stream are impressive examples. Under the conditions of many practical problems, however, such methods are excluded because the variations in these quantities are too small.

Despite their attractive lucidity, flow visualization techniques suffer from imperfections. They often yield only qualitative information and usually accumulate some of the history of the process under investigation. A smoke-line representation of a flow field gives the instantaneous distribution of tracers that have been introduced within a long time period prior to the instant of observation; it does not tell how these particles got to their present position or what their momentary velocity is. Furthermore, visualization usually demands an uneven distribution of matter in the flow to produce structures that can be identified in their motion. Thus we lack information from those flow regions that are vacant.

The need for quantitative analysis of flow fields has led to the development of various types of anemometers, instruments that measure flow velocity at a selected point of observation. Hot-wire or thin-film probes and laser–Doppler anemometers are very successful instruments that yield time records of the velocity at the measuring position. For a thorough analysis of the spatial structure of the flow field, a scan through different locations is necessary. In nonstationary fields, however, this is unsatisfactory because an instantaneous measurement at all locations would be required. Structural information about such fields is difficult to accumulate with anemometers and relies on spatial correlations of simultaneous measurements at two or more points, a costly technique of limited significance. Sometimes theoretical assumptions like the Taylor hypothesis of frozen turbulence help to establish space–time relationships

We thus need techniques that render the instantaneous velocity field over an extended region in space—quantitative whole-field methods. If such recordings are repeated at an appropriate frame rate, modern fluid dynamics is equipped with the tool to study the development of even nonstationary processes in space and time. This chapter is dedicated to the currently most promising approach to reaching this goal, particle image velocimetry (PIV). A recent review of this topic is given in Ref. 2.

Another method comparable in its intentions to PIV that is presently being explored is laser-induced fluorescence (LIF). Basically, a fluorescent pattern, for example a grid, is inscribed into a fluid by adding a fluorescent substance to the flow that is illuminated with an appropriate grid pattern produced by the laser beams. The flow pattern can be derived from the distortion of the pattern over the course of time.

6.1.2 Development of Particle Image Velocimetry

The solution to the problem of quantitative whole-field velocimetry is straightforward and not new. Assume we are standing on a bridge wondering about the velocity of the water underneath. We pick up a little piece of wood, throw it onto the water surface, and register its position at two instants of time that we take from our wristwatch. If the flow field is irregular it requires several floating objects at different positions for its survey. Finally, we could cover the entire surface with densely distributed tracers and take two photographs or, maybe better, a double-exposure photograph at the appropriate moments. Careful examination of this flow record, pairing the images of our floating objects and measurement of their separations, will yield the velocity distribution in space.

The picture presented also illustrates some conditions that have to be observed for a successful experiment. The density of the tracers has to be matched to the finest flow details, and the time between exposures has to be adjusted to ensure that pair separation is small compared to this structural scale. On the other hand, however, the displacement must be large enough to be measured accurately. Furthermore, the accuracy in position measurement depends on the size of the imaged tracers. These should be small, and the exposure time should guarantee an unblurred image.

We have thus introduced the concept of particle image velocimetry. Usually, the investigation concerns a volume flow instead of a two-dimensional motion. Thus, for successful photography of the tracing particles the analysis is restricted to a plane within the flow by illuminating it with a sheet of light, as shown in Fig. 6.1. The concept could also be extended to three dimensions by a holographic double-exposure record of the particles within the whole volume.

Double-exposure photography as outlined above asks for two closely spaced short pulses of light. The advent of lasers has supplied ideal light sources for such a task. In use prior to the development of lasers and still used sometimes today is a related technique that employs a single, but longer, exposure of the floating particles. Particle motion blurs the image into a streak line that reveals direction and magnitude of the motion. The visual inspection of such a photograph requires no pairing of images and gives a good indication of the velocity field. Many impressive flow studies are based on such pictures [3].

It is interesting to note that this rather traditional concept of flow diagnostics has been approached historically from still another direction.

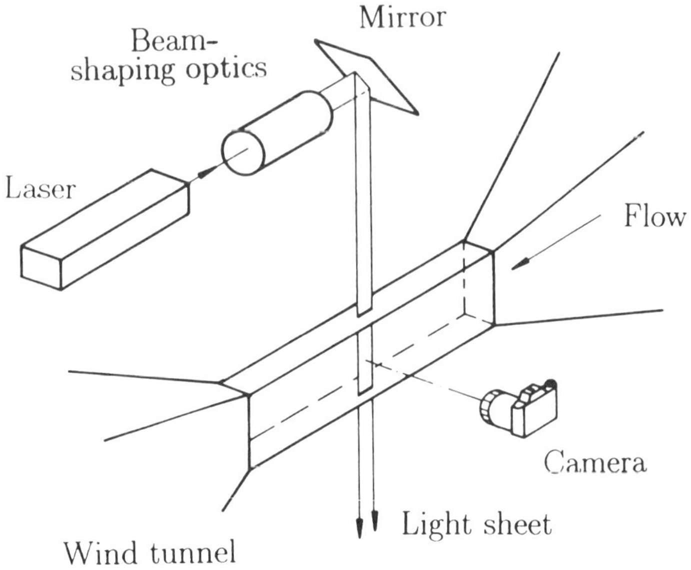

Figure 6.1 Schematic setup for PIV recording in a wind tunnel flow.

After the discovery of the speckled nature of scattered coherent light, measuring techniques developed that used the tiny speckles covering the image of a rough object as marks for displacement analysis. While originally conceived for surface studies, transparent models of engineering specimens were illuminated by a laser light sheet, and scattering from natural or artificially introduced inhomogeneities produced speckle in an image of this section [4]. The speckle displacement under load, for example, was used for strain analysis. It was obvious to extend this scheme from solid bodies to transparent fluids, and early flow analysis employed actual speckle photographs from coherently illuminated sections in flows densely seeded with scattering particles [5–8].

The development of particle image velocimetry can be traced by studying the different names this technique has been given over its short history. Because early measurements produced speckle images instead of single resolved particle images, the technique was assigned the term *scattered light speckle* (SLS) *metrology*, and the photographic records were called *specklegrams*. Generally the term *speckle velocimetry* was used for metrology employing the evaluation of the Young's fringes that the double-exposure speckle photographs yield. Their twin structure is difficult to identify in the photograph, so they are usually processed by examining their diffraction patterns (Young's fringe system). The fringe orientation is perpendicular to the direction of displacement, and the fringe spacing is inversely proportional to the magnitude of the displacement. This interrogation scheme was adopted in flow velocimetry. Gradually it was realized that speckles are not necessary for flow diagnostics. If particle images do not overlap (which depends on seeding density and imaging aperture), discrete image pairs are produced, and the records become multiple-exposure photographs and can be produced even with classical illumination [9]. There are good reasons to prefer this mode of recording because the speckle mode is more sensitive to decorrelation by out-of-plane particle motion. Yet a long time the terminology adhered to *laser speckle velocimetry* (LSV) or *photography* (LSP), mostly because the interrogation technique was identical to that used in speckle metrology. Finally researchers became more precise in their terminology, and today the term *particle image velocimetry* (PIV) is widely accepted [10, 11].

Particle image velocimetry in its essentials is so simple that the reader may ask: Why devote a complete chapter of this book to the topic? It often occurs in physics that the basic scheme of an experimental technique seems very straightforward but its realization encounters a variety of challenging problems. This is the case here. From the initial task of measuring precisely the distance between a pair of particle images, PIV has initiated or made use of a large number of fascinating investigations, many of them in optics. The immediate need for a well-functioning measurement technique in flow analysis has been the driving force for the experimenter; technological developments have set a series of milestones in progress. Instead of giving a detailed account of the history of PIV, which can be found elsewhere (see, e.g., Ref. 12), I will briefly outline these interrelations.

The recording of a PIV image reveals the technique's dependence on available technology. In low-speed flows, traditional flash lamps or modulated continuous-wave lasers give satisfactory records. As long as there were no powerful pulsed light sources, streak technology was the only way to produce particle tracks in higher speed flows. The pulsed ruby laser and then the Nd:YAG laser provide double or multiple pulses for particle image records in higher velocity fields. The coupling of two laser oscillators frees the method from almost all limits as to pulse separation and greatly extends the range of applicability. With ion lasers of up to 15-W power a scanning beam technology has become feasible that is free from some of the Gaussian beam limits of expanded beam illumination [13].

Limited laser light intensity has also restricted the kind of tracer particles that can be employed. Low illumination requires large particles that scatter enough light to expose the photographic film. Large particles, on the other hand, may not follow the flow faithfully. Conditions are easier to meet in water than in air. Thus, early PIV investigations were made in fluids; only later was the extension to gas flows demonstrated [14].

The history of evaluation techniques in PIV reveals constant competition between coherent optical and digital electronic processing techniques. The quite tedious task of pairing images and determining their separation was accomplished with the elegant method of optical Fourier transformation that is the basis for the formation of the Young's fringes. Now fringe orientation and spacing had to be determined. A manual approach to this task received assistance from video grabbing of the fringe pattern and computer-aided and operator-controlled fringe analysis. These strategies were in accord with the then existing capacity of digital computers. Relief was expected from various schemes of optical processing such as one-dimensional compression by cylindrical lenses [15], rotating prisms, anamorphic processing [16], or optical whole-field filtering [8] to generate isodeformation contours. It was a time of clever ideas being advanced by physicists and engineers as solutions to existing problems. Everybody knew, of course, that a two-dimensional Fourier transform of the fringe pattern would yield the desired information in the form of autocorrelation function peaks. Yet digital electronic techniques were too slow to allow consideration for routine applications. Thus all the other brilliant brainwork!

Computers, and especially digital image processing, have developed quickly, however, and all clever optical methods had to be examined in view of the competing digital evaluation. Special hardware encouraged digital Fourier transformation of the fringe system, and it was finally suggested that even the first optical processing step of fringe production be dropped: The photographic image of the interrogation area is directly input to the computer for complete digital treatment. Processing time per interrogation is still between 1/10 sec and 1 sec.

Presently, attractive developments in electro-optics tend to reestablish some optical priorities. They reflect efforts to build a new generation of processors, or even computers, on the basis of optical information transfer and handling. As an example, take the real-time storing and processing capability of electro-optical materials such as photorefractive crystals. Installed in an optical filtering setup they can perform a correlation operation in a fraction of a second [17]. Another example is that of electrically or optically addressed spatial light modulators, which can be employed to modulate a coherent light beam for fast optical processing [18]. Further, modern electron beam lithography allows the production of very precise arrays of a large number of holographic lenslets that can be employed for large-scale parallel interrogation of PIV records [19].

On the other hand, recent advances in video technology, the development of multi-element charge-coupled devices (CCDs) for video pickups, and the even faster image processing by computers allow some researchers to completely ignore optical processing aspects. Light is needed only to illuminate the particles; everything else is done electronically or digitally. We have arrived at digital particle image velocimetry (DPIV) [20].

Still it is difficult to say in which direction the method will develop. Thus a chapter on PIV at the moment reflects a state of the art that is in the midst of evolutionary new concepts. As this is a book on optics, we will concentrate on the role of optics in the progress of PIV, not forgetting, however, that all proposals are on trial in the competitive field of newly evolving technologies.

6.1.3 Speckle Velocimetry in Fluids

I have mentioned that the dominant optical whole-field technique today produces records that contain images of the flow-tracing particles. Before continuing with details of PIV, we should, however, dwell a little on the speckle mode of operation. The mode of operation must be fully understood before the setup parameters on an optical flow experiment are chosen.

In several investigations, general conditions for the distinction between particle image velocimetry and speckle velocimetry have been formulated [10]. Generally, speckles arise from the superposition of many coherent random phase waves. We must therefore answer the question: How many particles from the measuring volume scatter onto a certain location in the photograph? Obviously, this number increases with particle concentration C, depth of light sheet $\Delta z'$, and size D of the particle image. The latter is governed by the photographic magnification M and the f-number (F)-dependent diffraction image D_d:

$$D = [M^2D'^2 + D_d^2]^{1/2} \tag{6.1}$$

with particle diameter D' and

$$D_d = 2.4\lambda F(1 + M) \tag{6.2}$$

A quantity S called source density has been introduced [10]:

$$S = C\Delta z'D^2/M^2 \tag{6.3}$$

Roughly, S gives the number of visible particles within a volume that is cut from the light sheet by the cross section of a particle image in object space. Large S characterizes the speckle mode, small S the particle image mode.

Despite the speckle nature of early whole-field flow analysis, the speckle mode is rarely met in today's applications. Reasons are that the particle image diameter is made as small as possible to obtain high accuracy in particle positions and the concentration stays moderate owing to practical limitations in particle generation and optical penetration. Thus it is justified to keep the present chapter restricted to an analysis of the particle image mode. Readers should keep in mind, however, that the planning of a PIV experiment requires an estimation of S, because some interrogation schemes such as particle tracking will not operate on speckle images. Convenient estimates can be found in the literature [10]. Furthermore, there are some techniques in optical flow analysis that rely on the measurement of speckle displacements induced by the refraction of light by inhomogeneities in the fluid [21, 22].

6.2 THEORETICAL ANALYSIS OF PIV

Particle image velocimetry, as we have seen, consists of two steps: the recording of a suitably illuminated configuration of small particles within the flow, usually as a double-, rarely a multiple-exposure record, traditionally on a single photographic frame, and the interrogation of this record to obtain the spatial distribution of the velocity. Multiple-frame techniques are becoming popular with videoelectronic recording but are not typical of PIV. Much experimental skill goes into the production of the double-exposure photograph as explained in Section 6.4. A typical example from a flow is shown in Fig. 6.2; careful examination reveals the double structure of the particle images. Considerable development time has been invested in the processing of such PIV records, devising schemes to extract the desired displacement information from them and optimizing the signal-to-noise ratio. To understand more about these techniques and their limitations, we will proceed with a theoretical analysis of the data stored in the double exposure. To begin with, we assume ideal conditions, i.e., we have obtained undistorted images of the particles, identical over the entire record and in both exposures. Since velocity interests us as a space-dependent quantity, we select a small interrogation area from the record. Let us describe the image within this region.

6.2.1 Space-Domain Analysis

Let a particle image be given by its transmittance function $h(r)$ on the photograph. This can be the original negative or, as is often used, a positive contact copy thereof. The two-dimensional coordinate of position in this plane is denoted by r. In the first exposure, particles were caught at random positions r_l as described by Poisson statistics. During the time between

Figure 6.2 PIV double-exposure record of vortex ring interaction in an air jet of 20 mm diameter. (With permission from Ref. 14.)

exposures each particle traveled a distance d_l, so that the complete record within the interrogation region, when illuminated, produces the light distribution

$$f(r) = [i(r - r_l)]\left\{h(r) * \sum_{l=1}^{N} [\delta(r - r_l) + \delta(r - r_l - d_l)]\right\} \qquad (6.4)$$

Here the shift of the particle function $h(r)$ to its various positions has been expressed as a convolution $*$ with a Dirac delta function, and the illuminating light $i(r)$ has been placed at the interrogation position r_I. The sum is taken over all particles, N in number; the illumination, of course, cuts only a limited number of pairs from the complete record. Note that our representation allows for an arbitrary profile of the illuminating light and for a particle-position-dependent displacement. When, for example, the PIV record is viewed directly with a video camera through a microscope, $i(r)$ could be a constant illumination over a region defined by the field of view. On the other hand, when the photograph is scanned with a laser for Young's fringe analysis, $i(r)$ would be the Gaussian beam profile. As the whole record is scanned, r_I changes.

The relation can be simplified considerably, especially for further manipulations, when we consider that the size of a particle image is much smaller than the extent of the interrogation window. In this case we can assume $i(r)$ constant within a particle image to arrive at

$$f(r) = h(r) * \sum_{l=1}^{N} [i(r_l - r_I)\delta(r - r_l) + i(r_l + d_l - r_I)\delta(r - r_l - d_l)] \tag{6.5}$$

In the following calculations let us set $r_I = 0$ as we are studying conditions within a single interrogation area.

The PIV transparency is dominated by the random variable $\mathbf{r}$, the accidental initial position of the particle at the moment of exposure. Principally, this random nature is the origin of noise spoiling the signals. Let us, for example, attempt to evaluate the PIV record directly by searching for particle pairs, a philosophy that has guided various processing schemes. If particle density is low, which puts us at risk of encountering many areas that have no particle at all, we will be lucky and discriminate pairs because the partners are closer together than are other neighbors. If the average particle spacing is of the same order as the separation within a pair, however, the image looks like two interlaced skies full of stars, and it is difficult to identify pairs.

An appropriate way to describe a random process is by means of the autocorrelation function or power spectrum of the random variable, functions that are related by a Fourier transform. Since the power spectrum of a light distribution can be obtained easily just by generating its far-field diffraction pattern intensity distribution, its analysis has received considerable attention. It is the well-known Young's fringe pattern. Let us thus Fourier transform the light distribution derived above, apply statistical averaging, and also calculate the autocorrelation function. This treatment is equivalent to directly calculating the autocorrelation function on the basis of a Poisson distribution of particle positions [23].

6.2.2 Spatial Frequency Domain Analysis

We describe the Fourier transform $F(k)$ of $f(r)$ as a function of spatial angular frequency k, a quantity that in actual optical experiments is directly proportional to a spatial coordinate ρ. The coefficient contains the wavelength λ and geometrical parameters as shown later in Eq. (6.44).

$$F(k) = H(k) \sum_{l=1}^{N} \{i(r_l) \exp(-ikr_l) + i(r_l + d_l) \exp[-ik(r_l + d_l)]\} \tag{6.6}$$

All capital letters denote Fourier transforms of the lowercase functions. The calculation of the power spectrum, i.e., the fringe intensity, asks for the modulus squared

$$I_F(k) = |F(k)|^2 = F(k)F^*(k) \tag{6.7}$$

where the asterisk indicates the complex conjugate. This yields

$$I_F(k) = |H(k)|^2[P_1 + P_2] \tag{6.8}$$

with the abbreviations

$$P_1 = \sum_{l,m=1}^{N} [i(r_l)i(r_m) + i(r_l + d_l)i(r_m + d_m)] \times \exp[-ik(r_l - r_m)] \tag{6.9}$$

$$P_2 = \sum_{l,m=1}^{N} i(r_l)i(r_m + d_m)\{\exp[ik(r_l - r_m - d_m)] + \exp[-ik(r_l - r_m - d_m)]\} \tag{6.10}$$

We now have to introduce the random nature of the pattern. For a single realization of the process, that is, one special PIV record, the above formulas must be evaluated as they are. We are interested, however, in statistical quantities that are ensemble averages. Thus, we could consider repeating the experiment many times for an identical flow field and determining the average outcome. Thus, all expected values are conditional averages, given a specific flow field. In addition, we also want to know how the single realization result may differ from the expected values; we are interested, for example, in the higher moments of the observed light intensity. This is important in assessing the quality of our results. An experimental way to perform optical averaging over several experimental realizations has been to repeatedly double expose a film in a stationary flow with low particle concentration [24]. Successive realizations of the same flow field are each represented by a few particle pairs, whose superposition contains the statistics of the field at the locations in question.

Often it is difficult to record many realizations of particle distributions in the same flow to obtain the required ensemble average. When the number of particles within the interrogation region is very large, however, we can instead perform spatial averages. This smoothing can be achieved with a window of appropriate size, a procedure often applied in practice by scanning the light intensity with a pinhole of finite aperture or by spatially integrating parallel to the fringes.

To derive the expected value of the fringe intensity in Eq. (6.8) we separate the $l = m$ terms:

$$\langle I_F(k)\rangle = N|H(k)|^2\{\langle i^2(\mathbf{r})\rangle + \langle i^2(\mathbf{r} + \mathbf{d})\rangle + \langle i(\mathbf{r})i(\mathbf{r} + \mathbf{d})[\exp(ik\mathbf{d}) + \exp(-ik\mathbf{d})]\rangle\} + O(k) \tag{6.11}$$

where the bold characters **r** and **d** indicate the corresponding random variables, the brackets $\langle\ \rangle$ denote expected values, and $O(k)$ is an abbreviation for the averaged $l \neq m$ part of Eq. (6.8). We do this separation because $O(k)$ contributes to the fringe intensity only in a region very close to the axis. For the moment let us thus disregard this term.

In the discussion of the fringe formula let us distinguish two cases.

1. Constant displacement within the interrogation area. In the simple case that the interrogation region has been chosen properly small, we encounter only a single value d_0 of the random variable **d**

$$\langle I_F(k)\rangle = 2N|H(k)|^2[\langle i^2(\mathbf{r})\rangle + \langle i(\mathbf{r})i(\mathbf{r} + d_0)\rangle \cos kd_0] \tag{6.12}$$

$$= 2N|H(k)|^2\langle i^2(\mathbf{r})\rangle[1 + \tau_I(d_0)\cos kd_0] \tag{6.13}$$

where the normalized autocorrelation function $\tau_I(d)$ of the intensity profile in the interrogation area has been introduced as

$$\tau_I(d) = \frac{\langle i(\mathbf{r})i(\mathbf{r} + d)\rangle}{\langle i^2(\mathbf{r})\rangle} \tag{6.14}$$

Here we meet the familiar fringe system modulating a function $|H(k)|^2$, the spatial power spectrum of the particle image called a halo, often referred to as the pedestal component in analogy to laser–Doppler velocimetry [23]. It extends to a spatial frequency k_D that corresponds to a scale equal to $1/D$, where D is the diameter of the particle image. The fringe period is used to determine the displacement d_0.

Our analysis includes the finite cross-sectional intensity distribution in the interrogation area. As a consequence we observe a reduction in fringe visibility V,

$$V = \tau_I(d_0) \tag{6.15}$$

that is identical with the normalized autocorrelation of the interrogating beam evaluated at the particle displacement d_0. This relation is easy to understand. It expresses that partners within an image pair receive different intensities of light as the particle moved through the intensity profile between exposures, and the complete loss of partners as the separation approaches the size of the interrogation region.

2. Nonuniform displacement within the interrogation region. The spatial extent of the interrogation area may be comparable to the scales of the flow, thus including regions of different velocity. For such a distribution of displacements $\mathbf{d}$ within the interrogation region we decompose $\mathbf{d}$ into its mean value $\langle \mathbf{d} \rangle = d_0$ and a fluctuating part $\mathbf{s}$ with zero mean $\langle \mathbf{s} \rangle = 0$, that is,

$$\mathbf{d} = d_0 + \mathbf{s} \tag{6.16}$$

The statistics of $\mathbf{s}$ can be described by its probability density function (pdf) $p(s)$ or its characteristic function $\Phi(k)$ [25], where

$$\Phi(k) = \langle \exp(ik\mathbf{s}) \rangle = |\Phi(k)| \exp[i\phi(k)] \tag{6.17}$$

as decomposed into its modulus and argument. From its definition it is apparent that $\Phi(k)$ is related to $p(s)$ by Fourier transformation.

It is a little more difficult now to obtain the expected value in the latter part of Eq. (6.11) as we have two random variables $\mathbf{r}$ and $\mathbf{d}$. However, we can first average over $\mathbf{r}$ to arrive at

$$\begin{aligned}\langle I_F(k) \rangle &= N|H(k)|^2 \langle i^2(\mathbf{r}) \rangle \\ &\quad \times \{2 + \langle \tau_I(\mathbf{d})[\exp(ik\mathbf{d}) + \exp(-ik\mathbf{d})] \rangle\}\end{aligned} \tag{6.18}$$

and after decomposing $\mathbf{d}$ into its mean and fluctuating parts,

$$\begin{aligned}\langle I_F(k) \rangle &= N|H(k)|^2 \langle i^2(\mathbf{r}) \rangle \{2 + \langle \tau_I(d_0 + \mathbf{s}) \exp(ik\mathbf{s}) \rangle \exp(ikd_0) \\ &\quad + \langle \tau_I(d_0 + \mathbf{s}) \exp(-ik\mathbf{s}) \rangle \exp(-ikd_0)\}\end{aligned} \tag{6.19}$$

Later, in the derivation of the autocorrelation function, we will make use of this very general formula because it explains how a distribution of velocities tends to bias the mean value. Presently, let us make another assumption that is very reasonable for most applications: that the fluctuations in $\mathbf{s}$ given by its standard deviation $\sigma_s = \langle \mathbf{s}^2 \rangle^{1/2}$ are small in comparison with the mean value d_0. We can then approximate

$\tau_I(d_0 + \mathbf{s})$ by $\tau_I(d_0)$ and introduce the pdf $p(s)$ to evaluate the terms in brackets. We discover that these are equal to a Fourier or inverse Fourier transformation, respectively, of $p(s)$, resulting in

$$\langle I_F(k)\rangle = N|H(k)|^2\langle i^2(\mathbf{r})\rangle\{2 + \tau_I(d_0)\Phi(k)\exp(ikd_0) + \tau_I(d_0)\Phi^*(k)\exp(-ikd_0)\} \quad (6.20)$$

or

$$\langle I_F(k)\rangle = 2N|H(k)|^2\langle i^2(\mathbf{r})\rangle\{1 + \tau_I(d_0)|\Phi(k)|\cos[kd_0 + \phi(k)]\} \quad (6.21)$$

This is an expression for the fringe system accounting for displacement fluctuations within an interrogation region of finite extent. Visibility within such a fringe system becomes position-dependent,

$$V(k) = \tau_I(d_0)|\Phi(k)| \quad (6.22)$$

and fringe phase $\psi(k) = kd_0 + \phi(k)$ does not necessarily increase linearly with position any longer. Relation (6.22) was used to determine velocity statistics from visibility analysis [26].

In the discussion so far we have disregarded the term $O(k)$ containing the contributions $l \neq m$ in Eq. (6.19). For large values of k, i.e., far off-axis in the fringe system, where $k(r_l - r_m) > 2\pi$, phase values are distributed equally and the exponentials average to zero. In this region the formulas derived above are valid without any restrictions. When $k(r_l - r_m)$ is no longer smaller than 2π, we obtain

$$\langle\exp[-ik(r_l - r_m)]\rangle = \langle\exp(-ik\,\Delta\mathbf{r})\rangle = \Phi_{\Delta r}(k) \quad (6.23)$$

where $\Phi_{\Delta r}(k)$ is the characteristic function of the random variable $\Delta\mathbf{r}$, the separation between any two particle positions within the interrogation area. Once more we use the fact that $\Phi_{\Delta r}(k)$ is the Fourier transform of the probability density function $p_{\Delta r}(\Delta r)$ for these separations [25]. Since

$$p_{\Delta r}(\Delta r) \sim \langle i_I(r)i_I(r + \Delta r)\rangle \sim \tau_I(\Delta r) \quad (6.24)$$

the autocorrelation function $\tau_I(\Delta r)$ of the interrogation intensity, its Fourier transformation renders the power spectrum $|I_I(k)|^2$ of i_I:

$$O(k) \sim N^2|I_I(k)|^2 \quad (6.25)$$

This is a high peak closely concentrated around the axis with a width inversely proportional to the size of the interrogation area. For comparison, the extent of the halo, i.e., the size of the fringe pattern, is inversely proportional to the particle size. In many applications this bright so-called zero-order intensity is quite annoying, and often a stop is placed on-axis to remove it.

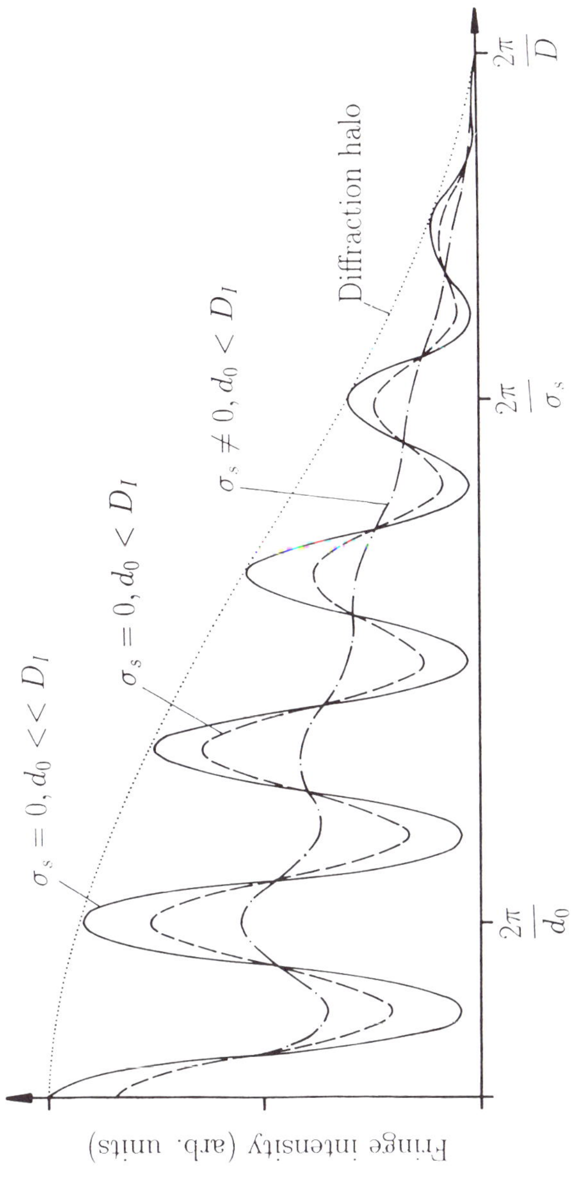

Figure 6.3 Characteristic Young's fringe systems from PIV records under different conditions governed by particle size D, size of the interrogating area D_I, mean displacement d_0, and standard deviation σ_s of the displacement within the interrogation area.

It is illustrative to interpret the zero-order contribution in terms of the random walk model of speckle formation. For phase angles evenly distributed over 2π, the walker's steps are oriented at random. When the possible phase angles are restricted to less than 2π, the walker prefers a certain general direction, thus increasing the rate at which he departs from his starting position. This implies that toward lower frequencies average fringe intensity will be greater than expected. When evaluating the fringe function in this region this should be taken into consideration. Obviously, these effects depend on the average particle separation, interrogation field size, and number of particles [27].

Let us summarize what we have learned about the fringe system. A decrease in visibility can be due to two effects: (1) The finite interrogation region produces a loss of partners, and (2) a distribution of displacements within the region results in a decorrelation of the patterns that are compared. An illustrative summary of the effects is obtained by looking at the fringe characteristics shown schematically in Fig. 6.3. High-visibility fringes result when all particles within the interrogation region have undergone the same displacement d_0 (standard deviation $\sigma_s = 0$) and when the size D_I of the region is large compared to the particle displacement d_0. With d_0 approaching D_I, the overall visibility is reduced. If, in addition, a distribution of $\mathbf{d}$ around its mean value d_0 is encountered ($\sigma_s > 0$), there results a position-dependent visibility.

6.2.3 Analysis by Autocorrelation

As will be mentioned in Section 6.3, the final interrogation evaluation often is not made in the fringe domain. Rather, a Fourier transform of the fringe intensity (done optically or by computer) produces the autocorrelation function $R(r)$ of the light distribution within the interrogation region. Performing this transformation on Eq. (6.20) results in

$$\begin{aligned} R(r) = {} & 2N\langle i^2(\mathbf{r})\rangle[h(r) \odot h(r)] \\ & * \{\delta(r) + \tfrac{1}{2}\tau_I(d_0)[p(-r) * \delta(r + d_0) \\ & + p(r) * \delta(r - d_0)]\} \end{aligned} \tag{6.26}$$

$$\begin{aligned} R(r) = {} & 2N\langle i^2(\mathbf{r})\rangle R_p(r) \\ & * \{\delta(r) + \tfrac{1}{2}\tau_I(d_0)[p(-r + d_0) + p(r - d_0)]\} \end{aligned} \tag{6.27}$$

where $\odot$ symbolizes a correlation operation and

$$R_p(r) = [h(r) \odot h(r)] \tag{6.28}$$

is the autocorrelation function of the single particle image.

Recall that we disregarded the $l \neq m$ terms contained in $O(k)$ that gave rise to the zero-order peak [Eq. (6.25)] in the fringe system. Its contribution in the autocorrelation domain is calculated by Fourier transformation, which brings us back to $\tau_I(r)$, a background intensity that decreases as the autocorrelation of the interrogation intensity. It is an expression of the random pairing of particles that creates the background noise responsible for false peak detection in the evaluation of the autocorrelation output.

Figure 6.4 shows a schematic diagram of the basic features of $R(r)$ to illustrate once more the fundamental conditions governing PIV. The different cases discussed above for the fringe system are found once again. The autocorrelation function, half of which is shown in the figure, consists of a central peak (pedestal) and two side peaks separated by the mean displacement d_0. The background due to random correlation of unpaired particles has been omitted. Let us once again study two cases.

1. Constant displacement within the interrogation area. In this case we have $p(s) = \delta(s)$, and each peak assumes the shape of the particle autocorrelation. This peak is of twice the particle width D; a clear separation of the peaks thus requires $d_0 > D$, that is, particle images should not overlap. The height of the side peaks depends on the ratio D_I/d_0. If the size of the interrogation region exceeds the particle separation by a large factor, the side peaks will be about half the central peak, which is equivalent to a visibility of $V = 1$. As D_I approaches d_0, the influence of the factor $\tau_I(d_0)$ reduces the peak height until a limit is reached at $D = d_0$ when the side peak vanishes. This illustrates the loss of partners with decreasing size of the interrogation region until finally it is impossible for a pair to fit into this region.
2. Nonuniform displacement within the interrogation region. Now each side peak is modified such that the particle image autocorrelation function has to be convolved with the probability density function of the displacement fluctuations within the interrogation area. With growing standard deviation σ_s it spreads and decreases in size. An evaluation of the peak shape can thus yield information on the flow statistics; if properly done, it could render the pdf of the velocity. A simple application would be to determine the degree of turbulence from the width of the peak [28]. More details will be found in Section 6.3.3 on large-area processing.

When the autocorrelation function is produced optically, the observed light intensity is its modulus squared. Such a function as calculated from the Young's fringe system obtained in a flow study is shown in Fig. 6.5. The photographic recordings of corresponding double-exposure transpar-

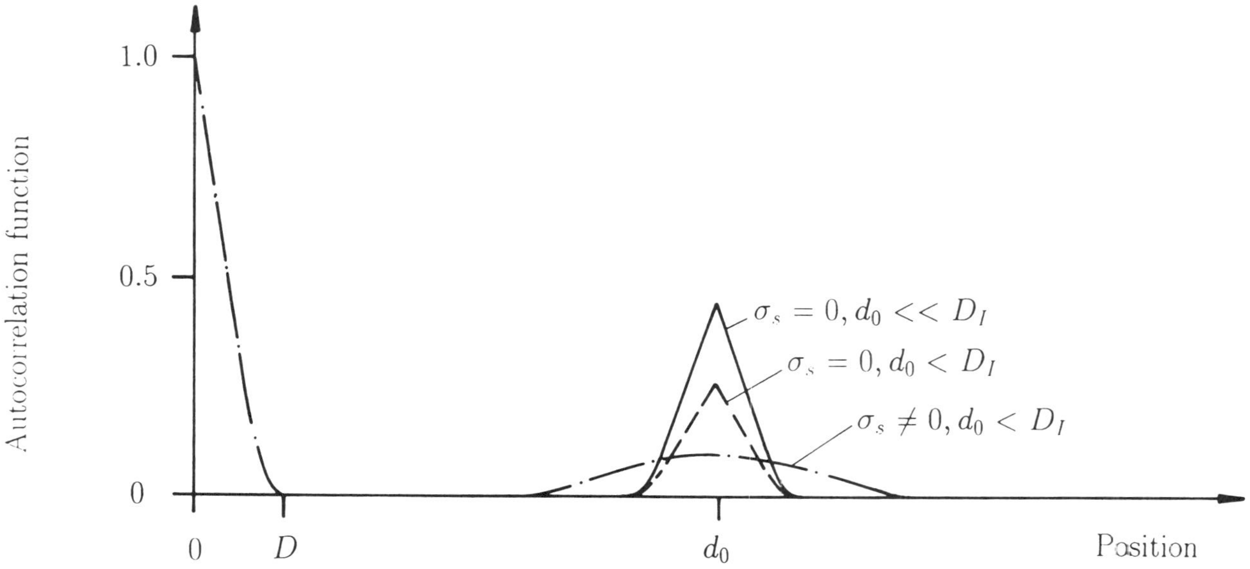

Figure 6.4 Characteristic autocorrelation functions of PIV records under different conditions governed by particle size D, size of the interrogating area D_I, mean displacement d_0, and standard deviation σ_s of the displacement within the interrogation area. Compare the corresponding fringe systems in Fig. 6.3.

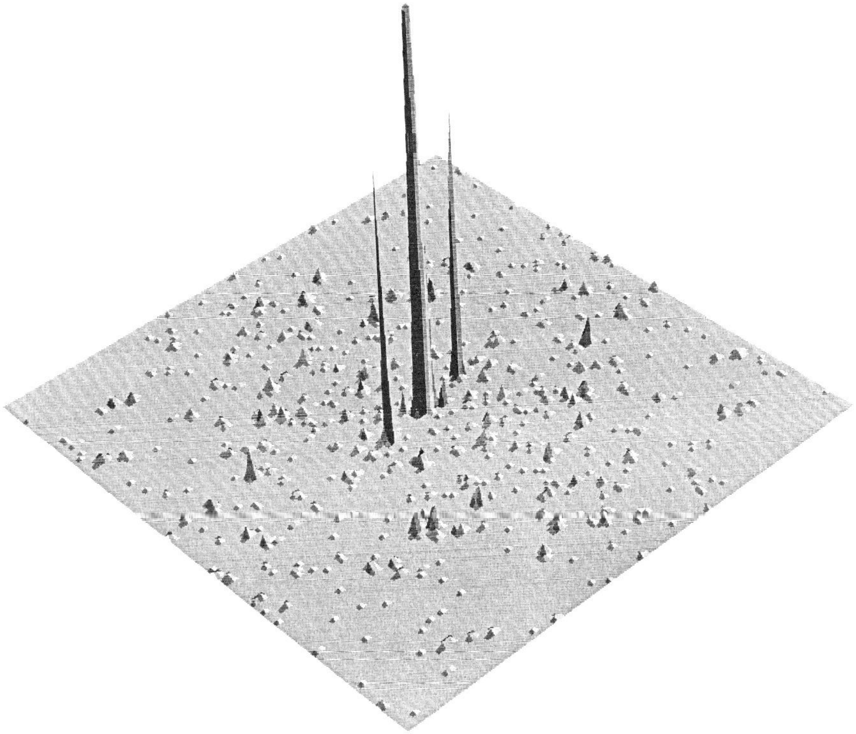

Figure 6.5 Three-dimensional representation of the modulus squared of the autocorrelation function of a PIV transparency. The corresponding light distribution is shown in Fig. 6.9.

ency, fringe system, and autocorrelation are shown later in Figs. 6.7 and 6.9. The central peak has been clipped to enhance the details of the two side peaks and the competing background peaks originating from the correlation of unpaired particles. Signal-to-noise considerations as in Section 6.2.4 aim at minimizing the disturbance from these peaks.

It should be noted that a double-exposure record of flowing particles does not reveal which image was made in the first exposure and which in the second. Thus, information about the sign of the flow direction gets lost. The velocity can be determined only to within a 180° ambiguity. Later in this chapter we consider modifications of the recording procedure to eliminate this ambiguity.

6.2.4 Noise in PIV

Up to now, we have calculated only averaged quantities characteristic of a PIV record either by assuming ensemble averages or by forming expected values with contributions from a very large number of particles. In practice, this is not the usual case. More common is the evaluation of a single realization of a flow field. Often, only a very limited number of particle pairs are involved in forming the signal from an interrogation area. Noise obscures the signal, and the useful information is difficult to extract and may even be distorted. Thus the performance of everyday PIV differs from that under the ideal conditions assumed here.

The useful information hidden in a PIV record is contained in the pairs of particle images. Among the large number of particles involved, however, pairing is not unambiguous, for any two particles can be grouped. The higher the ratio between the number of correct pairings and incorrect combinations, the better the signal-to-noise ratio (SNR). Quite obviously, the number of particles involved plays the decisive role. Optimum under these aspects would be a single particle pair; no erroneous competitive pairing would be possible. The occurrence of two particle pairs is disastrous. In addition to the correct combination of first-exposure with second-exposure particle images, the two first exposures and the two second exposures could also be paired. As each event has the same signal strength, a decision is impossible. With increasing particle density, SNR will again improve.

At first guess one might now try to achieve an average of one particle pair per interrogation area. Particles, however, are found at random because of the irregular addition to the flow. Besides the intended one-pair fields, the required low density will produce many interrogation areas without any particle at all and many others with the forbidding two pairs, a few with even more. At many sampling positions information will be absent or ambiguous. The aim, therefore, will be to increase particle density just as much as necessary for a reliable evaluation. This is sometimes called the high-image-density mode of PIV. The probability, for example, of finding two particles in an area that on average contains 15 particles can be calculated from Poisson statistics to be less than 3.5×10^{-5}. High concentrations are difficult to obtain, and the quality of the results has to be traded off against more relaxed particle production.

Another aspect must be considered in these efforts. We have seen that only paired particles add to signal strength. Any particle without a partner creates a background signal that tends to bury the useful information. The same holds if particle images that should form a pair are not identical. There are numerous causes for such a degradation of the PIV record. Our analysis has already included the effects of a nonuniform interrogation intensity $i_I(r)$, where particle images of a pair are illuminated with different brightness. Even earlier, in the photographic recording, the two pulsed light sheets may catch a particle at locations that differ in brightness. And furthermore, lens aberrations will produce particle images that vary with position.

A good estimate of the noise involved can be gained with the idealistic assumption that a large number of particle pairs contribute to the fringe system. Equations (6.9) and (6.10) have to be evaluated statistically, a task that has been covered extensively in the treatment of laser speckle [29]. These theories yield that the standard deviation of the fringe intensity at a given location equals its expected value, a considerable multiplicative noise that deteriorates the fringe system and gives rise to its speckled appearance. Another question of importance concerns the speckle size because it sets limits to the resolvable fringe spacing. As is well known, the size of objective speckles decreases as the extent of the scattering spot increases. In the present case, speckle size is inversely proportional to the diameter of the interrogation spot. Once again the coefficient of proportionality is the one relating the space domain to the Fourier plane. The requirement that the fringe spacing must exceed speckle size is thus equivalent to the quite obvious requirement that the particle displacement must fit into the interrogation spot.

The situation gets worse when we take into account that a single experiment may concern just 10 or so particle pairs, so we are far off the laws of large numbers. We are challenged with the problem of optimizing PIV parameters for reliable results. Let us do so, for example, for the derivation of displacement data from a measurement of the position of the side peak in the autocorrelation function. First of all we have to decide which is the correct peak. The previous discussion of the pairing possibilities has shown that this is a question that is not simple to answer. We then have to determine a suitable quantity giving the "position" of the successful peak.

Obviously, the highest peak will be selected for further evaluation. But how sure can we be that we have not chosen the wrong one? Maybe we had better not take a value at all when we cannot be sure! It has proved useful to introduce a reliability criterion by comparing the value of the tallest peak $R(r_1)$ with the second tallest $R(r_2)$ [58],

$$q = R(r_1)/R(r_2) \tag{6.29}$$

In some investigations a logarithmic quantity is used, the signal-to-noise ratio [30],

$$Q = 20 \log_{10} q \tag{6.30}$$

It is then decided that values will be accepted only if $q > q_0$, where q_0 is a suitably specified threshold; a typical value is $q_0 = 1.5$.

A possible way to treat the optimization of PIV parameters is to calculate the probability for peak heights starting out from a Poisson distribution of particle positions and determining the statistics for the random spacing between particle pairs. This result must be combined with data on the light sheet geometry and the interrogating beam intensity distribution to give final results. More common in today's age of computers is the Monte Carlo approach, in which a particle field is modeled on the computer, different realizations are created from random numbers, and the outcomes are averaged. Let us find optimum PIV parameters by presenting exemplary data from a recent investigation [58].

Particle images of Gaussian cross-sectional transmittance are Poisson distributed over a square interrogation window. The inspected flow volume originates from illumination with a light sheet of depth $\Delta z_I'$. Both the light sheet and interrogation areas are of constant cross-sectional intensity. Different displacement statistics were simulated, beginning with a constant purely transverse velocity v, adding a depth component w, and finally assuming different flow configurations within the interrogation area. The mean number N of particles within the window was changed, and an ensemble of 1200 realizations was evaluated for each flow configuration. A probability P for valid detection was derived that gave the percentage of correct pairings among the evaluations. An example for a flow field of constant in-plane velocity is shown in Fig. 6.6a, where P is plotted versus the relative in-plane displacement $\Delta x/D_I$ for various particle densities N. The calculations confirm that a successful completion of the interrogation is more probable when particle density is high and displacement is much smaller than the size of the interrogation spot. According to an old rule of thumb, the spot size should be at least five times the particle displacement. If there are an average of 15 particles within the area, a satisfactory 90% valid detection probability can be expected.

Let us take the curve for $N = 15$ and see how it is affected when an additional out-of-plane motion (normalized to light sheet depth) $\Delta z'/\Delta z_S'$ is allowed. As illustrated in Fig. 6.6b, additional particle loss reduces the number of possible pairings, and thus the probability decreases. In any case, all these curves saturate, and an incremental improvement in detection probability requires more and more particles, soon a gain not worth the effort. Additional simulations have been performed for more complicated flow configurations, a plane shear velocity, for example [58]. Basically, this type of analysis presents the data required to design a PIV configuration for successful operation. It must be remembered, however, that in real-world applications additional disturbances will cause a reduction in performance.

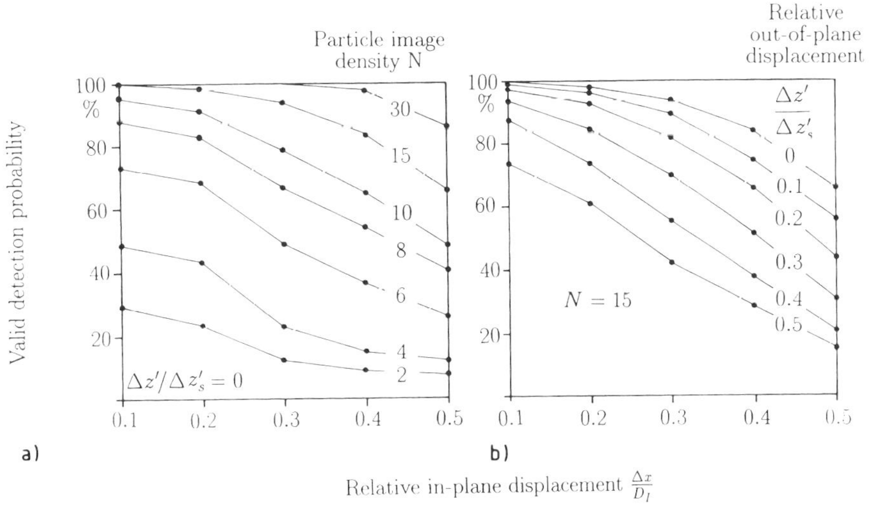

Figure 6.6 Valid detection probability for velocity measurements in a Monte Carlo simulation of a PIV experiment for different light sheet and interrogation parameters. (After Ref. 58.) (a) Probability versus in-plane displacement in units of interrogation area size for various numbers N of particles, no out-of-plane motion. (b) Probability versus in-plane displacement for $N = 15$ and different relative out-of-plane displacements in units of light sheet depth.

Once the tallest correlation peak has been identified, its position has to be determined. All discussions in this section have made clear that the correlation amplitude will be noisy and methods are required to extract a good estimate of peak position. It is agreed that the centroid of the peak serves this purpose well [58]:

$$r_P = \frac{\int_{A_i} sR(s)\, ds}{\int_{A_i} R(s)\, ds} \tag{6.31}$$

where integration is executed over an appropriate small area A_i containing the peak. Again, the noisiness of the signal limits the reliability of this result. Quite often, however, a resolution better than the interrogation sampling can be obtained. Let us postpone a discussion of velocity biasing of the peak position to Section 6.3.3.

Before its acceptance, a PIV velocity map undergoes a final correcting procedure. All sampling positions with missing or obviously incorrect data undergo a finishing process. During a validation step the evaluated velocity vectors are examined for consistence with their neighbors. For obviously incorrect peak identification a new try is made with the second tallest peak. Interrogation spots without a valid result receive interpolation from the neighbors.

Noise constraints can usually be relaxed at the expense of measuring time or spatial resolution; lock-in amplification is a good example. A similar strategy has been proposed and applied in PIV. To increase the number of favorable particle pairings, a multiple-exposure instead of double-exposure system has been employed in the recording of the flow field. Each particle is caught many times, resulting in a track of beads along its path. Spatial resolution, of course, decreases as the field of interrogation has to be increased. Furthermore, it must be assumed that within the much longer time there is no change in velocity to obtain equally spaced particle images along a straight line.

To illustrate the effects of a multiple exposure let us start with the familiar fringe system that is obtained from a constant-displacement double-exposure record. For simplicity let us assume constant interrogation intensity i_0 and displacement $d_0 \ll D_I$, so the difference in illuminating light of the partners in a pair can be omitted, and we arrive at the simplest formulation of the Young's fringe system:

$$I_F(k) = 2N|H(k)|^2 i_{0^2}[1 + \cos kd_0] \tag{6.32}$$

$$I_F(k) = 2N|H(k)|^2 i_{0^2} \cos^2 (kd_0/2) \tag{6.33}$$

Let us recall that the periodicity is essentially explained by diffraction from a double structure, thus the name *Young's fringes*, which originated from the classical double-slit experiment. Now recall what happens when a double slit is amended by adding more and more slits at the same spacing, finally resulting in a diffraction grating. The multiple-wave interference becomes more sensitive to diffraction angle changes; the diffraction orders sharpen, their width decreases, and the peak height increases in proportion to L^2, where L is the number of slits. This yields the better resolution of a diffraction grating with a large number of grating lines. The fringe formula must be modified accordingly to yield

$$I_F(k) = \frac{1}{2} N|H(k)|^2 i_{0^2} \left[\frac{\sin (Lkd_0/2)}{\sin (kd_0/2)} \right]^2 \tag{6.34}$$

which has two advantages over the ordinary cosine fringe system. The positions of the fringe maxima are much more pronounced, and the contrast is enhanced due to the improved signal strength. In the autocorrelation representation, instead of a single side peak, we arrive at a series of $L - 1$ peaks that decrease in height in inverse proportion to the order of the peak. The first side peak has improved by a factor $L - 1$ in height. All higher order peaks, of course, can be included in the evaluation.

The positive effect of multiple-exposure PIV was realized early [9, 31, 32] and is still applied regularly [33]. In a thorough analysis [34] the optimization of particle image velocimeters using multiple-pulsed systems was carried out to arrive at probabilities for valid velocity detection just as it was in double-pulsed systems. It was confirmed that conditions improve as the number of effective pairs that contribute to the signal strength increases. Even a triple- or quadruple-pulse system yields considerable improvement.

6.2.5 Exploration of Spatial Flow Structures by PIV

In PIV, the velocity of a flow field is determined from data that are sampled in time and space. We have to examine how this kind of data reduction affects the information about the flow configuration. Quite generally, the required sampling distance and time are governed by the highest spatial and temporal frequencies present in the flow. For sampling at regular separations, the Nyquist theorem states that the sampling frequency must be at least twice the highest signal frequency. Violation of this rule excludes a reproduction of the signal from the samples and results in aliasing phenomena. If the sampling separation cannot be reduced sufficiently, the signal must be frequency-limited prior to sampling by passing it through an appropriate filter.

The power of PIV lies in revealing the spatial structures of a flow field. Let us thus pass over the discussion of the signal properties in time and turn to space dependences. To begin with, let us determine the information contained in PIV records, independent of the configuration that is applied in the interrogation. Sampling positions are given by the random locations of particles at the instant of the exposure. Because of the statistical nature of this sampling, a study of the frequency response requires ensemble averaging. The result is quite interesting as there is no frequency limit imposed by random sampling. This becomes quite obvious when we examine the statistics of the sampling intervals as calculated from a Poisson distribution of positions. Many small separations occur, i.e., even the tiniest structures are sampled; and in the many realizations that form the basis of ensemble averaging these small intervals will be placed anywhere in the field of view, thus registering all details. In the case of PIV, of course, sampling is done by a pair of particles; thus a practical limit is imposed by the mean particle separation.

Reality, of course, differs from these ideal conditions. Often we have to rely on just a single sample from the ensemble of possible particle distributions in the flow field. Thus we are interested in the reliability of information that we derive about certain scales in the flow. A probability analysis must be made that calculates the variance in the Fourier coefficients describing a flow field. It has been shown [35] that this variance is independent of frequency. Thus, wherever in the spatial frequency spectrum of the velocity signal a component is small its uncertainty is the same. Take the usual case of a signal spectrum decreasing toward higher frequencies. Above some cutoff frequency the high statistical uncertainty rules out reliable measurements. Improvement, of course, can be found by increasing the particle density as the number of sampling positions increases.

For interrogation, the PIV record is usually subjected to procedures that impose additional restrictions on the transfer of spatial frequencies. If we assume that there are a sufficient number of particles present over the entire interrogation region, the interrogation process can be treated as follows.

First, the original velocity signal $v(r)$ is subjected to a smoothing operation expressed as convolution with the interrogation intensity $i_I(r)$,

$$v'(r) = v(r) * i_I(r) \tag{6.35}$$

In the spatial frequency domain this means a multiplication of their respective spectra,

$$V'(k) = V(k)I_I(k) \tag{6.36}$$

$I_I(k)$ usually decreases to a small value at a frequency inversely related to the size of the interrogation area. The remaining contribution beyond this frequency depends on the details of $i_I(r)$. If the signal spectrum is wider than the extent of $I_I(k)$, interrogation causes a high-frequency cutoff at structures comparable in size to the interrogation area. The sampling frequency required by the Nyquist theorem is then set by $I_I(k)$. Usually a sampling interval of half the interrogation spot size, i.e., overlapping interrogation, is found to be sufficient. In some applications even fourfold overlapping sampling has been used [20].

Another restriction on the obtainable information is imposed by the finite size L of the complete PIV record. We are unable to predict the flow field outside the photograph. Mathematically speaking, the velocity function is multiplied by the window function $w(r)$ representing the photograph geometry,

$$v''(r) = v'(r)w(r) \tag{6.37}$$

Usually all information within the field is passed and outside it is canceled; that is (in one dimension),

$$w(r) = \text{rect}\,(r/L) = \begin{cases} 1 & \text{for } |r| \le L/2 \\ 0 & \text{else} \end{cases} \qquad (6.38)$$

where the origin has been placed in the center of the field of view. In two-dimensional formulation, the size of the photograph in the transverse direction has to be added. In the spatial frequency domain, the multiplication turns into a convolution with the Fourier transform of rect (r), which utilizes sinc (x) = sin (x)/x:

$$V''(k) = V'(k) * L \text{ sinc } (Lk/2) \qquad (6.39)$$

This formula is an expression of the broadening and smoothing of a spectrum and yields the uncertainty in frequency resolution given by

$$L\,\Delta k \ge 2\pi \qquad (6.40)$$

Let us reflect upon the effect this finite viewing section has on measurements of turbulence intensity by PIV. Turbulence as a predominantly spatial phenomenon is a subject well suited for study by PIV. Usually, however, turbulent velocity fluctuations have been derived from a time record of the flow velocity at a single measuring site. Such equipment is characterized by a frequency transfer function determined by the sensor (hot wire or LDA, for example) and the electronic signal handling. In comparing data from such measurements with PIV, we can take a PIV record as a "frozen" sample of an evolving flow. When the mean flow is constant or a known function of the spatial coordinates, space coordinates can be converted to time coordinates by dividing by the mean velocity. Spatial and temporal frequencies are related accordingly.

When calculating the degree of turbulence from the ratio of the standard deviation of the velocity to the mean velocity, the spatial frequencies making up the velocity signal of a PIV record undergo a weighting due to the finite viewing area. Let us develop the argument for a single-frequency signal at $k = k_0$, that is, $v(x) = v_0 \exp(ik_0x)$ with zero mean. Usually, all power $P(k_0) = v_0^2$ within this frequency contributes to the turbulence. The size-limited PIV photograph, however, contains only a window of length L placed randomly in the signal. The mean within the window no longer vanishes; rather, it assumes a value dependent on the random phase at which our window catches the flow periodicity. The result is a transfer of signal power to the dc term that is interpreted as mean flow and thus is subtracted from the signal power. Disregarding for a moment the effects of the finite interrogation area treated initially, the dc power $P(0)$ can be derived from the $k = 0$ value of the broadened spectral line centered at $k = k_0$:

$$P(0) = v_0^2 \operatorname{sinc}^2 (Lk_0/2)\, \delta(k) \tag{6.41}$$

The weighting function thus assumes the form

$$G'(k) = 1 - \operatorname{sinc}^2 (Lk/2) \tag{6.42}$$

Low-frequency components in the turbulence spectrum thus do not receive the necessary attention in a PIV turbulence study. Combining all effects into one general weighting function, we arrive at

$$G(k) = [1 - \operatorname{sinc}^2 (Lk/2)] I_I^2(k) \tag{6.43}$$

a curve that is shown for Gaussian interrogation intensity in Fig. 6.7. This correction has been applied in the comparison of turbulence data from PIV and hot-wire studies, as will be explained in Section 6.3.3.

We should note that the two-dimensionality of the PIV photograph was disregarded in our arguments. Turbulence, of course, has to be determined

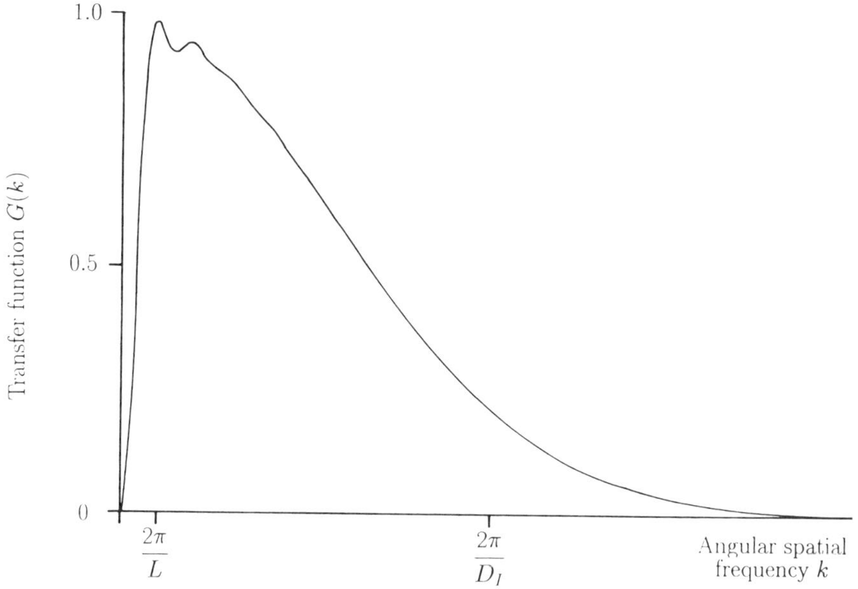

Figure 6.7 Weighting function for the calculation of turbulence intensity from PIV data obtained with a record of width L and a Gaussian interrogation intensity of diameter $d_I = 2r_I$ [r_I indicates the radius at which the intensity has dropped to exp (π) = 0.043].

for each flow velocity component; the present treatment covered the component in the mean flow direction. Even more important, we calculated the effects for just one scan lengthwise through the field. Depending on the correlation size of the turbulence, the PIV record may contain several such independent scans parallel to each other. In this case L should be replaced by appropriate multiples of L.

6.3 PROCESSING OF PIV RECORDS

A PIV photograph, as described previously, is a record of an extended region of the flow field that is characterized by paired images of tracer particles spread randomly over the field of view. The interpair spacing differs with position in the image and contains the information that is to be extracted from the picture. Thus small areas (the interrogation areas) are selected from the complete record, and the mean displacement values from the different areas are taken as samples of the velocity field. We have already learned how the transmittance of the PIV interrogation spot can be described in the space domain, the spatial frequency domain, or the autocorrelation representation, each of which can serve to deduce the displacement information. I have mentioned that a straightforward approach is to evaluate the PIV image directly. The feasibility of this method depends very much on low image density. Usually the image density is high enough to obscure the obvious pairing of particle images. Thus statistical analysis must be employed, preferably a two-dimensional autocorrelation of the image that renders, as we have seen, strong peaks at the displacement values. It may be of interest to know that the method of histogramming—extracting all existent separations between particles and distributing them into appropriate two-dimensional boxes—is equivalent to an autocorrelation. Autocorrelation, however, takes much time to compute, and the need for its repetition at up to thousands of sampling points initiated the search for more economical procedures. We will see that modern computer developments have again favored this approach. Yet in the meantime much effort has been invested in alternative evaluation procedures, most of it in spatial frequency domain processing, the study of the Young's fringes. Before treating this in more detail, let me, however, mention a strategy to reduce the number of steps required for the calculation of the two-dimensional autocorrelation by reducing the problem to one-dimensional tasks.

Let us produce projections of the PIV transmittance $f(x, y)$ onto the x and y axes by integrating $f(x, y)$ over either of its coordinates [15]. Each particle image will show up accordingly as a peak in this representation, and thus the pair separation will also show up in the x or y component.

Two one-dimensional autocorrelation functions, easy to calculate, could reveal these quantities. Obviously, and to the disadvantage of the method, background noise will increase in its contribution as it sums up during the integration. Yet the method has been applied successfully, especially as it can use optical processing to carry out effectively some of the tasks involved. To reduce the effects of background noise, an optical high-pass filter was inserted. Such a filter consists of a 1 : 1 symmetric two-lens imaging system of total length $4f$ (focal length f). In the plane of symmetry, the spatial frequency plane, the low-angle diffracted light is blocked by a stop on the axis of the optical system [36]. For further improvement, contact printing of the original PIV negative was proposed. All remaining background light must now pass through an area of very low transmittance, a technique that has been favored for different kinds of PIV processing [37]. One-dimensional integration was also executed optically by focusing the light onto a linear CCD array with a cylindrical lens oriented parallel to either of the axes. Meanwhile this technique of orthogonal compression has been improved by introducing small image windows that are positioned by estimates of the displacement [38]. It is interesting to note that similar ideas reappear in modern digital processing of PIV images, where the disadvantageous reduction in the signal-to-noise ratio can be avoided because digital preprocessing can remove most of the annoying background noise [33].

Optical processing is quite useful in the handling of large amounts of information as in a PIV transparency. Unfortunately, all methods for optical correlation are rather complicated for real-time implementation; we will return to more sophisticated techniques for this task in Section 6.3.4. Elementary optics, however, offers a method for efficient Fourier transformation of two-dimensional transparencies. Thus the production of the spatial power spectrum of a PIV record has been at the center of efforts for a long time, i.e., the study of the Young's fringe diffraction pattern. Let us therefore dedicate the next section to this technique, which is still quite popular.

6.3.1 Formation of Young's Fringes

It has been mentioned many times already that diffraction from a two-dimensional structure given by its amplitude transmittance $f(r)$ can be described by a Fourier formulation [36]. Thus the light amplitude in the far-field diffraction pattern is equal to $F(k)$, where the two-dimensional angular spatial frequency k is related to the two diffraction angles. To avoid the long distances required for far-field conditions, Fourier transformation of a PIV transparency is usually performed with the aid of a lens in the setup of Fig. 6.8. Here, a thin collimated laser beam is focused by a lens

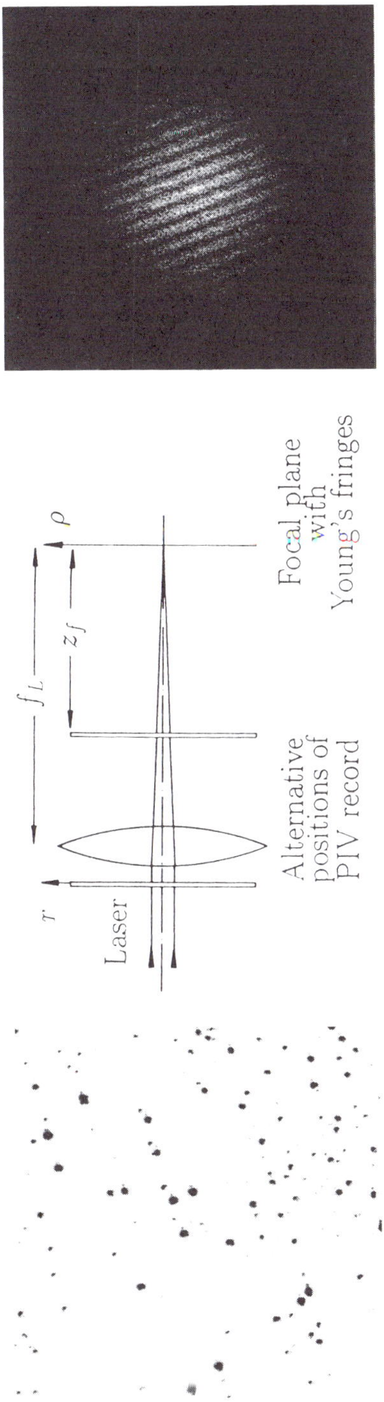

Figure 6.8 Formation of Young's diffraction fringes from a positive transparency of the original PIV double-exposure negative that is shown at left. Possible positions of the transparency are indicated.

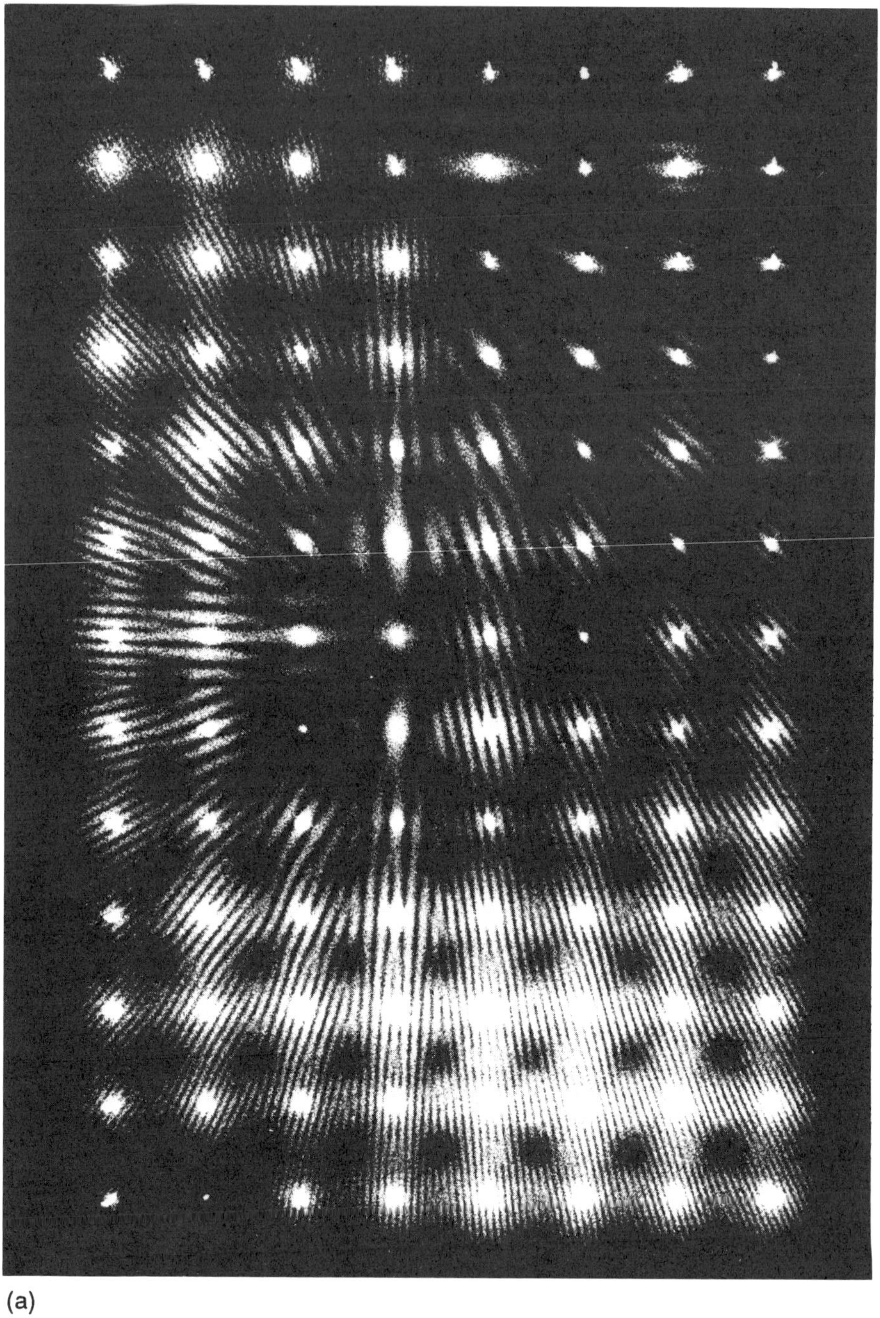

(a)

Figure 6.9 Parallel optical processing by an array of interrogating holographic lenslets. Speckle photography of the rotation of an aluminum plate impeded by a radial crack [19]. (a) Output array of Young's fringe patterns. (b) Output array of autocorrelation functions (Section 6.3.2).

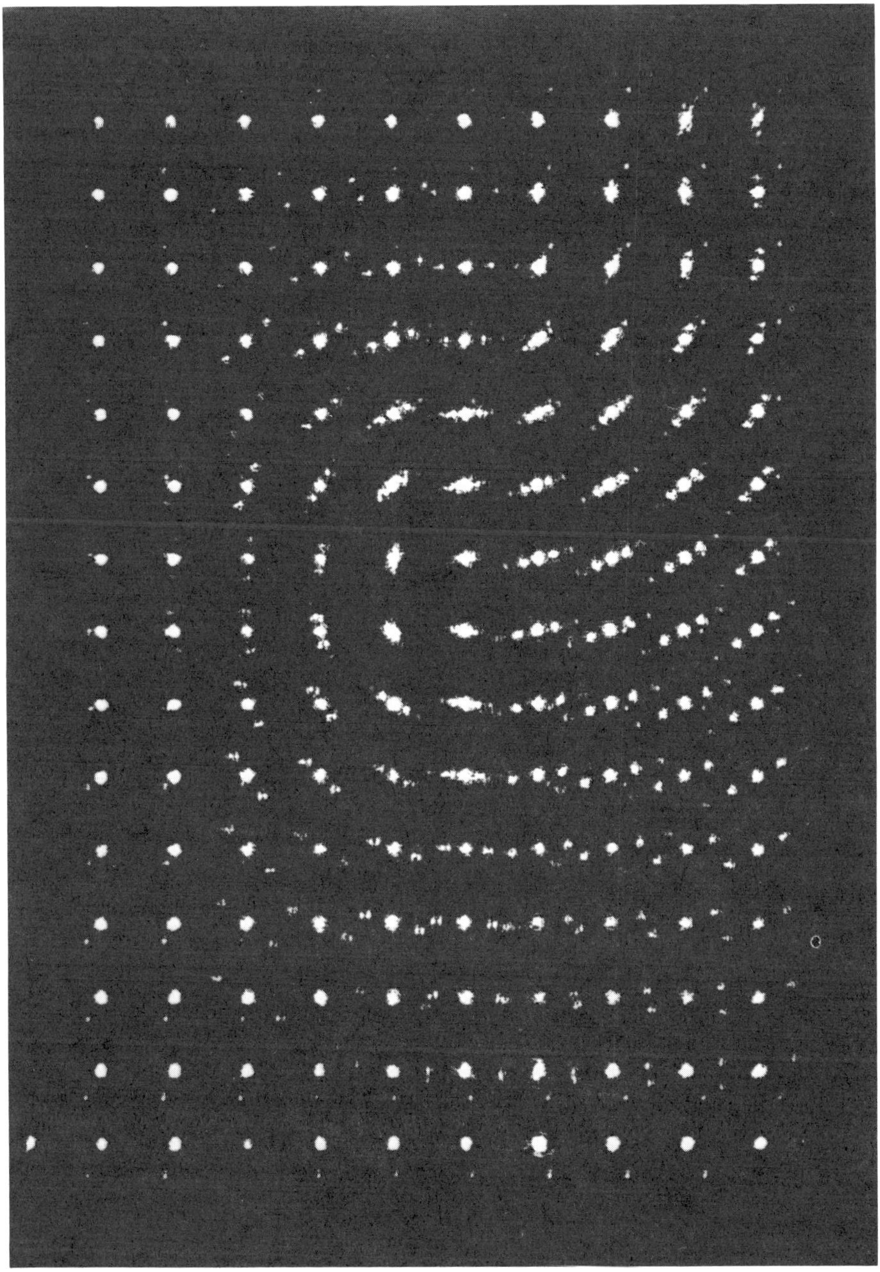

(b)

Figure 6.9 Continued.

of focal length f_L onto the so-called Fourier plane represented by the two-dimensional coordinate ρ. The PIV transparency can be introduced either directly in front of the lens or at any distance z_f between the lens and the Fourier plane. Spatial frequency k and coordinate ρ are then related by

$$\rho = (\lambda a/2\pi)k \tag{6.44}$$

where $a = f_L$ when the transparency is in front of the lens and $a = z_f$ when it is behind the lens.

The intensity at a location in the Young's fringe pattern thus indicates the contribution of the corresponding spatial frequency to the power spectrum of the input transparency. All power spectrum calculations in Section 6.2.2 thus yield the light intensity in the output plane of the optical spectral analyzer. It should be well noted here that the digital Fourier transformation of a laser-lit interrogation area that is picked up by a video camera operates on the intensity transmittance (transmissivity) of the PIV record and therefore is not identical with what is produced by coherent optical Fourier transformation (cf. Ref. 58).

Figure 6.8 also presents a typical fringe system obtained in this manner. Owing to all the effects discussed earlier, it is very noisy, but fringes are well resolved. Usually the system has an arbitrary orientation in space depicting the direction of the velocity. Fringe spacing and orientation are the primary data to be evaluated.

The evaluation of a flow-field record asks for the fringe systems at different subregions of the PIV record. Their size is determined by the diameter of the interrogating beam; their position is changed by moving the transparency on an xy translation stage to the desired coordinates. In addition to the data extraction, this subsequent movement of the photograph requires considerable time. Schemes have therefore been proposed to economize this step in the evaluation.

The effort can be reduced if a long strip of the PIV record that has the width of the usual interrogation spot is all illuminated at the same time. With the usual Fourier transform lens the different fringe systems would now overlap, and assignment to a definite position would be impossible. Therefore, an anamorphic processor has been devised that consists of two cylindrical lenses [16] configured in such a manner that from input to output plane a Fourier transformation is performed in one direction and there is imaging in the other direction. Thus, only one displacement component is evaluated; however, it is displayed simultaneously for all points along the interrogated strip.

Another concept simultaneously produces fringe systems at a grid of interrogation positions by employing an array of small lenslets, each lenslet producing an interrogating beam [19]. Such lenslets could be fabricated with micro-optical techniques. Here, an array of holographically produced focusing elements was used. The array is produced by sequential recording of small holograms of a suitably converging object beam at the grid points of the array. Producing phase holograms in dichromated gelatin gives a very high diffraction efficiency. An alternative is to produce tiny zone plates by electron beam etching, which results in very accurate and reproducible elements. If scaled properly, this holographic optical element produces an array of nonoverlapping Young's fringe patterns that give a very good overview of the velocity field under investigation. Many hundreds or even thousands of interrogations can be done at the same time with such an element. It serves for fast control of the success of an experiment or to prescreen large numbers of records for an early selection of those samples with valuable information. Figure 6.9a shows an output of fringe patterns in a double-exposure specklegram study of an aluminum plate undergoing rotational motion impeded by a radial crack. The deviation from rigid-body rotation (fringes in a radial direction and decreasing spacing with distance from the rotation axis) is obvious. The same type of outputs have greatly facilitated PIV evaluation.

6.3.2 Processing of Young's Fringe Systems

A human observer investigating the diffraction pattern in Fig. 6.8 has no problem recognizing fringes and placing a cursor line on crests or troughs, thus measuring fringe spacing and orientation. From a photoelectric scan normal to the fringes, however, it would be difficult to extract the fringe spacing accurately. The reason is that the observer includes the complete pattern in the analysis, averages, and thus utilizes data that come closer to the ensemble average formula [Eq. (6.21)], from which the fringe period is easy to extract. A fringe system can be smoothed by integration parallel to the fringes or by low-pass filtering that replaces original values by a weighted neighborhood average. Averaging can be done optically by compression with a cylindrical lens or by slight defocusing. It can be done digitally with great flexibility on a set of computerized image data. Only high-quality fringes from true speckle images have been successfully processed on the basis of a few single-line scans. In an example from flow analysis, the displacement of a speckle field due to light deflection by density structures in the flow was thus evaluated [39]. Generally, two-dimensional fringe information is used.

Let us recall the fringe formula [Eq. (6.21)], omitting constant factors and taking into account its two-dimensional character,

$$I_F(k) = |H(k)|^2\{1 + V(k)\cos[k_x d_0 + \phi(k)]\} \tag{6.45}$$

where k_x denotes the component of the angular spatial frequency k in the x direction. The coordinate system has been aligned such that the mean displacement d_0 points in this direction.

The halo function $|H|^2$ is an envelope to the sinusoidal fringe pattern of visibility V and phase ϕ; in some cases these quantities may vary with position. Mainly, orientation and period d_0 of the fringes are to be extracted. Most attempts in fringe processing tend to replace the human operator's abilities with more or less automated methods. Although details are covered in Chapter 9 of this book, let us briefly summarize the progress obtained in PIV applications where requirements usually are more severe.

The straightforward improvement is to perform an optical integration of the pattern parallel to the fringes, thus converting the originally two-dimensional task into a one-dimensional problem. A cylindrical lens focused on a one-dimensional array of photodiodes serves this purpose. First, however, an operator must rotate the input transparency to align the fringes normal to the axis of the lens [40]. If this step is to be automated also, a suitable criterion must signal the correct orientation of a rotation input. The degree of modulation of the photodiode output can be such an indicator; it is obtained easily from the signal amplitude after high-pass filtering. Maximum amplitude signals the correct orientation angle. Then a one-dimensional Fourier transform of the output at this angle yields the fringe period. Such a system has been implemented at our laboratory employing a rotating Dove prism [41] for image rotation, analog electronic circuitry to determine the correct angle, and a personal computer for the calculation of fringe frequency by fast Fourier transform and control of the translation stage. It can perform an interrogation within about 1 sec.

Similar operations are performed after video read-in and computer storage of the complete fringe pattern. Software-based manipulation of the data allows more flexibility than the optical hardware solution [42]. Take, for example, the need to align the fringes before integration. Columnwise addition of computerized image data would encounter the same problem that can be attributed to the change of fringe phase from one line to the next. If, however, one-dimensional autocorrelations are computed for each line, the phase information is eliminated, and averaging can be done without further complications by adding up these functions [43]. The fringe frequency is obtained from the periodicity of the autocorrelation function. Summing up the line autocorrelations is equal to a two-dimensional autocorrelation for a purely horizontal displacement, a technique that has also been implemented [44]. Orientation of the fringes can be calculated from the difference in position of the maximum of the cross-correlation between lines in the fringe pattern [45]. To remove ambiguity, correlation is repeated for the columns. Another proposal for determining orientation is the angle-dependent integrated intensity called the orientation function [46]. It has a maximum when parallel to the fringes but needs a good quality signal and has found little use in PIV.

Another problem in fringe analysis is the influence of the halo in Eq. (6.45). It is well known that such an envelope function shifts the positions of maxima or minima in the fringe system. The amount depends on halo shape and visibility and must be taken into account if extrema are to be evaluated [47]. Furthermore, the extent of the halo function limits the largest fringe spacing still resolvable and thus the lowest detectable displacement. Therefore many proposals have been made to eliminate the halo prior to further processing. Again, PIV data, because of their reduced quality, require such countermeasures. Often, due to aberrations in imaging a large field of very small particles or because a considerable fraction of particles are out of focus, the halo function differs within an interrogation area and is a superposition of several different contributions. This makes analytical consideration difficult and calls for on-the-spot determination. Fortunately, today's leading candidate in fringe processing, the two-dimensional Fourier transformation, provides direct possibilities to deconvolve with the halo.

Besides many innovative ideas for tricky evaluation schemes to avoid the halo influence on extrema positions (see references in Ref. 47), special consideration has been given to the limits imposed by the halo in the case of very few fringes [48]. The situation is best explained by examining the Fourier transform of the Young's fringe system, the autocorrelation function given by Eq. (6.27). The central and side peaks are each convolved with the Fourier transform of the halo function, i.e., the autocorrelation function of the particle image transmittance. Accuracy in measuring their separation decreases when the peaks overlap or even merge. Improvement can be expected if the dc peak is eliminated, which is equivalent to subtracting the term produced by the 1 in the brackets of Eq. (6.45). This was obtained from interpolation between the peaks of the fringe intensity profile. The execution of the procedure rendered fringe periods visible where they were buried by the halo otherwise.

Visibility or fringe phase as position-dependent quantities in the fringe pattern have been evaluated only rarely [26], and then by the classical definition of visibility that incorporates maximum and minimum intensity values of neighboring fringes. Since the information expected from such an analysis can be obtained more readily in the autocorrelation domain, there will be no further discussion here.

Let us progress to the presently most common method of fringe pattern evaluation, its two-dimensional Fourier transformation, which yields the autocorrelation function. This has been shown to be superior to most other methods concerning noise suppression and accuracy [42]. Omitting constant factors Eq. (6.27) can be simplified in accordance with the above representation of the fringe system to yield

$$R(r) = R_P(r) * \{\delta(r) + \tfrac{1}{2}\tau_I(d_0)[p(-x + d_0, y) + p(x - d_0, y)]\} \tag{6.46}$$

In the general case, the displacement must not be in the x direction, so the output has to be searched for the secondary peaks in both dimensions to determine peak positions. Usually, a Fourier transformation of the fringe system is done on the computer-stored video image of the pattern. The result may be passed through some thresholding to locate the peaks (cf. earlier considerations as to the validation of PIV results). Furthermore, various kinds of filtering operations may be employed prior to the Fourier transformation to smooth out the noise components in the fringe pattern. An example is shown in Fig. 6.10.

In a book on optics, let us not forget optical alternatives to the digital FFT of the fringe system. Once again, a Fourier transformation is needed, and we know how to perform this optically. Unfortunately, the process requires production of a coherent light distribution equal to the intensity of the fringe system; therefore, an intermediate recording is necessary to produce a transparency with a transmittance proportional to this intensity. It is the production of the power spectrum by a modulus squared operation that is responsible for this effort. Obviously, the Young's fringe pattern could be recorded on photographic film that is then developed and used as input to the optical spectrum analyzer. This photographic process, however, is too time-consuming for further application of the technique. We have to look for one of the many quasi-real-time optical recording media to take the place of the silver halide film. Recently photorefractive crystals that have been employed for holographic correlation (Section 6.3.4) have also been used for this purpose [49].

Presently, there is a group of candidates that are undergoing considerable development as they are intended for a big future in optical computing: spatial light modulators (SLMs) [50]. These are devices that contain a two-dimensional array of elements whose transmittance can be controlled either by light (optically addressed SLMs) or by an electric voltage (electrically addressed SLMs). Most of them utilize the properties of different types of liquid crystals to change transmittance or reflectivity in response to an electric field. Other principles such as magneto-optic or electro-optic effects in suitable materials are also exploited. The fringe pattern must be recorded as a video signal that can be fed to the light modulator, which in turn produces the coherent input to the optical Fourier transformer. In

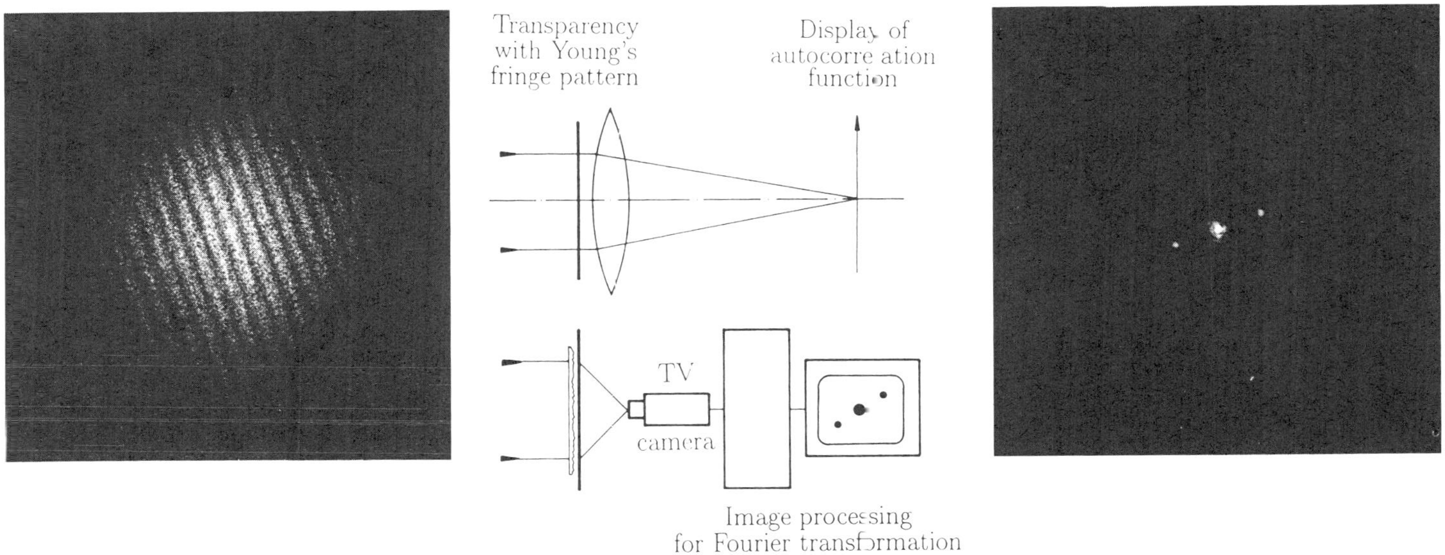

Figure 6.10 Formation of the autocorrelation function (right) of a PIV transparency by Fourier transformation of the fringe pattern (left). Shown is the modulus squared of the function as a result of digital electronic computing; an alternative is analog optical processing.

conjunction with an optoelectric sensor array that is sandwiched to the modulator plane, the device can be addressed directly by an incident light field. This saves the intermediate video recording. Commercially available equipment ranges from simple liquid crystal computer displays for overhead projection or TV pocket receivers to very sophisticated optically addressable devices. Current research is producing continued improvements. The size of the arrays ranges from a few hundred up to 1000 pixels per dimension. Frame rates are from 25 Hz video cycle up to a few kilohertz, and the modulation depth lies between 15:1 and 100:1. Unfortunately, not all the maximum performance specifications are combined in a single element, so compromises must be made. The required dynamic range, for example, can be greatly reduced by early elimination of the halo, as pointed out earlier.

An example of such SLM processing of fringe patterns is shown in Fig. 6.11, where a magneto-optic Faraday effect device has been used to modulate a He-Ne laser beam to produce the autocorrelation output [18]. The figure demonstrates the suitability of the SLM (640 × 400 pixels, 60 × 76 μm^2 pixel size, binary transmittance) for the evaluation of a PIV image. As their performance improves, employment of SLMs becomes even more promising. The output, it should be noted, is the modulus squared of the autocorrelation function because we register the light intensity in the Fourier plane. If quantitative use is to be made of the shape of the side peaks, this must be taken into account. In the detection of peak positions it is no disadvantage; on the contrary, the squaring operation sharpens the peaks.

We mentioned that for two-step optical Fourier transformation, photographic intermediate recording of the system of fringes is not appropriate because it has no real-time capability. The process regains attractiveness when it can be performed for many fringe patterns at the same time. Thus, let us try it on a photographic transparency containing the array of numerous fringe patterns produced in parallel with a holographic lenslet array (Fig. 6.9). A second array has to get into operation. Each beam has to be centered on one of the fringe patterns. In the output plane we will observe an array of autocorrelation functions, i.e., numerous small three-peak combinations, that just have to be connected to indicate the displacement. One might think of using the initially applied holographic optical element once again in the second step. In this case, however, when the focal lengths of the two elements are identical, the output will have the same scale as the input. The autocorrelation function will therefore extend over only twice the original displacement, which is small compared to the interrogation region and thus much less than the spacing between interrogating lenslets. No efficient use is therefore made of the space available. Consider an interrogating spot having a diameter 20 times the maximum displacement.

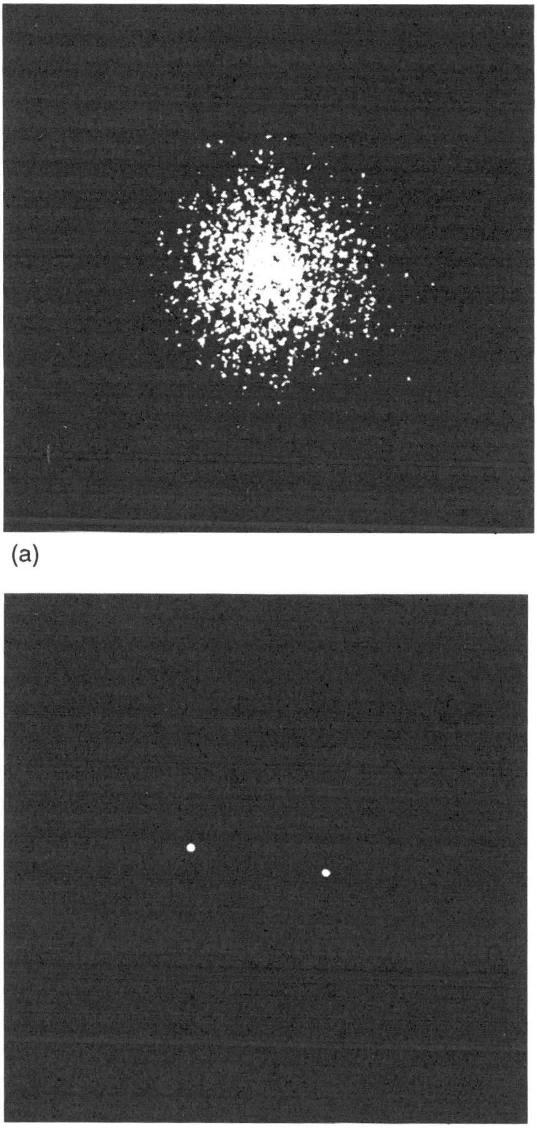

(a)

(b)

Figure 6.11 Autocorrelation function produced by optical Fourier transformation of a Young's fringe pattern written into a magneto-optic spatial light modulator (SLM) [18]. (a) Binary fringe pattern written on SLM; (b) autocorrelation output (the central peak was blocked). (Photos courtesy of DLR, J. Kompenhans.)

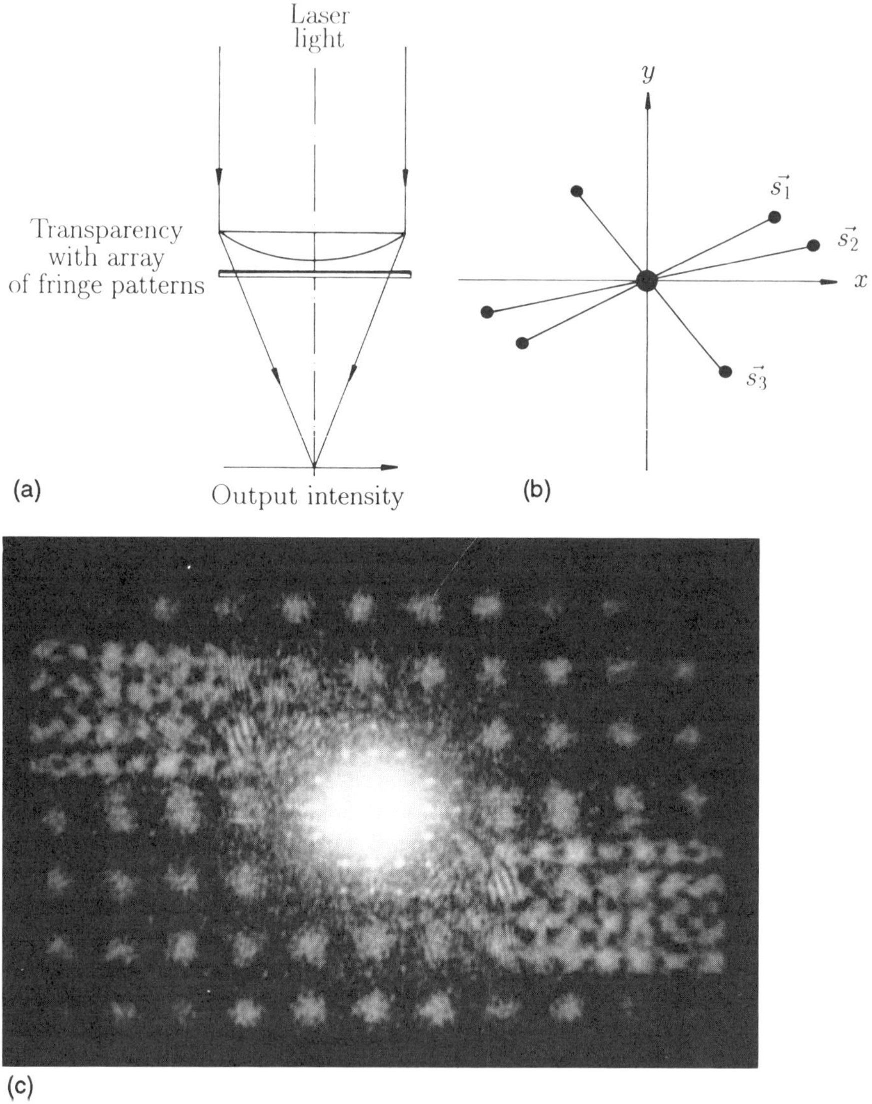

Figure 6.12 Accumulated displacement information (probability distribution) from optical whole-field Fourier transformation of an array of fringe patterns. (a) Optical setup; (b) principle of superimposed three-peak correlation functions; (c) application to the array of fringes in Fig. 6.9a (rotated defective plate [19]).

The focal length f_2 in step 2 should be 10 times the focal length in step 1, which is typically $f_1 = 5$ cm. Figure 6.9b shows the fringe systems of Fig. 6.9a thus processed. It presents an impressive picture of the overall displacement field and does not require thinking in inverse terms to convert fringe spacing to displacement.

Processing of the systems of Young's fringe patterns offers still another interesting application [51]. Let us take the photographic transparency with the array of fringe systems as input to an optical Fourier transformer, illuminating, however, not each pattern individually, but the whole array with an expanded beam (Fig. 6.12a). Each fringe pattern now produces its three-peak autocorrelation output. All of them are centered on the optic axis (Fig. 6.12b). The cloud of peaks thus produced gives the cumulative displacement in the interrogated field, i.e., the probability distribution of displacement. The size of such a cloud, for example, could be a measure of turbulence, and the existence of two distinct clouds an indication of separated regions within a flow. Figure 6.12c shows the displacement distribution of the rotated defective plate of Fig. 6.9. While a pure rigid-body rotation gives displacement values on a regular grid, the defect-induced irregular motion clearly shows up as disturbances of such a pattern. Additional proposals for further processing have been made. Position information that is usually lost in this kind of evaluation can be reintroduced by using a color-coded pattern in the illumination of the array. Certain color combinations in the output would signal the existence of specific structures in the motion field.

6.3.3 Large-Area Processing

All prior expositions concerned the derivation of flow data averaged over a small interrogation area. The main aim of the procedures was to extract the direction and magnitude of the mean displacement. The size of the interrogation area was made much smaller than the structural scales of the flow. Fluctuations in displacement **d**, though included in our theoretical analysis, have been disregarded so far. The present section deals with the situation when the size requirement is no longer fulfilled. This could be deliberate, to extract information from a larger field, or it could be unavoidable such as when flow properties and interrogating setup leave no alternative.

Omitting constant coefficients, let us recall Eq. (6.21):

$$I_F(k) = |H(k)|^2\{1 + \tau_I(d_0)|\Phi(k)| \cos [kd_0 + \phi(k)]\} \tag{6.47}$$

Recall that $\tau_I(r)$ is the autocorrelation function of the interrogating intensity profile. The fringe formula includes a fluctuating part **s** of the displacement

$\mathbf{d} = d_0 + \mathbf{s}$ expressed by its characteristic function $\Phi(k)$ that is the Fourier transform of the probability density function $p(s)$. It is based on ensemble averaging and statistical independence of random particle position $\mathbf{r}$ and displacement $\mathbf{d}$ and requires a large number of particles to approach these conditions by averaging within a single realization experiment. While effects are most easily understood in the autocorrelation domain, let us briefly start with the fringe pattern.

Any considerable fluctuating component within the interrogation area causes quite a pronounced reduction in visibility. This accounts for the severe loss in fringe quality when a small-scale turbulent fluid is observed. For fluctuations with a large variance, $\Phi(k)$ will be a rapidly decreasing function, and visibility can diminish even within the first fringe to such an extent that no fringe system is recognizable. This sets a strict limit to large-area interrogation. On the other hand, when the drop in visibility occurs gradually over the fringe system, it can be exploited to yield $|\Phi(k)|$ or $\phi(k)$ and thus information on $p(s)$. The space-dependent visibility has been exploited to obtain statistical information on a flow field [26, 52–54].

It is elucidating to note that the derivation of fringe visibility in this case is quite analogous to the formulation of the interference fringe visibility as a function of pinhole separation in a double-pinhole experiment with an extended incoherent light source (Van Cittert–Zernike theorem [55]). In that case the pinholes produce a pair of wavelets from each original point source differing in phase according to their position in the source and the aperture spacing. The strength of each point source is determined by the intensity distribution of the incoherent light source. Mutual interference between light from different sources is excluded. As a result, visibility as a function of aperture spacing is given by the modulus of the Fourier transform of the source intensity distribution. Similarly, in PIV we have to add pairs of wavelets that differ in phase according to their separation (the probability of which is governed by the pdf) and diffraction angle. Thus the visibility as a function of position in the fringe system is given by the modulus of the Fourier transform of the probability density function for the displacement.

Let us illustrate the method by two examples:

1. Computer-drawn model particle fields (small transparent disks) with well-defined displacement statistics were analyzed [26, 52]. Consider, for example, the one-dimensional exponential probability density function given by

$$p(s) = \begin{cases} \alpha \exp\{-\alpha[s - (1/\alpha)]\} & \text{for } s \geq -1/\alpha \\ 0 & \text{elsewhere} \end{cases} \tag{6.48}$$

The characteristic function $\Phi(k)$ is the Fourier transform of $p(s)$,

$$\Phi(k) = \exp\left(-\frac{ik}{\alpha}\right)\left[\frac{1 + ik/\alpha}{1 + (k/\alpha)^2}\right] \tag{6.49}$$

This results in visibility $V(k)$ being

$$V(k) = [1 + (k/\alpha)^2]^{-1/2} \tag{6.50}$$

Figure 6.13 illustrates the results of such an experiment. The input probability histogram of particle separations is shown in Fig. 6.13a. The fringe system in Fig. 6.13b displays a decrease in visibility with distance from the vertical axis (Fig. 6.13c).

2. In a wind tunnel experiment, large-area interrogation was made on PIV records of several low-turbulence airflows at $v = 10$ m/sec. The degree of turbulence was controlled by turbulence grids and thus was well known [28]. Figure 6.14 shows visibility plots for different degrees of turbulence. Quite obviously, the decrease in visibility accelerates with higher turbulence. The quantitative evaluation of such relationships was compared with hot-wire measurements. For this purpose the spatial frequency weighting characteristics of PIV that we derived in Section 6.2.5 have to be taken into account. Complementary measurements in the same flow were carried out by single-point interrogation [56]. When accounting for the two-dimensionality of the interrogated PIV record that renders several independent longitudinal scans through the flow field, the data can be interpreted correctly.

It must be noted that some of the assumptions in the above turbulence analysis are problematic [54]. First of all, only a single realization of the particle distribution is observed; thus visibility may be treated as a random variable. If the distance over which visibility drops to its $1/e$ value is taken as a measure of turbulence, then this distance too becomes a random variable. Its variance has been found by Monte Carlo simulations and should enter into reliability discussions. Second, the statistical relevance of the results depends on the size of the statistical sample that has been obtained from the interrogation area. If essential structures within the flow are comparable in size to the interrogated region, the statistical sample will be too small and the results may be biased and thus lose relevance.

All statistical information can be obtained also from the autocorrelation function of the PIV record, even more illustrative than from the pattern of fringes. Let us investigate one of the side peaks derived in Eq. (6.27),

$$R_+(r) = \tfrac{1}{2}\tau_I(d_0)\{R_P(r) * p(r - d_0)\} \tag{6.51}$$

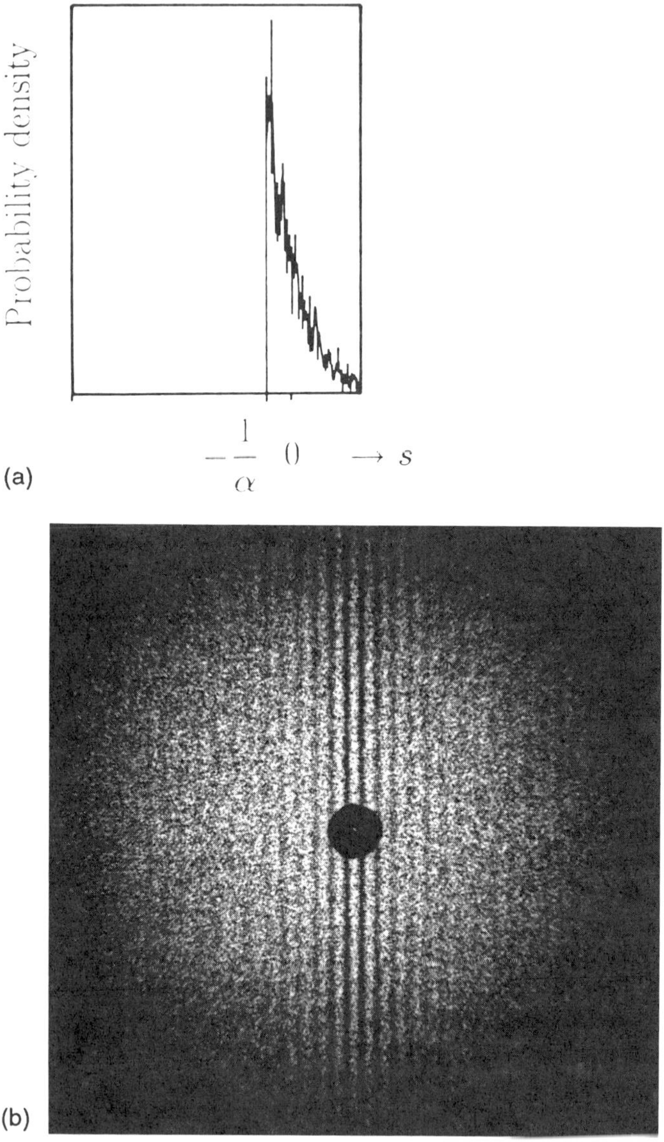

Figure 6.13 Restoration of the displacement distribution by large-area interrogation of a computer-produced PIV image [26]. (a) Probability density of particle separations for input plot; (b) fringe system; (c) fringe visibility; (d) autocorrelation function.

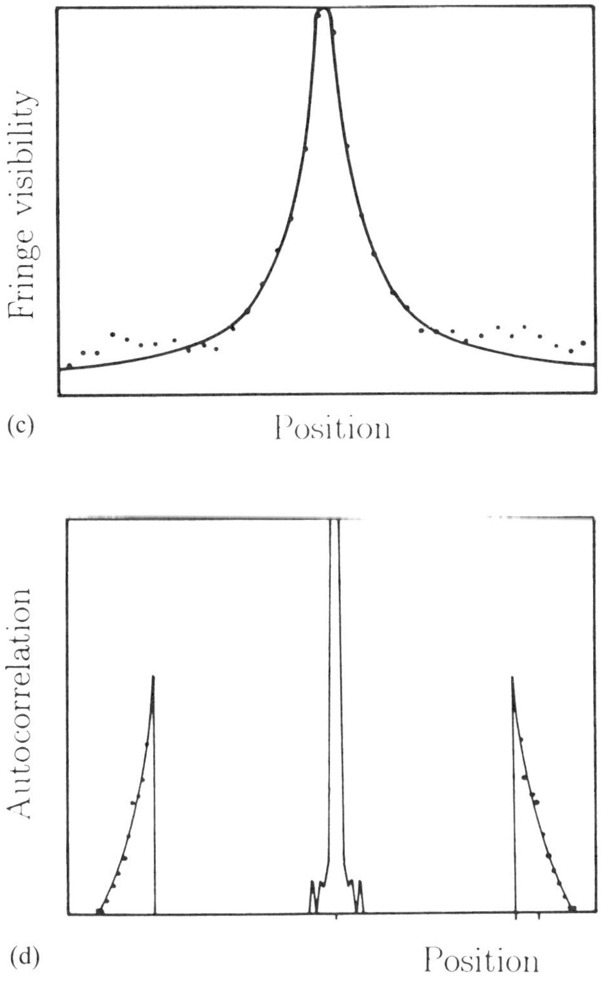

Figure 6.13 Continued.

This peak renders the mean displacement for $r = d_0$ but also contains the desired two-dimensional probability density function $p(s)$. We have to examine the way in which $R_P(r)$ influences these data.

Inspecting Eq. (6.51) we see that fluctuations widen and reduce the side peak. The probability density function that we would like to know is convolved by the particle autocorrelation function. A precise restoration of $p(s)$ thus requires deconvolution with $R_p(r)$. Fortunately, this function appears unspoiled in the central order of the correlation output given by

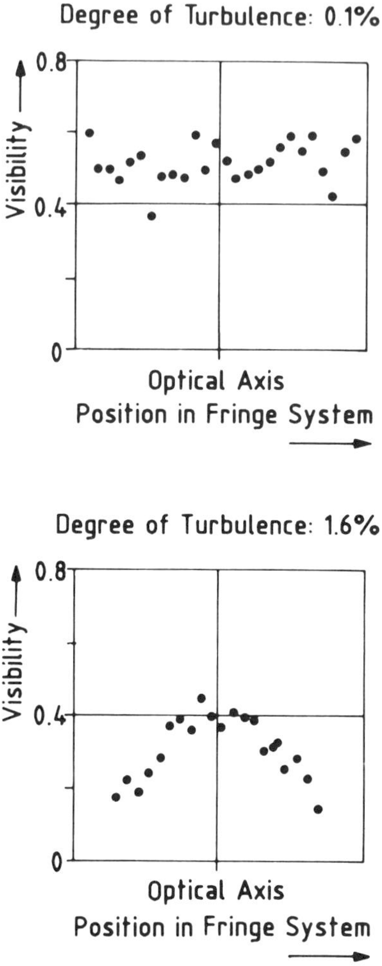

Figure 6.14 Visibility plots of grid turbulence in a laminar wind tunnel for different degrees of turbulence [28].

Eq. (6.27) and can be extracted to perform the necessary correction. In the space domain this means that from the knowledge of the particle image shape we could, for example, determine its center, reduce our data to just these coordinate values, and determine the separation of centers with high accuracy. Examples of the deconvolution have been shown [57]. Often this operation is done in the Fourier domain, which brings us back to the Young's fringe pattern. Thus we realize that deconvolution is equivalent to early elimination of the diffraction halo.

We should now amend Fig. 6.13 by the autocorrelation output that is obtained after digitizing the fringe pattern and executing a two-dimensional Fourier transformation. Obviously, the input pdf is fairly well reestablished in Fig. 6.13d. Particle images were tiny enough that deconvolution could be omitted.

Large-area interrogation usually operates on fields of a large number of particle pairs, and interexposure separation is then small compared to the field size. This greatly reduces the influence of the finite interrogation area, so it can often be disregarded. Furthermore, all SNR considerations are greatly relaxed because they yield improved conditions for high particle image densities. With increasing fluctuations, however, the autocorrelation side peak spreads and has to compete against the noise background from unpaired particles.

In the derivation of Eq. (6.20) we approximated $\tau_I(d_0 + \mathbf{s})$ by $\tau_I(d_0)$, forming the basis for the present discussion. If we abandon this condition it can be shown through a more detailed calculation of the expected value in Eq. (6.19) that there is no longer a constant factor $\tau_I(d_0)$ reducing the autocorrelation peak in Eq. (6.51), but instead we obtain

$$R_+(r) = \tfrac{1}{2}\{R_P(r) * [\tau_I(r)p(r - d_0)]\} \tag{6.52}$$

Since $\tau_I(r)$ is a gradually decreasing function of r, small displacements are weighted more heavily than large displacements. This enters into the calculation of the centroid according to Eq. (6.31) and has been realized as a biasing effect favoring small displacements [58]. This is easy to understand, as there is a growing probability for the loss of a partner when the separation approaches the interrogation size. The actual error must be evaluated for the specific interrogation intensity and particle image function.

It should be mentioned that large-area processing can be applied to problems other than flow analysis. A double-exposure microscope record of small organisms, for example, could identify directed or unorganized motion in the sample population and render data on the vitality of the organisms concerned.

To end our discussion on the techniques for fringe processing in PIV, let us mention the so-called whole-field Fourier plane processing of fringes. As in large-area processing, no small-diameter interrogation field is selected from the record for evaluation; rather the whole picture is illuminated with an expanded collimated laser beam. In the back focal plane of the Fourier transform lens there results a superposition of many fringe patterns of varying orientation and spacing, each particle pair of the large transparency represented by a small contribution. The spatial frequency spectrum of the complete input is produced. The idea of optical filtering is to do some operation on this spectrum and then recombine the altered spectrum to form a filtered image. This is a technique known from the early days of optical observation and measurement. The hundred-year-old schlieren technique, the Nobel prize-winning phase contrast method, and coherent optical correlators all fit into this category. A thorough theoretical formulation has come up with the improvement of coherence since the invention of the laser [36]. Let us recall the basics of optical filtering, as we will need them once more in the explanation of real-time optical correlators for PIV applications.

As mentioned, the filtered spatial frequency components have to be recombined into object space, which requires another Fourier transformation. So the setup for filtering consists, for example, of two identical lenses spaced twice their focal length f_L apart. The input transparency is placed a focal length in front of the first lens, the filter is introduced into the common focal plane in the center, and the filtered output is observed in the back focal plane. Since this arrangement is four focal lengths long, it is often called a $4f$ processor. If no manipulations are performed in the filter plane, the output is a diffraction-limited 1:1 inverted image of the input. Let us filter the light from an input region that is characterized by a definite displacement vector. This region produces in the filtering plane the familiar pattern of Young's fringes. We now insert a screen in the filter plane that contains a small aperture at a position corresponding to a spatial frequency of, say, $k_0 = (k_{0x}, 0)$; a second aperture may be placed symmetrically at $-k_0$ without changing our argument. If the fringe frequency $k_f = (k_{fx}, k_{fy})$ is such that $k_{0x} = nk_{fx}$ (n integer), diffracted light from this region in the input will pass through the filter to be combined for a contribution to the brightness in the respective output position. The filter thus draws contour lines into the flow-field record, lines that combine points where the x component of the velocity has a certain value or integer multiples thereof. If a similar aperture in another filter is placed on the k_y axis, isolines of the v_y component are drawn.

The filtering process could just as well be described in a systems theory formulation, which ends up relating the output to the input by a convolution with the Fourier transform of the filter function, the so-called point-spread function. This approach allows more detailed analysis of the effects of the filter, such as speckles in the output image or resolution.

Fourier plane filtering is an instructive application of optical data processing. Sophisticated proposals include the use of color filters in white light illumination that write colored isolines [59]. Apart from some impressive initial demonstrations [60], however, Fourier plane filtering has found no consequent application in PIV. To some extent this may be due to the difficulty in interpreting the isolines. Since only one component is mapped in such a representation, the type of displacement field is undefined unless the second component is included. An example are the parallel isolines that result from a purely rotational motion. In the visual inspection, the observer associates no rotational symmetry whatsoever with such a picture. Not until he learns that the orthogonal component produces just the same kind of system does he suspect some symmetry. Integration of the two isoline systems, however, calls for complicated image processing for interpolation onto a grid of sampling points. Furthermore, high seeding density is required to secure sufficient light after the filtering losses.

6.3.4 The Future of PIV Processing: Video and Computer Versus Optoelectronics

Previous analyses have shown that PIV has from the very beginning profited from a close symbiosis of optical and electronic processing. Ruler and pencil as tools for the evaluation of a fringe system had only a short lifetime. Both optics and electronic processing are in a steady state of fast development. On the one hand, the use of exotic materials for optical storage or the interaction of light with light in nonlinear materials is being investigated. On the other hand, video pickups get more powerful in sensitivity and resolution and computer-based image processing moves into the age of neural networks. It was pointed out earlier that with the advent of increasing computing capacity, optical processing steps were gradually abandoned in favor of digital processing. The original interrogation region from the PIV record was detected with a video system, and all evaluation was done by performing, for example, fast Fourier transformation on the data. At this stage of the report I wish to introduce the newest proposals in either direction, knowing that before long new inventions may shift the emphasis either way.

Let us begin with a completely digital PIV (DPIV) system [61]. The photographic record of the tracer particles in the flow is omitted in favor of a video-electric recording of the light sheet. Records are obtained with the frequency of the video system, i.e., some 30 Hz, and data are stored and then evaluated digitally. A variety of processing schemes are possible because the information exists in a sequence of single frames. The traditional scheme of autocorrelation is followed by digital (in some cases optical [62]) Fourier transformation of digitally mounted double- or multiple-exposure images [61, 63]. Alternatively, a cross-correlation can be performed between individual frames. Some recent work [20] is representative of the possibilities that digital processing offers.

Vortex ring formation in a water flow of about 3 cm/sec was studied using 80-μm phosphorescent polymer tracer spheres illuminated with 4-msec pulses from a 5-W Ar laser. The particle configuration was recorded with a 512 × 480 pixel color camera of 8-bit data depth. The frames, arriving at a rate of 30 Hz, were stored on a real-time digital video disk that allowed recording of a total of 750 images. Subsequent frames were processed on 32 × 32 pixel sampling windows that contained between 10 and 20 particles, each particle 2 or 3 pixels in diameter. Cross-correlations were computed via fast Fourier transforms, and peak detection was performed using a centroid method that allowed subpixel resolution. The evaluation of a velocity map on an 8 × 8 pixel grid, i.e., with overlapping interrogation regions, took about 2 hr but could be speeded up to a few minutes by implementation of an array processor. An impressive velocity map thus obtained is shown in Fig. 6.15.

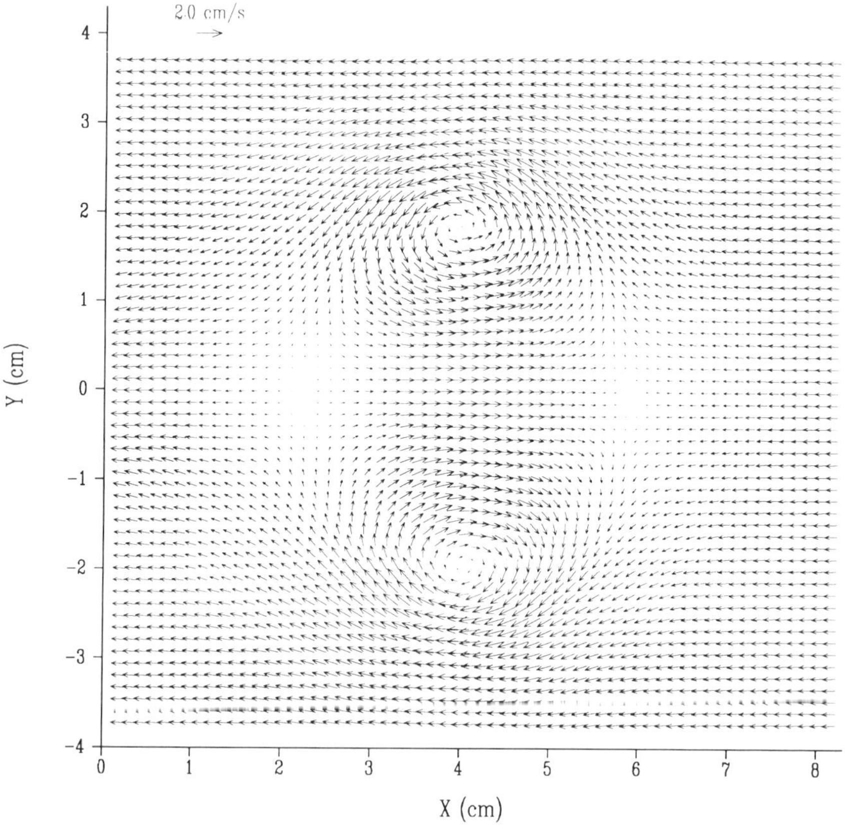

Figure 6.15 Velocity map from a purely digital PIV experiment on vortex ring formation in a water flow at $v = 3$ cm/sec. (With permission from Ref. 20.)

In regard to the relative size of the interrogation window, with respect to the whole field, DPIV is found to be still about a factor of 4 less efficient in spatial resolution than photographic PIV [20], but CCD arrays of larger dimensions and smaller element size are becoming available. DPIV offers almost real-time capability, has no directional ambiguity as long as single frames are recorded, and will profit from computer developments that permit a variety of processing schemes [64]. In high-speed flows, of course, double-exposure schemes must be employed.

On the other hand, consider the advances in modern optical processing. Optics has advantages because light is fast and light waves can be criss-crossed without disturbing each other. Disadvantages have been the lack of recording media that have the sensitivity and resolution of silver halide materials but do not require cumbersome processing. Furthermore, spatial light modulators (SLMs) as efficient as a photographic emulsion but with real-time addressability had to be developed. Both tasks have seen considerable progress recently. Upcoming nonlinear optical materials suggest that the application of some optical processing architectures should also be considered for PIV.

We have learned that profitable evaluation of a photographic PIV record requires a two-dimensional autocorrelation of the transmittance function of the transparency. One way to do this optically is to record the fringe intensity and Fourier transform this with the aid of a coherent light beam. However, an intermediate recording process that responds to the light intensity is required. It becomes feasible to apply one of the modern recording materials that allows on-site processing. Experiments have been carried out, for example, with photothermoplastic film that produces a phase-modulating record within a few seconds. Recently, however, some other correlating configurations have been exploited.

The old concept of coherent optical correlation by a matched holographic filter [65] experienced a revival with the advent of better filter materials. Let us illustrate the principle of operation for the present case with the aid of Fig. 6.16. The photographic PIV transparency serves as an optical input to a $4f$ correlator. The first lens produces the Young's fringe amplitude $F(k)$ of the original transmittance $f(r)$ in the spatial frequency plane. It is recorded as a Fourier transform hologram by adding a plane reference wave $R(k)$.

The principle of the correlation operation with the aid of this hologram serving as an optical filter can be understood by simple reasoning. The hologram is an interference pattern between object wave $F(k)$ and reference wave $R(k)$. Usually a hologram is used to reconstruct the object wave by illuminating it with the reference wave. For the present application let us exchange the roles of the two waves. The reference wave will be recon-

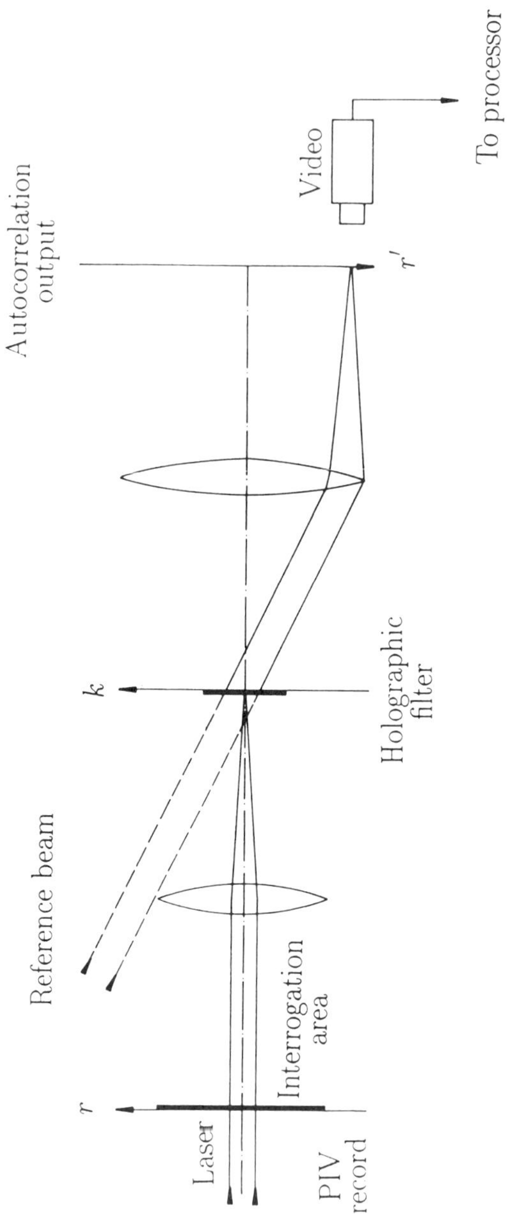

Figure 6.16 Setup for optical autocorrelation of a PIV transparency by holographic matched filtering.

structed by illuminating the hologram with the object wave. Thus a well-collimated reference wave leaves the filter to be focused into a bright light peak in the output plane of the correlator. The object wave can be recognized on account of this peak, and any changes degrade the quality of the peak. We have an apparatus that measures the similarity between two waves—a correlator.

In the above reasoning we were not quite accurate, as the interchanging of the two waves is not correct formally. The two waves differ in that the reference wave has constant intensity over the hologram plane but the object wave does not. The correct treatment employs Fourier optics. First we express the inclined plane reference wave as the Fourier transform of a (hypothetic or real) pinhole light source $\delta(r - r_R)$ in the input plane, where r_R is determined by the angle of inclination. When calculating the transmittance of the hologram there are several contributions that arise from the superposition of reference and object wave amplitudes and the quadratic recording. Upon illumination with $F(k)$, each term produces a wave field corresponding to the well-known direct and conjugate holographic image and the zero-order term. The wave of interest for our application is $E(k) = F(k)F^*(k)R(k)$ that leaves the holographic filter. It is Fourier transformed into the output plane to yield

$$e(r') = [f(r') \odot f(r')] * \delta(r' - r_R) \tag{6.53}$$

where r' is the spatial coordinate in the output plane, pointing in a direction opposite to r in the input plane (Fig. 6.16).

The equation confirms what we analyzed intuitively before. Centered at position r_R where we expected to see the focused reference beam we find the two-dimensional autocorrelation of $f(r)$, which is dominated by its central peak and contains the wanted information in the separation of the side peaks. What we actually observe is the light intensity, i.e., $|e(r')|^2$. A video pickup could feed the output to an evaluation system determining peak separation.

To operate this correlator to the advantage of PIV we should have a recording material for the holographic filter that allows a high-frequency throughput of writing and reading while the PIV record is scanned through the different interrogation areas. Recording media of choice are photorefractive materials [66] that exhibit nonlinear properties at milliwatt light power levels. Typical representatives of this family are ferroelectric lithium niobate ($LiNbO_3$) or bismuth silicon oxide ($Bi_{12}SiO_{20}$; BSO). When a photorefractive material is illuminated, photoexcited carriers migrate from bright regions to dark regions. This charge separation produces a space-charge field that gives rise to a refractive index modulation due to the electro-optic effect. The incident light field is thus stored by an appropriate phase pattern in the material. Dopants like Fe^{2+} are often responsible for the charge generation. In the dark regions electrons are caught by Fe^{3+} traps. $LiNbO_3$ has a crystallographic axis of preferential electric polarization that induces photovoltaic drift of the electrons. In BSO a drift may be induced by an external electric field. Some recently investigated photorefractive semiconductor materials have the advantage of high sensitivity [67]. Readout, of course, causes a gradual erasure of the stored pattern if no special fixation is carried out.

A BSO crystal was employed in the filter plane of Fig. 6.16 to correlate artificial inputs [17, 49] or real PIV transparencies [57]. Even with moderate light intensities in the milliwatt region, response times of about a second were obtained. Erasure of the stored holograms can be achieved by use of constant illumination, which preferably should not be coherent as irregularities in the crystal tend to produce irreproducible diffraction and the interference phenomena called parasite holograms. A somewhat different configuration, the joint transform correlator, has also been proposed in conjunction with photorefractive optical filtering [68], and both types have been assessed critically as to their suitability in PIV [57]. As interest in photorefractive crystals is very great, new and improved materials are expected that will also fertilize processing schemes in PIV.

The $4f$ correlator may also be used for cross-correlation [69]. The holographic filter is matched to a given particle image separation and yields output peaks wherever the corresponding velocity is found. If the filter has real-time capacity as in a magneto-optic spatial light modulator, it can be controlled for a search process. The use of an irregular multiple-pulse sequence even offers ambiguity removal.

6.4 EXPERIMENTAL PRACTICE

Up to now, our analysis has been concerned with the properties and evaluation of a record of particle images. The practical problems begin much earlier when an experimental setup has to be designed that allows faithful mapping of the flow through optical recording of particle positions. Let us study in detail the various obstacles on this path.

6.4.1 Particles for Faithful Flow Tracking

Since the flowing fluid is studied via tracer particles, it is essential to check whether fluid and particle velocity are truly identical (tracking fidelity [2]).

This requires a thorough investigation of particle dynamics. Mainly there occur two interferences

1. A difference in particle density ρ_p and fluid density ρ_f causes a gravitationally induced velocity v_g that is calculated from an appropriate drag law. For particle Reynolds number

   $$\mathrm{Re} = v_g D' \rho_f / \mu_f \tag{6.54}$$

   smaller than unity, Stokes's law for the drag force yields

   $$v_g = (\rho_p - \rho_f) g D'^2 / 18\mu_f \tag{6.55}$$

   Here g is the acceleration of gravity and μ_f the dynamic viscosity of the fluid. Obviously, this effect interferes with flow measurement when v_g cannot be disregarded with respect to the flow velocity. It can be minimized by selecting small particles and closely matching densities. The latter condition is easy to observe in liquid flows but difficult to observe in gaseous flows.
2. If the flow is accelerated, the particles respond with an inertial delay. Assuming once more the validity of Stokes's law, we replace g in Eq. (6.55) with the local acceleration a to obtain the slip velocity

   $$v_s = v_p - v_f = (\rho_p - \rho_f) a D'^2 / 18\mu_f \tag{6.56}$$

The slip velocity in relation to the flow velocity indicates the measurement error. A flow velocity v_f will be accepted only gradually by a particle that is originally at rest, following the exponential law [1]

$$v_p(t) = v_f[1 - \exp(-t/\tau_s)] \tag{6.57}$$

which is governed by a relaxation time τ_s given by

$$\tau_s = \rho_p D'^2 / 18\mu_f \tag{6.58}$$

These simple relations assume a pure drag force obeying Stokes's law. For 10-μm particles in water, Stokes's law holds up to velocities of about 0.1 m/sec. In air, the limit is 1 m/s. A more elaborate analysis for nonstationary flows has to include three additional force terms accounting for a pressure gradient, an added fluid mass, and the history of acceleration [70].

Let us illustrate the consequences of the equations by introducing values for a water droplet of 1-μm diameter in air. The gravitational falling velocity is found to be $v_g = 3 \times 10^{-5}$ m/sec; the relaxation time $\tau_s = 3 \times 10^{-6}$ sec. Note that a factor of 10 in size multiplies these values by 100. A 30-μm droplet will fall at about 3 cm/sec, a disturbance that cannot always be disregarded.

The comprehensive treatment of particle dynamics has been employed in a numerical calculation of the frequency response of particles to flow oscillations [70]. For a simpler estimate we develop the following approach describing the system by its transfer function $H(\omega)$. Equation (6.57) is interpreted as the response $\sigma(t)$ of the particle to a unit step change $s(t)$ in flow velocity. The relation between the respective Fourier transforms $\Sigma(\omega)$ and $S(\omega)$ is then given by $H(\omega)$ as

$$\Sigma(\omega) = H(\omega)S(\omega) \tag{6.59}$$

$H(\omega)$ is determined by introducing the required Fourier transform [see Ref. 71], which ends up with an amplitude transfer function

$$|H(\omega)| = 1/[1 + (\omega\tau_s)^2]^{1/2} \tag{6.60}$$

In Ref. 70 a frequency limit is derived for 7-μm water droplets in air. It is found that $|H(\omega)|$ decreases to a value of about 0.75 at 812 Hz. Introducing the appropriate values into Eq. (6.58) yields $\tau_s = 0.15 \times 10^{-3}$ sec. The respective frequency of 1.06 kHz compares favorably with the literature value. Another example from the same reference concerns 0.5-mm sand particles in water, for which a frequency limit of 0.16 Hz is given. Introducing the appropriate data into Eq. (6.58) and evaluating the amplitude transfer function yields a frequency limit of 5.0 Hz. The discrepancy with the cited value is explained by a thorough analysis of Ref. 70, which reveals that the indicated frequencies are not comparable. In the case of sand particles, a phase angle criterion was used in the derivation of the frequency limit. When the same factor 0.75 is applied for the amplitude transfer, the data in Ref. 70 yield 4.0 Hz, now in good agreement with our value.

In summary, appropriate particle selection depends very much on the flow details. Acceptable particles in air typically are less than 10 μm in diameter. Due to the much greater density and viscosity, water flows sometimes allow larger particles.

A variety of particles, many of which are well known from LDA applications, have been employed in PIV. Cigarette smoke, oildrops, helium bubbles, and glass beads have been used in air, and conifer pollen, polystyrene particles, and hydrogen bubbles in water, to name a few. An analysis of the flow characteristics under study will determine the final selection. Additional considerations concern the mechanisms of particle generation and distribution in the flow. For the production of small droplets a liquid is often passed through appropriate nozzles. Precautions must be observed to avoid any possible health risks associated with the particle material. Many small particles are hazardous when inhaled, and some exhibit additional toxicity.

6.4.2 Light for Particle Imaging

In addition to their fluid dynamical behavior, particles are selected on the basis of their optical properties. They must scatter light sufficiently to expose the photographic material. We have learned that usually the particles are rather small. Light scattering depends on particle shape and size, the refractive index of particles and fluid, and the polarization and wavelength λ of the radiation. The latter enters by a normalized diameter $q = \pi D'/\lambda$. PIV conditions usually require particles that are between the limiting cases of $q \ll 1$ (Rayleigh scattering) and $q \gg 1$ (geometrical scattering). In the intermediate range, scattering depends in an irregular manner on the direction between incident light and observation that is described for spherical particles by Mie's theory. There exists an abundant literature on this topic (see, e.g., references in Ref. 55). Frequently encountered particles such as spherical water droplets have been treated extensively.

As a disadvantage of PIV it is found that scattering in the Mie region is mainly in the forward direction. In the 90° direction that is characteristic of PIV configurations, scattering intensity has typically decreased by a factor of 10^{-3}. On average the scattered light increases with increasing values of the parameter q. Larger particles and smaller wavelengths are thus desirable.

In the final assessment of a particle's appropriateness for a flow study, the imaging process must be included. Lens aperture (i.e., f number) and magnification M determine how much of the scattered light is collected and onto how large an area on the photographic emulsion it is focused. A large aperture ensures much light and small diffraction-limited particle images. The finite depth of the light sheet, however, needs an appropriate depth of focus, which may require that the lens be stopped down.

Let us study a typical particle of diameter less than 10 μm at $M = 0.5$ in ruby laser light ($\lambda = 694.3$ nm). The lens is set at $f/5.6$. According to Eq. (6.2) the particle image assumes a size of 15 μm fairly independent of its diameter within a depth of focus of about 1 mm, just sufficient to cover the light sheet in depth. Any increase in aperture would result in defocused images of larger diameter. Provided that the nature of the flow allows it, an additional increase in particle size up to the resolution size would increase the scattered light but not enlarge the particle image much. A further increase brings no benefit as the light is now imaged onto a larger area.

The type of analysis just conducted is typical of a system study that results in the selection of particles and designs the appropriate illumination. Numerous examples that lead to minimum detectable particle diameters for given light sheet dimensions and laser power have been calculated. The largest particle should be chosen that follows the flow and has an image (at unit magnification) not significantly greater than its physical size [72]. Special particles have been used to increase the scattering intensity (polycrystalline Vestamid [73] or photochrome particles). Finally, forward-scatter configurations with a much higher efficiency have been devised [74]. These, however, need blocking of the unscattered light by spatial filtering and careful exposure adjustment to minimize noise from out-of-focus particles.

The complete design of the PIV system now requires an appropriate light source and the formation of a light sheet. The illuminating light need not be coherent; however, lasers have become the preferred light sources in PIV. The short-exposure illumination can be achieved in two ways. (1) The laser beam is shaped into a light sheet and turned on for the necessary exposure time, or (2) the unexpanded laser beam is scanned through the region of interest.

In small viewing fields or low-velocity flows a CW laser output can be shuttered to produce double or multiple pulses or even pulse sequences for time-resolved PIV [73]. Argon lasers are popular candidates in these applications. When the area under investigation or the flow velocity is large, the necessary sequence of short pulses (typically about 10 nsec in duration) of sufficient energy is produced from *Q*-switched pulsed lasers. Early PIV was done entirely with ruby lasers ($\lambda = 694.3$ nm), but recent developments use the more suitable frequency-doubled Nd:YAG laser ($\lambda = 532$ nm). Despite its lower energy (about 100 mJ as compared to a few joules), the performance is comparable. This is due to the shorter wavelength for more efficient scattering and photographic recording. Furthermore, Nd:YAG lasers allow a repetition rate of about 10 Hz while a ruby laser needs about 10 sec of recharging before it can be fired again. This greatly facilitates alignment operations. The optimum setup today consists of a combination of two Nd:YAG oscillators that allows almost any separation between the two pulses, extending PIV from very low speeds to flows at Mach numbers around $Ma = 1$.

Light sheet formation in this case is usually done by expanding the output beam of the Gaussian intensity profile with the aid of appropriate optics into the desired shape. In the simplest configuration a glass rod fans out the beam. A sophisticated and very versatile combination of a telescope with a cylindrical lens [75] allows the dimensions and position of the light sheet to be adjusted. Suitable additions even furnish movable sheets for three-dimensional investigations. For creation of a very wide sheet, an expansion lens can be combined with a two-dimensional parabolic mirror of appropriate dimensions.

The disadvantage of these light sheets is that the Gaussian profile is imprinted onto the sheet not only in depth but also across the field of view.

Thus, regions far off the axis get a reduced amount of illuminating light. While proposals exist for a reshaping of the intensity profile, the following alternative solution by beam scanning is appropriate in certain cases.

In the second type of configuration for achieving short-exposure illumination, a continuous-wave laser beam (an argon laser, for example) is directed onto a rapidly rotating polygonal mirror that scans the beam through the field of view. When the polygon faces are placed in the focus of a parabolic mirror (Fig. 6.17), the output beam passes over the field in a parallel scan at almost constant speed [13]. Thus all particles receive the same amount of light. The exposure time is determined by the time it takes the unexpanded beam to pass over a particle. The separation between exposures is given by the rotation rate and the number of faces. Double or multiple exposures are possible depending on the number of scans for which an additional shutter opens the laser output. The registration of the whole particle field, of course, takes longer than in a chopped expanded beam. Furthermore, not all particles are exposed at the same instant. Usually, however, the errors thus induced are not essential, and the better utilization of the available light is a big advantage. In a wave tank study it was concluded that the scanning method was superior to the expanded beam method by a factor of 3 in velocity and a factor of 4 in area observed [13]. With a 15-W Ar laser it can be applied to velocities up to 10 m/sec.

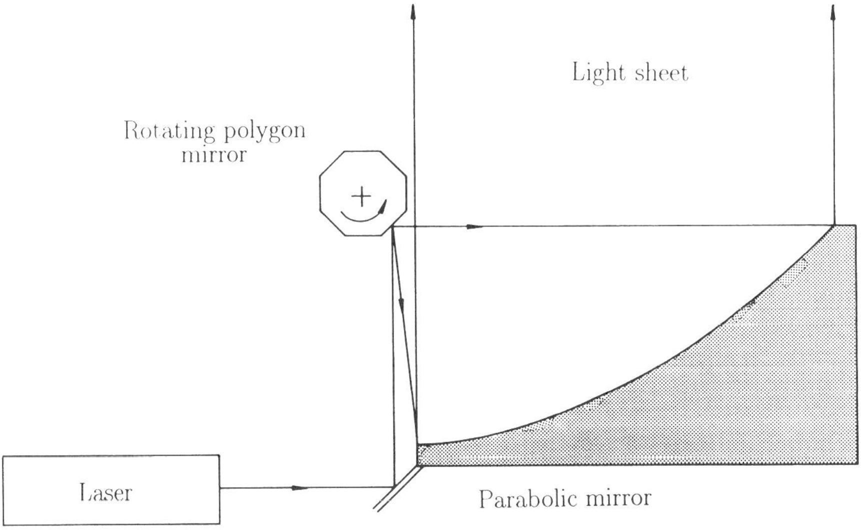

Figure 6.17 Scanning laser beam configuration for the formation of a light sheet.

6.4.3 Photographic Recording of Particle Images

In many PIV configurations the particle images are recorded on photographic film. As this record is the basis for all subsequent evaluation, its quality should be optimized.

To begin with, the camera lens must be well corrected for PIV conditions, which often operate in a region of magnification between 0.1 and 1. To obtain a minimum size particle image, focusing must be well controlled. Since the commonly used f numbers limit the depth of focus to roughly 1 mm, this adjustment is critical. To avoid long-lasting photographic experiments with changing object distance and focus evaluation, a video-operated focus control in the image plane of the camera has been found to be very useful [76].

Relevant properties of the photographic emulsion are spectral sensitivity, speed, resolution, photographic gamma, and noise. The final specification depends on the emulsion as well as on the chemical processing. Panchromatic films are sensitive over the entire visible range of the spectrum, but the long-wavelength region is usually at a disadvantage. The very important quantities speed and resolution, unfortunately, oppose each other; thus a compromise must be found. A holographic plate such as Agfa Holotest 10E75 has a resolution of more than 2000 line pairs per millimeter. To obtain an optical density of 1, this film requires an exposure of about 20×10^{-3} J/m^2 at ruby laser wavelength. Kodak Technical Pan 2451, on the other hand, resolves just 300 lp/mm but needs only 1.5×10^{-3} J/m^2 for the same density. Since the smallest particle images are determined by diffraction and range no smaller than 10 μm, the latter type of film is a good choice.

A double-exposure particle record is usually employed for Young's fringe analysis. It is illuminated with laser light to evaluate the diffraction pattern thus produced. Besides the useful particle image pairs, however, the photographic negative contains many other structures that may spoil the quality of the fringe system. Due to its granular composition a photographic emulsion contains sources of noise. Light is scattered by tiny silver particles (grain noise) and refractive index variations or emulsion-thickness variations (phase noise). At low transmittance, i.e., in a heavily exposed, rather dense negative, grain noise is the dominant source, whereas phase noise prevails in highly transparent negatives.

Principally, noise can be reduced by immersing the negative in an index-matching liquid called a *liquid gate*. Xylene has been found to match well to the refractive index of the emulsion gelatin (about 1.5). The improvement is most pronounced at high transmittances where the effect from the surface profile is reduced. Further improvement can be gained through a two-step process in which a positive contact transparency is made of the original negative particle record [11]. Regions of high transmittance responsible for a great portion of the noise are thus turned into low-transmittance areas with little effect. The gain, of course, will be most pronounced when the image density on the photograph is low. In records with high image density, the additional effort does not pay off [77].

Let us illustrate the possible improvement by results from appropriate operations on a model particle record containing four pairs of black dots resembling particle images on a transparent background (Fig. 6.18) [11]. Whereas the original negative produces hardly any fringe modulation, the liquid gate results in noticeable improvement, and the positive copy in high-visibility fringes. The two-step process has become quite popular in PIV. When it is employed properly the quality can be improved further because of the considerable reduction of the zero-order peak by appropriate selection of the copying emulsion (hard clipping of the images).

Often the brightness of particle images or the particle image density varies with position on the PIV record. In this case the brightness of the fringe system will be affected, and the registering device (often a video camera) may not operate under optimum conditions. Automatic brightness control by adjustable filters has proved to be very useful in this case [76].

Keeping all previous considerations in mind, let us outline a typical PIV recording system for the study of wind tunnel flows. A Nd:YAG laser of 100 mW per pulse serves as light source. Oil droplets 1 μm in diameter are introduced along a line upstream from the measuring section to evenly fill the light sheet cross section about 75 mm high and 1 mm deep. At a

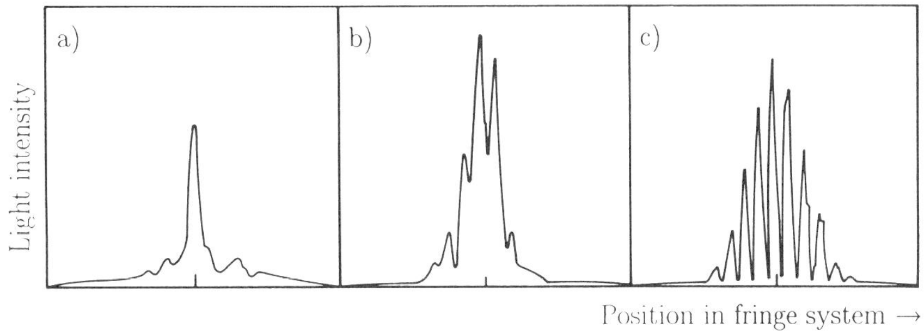

Figure 6.18 Improvement of the quality of Young's fringes by noise reduction. Shown is an intensity scan across the fringe system of (a) an original negative, (b) the same negative in a liquid gate, and (c) a positive copy. (After Ref. 11, with permission.)

magnification of M-0.3, a 35 mm camera system of 60 mm focal length allows imaging the flow field at $f/2.8$ onto Kodak Technical Pan 2415 film. Velocities up to several tens of meters per second can be studied with this setup.

6.5 ADVANCED OPTICS FOR PIV PERFECTION

6.5.1 The Problem of Directional Ambiguity

In particle image velocimetry, flow velocity is determined from the positions of tracer particles at certain instants of time. The typical double-exposure record, for example, contains pairs of particle images. In the evaluation it is not known which particle was imaged during the first exposure and which one was second; there is 180° ambiguity of the velocity. Fringe processing yields only the magnitude of the velocity. In many applications additional knowledge about the flow resolves this ambiguity. In complex flow fields like recirculating flows, however, this is not possible. Solutions to the problem are manifold.

Directional ambiguity arises from the inability to distinguish the two images. A straightforward approach is to obtain single-exposure multiframe records where the first and second images are separated. In fast flows this requires high-speed cinematographic equipment. An alternative is to record a double-exposure hologram and shift the reference beam direction between exposures [78]. The principle has been demonstrated with image-plane holography of a simulation of particles by a spray of 100-μm dots on a glass substrate. Simultaneous illumination with the two reference beams produces the ordinary double-exposure record, while illumination with a single reference beam enables the direction to be discriminated. In pulsed-laser applications, reference beam shifting can be achieved by polarization switching or acousto-optical beam deflection, to name two possibilities.

Another approach with single-frame recordings is to label the images by special characteristics of the train of light pulses. Various schemes have been proposed. The most obvious is to employ light pulses of different intensities, rendering a fainter image of, say, the second exposure. Usually, however, this interferes with the spatially inhomogeneous illumination in the light sheet. A triple-pulse sequence of different pulse intervals is another example that permits the direction of motion to be obtained from the succession of particle separations. The same objective is achieved with a low-intensity tag pulse prior to each main illuminating pulse of the pulse train [79]. In this case the particle images are slightly deformed in the direction of motion. Similarly, phosphorescent particles will produce images with a gradually fading tail. Furthermore, the light pulses can be made to differ in color. A green 532-nm pulse from a frequency-doubled Nd:YAG laser was followed by the red 607-nm pulse from a Nd:YAG pumped dye laser [80]. The resulting color photograph was digitized and processed, using the direction from green to red to resolve ambiguities.

All solutions mentioned have the disadvantage that they do not allow the familiar evaluation procedures of PIV. Basically, these perform only an autocorrelation analysis, in which case asymmetry is lost. Detailed particle identification and tracking are therefore required. To avoid this the philosophy of image shifting has been introduced [81]. A bias displacement is performed between exposures of such magnitude that the largest negative displacement that occurs is fully compensated. The true particle displacement, including its direction, is regained from the experimental data later by subtracting the bias value.

The experimental realization of this philosophy asks for a device that shifts the image within the short time between the two light pulses. In flow velocities below about 10 m/sec the image offset can be produced by a mirror mounted on an oscillating galvanometer scanner or a rotating mirror and placed in front of the imaging lens [38]. Synchronization between mirror movement and light pulses must be provided. Recently this mechanical image shifting was extended to flow velocities of about 100 m/sec. To eliminate errors due to inaccuracies of the mirror motion, a stationary marker in the field of view can provide calibration data.

For higher flow velocities and better accuracy, electro-optical image shifting has been introduced. The two light pulses are labeled by orthogonal directions of polarization. An electro-optical device such as a Pockels cell rotates the polarization within the required short time between exposures. A birefringent crystal placed in front of the imaging optics induces a displacement between images in differently polarized light [82]. Three aspects must be considered in such a system:

1. Scattering of the incident light by the particles may cause depolarization. By applying Mie's theory to commonly used combinations of particle and fluid, it was shown that the incorrectly polarized intensity is far below the threshold for photographic registration.
2. The displacement is nonuniform across the field of view. This effect depends on the orientation of the optic axis of the crystal and can be minimized. Furthermore, its amount is known and can thus enter into correcting computations during the evaluation.
3. The image shift must be adjusted to the specific flow characteristics. This is done by choosing an appropriate thickness of the crystal. This parameter, however, is not very easy to change.

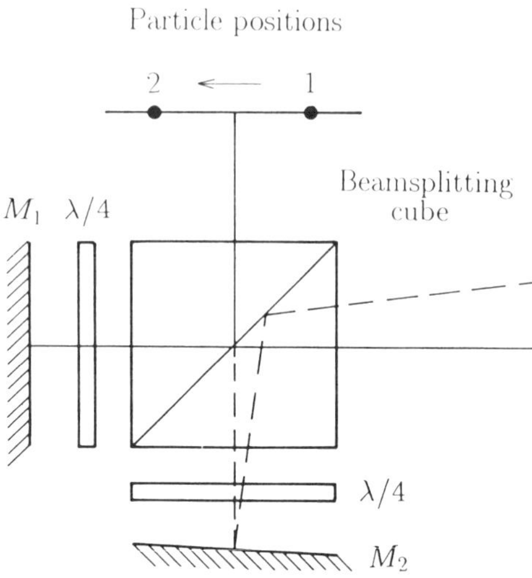

Figure 6.19 Beamsplitting device used in two configurations for different purposes: (1) Electro-optical image shifting with a polarizing beamsplitter and quarter-wave plates produces the second image of the particle at a position *2′* displaced from its regular position *2*. (2) Image doubling with an ordinary beamsplitter and without quarter-wave plates produces a regular image pair (*1*, *2*) and a shifted image pair (*1′*, *2′*).

In Fig. 6.19 another setup is shown that shifts the images from light pulses of orthogonal polarization. The particle field is imaged by a beamsplitter cube and two mirrors in a configuration reminiscent of a Michelson interferometer. Originally, this arrangement was introduced for the purpose of image doubling [83] (see Section 6.5.2). It has been modified for electro-optical image shifting by using a polarizing beamsplitter and inserting two quarter-wave plates [84]. Linearly polarized light passes a quarter-wave plate, is totally reflected, and leaves the plate with a 180° rotation of the polarization. In the final outcome the two orthogonal polarization states are imaged over different mirrors. They coincide when the mirrors are aligned normal to the optical axes. They are displaced by tilting one of the mirrors. Thus the second image of the particle is not found at its normal position 2, but at the displaced location 2′.

6.5.2 Extending the Range of Application by Image Shifting

The analysis of an interrogation region may serve two purposes. First, the mean separation of paired particles yields the mean flow velocity. Second, by determining the distribution of particle separations, information is obtained about the statistics of the flow. The variance, for example, is an important quantity in describing turbulence. We have seen that successful measurement in both these cases is limited by certain detection criteria that can be summarized by four rules:

1. The mean displacement d_0 must be greater than the particle image size D: $d_0 \geq D$.
2. The size D_I of the interrogation region should be much larger than d_0; typically: $D_I \geq 5d_0$.
3. For successful measurement of d_0, the displacement fluctuations σ_s must be small enough, typically $5\sigma_s \leq d_0$.
4. If we want to determine the value of σ_s, this quantity should be larger than the particle image size: $\sigma_s \geq D$.

Usually all displacement quantities d are related to velocity values v by $d = M\,\Delta t\,v$, where M and Δt are the magnification and the time between exposures, respectively. In this way the above rules set limits also to the mean value v_0 of the velocity and its variance σ_v^2.

In certain cases the limits can be extended by introducing an appropriate bias displacement d_b by image shifting. Let us illustrate this for the two types of investigations.

For the measurement of the mean velocity in the presence of a high degree of turbulence, rules 2 and 3 determine the maximum pulse separation

$$\Delta t = 0.04 D_I / M\sigma_v \tag{6.61}$$

which decreases to a rather small value for large velocity fluctuations. Remembering that $d_0 = M\,\Delta t\,v_0$, we see, however, that rule 1 sets a limit to the required reduction of Δt: The mean displacement must still exceed the particle image diameter. By adding a bias displacement d_b, on the other hand, we can circumvent this rule and arrive at an effective mean displacement $d_0^1 = d_0 + d_b$. Now all rules will be met when

$$d_b = 0.2 D_I (1 - 0.2 v_0/\sigma_v) \tag{6.62}$$

For the determination of low turbulence, rule 4 asks for a rather long pulse separation

$$\Delta t = D / M\sigma_v \tag{6.63}$$

which brings us into conflict with rule 2, because d_0 will become too large. Both particles must still lie well within the interrogation region. Again the bias helps; in this case a negative shift d_b lowers the mean spacing $d_0' = d_0 - d_b$,

$$d_b = Dv_0/\sigma_v - 0.2D_I \tag{6.64}$$

The bias image shift can be implemented by any of the means mentioned previously. As there is no need, however, to tag pulses, it is also possible to manage without rapidly moving parts or high-speed switches. The required particle configuration can be created simply by image doubling [83].

A version of the device in Fig. 6.19 with an ordinary beamsplitting cube and without quarter-wave plates is placed in front of a photographic lens. When the mirrors are properly tilted with respect to each other, the double-exposure image can be repeated at a distance d_r. The repetition of the first-exposure image can thus be brought into the neighborhood of the original second-exposure image at a mean distance $d_0' = d_0 - d_b$ as required by the above considerations. These two images then form the particle pair for evaluation. Other particle pairings will yield separations outside the size of the interrogation region. The increase in the number of single particles, however, contributes to the overall noise.

Conditions are easily illustrated in the autocorrelation domain. We calculate the autocorrelation function $R'(r)$ for the doubled particle record from $R(r)$, the autocorrelation function for the single record.

$$R'(r) = R(r) * [2\delta(r) + \delta(r - d_r) + \delta(r + d_r)] \tag{6.65}$$

For reasons of clarity let us simplify $R(r)$ as given by Eq. (6.27) by disregarding the finite particle size; that is, we set the autocorrelation of a particle $R_p(r) = \delta(r)$. For the moment we also disregard the effect of

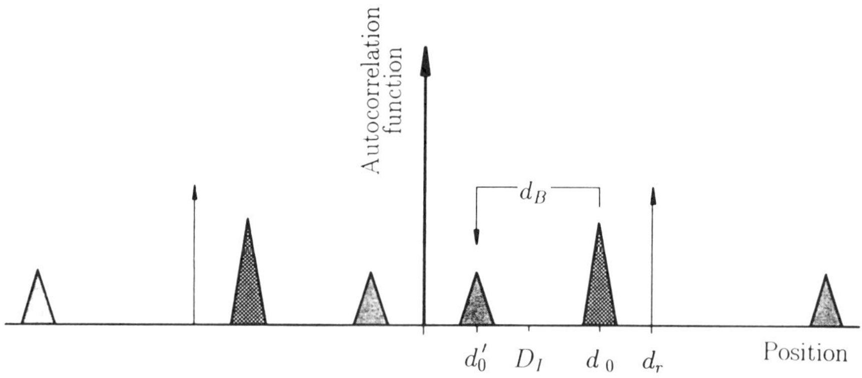

Figure 6.20 Autocorrelation function for a PIV record with image doubling. Particle images are approximated by δ pulses.

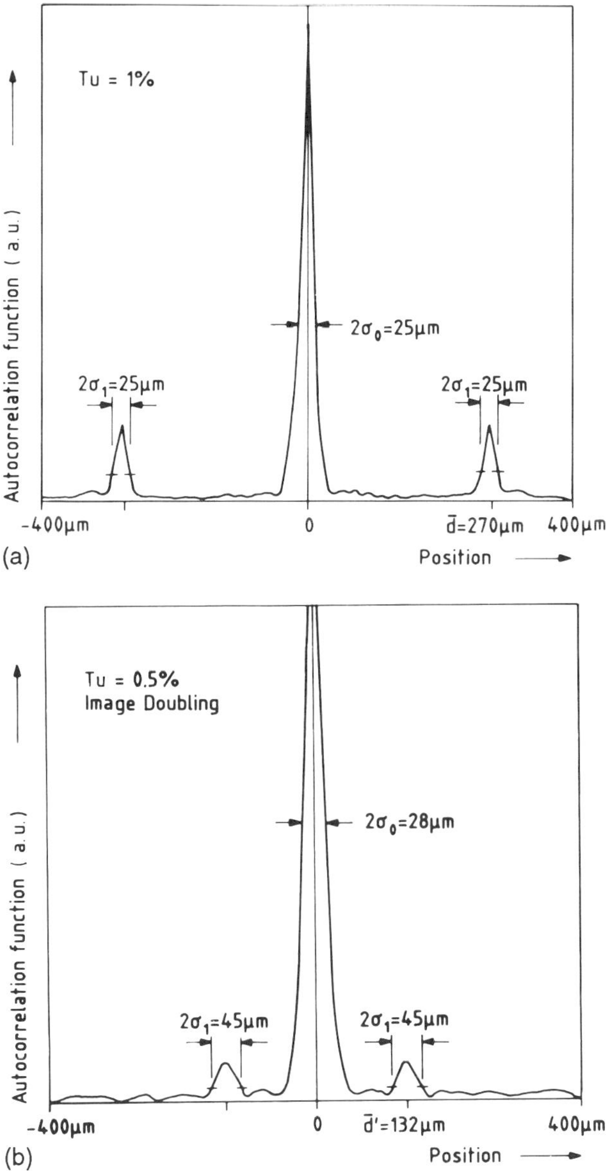

Figure 6.21 Autocorrelation function output from flow measurements in a low-turbulence wind tunnel using a large interrogation area [83]. Image doubling (b) allows measurement of the degree of turbulence Tu from broadening of the side peak that is not detectable otherwise (a).

the finite interrogation area. The darkly shaded contributions in Fig. 6.20 thus represent $R(r)$, where the triangular function at position d_0 symbolizes the probability density function of the particle separations. Let us study the example of a low-turbulence measurement (outlined above). To obtain a pdf wide enough, the time Δt has been chosen large; thus d_0 is well outside the diameter D_I of the interrogation region. We recall that the interrogation area size induces a gradual decrease in the correlation peak until it vanishes at D_I. In the present case one partner of a pair of particles is always outside this area. The purpose of the image doubling is to change this, i.e., to place a pdf peak within the limit set by D_I. According to Eq. (6.65) there are two shifted versions of $R(r)$ that have been added in Fig. 6.20 by the smaller δ pulses and the lightly shaded triangles. They are centered at d_r and $-d_r$. If d_r is selected properly we can shift the pdf by the amount $d_b = 2d_0 - d_r$ into the favorable region.

This strategy has been used for studies in a low-turbulence wind tunnel [83]. Ordinary conditions required a short pulse separation of 80 μsec so that at a low turbulence of Tu = 1% there is no widening of the side peaks (Fig. 6.21a). When a shift of 1800 μm was introduced by image doubling, the pulse separation could be extended to 1 msec. Even at a degree of turbulence of only 0.5% there resulted a well-resolved peak broadening that was used to measure the turbulence (Fig. 6.21b).

6.5.3 Recording of Three-Dimensional Flow Fields

The two-dimensional information obtained by particle image velocimetry represents great progress in comparison with the single-point measurements of conventional anemometry. The increasing importance of studies in nonstationary flows, however, suggests extensions even in the third dimension that serve two purposes: (1) determination of the depth component of the velocity vector and (2) covering a three-dimensional measuring volume.

To cover a three-dimensional field of view, the standard PIV setup can be extended by successive movements of the light sheet through the depth of the measuring volume. Light sheet configurations for this purpose have been proposed [75], and applications are known from ordinary flow visualization [85]. The different layers, however, are not recorded strictly simultaneously because of the finite scan time.

For determination of the depth component of the velocity, a light sheet may be composed of two parallel and adjacent sheets of different colors (green and blue from an Ar ion laser) to color code the depth position of a particle [86]. The position is obtained from a color analysis of the red, green, and blue video signals of a camera that is employed to analyze a photographic multiexposure record of the particles. The depth component has been found to be less accurate than the transverse components by an order of magnitude. Another possibility is to use the degree of defocusing of the particle image as a measure for the depth coordinate. As this is difficult to quantify and there is ambiguity as to positions in front of or behind the plane of maximum focus, operation with a coded aperture has been demonstrated in a digital application [87]. A diaphragm with three circular holes at the corners of an equilateral triangle is placed in the aperture of the imaging lens. Out-of-focus particles are imaged as replicas of this aperture pattern; the farther away from the focus plane, the larger the pattern. The front and back regions are discriminated by the orientation of the pattern.

A straightforward solution for the examination of three-dimensional particle motions is to view the illuminated measuring volume from two directions at right angles to each other. The optical parameters must allow sufficient focusing depth. Particles are identified and tracked in both images, and the resulting projections must be matched for three-dimensional information. This is basically a task for digital evaluation; the well-established PIV processing schemes cannot be applied. Multi-exposure tracks of seed particles, for example, were recorded with two standard-rate video cameras, and the outputs were mixed to occupy the two halves of a single video frame [88]. Digital processing relied on images of calibrating gridlike test targets. A large increase in the framing rate can be achieved by applying a high-speed 16-mm movie camera (up to 5000 frames per second, for example) and recording through an appropriate mirror arrangement that allows side-by-side orthogonal views [89]. In either case it was realized that correct matching required a low seed density, which resulted in incomplete spatial flow data.

Let us for a moment return to the familiar PIV configuration to study a two-dimensional cross section of the flow. The depth displacement of particles in a light sheet can be recovered from a stereo analysis of pictures obtained at different angles. The separation between particles within a pair is projected in the viewing direction and thus depends on the angle. The larger the angle between the viewing directions, the better the resolution. It should be noted that even in ordinary PIV records the stereoscopic effect must be included as a correction in the quantitative evaluation. Particles at the edge of the field of view are seen under larger angles, and any depth component of the displacement is contained in the projected separation [90].

Principal stereoscopic recording setups that also illustrate the problems encountered are shown in Fig. 6.22. When the optical axes of the cameras point at the center of the field of view, only a limited region will be in

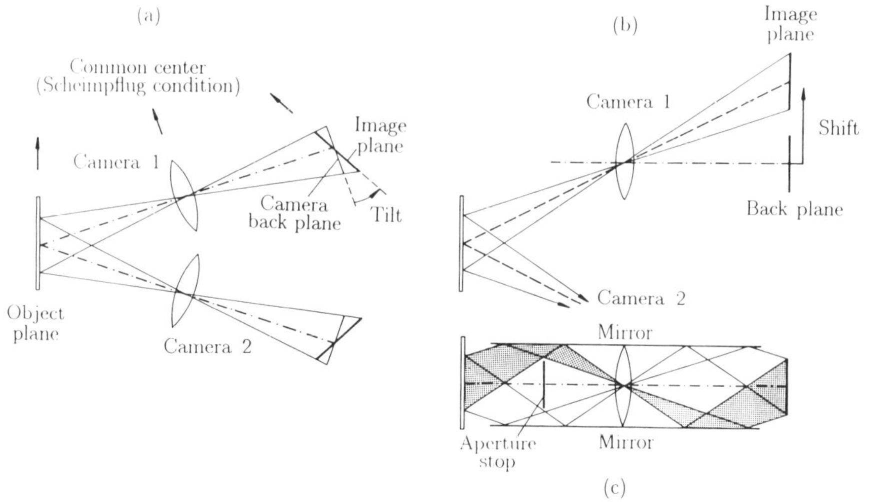

Figure 6.22 Stereoscopic recording of particles in a light sheet. (a) Oblique axis viewing requires Scheimpflug condition for object, lens, and camera back plane and results in distortion of the image field. (b) Normal axis viewing preserves focusing but requires back plane shifting and far-off-axis imaging. (c) Introduction of mirrors [91] condenses the setup of (b); the mirrors can be arranged for superimposed or separated images.

focus (Fig. 6.22a). The small f numbers needed for small-particle images also result in a small depth of focus. A whole-field focused image requires that the back plane of the camera be tilted in such a way that the planes of object, lens, and image meet in a common line (Scheimpflug condition). Now, however, magnification becomes position-dependent, and the image of the light sheet is thus distorted. Furthermore, the distortion in the two images to be compared is different. Alternatively, the optical axes of both cameras are aligned perpendicular to the plane of the sheet (Fig. 6.22b). In this case, the region in the image plane is well in focus but displaced far off the axis. In its ordinary position the camera back plane will record only a small section, if any, of the field of view. The second camera will do the same, but symmetrically shifted so that at best only a small region of the field of view overlaps. This problem can be solved by displacing the back plane accordingly. The necessary unwieldy setup has been condensed to a single-lens system with the aid of mirrors (Fig. 6.22c) [91]. Imaging very far off the axis, however, results in large aberrations.

In some practical applications we do not worry much about defocusing. In an elaborate unit for stereoscopic photogrammetry, for example, a digital particle tracking system was based on the images from three synchronized CCD cameras [92]. For intermediate storage the data arriving at a rate of 25 Hz were kept on video recorders. Digital processing determined the centroid of each particle image for further evaluation.

The distortion problem of the Scheimpflug arrangement was solved by a backprojection technique and optical evaluation of the superimposed particle images from both viewing angles [93]. Positive copies of the particle images recorded as in Fig. 6.22a were repositioned in the image plane, illuminated from the back, and imaged onto a photographic plate placed in the plane of the former light sheet. Each interrogation region now contains two kinds of particle pairs; their mean separation measures the in-plane displacement, their difference in separation the out-of-plane displacement. When the autocorrelation domain processing as described in Section 6.3.2 is applied, twin peaks are obtained that contain just this information (Fig. 6.23). This is easy to understand by applying the large-area considerations (Section 6.3.3) to an interrogation area that contains two distinct displacements. The normal angle viewing arrangement with mirrors (Fig. 6.22c) that has been employed to produce two separate images [91] could also be adjusted properly for superimposed images.

It should be noted that studying a flow with a considerable component normal to the light sheet requires thoughtful consideration as to the depth of the illumination. It was mentioned earlier that all unpaired particles produce noise and reduce the reliability of the measurement. Again image shifting could relax the conditions, as transverse and longitudinal displacements, which are usually both proportional to the time separation between exposures, could be decoupled.

The ideal method for preserving all three-dimensional information from a field of small particles is holography. A carefully produced short-exposure hologram can serve as storage for a huge number of particle images. Original positions and sizes of the individual particles can be retrieved at a later time and at leisure. From its very beginning holography has been employed in particle analysis [94, 95]. For small particles preference has been given to in-line holography, where the large amount of unscattered light serves as the reference background. Intensity ratios can be optimized with an auxiliary illumination beam [96] or in off-axis configurations where a separate reference beam is used. In double-pulse applications this has the additional advantage that the images can be distinguished by separate reconstruction with the appropriate reference wave (cf. Section 6.5.1). Usually real images of the particles are reconstructed for detailed investigation. Determining flow velocities from a double-exposure hologram requires particle pairing and the precise extraction of particle positions. This is done by computer-controlled video scanning of the field of particle

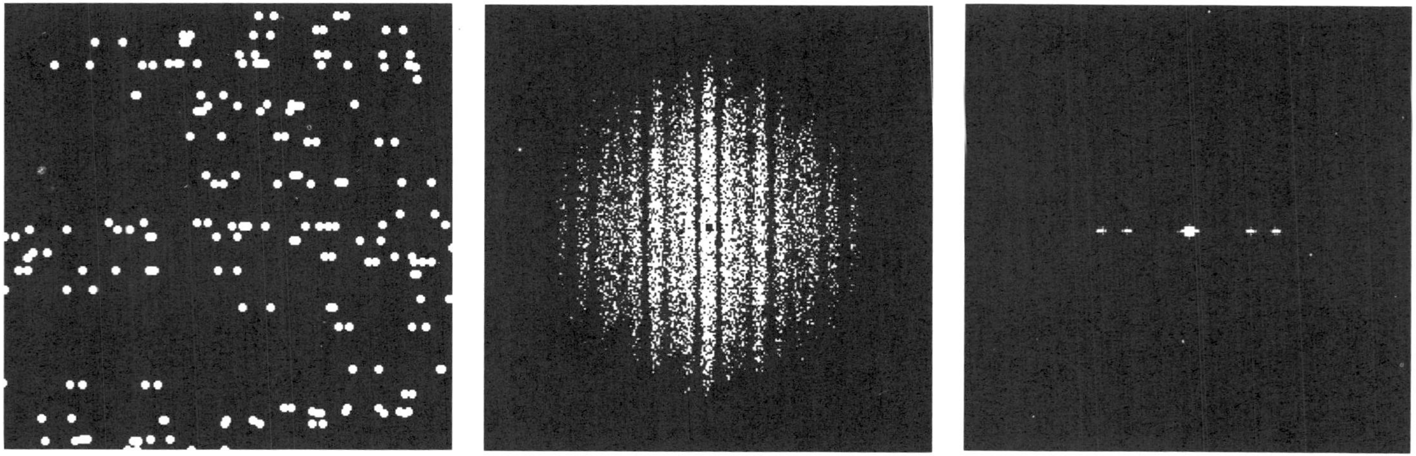

Figure 6.23 Computer simulation of the processing of stereoscopic PIV images obtained with the backprojection technique [93]. (a) Input PIV record containing particle pairs from both viewing angles. (b) Fringe system with reduced visibility due to the two discrete particle separations. (c) The twin peak in the autocorrelation function contains in-plane and out-of-plane displacements.

images in three dimensions. Again the depth component is obtained with much reduced accuracy, which has inspired proposed setups with holography at two orthogonal directions.

It is not appropriate to go into the details of particle holography here. Let us rather examine how the combination of PIV concepts with holography will advance the technique toward the third dimension. We would like to perform particle image holography by preserving the basic advantage of PIV interrogation, i.e., the simultaneous use of many particle pairs for the determination of displacement.

Let us retain the familiar light sheet configuration, sufficiently deep to catch enough particle pairs in their depth motion. Instead of the ordinary photograph, a double-exposed off-axis hologram of the particle field is recorded (Fig. 6.24a). We look for the three-dimensional equivalent to Young's fringe formation and autocorrelation analysis [97]. As position in the hologram is not assigned to position in the object space, the interrogation region must be selected by an aperture in the reconstructed real image of the particles (Fig. 6.24b). Scanning can be performed by moving the hologram in front of a stationary aperture, for example. Light from

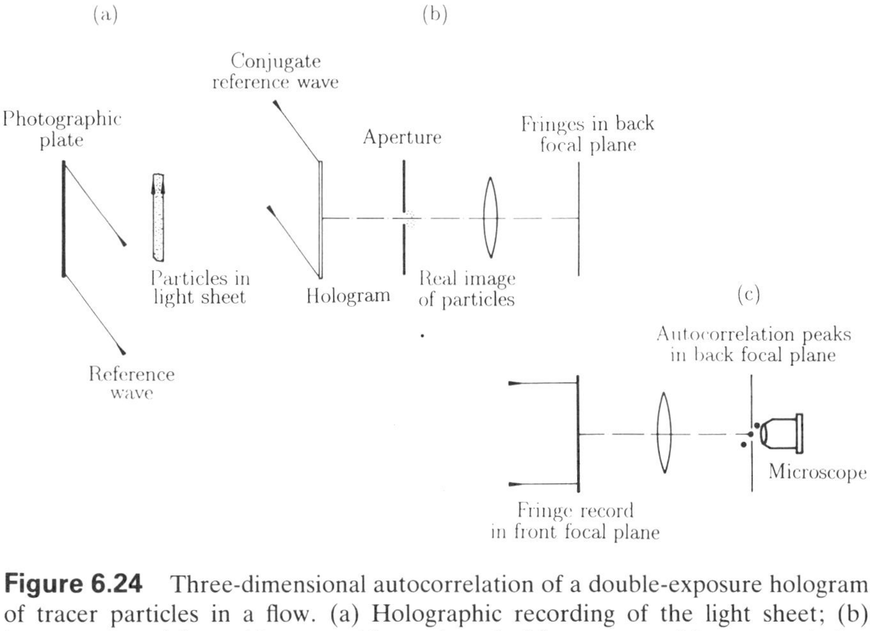

Figure 6.24 Three-dimensional autocorrelation of a double-exposure hologram of tracer particles in a flow. (a) Holographic recording of the light sheet; (b) interrogation of the real image and formation of a fringe system; (c) autocorrelation by Fourier transformation of the fringe system. (After Ref. 97.)

the particle pairs is recorded in the back focal plane of a lens. A transparency with this pattern is Fourier transformed optically, and the output is observed with a microscope (Fig. 6.24c).

Instead of a mathematical analysis [97] we present an illustrative approach with the aid of Fig. 6.25. We take a single particle pair and place one partner on-axis in the input focal plane. This is no restriction of generality as a transverse (axial) displacement adds only a linear (quadratic) phase factor that will not show up in the light intensity. The output in the back focal plane can be interpreted as the superposition of a convergent object wave from particle 2 with a parallel reference wave from particle 1, i.e., formation of a hologram of particle 2. It is made up of circular fringes of the Fresnel zone type, analogous to the Young's fringes in a two-dimensional input. When such a hologram is illuminated by a collimated beam (the reference wave), three waves are generated by diffraction, the direct and conjugate images and the zero-order wave. Each will be focused to a peak in the output. The peaks are arranged in space just as the original particle sources were. We have obtained the three-dimensional autocorrelation of the input. Its orientation and spacing is examined with a mi-

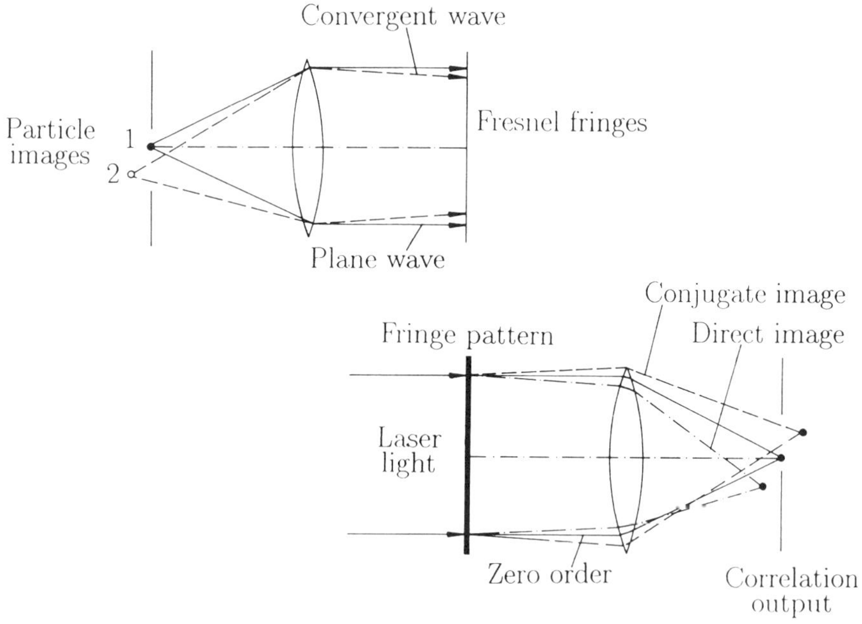

Figure 6.25 Explanation of three-dimensional autocorrelation of a particle pair in space by means of a holographic analogy.

croscope. An extension of this analysis to many particle pairs does not change anything in principle. It results in the addition of speckle noise from the random positions of the particles, as was the case in ordinary PIV also.

Expected results were obtained experimentally on a sample of small particles embedded in a transparent solid [97]. The method fails when the flow has only a depth component, because the collinear peaks will obscure each other. Since the sheet is usually placed parallel to the dominant flow direction, this case hardly ever occurs. However, it is possible to use separate reference beams for both exposures to obtain a cross-correlation with a single peak and the additional possibility to resolve directional ambiguity (cf. Section 6.5.1).

The above technique is reminiscent of earlier proposals to derive transverse displacements from direct Fourier transforms of a double-exposure in-line hologram [98] where, however, no use was made of the out-of-plane term.

It seems easy to extend holography of particle fields beyond the limited depth of a light sheet simply by expanding the illuminating light in depth. In this way an extended volume of the flow is available for inspection. Difficulties arise, however, if one likes to retain evaluation strategies from PIV and not go into single-particle tracking, which is very cumbersome in a large volume. A strategy is needed that allows us to choose a light sheet from a holographic record that originally covers the entire volume. Making this selection merely by focusing has been found to be unsatisfactory, as the many defocused particle images contribute considerable noise to the background. Various optical filtering techniques are pursued that should select particles in a given interrogation volume from the three-dimensional test region. Nonlinearities in recording materials are employed as well as spatial filtering [99], which was long ago proposed as a holographic matched filter for depth selection of small particles from in-line Fraunhofer holograms [100].

A method has been demonstrated that makes use of the limited coherence of laser light to select a single sheet in depth in the evaluation [101]. It is related to "light-in-flight" holography, in which the wave front of a propagating light wave can be visualized by holographic recording of only a very narrow spatial region around the location of equal path lengths. The basic idea is illustrated in Fig. 6.26 for a wind tunnel study. A single light sheet is redirected repeatedly through the flow at a spacing smaller than the flow scales to be resolved. Additional beam-shaping optics may be required to restore the quality of the light sheet whenever it passes the flow. The double-pulsed laser thus simultaneously illuminates an entire batch of sheets filled with particles. All scattered light reaches a holographic

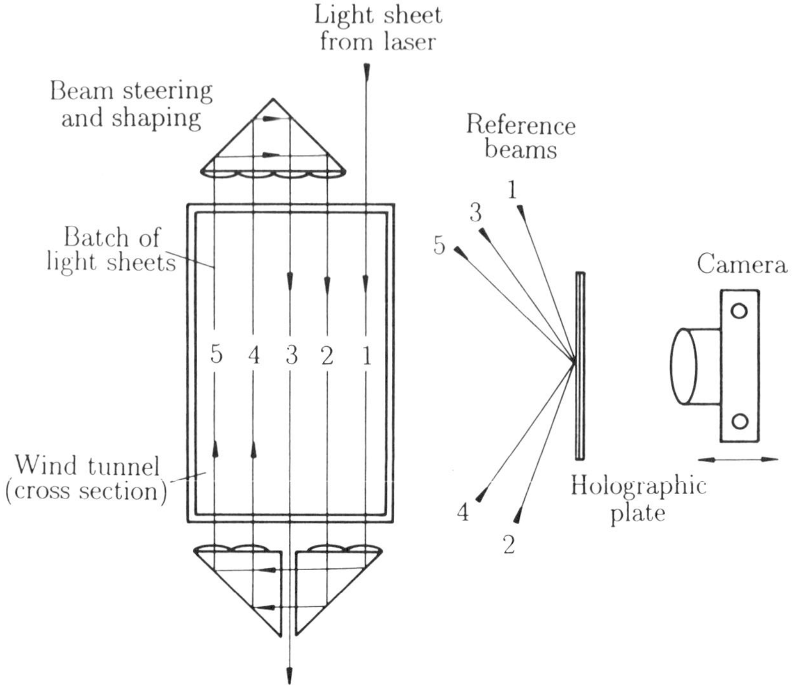

Figure 6.26 Particle image holography of a batch of light sheets by selective coherence matching with an appropriate set of reference beams.

plate. A number of reference beams arrive at the plate at different angles. The optical path of each reference beam is such that each beam produces interference with light from only a single sheet. This is guaranteed by appropriate selection of the coherence length of the laser and the optical detours between sheets. By shifting through the set of reference beams upon reconstruction, particles in one sheet can be evaluated without disturbing information from any other plane in depth. All PIV techniques explained so far, ranging from ordinary photography to three-dimensional autocorrelation, can be applied to the various planes successively. Finally, a complete set of three-dimensional information about the flow is available.

The outlined scheme cannot be extended to an arbitrary number of sheets. First of all, the quality of reconstructed images in holographic multiplexing decreases with the number of holograms. Second, the coherence function of a laser emitting in several longitudinal modes repeats with a period equal to twice the resonator length. For the maximum number of light sheets we have to divide this length by the coherence length. If more sheets are needed, the two (or more) light sheets that reconstruct with the same reference wave should be spaced far enough apart in depth to minimize cross talk. For pulsed holography a Q-switched ruby laser is usually used. With a single intracavity etalon its coherence length can be adjusted to measure about 10 cm.

6.6 STATE OF THE ART OF PIV

Researchers in fluid dynamics have become aware of the existence of PIV as a powerful method for quantitative whole-field recording of flow fields. Wherever flows are investigated the simultaneous registration of the spatially extended velocity field either offers scientific data otherwise not obtainable or provides better economy in collecting flow data. The traditional pointwise measuring facilities are supplemented by equipment for large-area velocity mapping. We have shown that there is still much development in this field, profiting from modern technological progress. It is difficult to predict what the typical PIV system of the future will be like. Will it be completely digital, or will analog optical processing play a role?

Today, most laboratories design their own equipment to satisfy the specific needs of their research projects. In addition, several companies offer equipment commercially, mostly the processing components. This includes the translation stage for the input transparency and the optical components for read-in of the PIV record and eventually Young's fringe formation. Data are recorded by a video system and framegrabber and are digitized and stored for further computer processing. Computer power depends on the amount of data and the speed required, ranging from PC-operated systems to high-speed array processors. Software for evaluation and validation and the assembly and graphical presentation of a velocity map is an important part of the systems.

A systematic experiment with PIV produces a huge amount of spatial flow information. The PIV system of the future could comprise all three-dimensional features we have introduced and operate at a given sampling rate for the length of an experiment. Such a flood of data must be presented properly for interpretation. This problem is not unique to flow measurements but arises in many fields where multidimensional data must be handled. A crew of computer science experts have joined the physicists to invent tools for the visualization and animation of such data fields [102]. Stereoscopic TV presentation is now possible, and holographic display may become feasible some day.

Let us conclude this chapter with a few exemplary applications that demonstrate the range of questions that profit from PIV. Furthermore, I would like to show some frontiers that have been reached. We will not

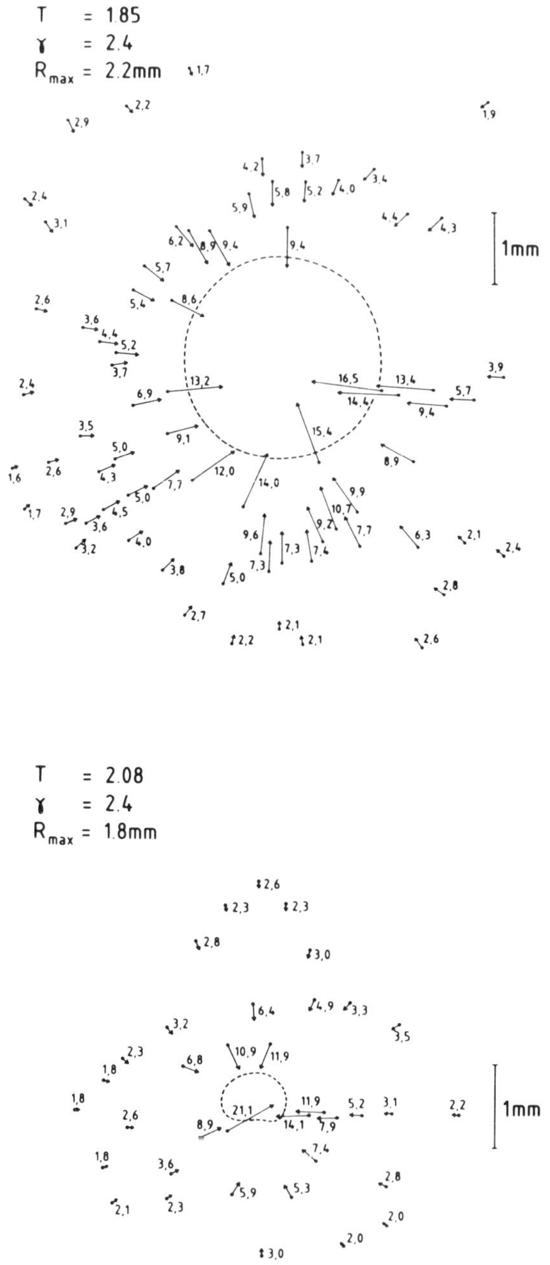

Figure 6.27 Flow fields around laser-produced cavitation bubbles collapsing in the vicinity of a solid wall; velocities are in m/s. (With permission from Ref. 73.)

cover the vast range of fluid dynamical problems that have made use of PIV. Interested readers are referred to work that is oriented accordingly [2].

Figure 6.27 shows flow fields during two phases of collapsing cavitation bubbles that have been produced by focusing a high-intensity ruby laser pulse into water [73]. Here, the complex task of describing the nonlinear dynamic behavior of a bubble collapse close to a solid wall benefited from the ability to obtain experimentally a velocity map in the vicinity of the bubble. Obviously, there is no time for pointwise velocity measurement in this case. The rapid motion of the bubble required high-speed, time-resolved recording. The continuous light from an argon ion laser was externally chopped by an acousto-optical deflector, and frames containing five pulses each were recorded on a rotating drum camera. Vestamid particles 5 μm in diameter and of a density very close to that of water allowed low-error registration even at the high accelerations of more than 10^6 m/sec^2 in the final stage of the collapse. Microstreaming around the bubble could be compared with theoretical predictions.

A device for PIV recording of velocity fields in water waves was installed in a wave tank [13]. Figure 6.28 gives the flow field in a breaking wave.

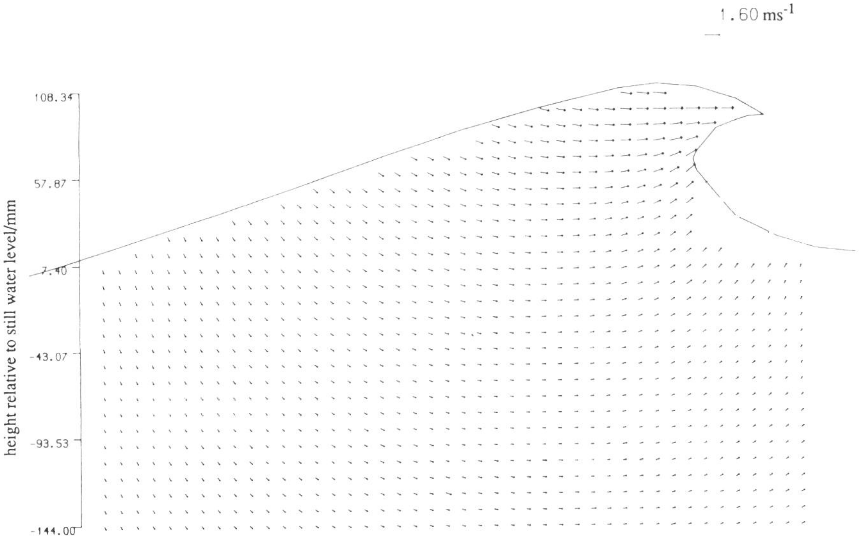

Figure 6.28 Velocity map from a breaking water wave obtained with the scanning mirror illumination. (With permission from Ref. 13.)

Conifer pollen served as tracer particles. To achieve a light sheet of 50-cm width, the scanning beam technique was applied. A 15-W argon ion laser beam was directed into the flow from underneath through the glass bottom of the tank. The section containing the rotating and parabolic mirrors was mounted on a sliding carriage so that the distance between the inspected region and the wave paddle could be changed easily. Such a laboratory setup provides a versatile complementary tool in routine wave studies.

Flow investigations with particle image velocimetry have been pushed to velocities in the transonic region [103]. Here, measurements in a high-speed wind tunnel had to be completed within 20 sec, the time needed to fill the vacuum tank that sucked air through the tunnel. This is an excellent example of the superiority of a whole-field method. A time delay of 2 μsec between illuminating pulses from a two-oscillator Nd:YAG laser allowed

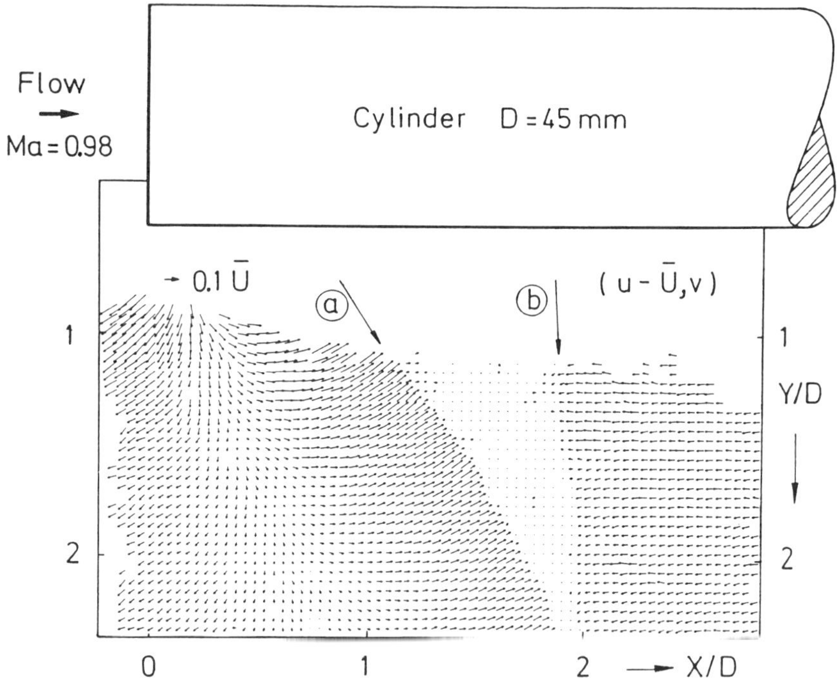

Figure 6.29 High-speed wind tunnel flow at Mach number Ma = 0.98 around a cylinder of 45-mm diameter. The mean velocity equal to the speed of sound has been subtracted. Supersonic regions (vectors pointing toward the right), sonic regions (vectors almost zero), and subsonic regions (vectors toward the left) are separated by shocks. (With permission by DLR from Ref. 103.)

the study of flow velocities up to 400 m/sec. Figure 6.29 shows the flow configuration around a bluff cylinder at a free-stream Mach number of Ma = 0.98. To enhance the impression of flow structures, the mean velocity $\langle U \rangle$ = 310 m/sec has been subtracted; the length of the small arrow indicates $0.1\langle U \rangle$. Since $\langle U \rangle$ is almost identical to the speed of sound, subsonic regions are distinguished by arrows pointing to the left, while vectors pointing to the right indicate supersonic regions. Clearly seen in this picture are an oblique shock that separates regions of supersonic and sonic flow and another shock normal to the wall of the cylinder that separates the sonic region from the following subsonic region. It is obvious that velocity information in the immediate neighborhood of the shock is subject to systematic error because even the 1-μm oil droplets will not follow the flow reliably.

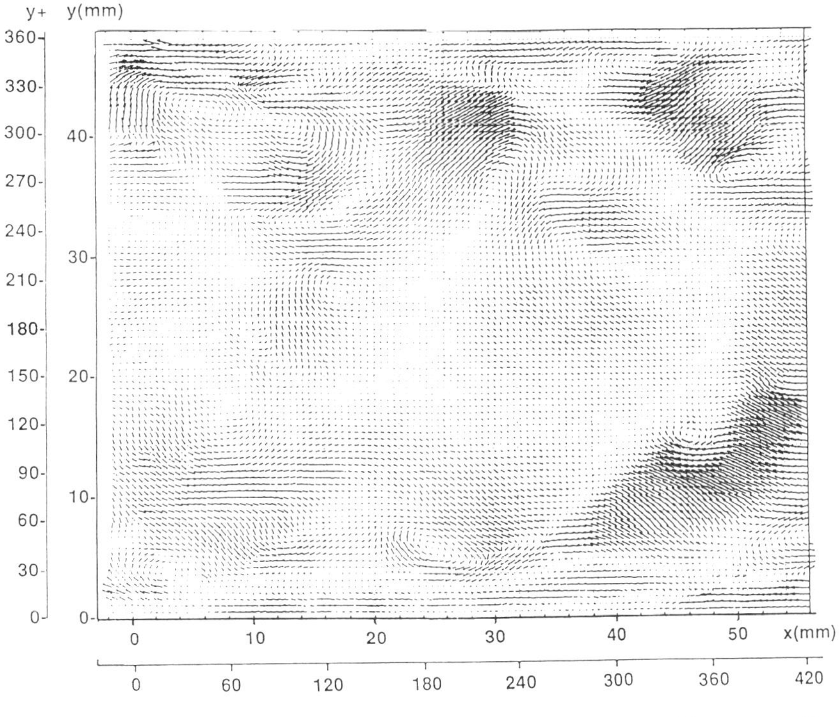

Figure 6.30 Structure of fully developed turbulence in a 50 mm × 60 mm downstream test field of a water channel flow. Shown is the fluctuating velocity obtained by subtracting the ensemble average value calculated from 11 experiments. (With permission from Ref. 104.)

Finally, let us demonstrate the capability of a current sophisticated PIV system [104]. Fully developed turbulent flow was studied in a water channel at a velocity of about 100 mm/sec. Double pulses from a ruby laser with 2.5-msec separation illuminated 5-μm aluminum oxide particles within a 50 mm × 60 mm field of view. Interrogation was performed within 0.3 sec per point by a Fourier transformation of the Young's fringe system on a fast array processor. Figure 6.30 gives the impressive result of such an investigation where the ensemble-averaged velocity value obtained from 11 such records has been subtracted. The existence of coherent structures, ejections from the wall, and separating shear layers is obvious from such whole-field visualizations. This emphasizes the power of PIV. How could a single-point technique accumulate this information from some 12,000 points in space? Furthermore, PIV results are the starting data for the computation of many other parameters relevant to flow. Impressive examples are color-coded representations of vorticity.

Even these few examples show that particle image velocimetry is no longer a technique in the initial state of laboratory experimentation. It has been employed over a wide range of velocities up to supersonic conditions and has found application in studies of gas and liquid flows. Well-established interrogation routines and sophisticated equipment yield spatial flow information unattainable with any other method. In addition, the technique benefits from many current technological developments that promise further progress. It is a good example for the synthesis of optical design and architecture with electronics and computers for a system of high performance.

REFERENCES

1. W. Merzkirch, *Flow Visualization*, 2nd ed., Academic, Orlando, 1987.
2. R. J. Adrian, Particle-imaging techniques for experimental fluid mechanics, *Annu. Rev. Fluid Mech.*, *23*, 261–304 (1991).
3. M. Van Dyke, *An Album of Fluid Motion*, Parabolic, Stanford, Calif., 1982.
4. D. B. Barker and M. E. Fourney, Displacement measurements in the interior of 3-D bodies using scattered-light speckle patterns, *Exp. Mech.*, *16*, 209–214 (1976)
5. D. B. Barker and M. E. Fourney, Measuring fluid velocities with speckle patterns, *Opt. Lett.*, *1*, 135–137 (1977).
6. T. D. Dudderar and P. G. Simpkins, Laser speckle photography in a fluid medium, *Nature*, *270*, 45–47 (1977).
7. R. Grousson and S. Mallick, Study of flow patterns in a fluid by scattered laser light, *Appl. Opt.*, *16*, 2334–2336 (1977).

8. K. Iwata, T. Hakoshima, and R. Nagata, Measurement of flow velocity distribution by multiple exposure speckle photography, *Opt. Commun.*, *25*, 311–314 (1978).
9. R. Meynart, Equal velocity fringes in a Rayleigh-Bénard flow by a speckle method, *Appl. Opt.*, *19*, 1385–1386 (1980).
10. R. J. Adrian, Scattering particle characteristics and their effect on pulsed laser measurements of fluid flow: speckle velocimetry vs particle image velocimetry, *Appl. Opt.*, *23*, 1690–1691 (1984).
11. C. J. D. Pickering and N. A. Halliwell, Laser speckle photography and particle image velocimetry: photographic film noise, *Appl. Opt.*, *23*, 2961–2969 (1984).
12. T. D. Dudderar, R. Meynart, and P. G. Simpkins, Full-field laser metrology for fluid velocity measurement, *Opt. Lasers Eng.*, *9*, 163–199 (1988).
13. C. Gray, C. A. Greated, D. R. McCluskey, and W. J. Easson, An analysis of the scanning beam PIV illumination system, *Meas. Sci. Technol.*, *2*, 717–724 (1991).
14. R. Meynart, Speckle velocimetry study of vortex pairing in a low-Re unexcited jet, *Phys. Fluids*, *26*, 2074–2079 (1983).
15. C. S. Yao and R. J. Adrian, Orthogonal compression and 1-D analysis technique for measurement of 2-D particle displacements in pulsed laser velocimetry, *Appl. Opt.*, *23*, 1687–1689 (1984).
16. S. H. Collicot and L. Hesselink, Anamorphic optical processing of multiexposure speckle photographs, *Opt. Lett.*, *11*, 410–412 (1986).
17. J. M. Coupland and N. A. Halliwell, Particle image velocimetry: rapid transparency analysis by optical correlation, *Appl. Opt.*, *27*, 1919–1921 (1988).
18. J. Kompenhans, J. Reichmuth, and R. Höcker, Data evaluation in particle image velocimetry by means of a spatial light modulator, ICALEO '89, *LIA Proc.*, *68*, 121–129 (1989).
19. K. Hinsch and W. Arnold, Spatially multiplexed optical correlation for displacement mapping in speckle metrology, *Proc. SPIE*, *1136*, 327–334 (1989).
20. C. E. Willert and M. Gharib, Digital particle image velocimetry, *Exp. Fluids*, *10*, 181–193 (1991).
21. U. Wernekinck and W. Merzkirch, Speckle photography of spatially extended refractive-index fields, *Appl. Opt.*, *26*, 31–32 (1987).
22. R. Erbeck and W. Merzkirch, Speckle photographic measurement of turbulence in an air stream with fluctuating temperature, *Exp. Fluids*, *6*, 89–93 (1988).
23. R. J. Adrian, Statistical properties of particle image velocimetry measurements in turbulent flow, Proc. 3rd Int. Symp. Applications of Laser Anemometry to Fluid Mechanics, Lisbon, 1986, Paper 17.6.
24. R. J. Adrian, Double exposure, multiple-field particle image velocimetry for turbulent probability density, *Opt. Lasers Eng.*, *9*, 211–228 (1988).
25. A. Papoulis, *Probability, Random Variables, and Stochastic Processes*, McGraw-Hill, New York, 1965.

26. K. Hinsch, W. Arnold, and W. Platen, Flow field analysis by large-area interrogation in particle image velocimetry, *Opt. Lasers Eng.*, *9*, 229–243 (1988).
27. C. S. Moraitis and M. L. Riethmuller, Stochastic modelling and performance prediction of PIDV experiments, Proc. 4th Int. Symp. Applications of Laser Anemometry to Fluid Mechanics, Lisbon, 1988, Paper 5.14.
28. W. Arnold, K. Hinsch, and D. Mach, Turbulence level measurement by speckle velocimetry, *Appl. Opt.*, *25*, 330–331 (1986).
29. J. W. Goodman, Statistical properties of laser speckle patterns, in *Laser Speckle and Related Phenomena*, J. C. Dainty, Ed., Springer-Verlag, Berlin, 1975, pp. 9–75.
30. J. M. Coupland and C. J. D. Pickering, Particle image velocimetry: estimation of measurement confidence at low seeding density, *Opt. Lasers Eng.*, *9*, 201–210 (1988).
31. L. M. M. Lourenco and M. C. Whiffen, Laser speckle methods in fluid dynamics applications, Proc. 2nd Int. Symp. Applications of Laser Anemometry to Fluid Mechanics, Lisbon, 1984, Paper 6.3.
32. M. P. Arroyo, T. Tonte, M. Quintanilla, and J. M. Saviron, Particle image velocimetry in Rayleigh-Bénard convection: photographs with a high number of exposures, *Opt. Lasers Eng.*, *9*, 295–316 (1988).
33. A. Cenedese and A. Paglialunga, Digital direct analysis of a multiexposed photograph in PIV, *Exp. Fluids*, *8*, 273–280 (1990).
34. R. D. Keane and R. J. Adrian, Optimization of particle image velocimeters. Part II. Multiple pulsed systems, *Meas. Sci. Technol.*, *2*, 963–974, (1991).
35. R. V. Edwards, Some practical considerations of the resolution and accuracy of particle image velocimetry, Proc. 4th Int. Symp. Applications of Laser Anemometry to Fluid Mechanics, Lisbon, 1988, Paper 2.16.
36. J. W. Goodman, *Introduction to Fourier Optics*, McGraw-Hill, New York, 1968.
37. C. J. D. Pickering and N. A. Halliwell, Speckle photography in fluid flows: signal recovery with two-step processing, *Appl. Opt.*, *23*, 1128–1129 (1984).
38. C. C. Landreth, R. J. Adrian, and C. S. Yao, Double pulsed particle image velocimeter with directional resolution for complex flows, *Exp. Fluids*, *6*, 119–128 (1988).
39. R. Erbeck, Fast image processing with a microcomputer applied to speckle photography, *Appl. Opt.*, *24*, 3838–3841 (1985).
40. G. H. Kaufmann, A. E. Ennos, B. Gale, and D. J. Pugh, An electro-optical read-out system for analysis of speckle photographs, *J. Phys. E: Sci. Instrum.*, *13*, 579–584 (1980).
41. B. Ineichen, P. Eglin, and R. Dändliker, Hybrid optical and electronic image processing for strain measurements by speckle photography, *Appl. Opt.*, *19*, 2191–2195 (1980).
42. J. M. Huntley, Speckle photography fringe analysis: assessment of current algorithms, *Appl. Opt.*, *28*, 4316–4322 (1989).
43. R. Meynart, Instantaneous velocity field measurements in unsteady gas flow by speckle velocimetry, *Appl. Opt.*, *22*, 535–540 (1983).

44. G. A. Reynolds, M. Short, and M. C. Whiffen, Automated reduction of instantaneous flow field images, *Opt. Eng.*, *24*, 475–479 (1985).
45. A. Kirita, C. J. D. Pickering, and N. A. Halliwell, Particle image velocimetry: automatic fringe analysis by cross-correlation, *Opt. Eng.*, *27*, 188–192 (1988).
46. D. W. Robinson, Automatic fringe analysis with a computer image-processing system, *Appl. Opt.*, *22*, 2169–2176 (1983).
47. K. Hinsch, Fringe positions in double-exposure speckle photography, *Appl. Opt.*, *28*, 5298–5304 (1989).
48. C. J. D. Pickering and N. A. Halliwell, Particle image velocimetry: Fringe visibility and pedestal removal, *Appl. Opt.*, *24*, 2474–2476 (1985).
49. J. M. Coupland and N. A. Halliwell, Automated optical analysis of Young's fringes—optical autocorrelation, *Opt. Lasers Eng.*, *14*, 351–361 (1991).
50. J. A. Neff, R. A. Athale, and S. H. Lee, Two-dimensional spatial light modulators: a tutorial, *Proc. IEEE*, *78*, 826–855 (1990).
51. W. Arnold and K. Hinsch, Purely optical parallel processing in particle image velocimetry and the study of flow structures, *ICALEO '88 LIA Proc.*, *67*, 157–165 (1988).
52. K. Hinsch, W. Schipper, and D. Mach, Fringe visibility in speckle metrology and the analysis of random flow components, *Appl. Opt.*, *23*, 4460–4462 (1984).
53. I. Grant, E. H. Owens, and G. H. Smith, Particle image velocimetry fringe visibility as a means of turbulence estimation, *Appl. Opt.*, *28*, 2982–2984 (1989).
54. R. Rubinstein and P. S. Greenberg, Young's fringe particle image velocimetry in the presence of random particle motions, *Proc. ASME*, *Laser Anemometry*, *Vol. 1*, 337–340 (1991).
55. M. Born and E. Wolf, *Principles of Optics*, Pergamon, Oxford, 1980.
56. J. Kompenhans and R. Höcker, Investigation of turbulent flows by means of particle image velocimetry, *Flow Visualization V*, 5th Int. Symp. on Flow Visualization, Prague, 1989.
57. M. L. Jakobsen and P. Buchhave, PIV-processing: parallel processing with optical correlators, Proc. 5th Int. Symp. Applications of Laser Anemometry to Fluid Mechanics, Lisbon, 1990, Paper 12.2.
58. R. D. Keane and R. J. Adrian, Optimization of particle image velocimeters. Part I. Double pulsed systems, *Meas. Sci. Technol.*, *1*, 1202–1215 (1990).
59. Y. W. Zhang, Optical measurement of velocity and acceleration with colour coding, *Opt. Laser Tech.*, *18*, 256–258 (1986).
60. R. Meynart, Convective flow field measurement by speckle velocimetry, *Rev. Phys. Appl.*, *17*, 301–305 (1982).
61. Y.-C. Cho, Digital image velocimetry; DIV, *Appl. Opt.*, *28*, 740–748 (1989).
62. E. Okada, H. Enomoto, Y. Fukuoka, and H. Minamitani, Instantaneous imaging velocimeter with electronic specklegram, Proc. 5th Int. Symp. Applications of Laser Anemometry to Fluid Mechanics, 1990, Paper 18.2.
63. J. Westerweel and F. T. M. Nieuwstadt, Measurement of the dynamics of coherent flow structures using particle image velocimetry, Proc. 5th Int.

Symp. Applications of Laser Anemometry to Fluid Mechanics, 1990, Paper 18.3.
64. B. R. Frieden and C. K. Zoltani, Fast tracking algorithm for multiframe particle image velocimetry data, *Appl. Opt.*, *28*, 652–654 (1989).
65. A. Van der Lugt, Signal detection by complex spatial filtering, *IEEE Trans. Inf. Theory*, *IT-10*, 139–145 (1964).
66. P. Günther and J.-P. Huignard, *Photorefractive Materials and Their Applications*, Vols. I and II, Springer-Verlag, Berlin, 1988.
67. G. Gheen and L.-J. Cheng, Optical correlators with fast updating speed using photorefractive semiconductor materials, *Appl. Opt.*, *27*, 2756–2761 (1988).
68. C. S. Moraitis and M. L. Riethmuller, A real time holographic autocorrelator for the analysis of pulsed laser velocimetry images based on a $Bi_{12}SiO_{20}$ spatial light modulator, *ICALEO '89, LIA Proc.*, *68*, 112–120 (1989).
69. M. P. Wernet and R. V. Edwards, Real time optical correlator using a magnetooptic device applied to particle imaging velocimetry, *Appl. Opt.*, *27*, 813–815 (1988).
70. A. T. Hjemfelt and L. F. Mockros, Motion of discrete particles in a turbulent fluid, *Appl. Sci. Res.*, *16*, 149–161 (1965).
71. R. N. Bracewell, *The Fourier Transform and Its Applications*, McGraw-Hill, New York, 1978, p. 398.
72. R. J. Adrian and C.-S. Yao, Pulsed laser technique application to liquid and gaseous flows and the scattering power of seed materials, *Appl. Opt.*, *24*, 44–52 (1985).
73. A. Vogel and W. Lauterborn, Time-resolved particle image velocimetry used in the investigation of cavitation bubble dynamics, *Appl. Opt.*, *27*, 1869–1876 (1988).
74. A. Vogel and W. Lauterborn, Time resolved particle image velocimetry, *Opt. Lasers Eng.*, *9*, 277–294 (1988).
75. J. P. Prenel, R. Porcar, and A. El Rhassouli, Three-dimensional flow analysis by means of sequential and volumic laser sheet illumination, *Exp. Fluids*, *7*, 133–137 (1989).
76. R. Höcker and J. Kompenhans, Some technical improvements of particle image velocimetry with regard to its application in wind tunnels, Proc. Int. Congr. on Instrumentation in Aerospace Simulation Facilities—ICIASF '89, 1989, pp. 545–554.
77. C. J. D. Pickering and N. A. Halliwell, Particle image velocimetry: improving fringe signal-to-noise ratio with a two-step photographic process, *J. Opt. Soc. Am.*, *A2*, 1721–1724 (1985).
78. J. M. Coupland, C. J. D. Pickering, and N. A. Halliwell, Particle image velocimetry: theory of directional ambiguity removal using holographic image separation, *Appl. Opt.*, *26*, 1576–1578 (1987).
79. I. Grant and A. Liu, Directional ambiguity resolution in particle image velocimetry by pulse tagging, *Exp. Fluids*, *10*, 71–76 (1990).
80. L. P. Goss, M. E. Post, D. D. Trump, and B. Sarka, Two color particle velocimetry, *ICALEO '89, LIA Proc.*, *68*, 101–111 (1989).

81. R. J. Adrian, Image shifting technique to resolve directional ambiguity in double-pulsed velocimetry, *Appl. Opt.*, *25*, 3855–3858 (1986).
82. C. C. Landreth and R. J. Adrian, Electrooptical image shifting for particle image velocimetry, *Appl. Opt.*, *27*, 4216–4219 (1988).
83. K. Hinsch, W. Arnold, and W. Platen, Turbulence measurements by particle imaging velocimetry, *ICALEO '87, LIA Proc.*, *63*, 127–134 (1988).
84. L. M. Lourenco, Particle image velocimetry: photographic and video techniques, in *Laser Velocimetry*, von Karman Institute for Fluid Dynamics, Brussels, Lecture Series 1991-05, 1991.
85. T. Utami and T. Ueno, Visualization and picture processing of turbulent flow, *Exp. Fluids*, *2*, 25–32 (1984).
86. A. Cenedese and A. Paglialunga, A new technique for the determination of the third velocity component with PIV, *Exp. Fluids*, *8*, 228–230 (1989).
87. C. E. Willert and M. Gharib, Three-dimensional particle imaging with a single camera, *Exp. Fluids*, *12*, 353–358 (1992).
88. A. A. Adamczyk and L. Rimai, Reconstruction of a 3-dimensional flow field from orthogonal views of seed track video images, *Exp. Fluids*, *6*, 380–386 (1988).
89. R. G. Racca and J. M. Dewey, A method for automatic particle tracking in a three-dimensional flow field, *Exp. Fluids*, *6*, 25–32 (1988).
90. S. K. Sinha, Improving the accuracy and resolution of particle image for laser speckle velocimetry, *Exp. Fluids*, *6*, 67–68 (1988).
91. M. P. Arroyo and C. A. Greated, Stereoscopic particle image velocimetry, *Meas. Sci. Technol.*, *2*, 1181–1186 (1991).
92. H.-G. Maas, Digital photogrammetry for determination of tracer particle coordinates in turbulent flow research, *Photogr. Eng. Rem. Sens.*, *57*, 1593–1597 (1991).
93. F. Dreesen, H. Hinrichs, K. Hinsch and A. Roshop, Holographic particle image velocimetry (HPIV)—an extension of PIV to a 3-D measuring method, in *Flow Visualization VI*, Y. Tanida and H. Miyashiro, Eds., Springer-Verlag, Berlin, 1991, 822–826.
94. C. S. Vikram, *Holographic Particle Diagnostics* (SPIE Milestone Ser. MS21), SPIE, Bellingham, Wash., 1990.
95. B. J. Thompson, Holographic methods for particle size and velocity measurement—recent advances, *Proc. SPIE*, *1136*, 308–326 (1989).
96. H. Royer, Holographic velocimetry of submicron particles, *Opt. Commun.*, *20*, 73–75 (1977).
97. J. M. Coupland and N. A. Halliwell, Particle image velocimetry: three-dimensional fluid velocity measurements using holographic recording and optical correlation, *Appl. Opt.*, *31*, 1004–1008 (1992).
98. P. H. Malyak and B. J. Thompson, Particle displacement and velocity measurement using holography, *Opt. Eng.*, *23*, 567–576 (1984).
99. C. M. Moraitis and M. L. Riethmuller, Real time optical processing of holographically recorded particle images for 3D whole field velocimetry, Proc. 5th Int. Symp. Applications of Laser Anemometry to Fluid Mechanics, Lisbon, 1990, Paper 12.3.

100. F. M. Shofner, R. W. Menzel, T. H. Gee, and R. O. Webb, Processing holographic velocimetry data, *ICIASF '69 Record*, 136–144 (1969).
101. K. Hinsch, H. Hinrichs, G. Kuhfahl, and P. Meinlschmidt, Holographic recording of 3-D flow configurations for particle image velocimetry (PIV), *ICALEO '90*, *LIA Proc.*, *72*, 121–130 (1991).
102. L. Hesselink, Digital image processing in flow visualization, *Ann. Rev. Fluid Mech.*, *20*, 421–485 (1988).
103. R. Höcker and J. Kompenhans, Application of particle image velocimetry to transonic flows, Proc. 5th Int. Symp. Applications of Laser Anemometry to Fluid Mechanics, Paper 15.1, Lisbon, 1990.
104. Z.-C. Liu, C. C. Landreth, R. J. Adrian and T. J. Hanratty, High resolution measurement of turbulent structure in a channel with particle image velocimetry, *Exp. Fluids*, *10*, 301–312 (1991).

7

White Light Speckle Metrology

Anand K. Asundi

University of Hong Kong
Hong Kong

7.1 INTRODUCTION

The white light speckle method relies on the random pattern, either created artificially or present naturally, on the surface of the test object for deformation gauging. Thus in many respects this method is closely akin to the moiré methods. Whereas in the moiré methods a regular pattern of dots is imprinted on the specimen, the white light speckle method makes use of a random array of dots. Practically, it is much simpler to generate a random pattern than it is to obtain a regular pattern. Furthermore, with the speckle being physically attached to the specimen, analysis of speckle movements is much simpler and more straightforward than in the laser speckle method. Like the moiré method, this makes such applications as the measurement of large object deformation and the analysis of curved surfaces and interior displacement measurement a cinch. However, like its laser speckle counterpart, postprocessing in the form of optical filtering is required to delineate the deformation information.

Indeed, one of the first applications of random patterns, as used in the white light speckle method, was to show that the human visual system can detect correlations from noisy patterns [1]. As seen in Fig. 7.1, one can readily visualize "the wheel of fortune," "the starburst (whirler)," and "the big bang," patterns obtained by superposition of two slightly different

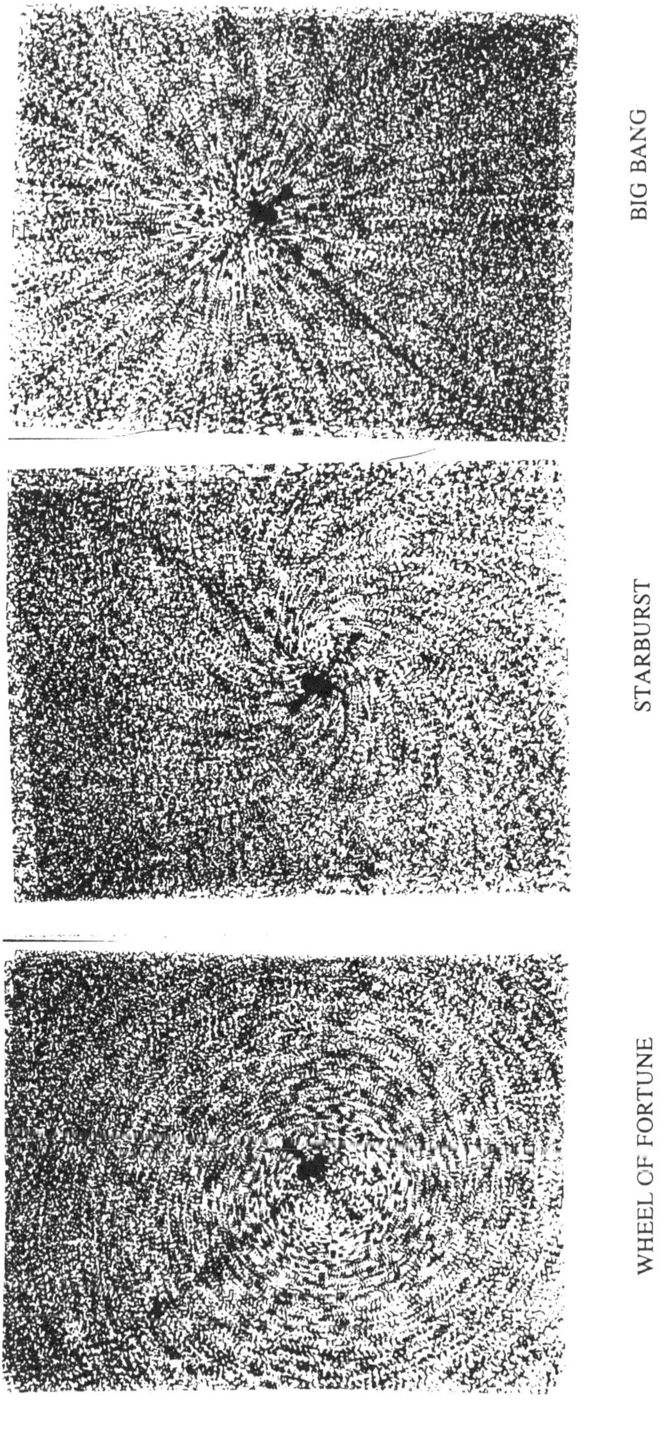

Figure 7.1 Visual ccrrelations from random patterns.

speckle patterns. For example, the wheel is obtained simply by rotating two identical speckle patterns, while the big bang corresponds to superposition of two speckle patterns, one of which is magnified by 5% compared to the other. Such qualitative interpretation can also be used to good effect in flow visualization [2] and vibration analysis [3].

White light speckle metrology, however, evolved from the two techniques most closely akin to this method: moiré and laser speckle. Burch and Forno [4] had developed a modified camera to record only a select band of frequencies present on the object. Using grids on the object and suitable demagnification, they recorded spatial frequencies of 300 lines/mm on high-resolution film. This was taken a step further by Forno [5], who used a random pattern instead of the regular grid to simplify the recording. The random pattern was created by coating the object with retroreflective liquid, thus creating a range of spatial frequencies, of which the multiaperture camera chose a narrow bandwidth. This was a step better than the moiré methods in that the frequency (i.e., sensitivity) was chosen at the recording stage. Around the same time, Boone and De Backer [6] generated white light speckles in four different ways and used the entire aperture for recording. Optical filtering, as for the laser speckle method, which was their guiding light, was demonstrated for pointwise and whole-field visualization of deformation. Both these researchers were looking to apply speckle techniques to large objects and objects outside the laboratory environment, applications that the laser speckle method could not readily handle. Some applications of the white light speckle method for large-object deformation measurement are discussed in Chapters 3 and 5. Chiang and Asundi [7] provided quantitative evaluation of the method and also demonstrated the ability to select the operating frequency at the processing stage. Jacquot and Rastogi [8] developed procedures to eliminate the out-of-plane component contribution from that of the in-plane component. Since then white light speckle metrology has undergone significant development, living up to its claims of simplicity and varied applicability. The method has been used for the measurement of small and large deformations on the surface and in the interior of both flat and curved, small and large objects [9–21]. White light speckle photography has also been applied to fluid mechanics [22, 23] and glaciology [24, 25]. The influence of intervening media in long-distance speckle photography has also been investigated [26]. Benefits of aperturing the imaging lens have also been reported by a number of researchers [27–29].

In this chapter, the theorctical equations governing fringe formation and interpretation in the white light speckle method will be detailed followed by applications of the method to engineering metrology.

7.2 THE WHITE LIGHT SPECKLE METHOD

The white light speckle method is similar to the laser speckle method in that it is a two-step process; recording the specklegram and reconstructing or filtering this pattern to give fringes that represent the displacement undergone by the object between the two exposures. However, as the name suggests, a white light source is used in the recording stage instead of a laser. Reconstruction could also be effected by using a white light source, but the advantage gained by using a noncoherent source for filtering is not significant compared to the versatility of laser-based filtering techniques. Furthermore, other noncoherent radiation sources such as the electron beam in an electron microscope have also been used in the recording stage [30]. However, in this chapter, discussions are restricted to cases wherein the recording is done using a white light source and reconstruction is accomplished with laser-based optical filtering schemes.

7.2.1 Creation of White Light Speckles

As the white light source is not coherent, no speckles due to random interference of scattered light are observed. It is thus necessary to create speckles on the object. Figure 7.2 shows enlarged speckle patterns created by various methods on the surface of the test object. The simplest and most straightforward is to first paint the surface white and then spray it with a fine mist of black paint to give the speckled appearance as in Fig. 7.2a. A retroreflective paint consisting of small glass beads suspended in liquid was the first choice of early researchers for generating the speckled appearance (Fig. 7.2b). For most applications, this is still the preferred choice for creating artificial speckles. Of course, as with moiré gratings, speckle patterns created on film can be adhered onto the specimen surface. A speckle pattern mold can be readily created by exposing film to a laser-generated speckle pattern obtained by illuminating a diffuse glass plate with a laser beam. Figure 7.2c is a photograph of such a pattern. Instead of creating these speckles artificially, the natural roughness of the object, or in some cases naturally occurring random patterns, can also be used. Figures 7.2d–f are photographs of a portion of a concrete wall, a machined aluminum block, and a wooden block. The speckled appearance is quite apparent in these examples. An example of naturally occurring random patterns is dew on grass with sunlight illumination.

Illumination of these artificially created, or naturally occurring, random patterns is accomplished using any white light source. However, in certain cases, to enhance the speckled features on the object, the angle of illuminating light has to be in preferred directions. For example, with retroreflective liquid–generated speckles, the directions of illumination and

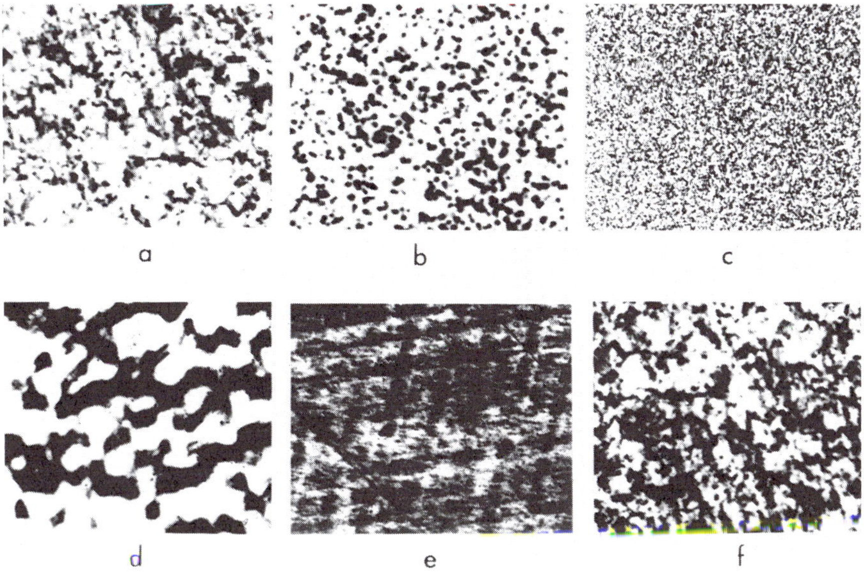

Figure 7.2 Enlarged white light speckle patterns due to (a) black paint sprayed on white surface; (b) retroreflective liquid coating; (c) laser-generated speckles recorded on stripping film; (d) concrete wall; (e) machined aluminum block; (f) wooden block.

viewing should be coincident for best results, although a small angle between them still provides very good contrast speckles. On the other hand, to enhance speckles due to surface roughness, the light should illuminate the surface at a very shallow angle. Finally, for black-and-white painted speckles or stuck-on speckles, a diffuse light source might provide better illumination, these techniques being closest to moiré methods.

7.2.2 Recording of White Light Speckles

The recording of the white light speckle patterns requires high quality optics and precise focusing if all the frequencies present on the objects are to be recorded. Thus, unlike laser speckles where the maximum frequency is limited by the aperture of the recording lens, in the white light speckle methods lens aberrations also influence the maximum frequency that can be imaged. It is no trivial matter to take into account all aberration effects, nor is it really essential in white light speckle metrology. However, one form of aberration caused by a defect in focus is quite interesting and will be considered in depth. It is worth mentioning at this stage that a number of researchers in this area opt for reduced sensitivity by operating at larger f numbers. Exceptions can be found in the modified camera system of Forno [5] and the high magnification recording initiated by Chiang and Asundi [12] and extended to very large magnification using electron microscopy by Chiang and Li [30]. A few researchers have used a multi-aperture mask in front of the imaging lens to obtain prefiltering sensitivity and other benefits [27–29].

7.2.3 Single-Exposure Recording

Consider the schematic diagram of a typical white light speckle recording setup shown in Fig. 7.3. The amplitude distribution on the image plane

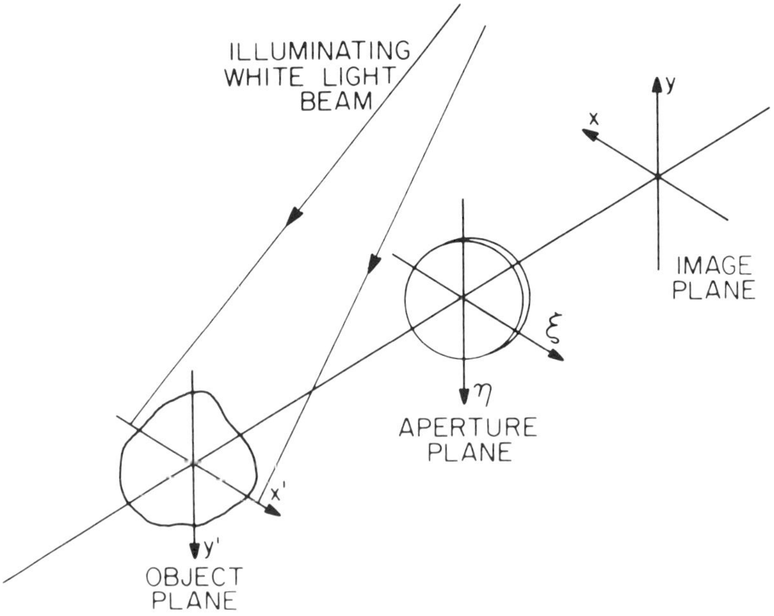

Figure 7.3 Schematic diagram of setup for recording white light speckle patterns.

$V(x, y; t)$ can be readily written in terms of the field on the object plane $V'(x', y'; t)$ and the amplitude impulse response of the imaging system $K(x - x', y - y')$ as

$$V(x, y; t) = \iint_{-\infty}^{\infty} V'(x', y'; t)K(x - x', y - y')\, dx'\, dy' \tag{7.1}$$

This expression is valid for a linear space-invariant imaging system, which usually is the case. Equation (7.1) is the well-known superposition integral.

The intensity on the image plane is the time-averaged product of the amplitude and its conjugate and can be expressed as

$$I(x, y) = \langle V(x, y; t)V^*(x, y; t)\rangle \tag{7.2}$$

where the angled brackets indicate an infinite time average and the asterisk denotes the complex conjugate. From Eqs. (7.1) and (7.2) and noting that the impulse response is time-independent, the intensity on the image plane can be written as

$$I(x, y) = \iiiint_{-\infty}^{\infty} \langle V'(x_{1'}, y_{1'}; t)V^{*\prime}(x_{2'}, y_{2'}; t)\rangle K(x - x_{1'}, y - y_{1'}) \times K^*(x - x_{2'}, y - y_{2'})\, dx_{1'}\, dx_{2'}\, dy_{1'}\, dy_{2'} \tag{7.3}$$

When the object is incoherently illuminated, each point on the object is statistically independent of every other point. Hence,

$$\langle V'(x_{1'}y_{1'}; t)V^{*\prime}(x_{2'}, y_{2'}; t)\rangle = I'(x_{1'}, y_{1'})\delta(x_{1'} - x_{2'}, y_{1'} - y_{2'}) \tag{7.4}$$

where the use of the delta function is an idealization but is valid for most practical purposes. Substituting Eq. (7.4) into (7.3) gives the intensity on the image plane as

$$I(x, y) = \iint_{-\infty}^{\infty} I'(x', y')Q(x - x', y - y')\, dx'\, dy' \tag{7.5}$$

where $Q(x - x', y - y') = |K(x - x', y - y')|^2$ is the point-spread function of the imaging system.

The properties of $I(x, y)$ can be analyzed by studying its frequency spectrum. In the Fourier plane, the convolution integral is replaced by a simple product. The intensity distribution in the transform plane is thus

$$\tilde{I}(\mu, \nu) = \tilde{Q}(\mu, \nu)\tilde{I}'(\mu, \nu) \tag{7.6}$$

where (μ, ν) are the spatial frequencies on the Fourier plane and $\tilde{I}(\mu, \nu)$, $\tilde{Q}(\mu, \nu)$, and $\tilde{I}'(\mu, \nu)$ are the Fourier transforms of $I(x, y)$, $Q(x, y)$, and $I'(x', y')$, respectively. The function $\tilde{Q}(\mu, \nu)$ is called the *optical transfer function* (OTF) of the imaging system. Thus frequencies present on the image plane depend on the frequencies on the object plane and the optical transfer function. For an aberration-free system, the OTF is also given by the autocorrelation function of the pupil function. The pupil of the lens therefore controls the spatial frequencies transmitted by the lens. In white light speckle photography the spatial frequencies present in the image may be limited by the object spectrum at low f numbers and by the aperture itself at high f numbers of the imaging lens. The influence of lens aberration has been emphasized by several researchers [31, 32]. However, one may also encounter in practice the influence of defocus. To study this we consider a point object represented by a delta function and write for the intensity distribution

$$I(x, y) = \iint_{-\infty}^{\infty} \delta(x', y')Q(x - x', y - y')\, dx'\, dy' = Q(x, y) \qquad (7.7)$$

Taking the Fourier transform of both sides results in

$$\iint_{-\infty}^{\infty} I(x, y) \exp\{2\pi i(\mu x + \nu y)\}\, dx\, dy = \tilde{Q}(\mu, \nu) \qquad (7.8)$$

We can now assume that defocus lends to smearing of the intensity distribution; the image of the point may be taken as a disk of radius $z \tan \alpha$ and of uniform intensity. Δz is the distance between the image plane and the defocus plane, and α is the semiangular subtense of the lens aperture at the image plane. Assuming symmetry, we can express the OTF of the defocused system as

$$\tilde{Q}(\mu, \nu) = \tilde{Q}(\rho) = \int_0^{2\pi} \int_0^{\Delta z \tan \alpha} \exp\{2\pi i \rho r \cos(\theta - \phi)\} r\, dr\, d\phi$$

On integration we obtain

$$\tilde{Q}(\rho) = c\, \frac{2J_1(2\pi\rho\, \Delta z \tan \alpha)}{2\pi\rho\, \Delta z \tan \alpha} \qquad (7.9)$$

where $\rho^2 = \mu^2 + \nu^2$, c is a constant, and $J_1(\cdot)$ is a first-order Bessel function of the first kind.

For a well-focused speckle pattern whose spectrum is limited by the aperture of the imaging lens, the halo distribution is circular. Due to defocus, rings appear in the halo, as a result of the modulation of the OTF by the Bessel function. The higher order rings are visible only with non-linear processing while the first zero is overwhelmed by the undiffracted beam. Hence with conventional recording at a defocused plane, all one would get is a halo of reduced size corresponding to the reduced spatial frequency content in the image. Of course, as the defocus becomes very large the spatial frequency content of the image becomes very small, and secondary diffraction due to apertures used in the filtering comes into prominence.

7.2.4 Double-Exposure Recording

For displacement and strain analyses, a double exposure of the speckled intensity distribution is recorded on film, with one exposure before and one after deformation. If the in-plane displacement components along two perpendicular directions are u and v, the total exposure E on the film will be

$$E(x, y) = [I(x, y) + I(x + u, y + v)]T$$

where T is the duration of each exposure. The amplitude transmittance of the film, assuming linear recording, is written as

$$t(x, y) = t_0 - \beta T[I(x, y) + I(x + u, y + v)] \tag{7.10}$$

with t_0 and β as constants.

These double-exposure patterns are not unlike those in Fig. 7.1. To delineate the displacement information from such patterns, optical filtering is done on a whole-field or pointwise basis. The use of digital processing to correlate the speckle patterns corresponding to undeformed and deformed states of the object is being considered [33]. However, due to the lower sensitivity of digital systems and slower speeds, optical filtering is still the preferred choice.

7.2.5 Whole-Field Filtering Technique

In the whole-field approach, the processed double-exposure specklegram is placed in a convergent beam of laser light as shown in Fig. 7.4a. The Fourier transform of the amplitude transmittance of the film is obtained at the transform plane. The intensity at the transform plane can thus be written as

$$I_1(r_x, r_y) = \left| \iint_{-\infty}^{\infty} t(x, y) \exp\left[ik \frac{xr_x + yr_y}{L}\right] dx\, dy \right|^2 \tag{7.11}$$

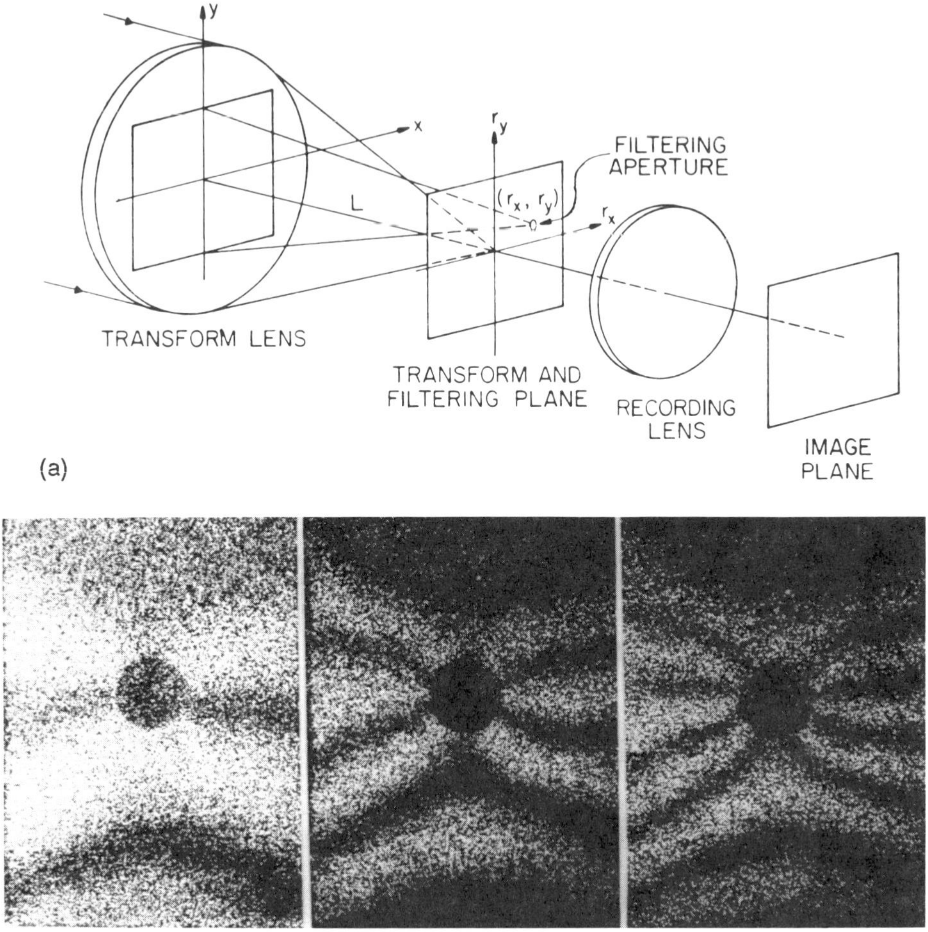

Figure 7.4 (a) Schematic of whole-field filtering setup. (b) Isothetic patterns with varying sensitivity for a tensile specimen with a hole.

Substituting for $t(x, y)$ from Eq. (7.10) into Eq. (7.11), ignoring the undiffracted d.c. term due to the constant t_0, and applying the shifting theorem, we obtain

$$I_1(r_x, r_y) = \beta^2 T^2 \left| \iint_{-\infty}^{\infty} I(x, y) \exp\left[ik \frac{xr_x + yr_y}{L}\right] dx\, dy \cdot \left\{1 + \exp\left[ik \frac{ur_x + vr_y}{L}\right]\right\}\right|^2 \qquad (7.12)$$

With u, v being slowly varying functions of x, y, this equation can be simplified to

$$I_1(r_x, r_y) = \tilde{I}(r_x, r_y)\cos^2\left(k\,\frac{ur_x + vr_y}{2L}\right) \tag{7.13}$$

where $\tilde{I}(r_x, r_y)$ is the Fourier transform of a single-exposure speckle pattern; it represents the diffraction halo. Thus the diffraction halo is modulated by the cosine squared distribution characterizing the deformation of the object. If the deformation is the same throughout, fringes will be visible corresponding to the cosine squared variation. The fringe spacing and orientation fully determine the displacement. In general, however, the deformation varies from point to point. Thus the spacing and orientation of fringes will be different, resulting in no visible fringes in the halo. However, by selecting a particular band of frequencies, by placing an aperture within the halo at a distance $\bar{r}(r_x, r_y)$ on the Fourier plane, isothetics representing the displacement component in the $\bar{r}$ direction are observed (Fig. 7.4a). Thus if the aperture is placed at $r(r_x, 0)$, the u-isothetics are revealed, which are governed by

$$u = n\lambda L/r_x \tag{7.14}$$

where n is the fringe order, which takes integral values for bright fringes. Similarly, the v-isothetics are revealed by placing the aperture along the r_y axis on the Fourier plane. The fringe patterns observed in these cases are identical to those that would be obtained if a grating of pitch $p = \lambda L/r_x$ were used. However, in the white light speckle method, since r_x, or for that matter $\bar{r}$, can be continuously varied up to the limit of the halo, the method provides variable sensitivity. Furthermore, the circular halo enables displacement components along any direction to be obtained, providing greater versatility than the moiré method. However, the price to be paid for this convenience is the relatively poor fringe contrast, as seen in Fig. 7.4b. The specimen in this case is a tensile coupon with a central hole, and the three isothetic patterns were obtained from one double-exposure specklegram by varying the sensitivity [7].

7.2.6 Pointwise Filtering Technique

The whole-field method provides an overall view of the deformation, but speckle noise limits the number of fringes that can be recorded. Thus quantitative evaluations based on these fringes are generally not accurate. The pointwise filtering approach improves sensitivity by analyzing the speckle pattern point by point. A narrow laser beam illuminates the region of interest on the double-exposure speckle pattern. The diffracted halo observed at a distance L from the film plane is modulated by the same cosine squared fringes as for the whole-field method. To make it more formal, a lens is generally inserted to transform the amplitude transmittance of the illuminated region. Since the deformation at this region is uniform, the cosine squared fringes are observed at the Fourier plane. The same governing equations apply as for the whole-field approach. However, since the fringes are equispaced, the magnitude of displacement at the point of interest can be obtained by using the relationship

$$(u^2 + v^2)^{1/2} = \lambda L/s \qquad (7.15)$$

where s is the fringe spacing normal to the fringe orientation, which is also the direction of displacement. To get the displacement map over the entire specimen, the laser beam has to be scanned over an array of points on the film. This process is amenable to computerization to speed up the data acquisition and processing (see Chapter 9).

7.3 IN-PLANE DISPLACEMENT AND STRAIN ANALYSIS

The white light speckle method has the capability of full-field displacement and strain analysis with variable sensitivity. A double-exposure specklegram can be optically filtered using the pointwise or whole-field method to obtain components of displacement in two mutually perpendicular directions. For strain analysis, these two directions are generally chosen as the Cartesian axes x and y. In the pointwise filtering approach, the displacement components in these two directions can be obtained by measuring the fringe spacing (s_x and s_y). Thus

$$u = \lambda L/s_x \qquad (7.16a)$$

and

$$v = \lambda L/s_y \qquad (7.16b)$$

The displacements on the object plane along the (x', y') directions can be readily deduced from the above equations by incorporating the magnification M of the imaging system, as

$$u' = u/M \qquad (7.17a)$$

and

$$v' = v/M \qquad (7.17b)$$

Whole-field displacement distribution can be obtained by fitting curves or surfaces through the array of data points. These can then be differentiated to obtain the strain components

$$\epsilon_{x'} = \frac{\partial u'}{\partial x'} \qquad (7.18a)$$

$$\epsilon_{y'} = \frac{\partial v'}{\partial y'} \qquad (7.18b)$$

and

$$\tau_{x'y'} = \frac{\partial u'}{\partial y'} + \frac{\partial v'}{\partial x'} \qquad (7.18c)$$

The verification of these expressions and their application to stress intensity factor determination with enhanced sensitivity are detailed in this section.

7.3.1 Experimental Verification

A perspex cantilever beam, 210 mm long, under tip loading was chosen as the specimen. The artificial speckle pattern was created by sticking on the surface of the specimen a laser-generated speckle pattern formed on stripping film. The specimen, being transparent, was illuminated by diffuse light from behind. The object was imaged by a 135 mm lens operating at $f/22$ and 0.75 × magnification. A double exposure of the speckle pattern was recorded on film, with a tip deflection of 0.78 mm applied between exposures. Figure 7.5a shows the pointwise-filtered Young's fringes at selected points along the neutral axes. The fringe density and orientation of fringes clearly show that the deflection is vertical and increasing as the loaded end is approached. The whole-field isothetics representing contours of the displacement component in the vertical direction are shown in Fig. 7.5b. The fringe sensitivity in this case was 0.04 mm/fringe on the image plane. The whole-field fringes with reduced sensitivity were also analyzed, and the resulting plots are also shown in Fig. 7.5b. Good agreement with the theoretical distribution is evident. Derivatives of displacement, which in this case corresponds to the rotational/shear component ($\partial v'/\partial x'$) can be determined from this plot. The normal strain ($\partial v'/\partial y'$) is fairly small, as can be seen from the nearly straight whole-field fringes.

7.3.2 Sensitivity Enhancement

As shown by the example in Section 7.3.1, the sensitivity of the white light speckle method is lower than that of its laser counterpart. There are two

(a)

Figure 7.5 (a) Pointwise-filtered Young's fringes at selected points on the neutral axis of a cantilever beam specimen. (b) Whole-field fringes for the cantilever beam specimen and plot of experimental and theoretical vertical displacement components.

reasons for this. First, artificially created speckles are generally larger than laser speckles. For example, the glass beads in the retroreflective liquid have a smallest diameter of 20 μm, which corresponds to a maximum frequency of 50 lines/mm. Finer speckle patterns can be generated without too much difficulty by alternative means. However, this introduces the second problem, that of resolving this fine structure. Defocusing, even by a small amount, substantially reduces the high-frequency content of the recorded image, as explained in Section 7.2.3.

One possible solution to this is to record the speckle pattern at high magnification. Of course, the higher the magnification, the smaller the depth of field and thus the greater the problem of defocus. However, higher magnification enables better visual resolution of the speckles, improving

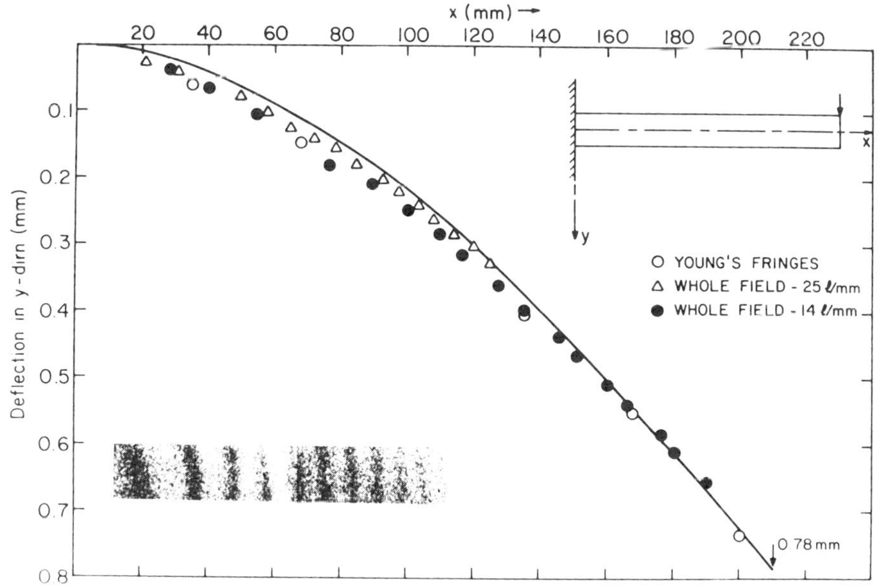

(b)

the possibility of accurate focusing. Although electron microscopy has been used to magnify speckle pattern and enhance sensitivity, in this example, only optical microscopes using white light illumination will be discussed. The principle can be readily extended to electron beam imaging.

An optical comparator capable of a maximum magnification of 450× was used in this demonstration. The speckles were created using three different schemes and were observed in the comparator in both the reflection and transmission modes. For a glass specimen, speckles were created by sandblasting the surface using particles approximately 1 μm in diameter. For an aluminum specimen, the surface was first cleaned using No. 600 emery paper and then polished using a belt sander and a liquid containing alumina particles of 1 μm diameter. The third method involved painting, using white paint overcoated with sprayed black paint. To the naked eye the surface appeared totally black, but magnification revealed the fine speckled pattern. The specimens were illuminated using the white light source present in the comparator. The ground-glass specimen was viewed in transmitted light. Double-exposure specklegrams, with a rigid-body in-plane displacement of the objects between exposures, were recorded at magnifications of 10×, 20×, and 50×. The resulting Young's fringes for the three specimens are shown in Fig. 7.6. The displacement

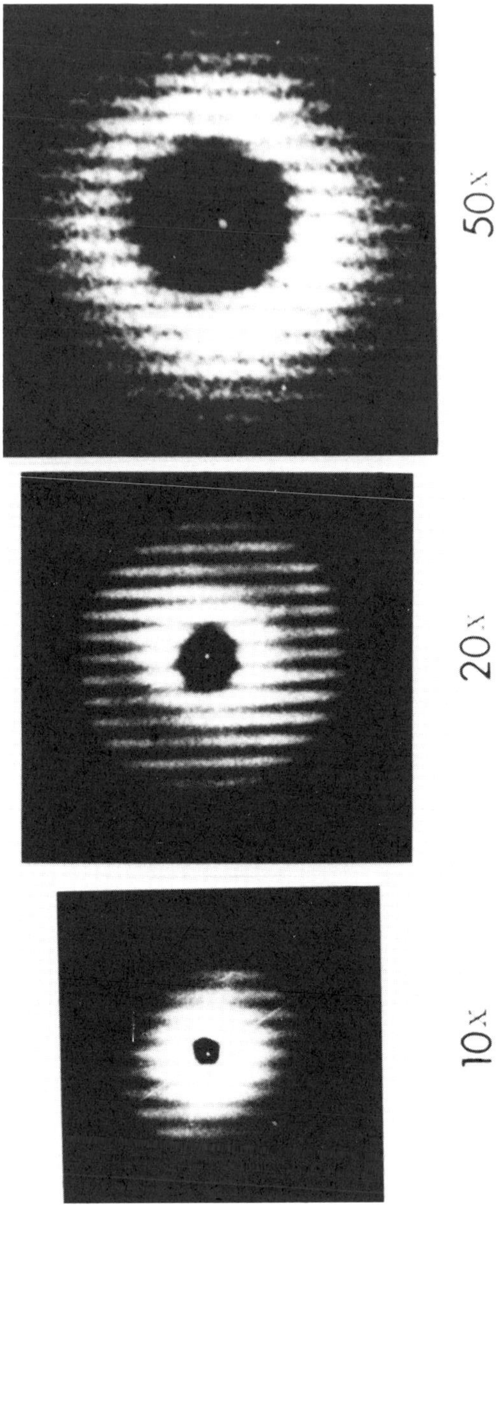

Figure 7.6 Young's fringes due to rigid-body translation with different recording magnifications for three different specimen surfaces.

Table 7.1 Effect of Magnification on Speckle Sensitivity

	Ground glass		Polished aluminum		Black paint	
Magnification	Frequency on object (lines/mm)	Speckle size (μm)	Frequency on object (lines/mm)	Speckle size (μm)	Frequency on object (lines/mm)	Speckle size (μm)
10×	280.6	3.60	322.0	3.1	301.5	3.30
20×	415.8	2.40	436.6	2.3	395.0	2.53
50×	571.7	1.75	623.8	1.6	571.7	1.75

calculated from the fringe pattern agrees well with the given displacement of 0.0127 mm. More important, however, the increased sensitivity, characterized by the higher spatial frequencies recorded as the magnification is increased, is evident from the halo size. The diffraction halos shown in Fig. 7.6 correspond to the spatial frequencies present on the object, as noted by the same fringe spacing (same displacement of object). Table 7.1 gives the frequencies on the object plane along with the speckle size for the three specimens at the three magnifications. Compare this with the speckle size on the object plane obtained using the laser speckle method; the speckle size is given by [34]

$$\sigma = 1.2\lambda F \frac{1 + M}{M} \tag{7.19}$$

It is evident that whereas for the laser speckle increasing magnification does not significantly increase the sensitivity, for the white light speckle method there is a 50% reduction in speckle size due to finer speckles being resolved. Higher magnifications in the present case would not significantly increase the maximum frequency because the smallest speckle has already been resolved at 50× magnification. It is thus necessary to generate even smaller speckles on the object if higher magnifications are to be useful.

7.3.3 Application to Fracture Mechanics

The enhanced sensitivity combined with a magnified image makes the white light speckle method particularly suitable for analyzing deformations around crack tips. Furthermore, decorrelation is not a serious problem with white light speckles as it is with laser speckles. As a demonstration,

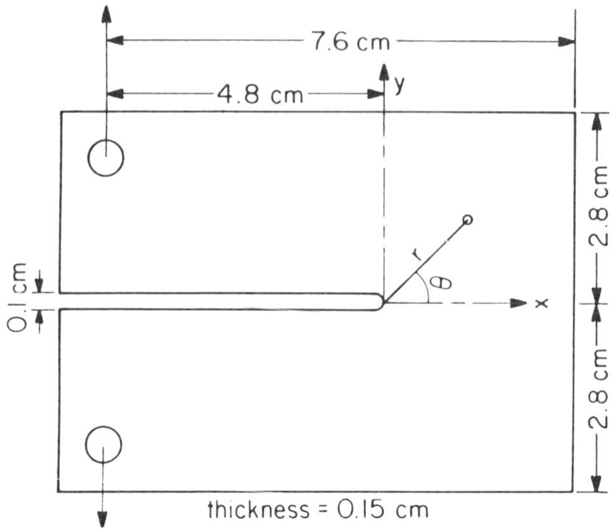

Figure 7.7 Schematic diagram of compact tension specimen.

the stress intensity factor (SIF) and displacement field around a crack tip in a perspex compact tension specimen (Fig. 7.7) were determined. Speckles were created using black on white paint, and the area around the crack tip was recorded at 10× magnification. A special loading jig was used to apply constant displacement to the two legs within the confines of the comparator. Double exposures were recorded, and the specklegram was analyzed using the pointwise and whole-field methods. The displacements, in a mode I loading configuration under plane stress, are proportional to $K_1\sqrt{r}$, where K_1 is the mode I SIF and r is the radial distance from the crack tip. Thus a plot of the displacement component as a function of $\sqrt{r}$ will directly provide the SIF as proportional to the slope. The ability to perform full-field displacement mapping makes possible numerous approaches to determining the SIF. Indeed, one can better estimate the SIF by averaging the redundant data. Figure 7.8a is the displacement component in the direction of load plotted along the crack line, with data obtained from the whole-field isothetics. Figure 7.8b plots the total displacement magnitude obtained from the Young's fringes. The SIFs obtained from these plots are 55.4 and 59.2 (kg/cm^2) cm$^{1/2}$ and compare well with the handbook value of 57.2 (kg/cm^2) cm$^{1/2}$. White light speckle photography has also been used to study dynamic fracture [35].

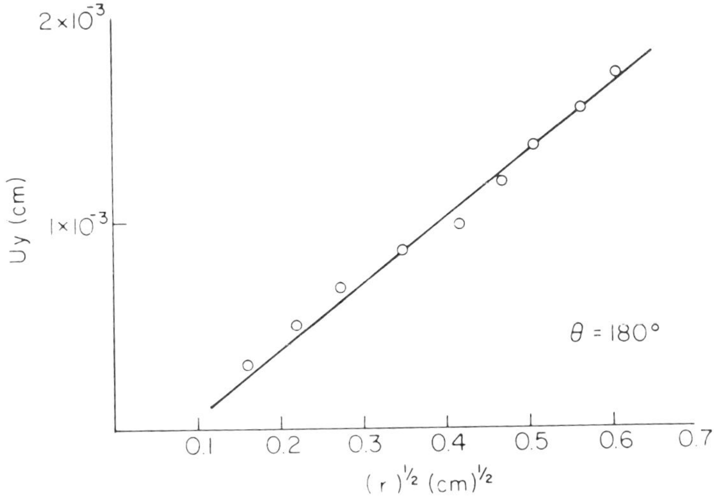

(a)

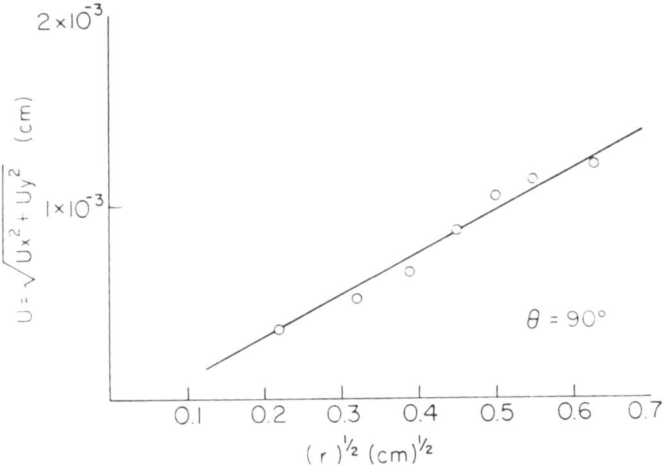

(b)

Figure 7.8 (a) Displacement component in the load direction for a compact tension specimen from whole-field data. (b) Total displacement magnitude obtained by pointwise filtering.

7.4 OUT-OF-PLANE DISPLACEMENT MEASUREMENT

The similarity of the white light speckle method to conventional moiré methods extends to the use of similar techniques for out-of-plane displacement measurement. Hence gap effect fringes, projection moiré, and shadow moiré—three of the more common out-of-plane displacement and contouring techniques in moiré—have corresponding counterparts in the white light speckle method. In addition, the defocus effect detailed in Section 7.2.3 can also be used for out-of-plane displacement measurement and contouring. In this section, only the gap effect for out-of-plane displacement measurement will be discussed. In Section 7.9, which deals with other applications, other techniques will be discussed to emphasize similarity with moiré methods.

7.4.1 Analysis of Fringe Formation

A double-exposure specklegram of a speckled surface illuminated in white light is recorded with an out-of-plane displacement z between exposures. A point at a distance R from the optical axis experiences an apparent in-plane displacement u_o that can be written as [16]

$$u_o = \beta \, \Delta z \tag{7.20}$$

with $\beta = R/p = r/q$, p and q the object and image distances, respectively, and r the radial distance on the image plane.

The fictitious in-plane displacement u_i on the image plane is thus

$$u_i = u_o M = \frac{MR \, \Delta z}{p} = \frac{Mr \, \Delta z}{q} \tag{7.21}$$

These strains exist even for rigid-body out-of-plane translation, and the speckles cannot differentiate it from in-plane displacement—thus the term "fictitious."

Pointwise and whole-field filtering can be performed to delineate this fictitious displacement component and hence deduce the out-of-plane displacement. For the pointwise method, the resulting equation is

$$\Delta z = (\lambda L/Ms)(p/R) \tag{7.22a}$$

and for the whole-field fringes,

$$\Delta z = (n\lambda L/Mr_x)(p/R), \qquad n = 0, 1, \ldots \tag{7.22b}$$

where r_x is the distance of the filtering aperture from the optical axis on the Fourier plane. The double-exposure specklegram containing axial displacement can also be Fourier filtered [36]. This results in a circular fringe

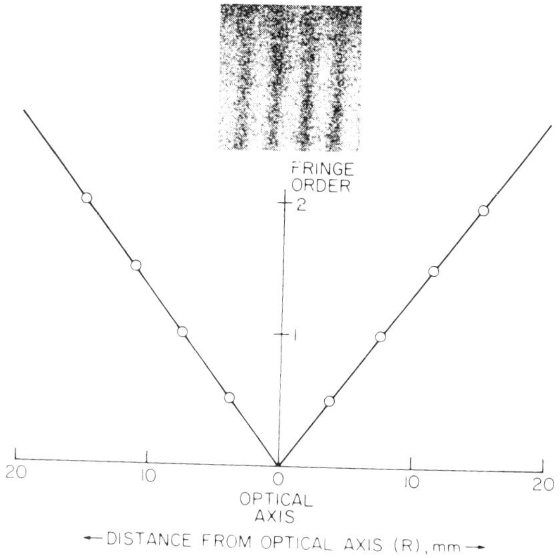

(a)

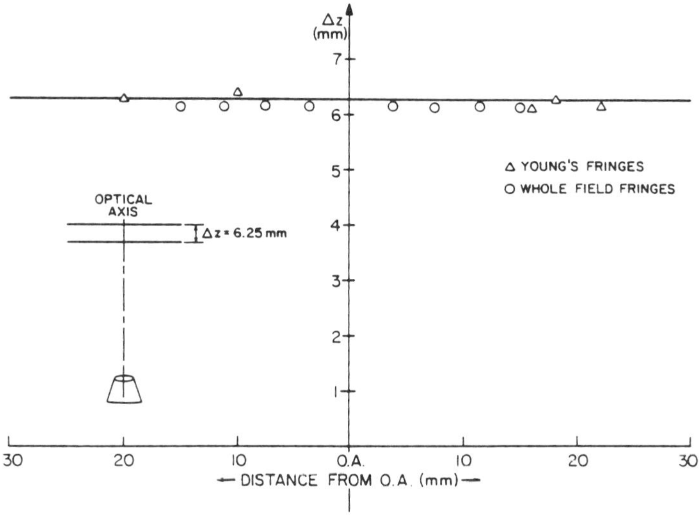

(b)

Figure 7.9 (a) Whole-field fringes and fictitious in-plane displacement due to rigid-body out-of-plane translation. (b) Comparison of experimentally derived out-of-plane displacement with actual given value.

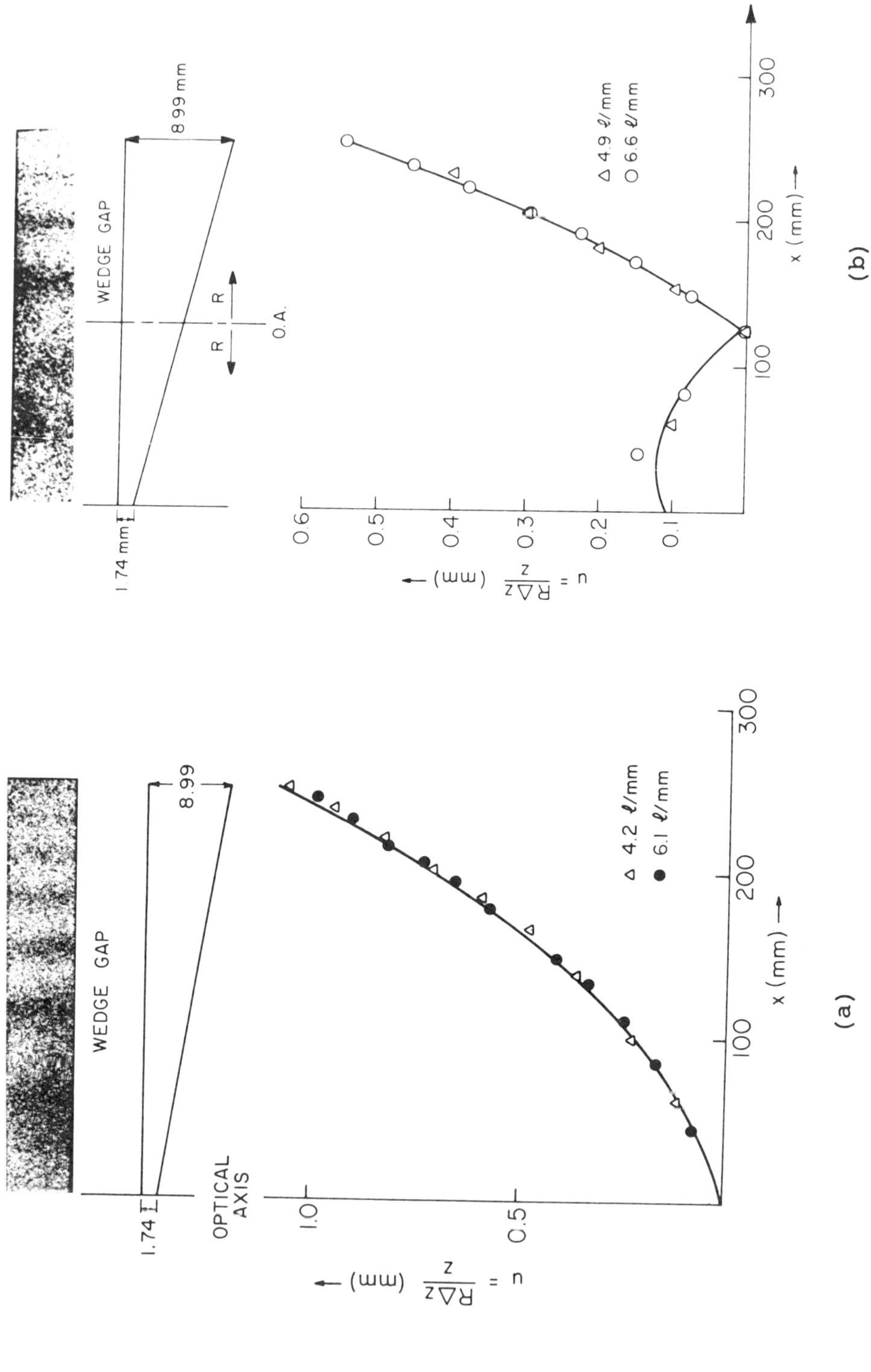
WEDGE GAP
8.99 mm
1.74 mm
R
R
O.A.
u = RΔz/z (mm)
Δ 4.9 ℓ/mm
○ 6.6 ℓ/mm
x (mm)
(b)
WEDGE GAP
8.99
1.74
OPTICAL AXIS
Δ 4.2 ℓ/mm
● 6.1 ℓ/mm
x (mm)
(a)

pattern; the displacement can be calculated from the diameter of the circular fringes. However, the plane of localization of fringes will vary depending on the experimental conditions [37].

Figure 7.9a depicts a typical whole-field pattern corresponding to a rigid-body out-of-plane displacement of 6.25 mm between exposures. The fictitious in-plane displacement obtained from these fringes as a function of R gives the straight-line distribution predicted by Eq. (7.21), while the measured out-of-plane displacement agrees well with the given displacement seen in Fig. 7.9b. If the out-of-plane displacement is linearly varying, then the fictitious in-plane displacement curve is quadratic as shown in Fig. 7.10a. By moving the optical axis, the fictitious in-plane displacement can be varied as shown in Fig. 7.10b. This provides a means of separating in-plane and out-of-plane displacement components. Note that zero fictitious displacement at the optical axis can be observed in the whole-field filtering as one continuously varies the sensitivity by moving the filtering aperture.

7.5 SEPARATION OF THREE-DIMENSIONAL DISPLACEMENT COMPONENTS

The total displacement of a typical speckle on the image plane along the x direction is composed of the real in-plane displacement u and the fictitious component u_i attributed to the out-of-plane displacement w. Thus one can write

$$u_t = u \pm MRw/p \tag{7.23a}$$

Noting that the fictitious component, owing to the perspective effect [14], emanates radially from the optical axis, the resulting displacement in the y direction can similarly be written as

$$v_t = v \pm MRw/p \tag{7.23b}$$

The double-exposure specklegram thus records these total speckle displacements, which can be determined by using the pointwise or whole-field techniques. However, this results in only two independent equations in terms of the three unknown displacements u, v, w. Thus one additional equation is required to separate out the displacement components.

Various schemes are available for separation of displacement components [8, 38–40]. In cases where the out-of-plane displacement is dominant,

Figure 7.10 Fictitious displacement fringes for linearly varying out-of-plane displacement with optical axis (a) at one end of beam, (b) at center of beam.

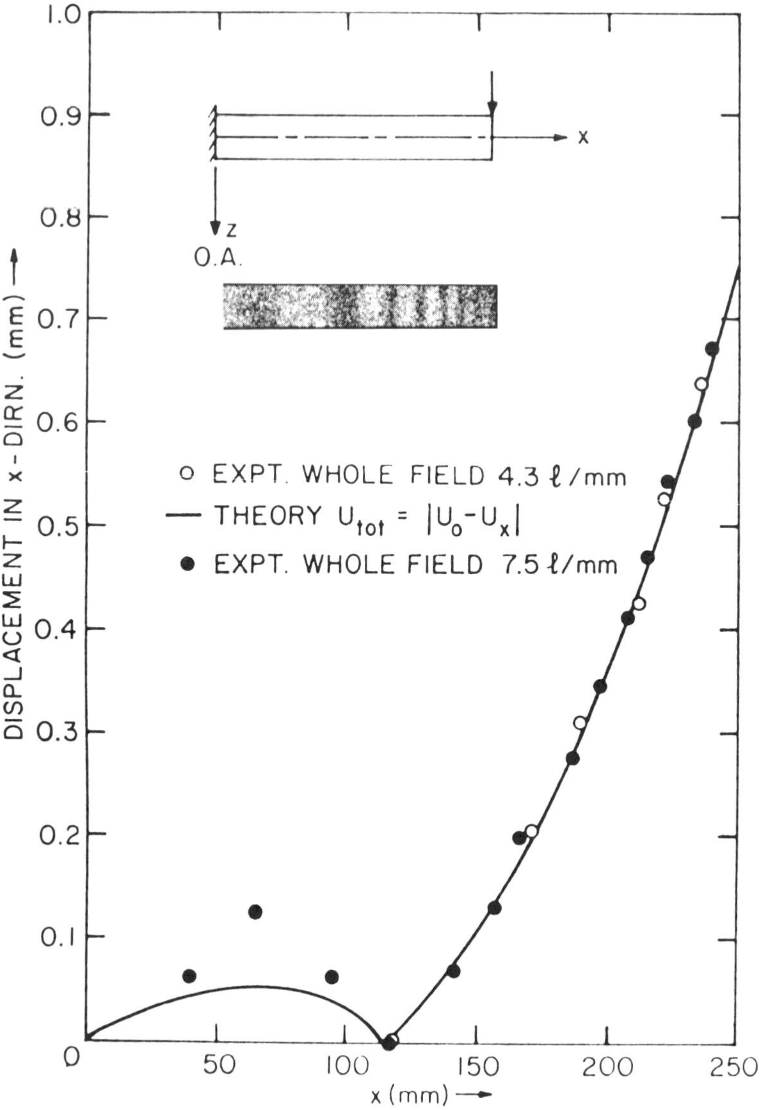

(a)

Figure 7.11 Whole-field fringes on a cantilever beam specimen loaded in the z direction with optical axis (a) at fixed end, (b) at center of beam.

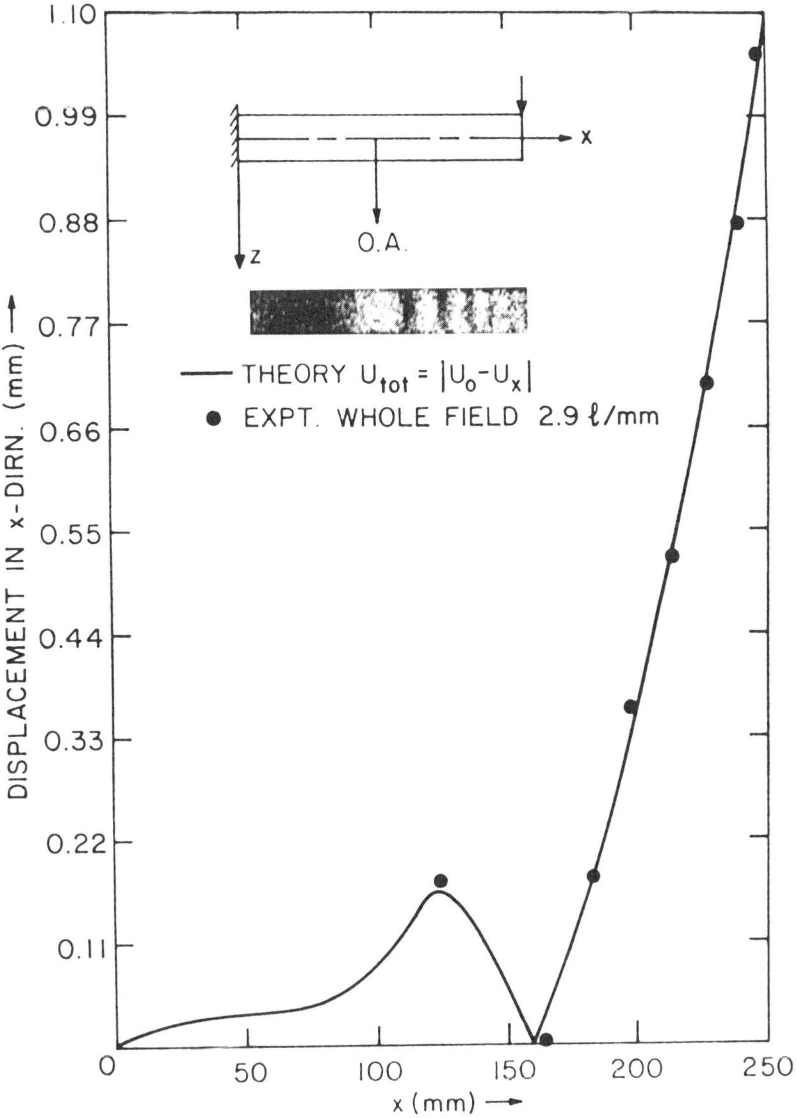
1.10
0.99
0.88
0.77
0.66
0.55
0.44
0.33
0.22
0.11
0
50
100
150
200
250
x (mm)
DISPLACEMENT IN x-DIRN. (mm)
x
z
O.A.
THEORY $U_{tot} = |U_0 - U_x|$
EXPT. WHOLE FIELD 2.9 ℓ/mm

(b)

as in plate-bending problems, one can devise the setup so as to enhance the out-of-plane displacement, thus making its effect dominant. Using the lens law, Eq. (7.23a) can be rewritten as

$$u_t = u \pm M^2Rw/(1 + M)f \tag{7.24}$$

where f is the focal length of the lens. For the same magnification, a shorter focal length lens would increase the sensitivity of the fictitious component. Thus for a suitable choice of lens and magnification, the fringes represent the out-of-plane displacement component within acceptable limits of error. On the other hand, when both the in-plane and out-of-plane displacements are similar, as in a general three-dimensional problem, one additional equation has to be obtained to separate out the fictitious in-plane component from the real in-plane component. This can be achieved in a variety of ways, two of which are exemplified below.

The first scheme for separation of displacement components takes advantage of the fact that the fictitious displacement due to the out-of-plane component depends on the location with respect to the optical axis. Thus by using two cameras with parallel optical axes, two double-exposure speckle patterns can be recorded [39]. The contributions of the real in-plane components are the same in the two patterns. However, the fictitious displacements are different, thus producing different fringe densities when optically filtered. Separation of displacement components then follows readily. To demonstrate this effect, the ubiquitous cantilever beam is chosen as the specimen. Loading is applied at the free end in the z direction. Thus the out-of-plane displacement is much greater than the in-plane component. However, sensitivity to the out-of-plane displacement is reduced by using a large focal length (485 mm) lens and setting the optical axis toward the loaded end of the specimen. For lack of two cameras, the two double-exposure specklegrams were recorded sequentially using one camera. The optical axes were located at the fixed end and the center of the beam. The resulting whole-field isothetics for the two cases are shown in Figs. 7.11a and 7.11b. The fact that the real in-plane displacements, despite being small, cannot be neglected is borne out by these fringes. For the case when the optical axis is at the fixed end, consideration of only fictitious displacement would result in only one zero displacement point, i.e., at the fixed end. However, the fringe pattern shows two zeros. The total displacement calculated using Eq. (7.23) exemplifies the discrepancy as shown in the plot of Fig. 7.11a. Similarly, for the case when the optical axis is at the center, the points of zero displacement, based on purely fictitious displacement analysis, should be at the fixed end and the center of beam. However, the real in-plane displacement shifts the central zero to the right (Fig. 7.11b). These two curves provide the two equations needed to separate the displacement components. This approach appears analogous to stereovision techniques [40] and photogrammetry, wherein two views of the object from different perspectives provide depth information on the object.

The second scheme uses only one double-exposure specklegram. The second equation is obtained by an alternative method. The projection moiré method is suitable for this, for generally a slide projector is used to illuminate the speckles in the white light speckle method. Thus one can readily project a grating also. However, if the speckles are created using retroreflective liquid, the projection angle has to be necessarily small, resulting in reduced sensitivity for the projection moiré. Double exposures are recorded in the usual way. However, in addition to the speckles, the projected grating is now also recorded in the undeformed and deformed states. But because of speckles, projection moiré fringes are not immediately apparent. If, however, this double-exposure specklegram is placed in the whole-field setup, the diffraction dot corresponding to the grating can be filtered to enhance the moiré fringe as shown in Fig. 7.12a. These fringes are governed by the familiar equation [39]

$$w = NP/(\sin \tau) \tag{7.25}$$

where N is the fringe order, P the pitch of the grating on the object, and τ the angle of projection. The out-of-plane displacement can be converted to fictitious displacement using Eq. (7.21), and the result is plotted in Fig. 7.12a along with the theoretical line. Filtering elsewhere in the halo gives the total displacement component, as shown in Fig. 7.12b, from which the real in-plane displacement component can be extracted.

7.6 INTERIOR DISPLACEMENT MEASUREMENT

Displacement measurement in the interior of three-dimensional objects is important for overall strain and displacement analysis. Three-dimensional photoelasticity is the only method widely used. Laser speckle offered an exciting alternative, but decorrelation restricted widespread application. The white light speckle method was thus explored. Areas of application that are receiving widespread attention include flow visualization [2] and velocimetry [22]. A detailed description of the use of speckles for flow measurement is given in Chapter 5.

7.6.1 Basic Principle

The basic procedures of the white light speckle method for interior displacement measurement are as follows. Small particles are embedded in the solid or mixed with the fluid flow. Like surface speckles, embedded

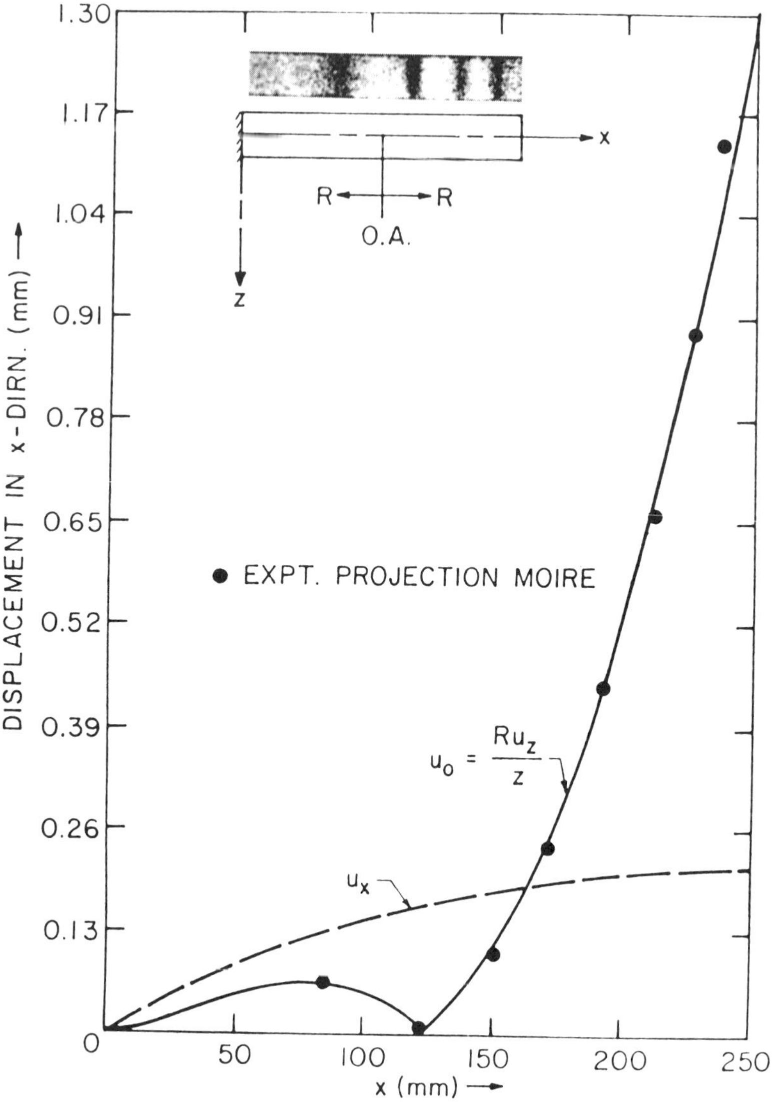

(a)

Figure 7.12 (a) Projection moiré fringes contouring out-of-plane displacement component of a cantilever beam. (b) Whole-field speckle fringes with optical axis at center of beam.

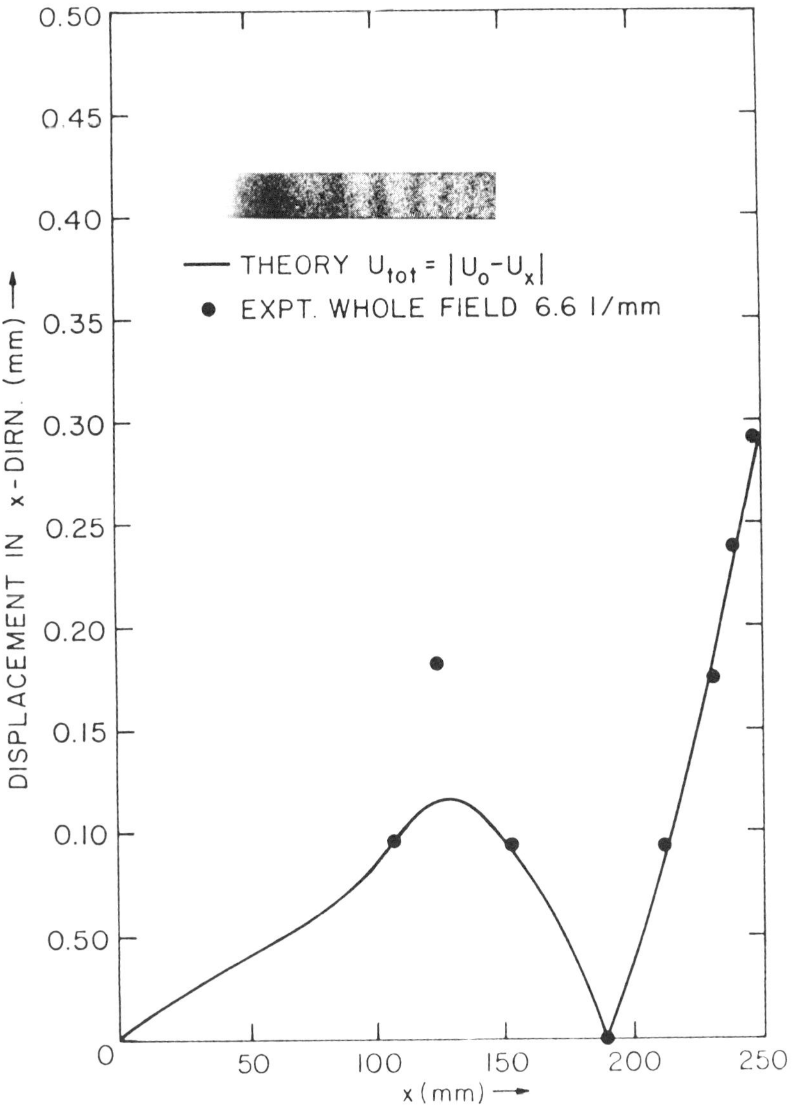
0.50
0.45
0.40
0.35
0.30
0.25
0.20
0.15
0.10
0.50
0
50
100
150
200
250
x (mm) →
DISPLACEMENT IN x-DIRN. (mm) →
— THEORY $U_{tot} = |U_0 - U_x|$
● EXPT. WHOLE FIELD 6.6 l/mm

(b)

speckles can be generated in a variety of ways. Specimens could be prepared with embedded glass beads, or flow could be seeded with aluminum or PVC particles. Indeed, air bubbles generated by improper casting of epoxy could themselves be a source of random speckles. Individual planes of these speckles are isolated by using either a sheet of light or a narrow beam of white light [10, 16, 20]. For illumination by a sheet of light, the scattered light from the speckles in this plane is recorded. The usual double-exposure recording of these speckles provides in-plane movement of the speckles using either pointwise or whole-field filtering methods. The plane of illumination can then be made to traverse the entire thickness of the specimen to build a composite of the overall three-dimensional deformation from individual slices. For flow visualization, one can record multiple exposures as explained in Chapter 5.

The use of a narrow beam to illuminate a line of speckles makes it possible to record speckle movement from two mutually perpendicular directions simultaneously [10]. All three displacement components can be determined from these two speckle patterns for different points on the line. Only pointwise filtering is possible due to the narrow width of the line, and a greater number of photographs must be analyzed to cover the entire object in three dimensions. However, this approach can readily determine all six strain components from the three displacement components at four neighboring corners of an internal cubic element [41].

7.6.2 Torsion of Shafts

For torsion of shafts with circular cross section, the displacement components at any plane are given as [9]

$$u = -\Omega zy \tag{7.26a}$$

$$v = \Omega zx \tag{7.26b}$$

$$w = 0 \tag{7.26c}$$

where Ω is the twist per unit length and u, v, w are the displacement components in the x, y, and z directions, respectively.

A 68-mm-diameter circular shaft made of urethane rubber epoxy with embedded glass beads was cast. The shaft was fixed at one end, and a twisting moment was applied at the other end via two micrometers. The shaft was illuminated by a sheet of white light normal to its axis. Two planes, at 54 mm and 67 mm from the fixed end, were investigated. The twist per unit length was 0.17×10^{-3} rad/mm. Double-exposure specklegrams were analyzed using the whole-field method. Good comparison with Eq. (7.26) was obtained, as can be seen from Fig. 7.13.

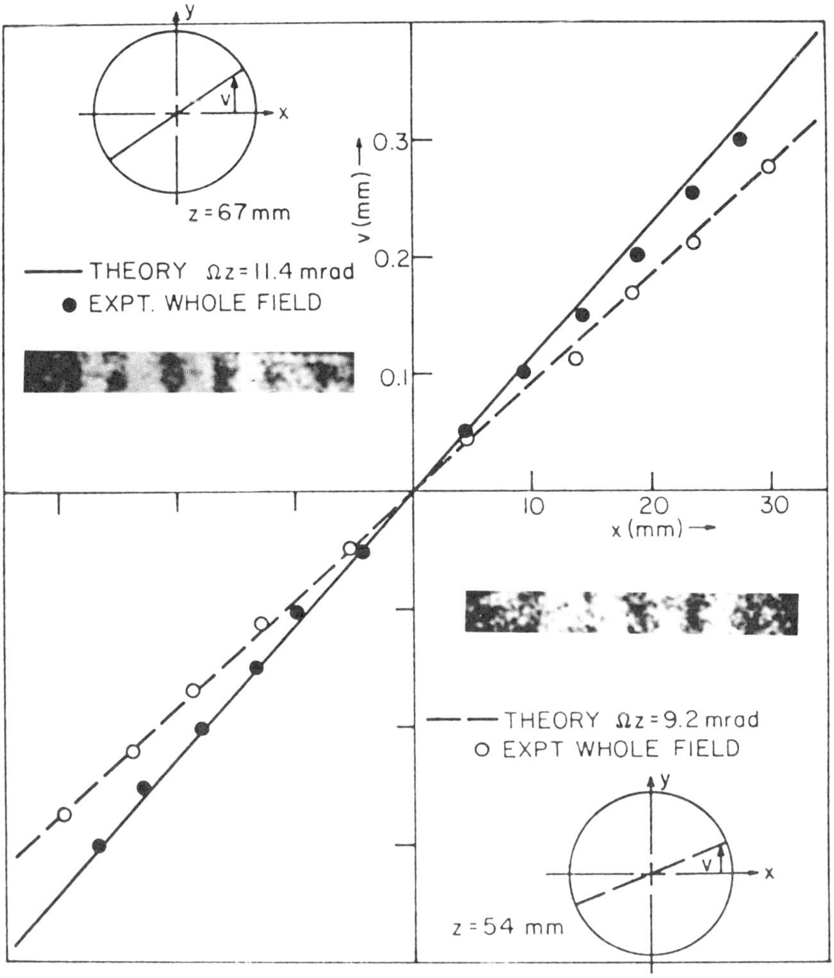

Figure 7.13 Whole-field fringes at two planes of a twisted circular shaft and corresponding displacement.

Torsion of cylindrical shafts has been analyzed using the scattered laser speckle method to yield better fringe quality [42]. However, for torsion of shafts with square cross section, out-of-plane warping of sections gives rise to laser speckle decorrelation despite the fact that the out-of-plane displacement is two orders of magnitude smaller than the in-plane displacements. Using the white light speckle method, this problem does not arise, and one can readily analyze shafts of any cross section. The displacement

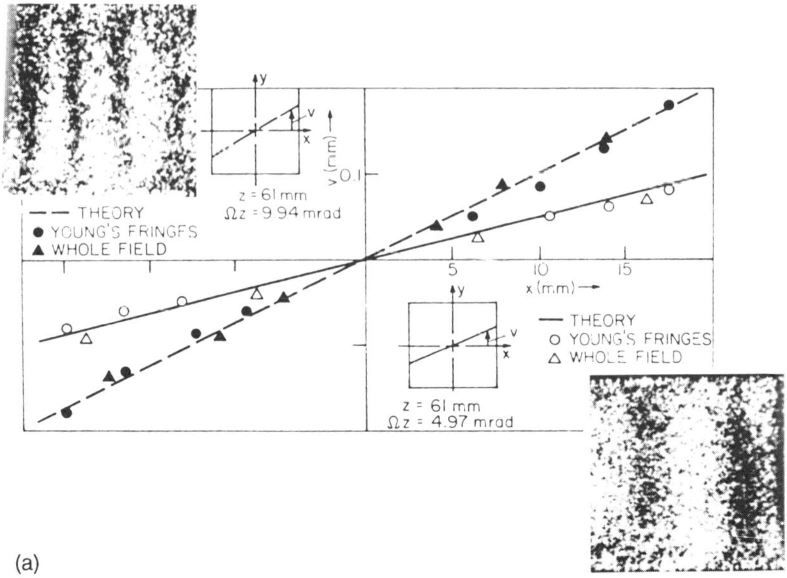

(a)

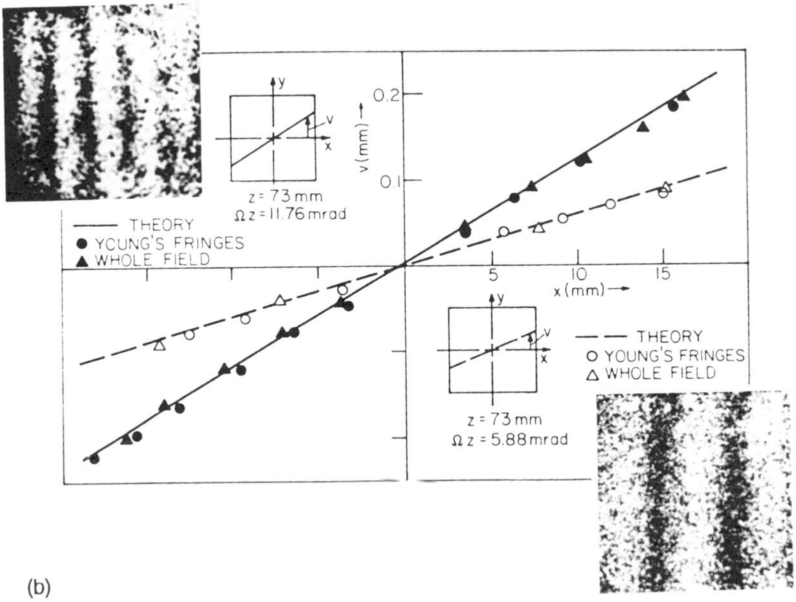

(b)

Figure 7.14 Whole-field fringes and data comparison for two isolated planes and two amounts of applied twist: (a) $z = 61$ mm; (b) $z = 73$ mm.

equations for torsion of a shaft of square cross section are identical to Eq. (7.26) except that the out-of-plane displacement w is now nonzero and is given as [20]

$$w = \Omega Q(x, y) \tag{7.27}$$

where $Q(x, y)$ is the warping function.

A shaft of square cross section was cast using urethane epoxy with embedded glass beads, and planes were isolated using sheets of light. Both the whole-field and pointwise filtering methods were adopted to analyze double-exposure speckle patterns for two values of twist and two locations along the length of the beam. Figure 7.14 gives the experimental results along with the whole-field fringes. The out-of-plane displacement is too small to be resolved, and thus the in-plane components show good agreement with theory. However, unlike the laser speckle, the problem of decorrelation does not exit, and good quality fringes are obtained.

7.7 CURVED SURFACE DEFORMATION

The white light speckle method provides the simplest approach to measuring the surface deformation of curved objects from both the experimental and analytical points of view [11]. Surface preparation is much simpler than for the moiré methods, and data analysis is more straightforward than with the laser speckle or holographic methods.

7.7.1 Basic Principle

The principle of this method is similar to that for the flat surfaces discussed before. Artificial speckles are created as before; these are then illuminated in white light and recorded using a camera. However, since the surface is curved, only certain regions of the object will be in focus and thus provide distinctive speckles with a discernible halo, and the region of the object that is not in focus cannot be used for analysis. Hence, in order to view the entire object at one setting a tandem-plate approach is proposed. As schematized in Fig. 7.15, photographic plates are arranged in tandem in the image space. Each of the plates would have a certain region of the object and hence speckles in best focus. Figure 7.15 also shows the enlarged speckles recorded for a spherical object using two plates in tandem. Note that the enlarged speckles shown correspond to only one half of the sphere. Plate (a) has distinctive speckles around the center of the sphere, which is its focus plane. The speckles progressively degrade as we move away from the center and are indistinguishable at the periphery. The second plate, which is focused at the periphery, has virtually no speckles at the center but sharp speckles at the edges.

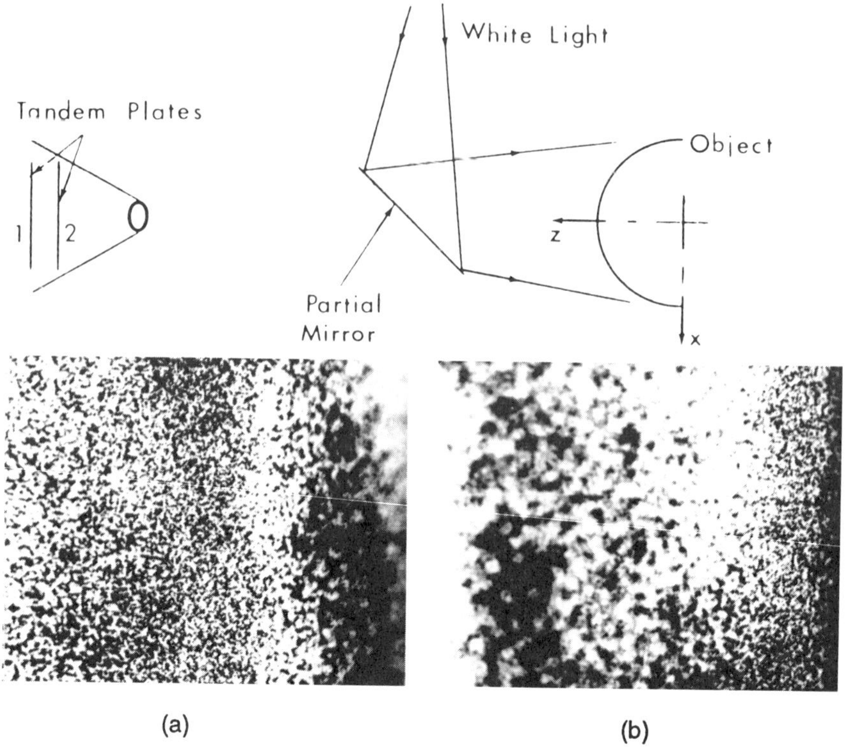

Figure 7.15 Tandem-plate arrangement for curved surface deformation measurement. See text.

Analysis of double-exposure specklegrams also proceeds in much the same way as discussed for flat surfaces. Points on the curved object are transformed to a flat image on the film plane. Hence, one can initially interpret the pointwise or whole-field fringes in terms of Cartesian displacement components u, v, w. The out-of-plane displacement manifests itself as fictitious in-plane components and has to be separated out by following the techniques described in Section 7.5. Having obtained the image plane displacement, the data have to be transformed to the object plane and appropriate spherical or cylindrical coordinates. Although the procedure is time-consuming, it is much simpler than trying to interpret complex three-dimensional speckle movements as for the laser speckle method.

7.7.2 Experimental Demonstration

A variety of experiments involving both rigid-body motion and strain-induced deformation were performed to demonstrate the use of the white light speckle method to determine the deformation of a curved surface. First, a spherical rubber ball 230 mm in diameter (with its surface painted with retroreflective liquid) was given a rigid-body shift of 0.382 mm along the x direction. Double-exposure specklegrams with two tandem plates were recorded. The resulting Young's fringes at different points along the meridian section are shown in Fig. 7.16a. The first two patterns were from one plate and the last two from the second plate. The calculated displace-

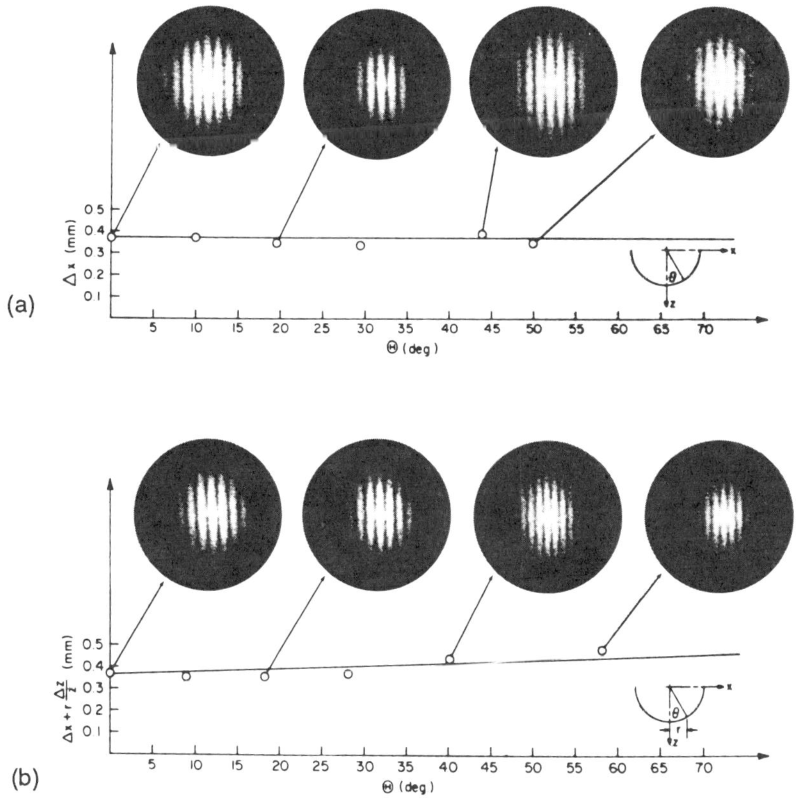

Figure 7.16 (a) Rigid body in-plane displacement fringes on a spherical ball using the tandem-plate setup. (b) Young's fringes due to rigid-body in-plane and out-of-plane displacements of a spherical ball.

ment from these fringes agrees fairly well with the actual displacement shown in the same figure. To show the effect of out-of-plane movements, a second experiment was performed with a rigid-body shift of 0.382 mm given in both the x and z directions. The resulting Young's fringes obtained from the two plates are shown in Fig. 7.16b. At $\theta = 0$, where $R_z = R_y = 0$ (i.e., the optical axis), the effect of the out-of-plane movement is zero, as expected. As θ increases, the effect becomes more pronounced, as is apparent from the figure.

For strain analysis, a curved beam with a radius of 100 mm and 0.31 mm thick was first coated with retroreflective paint. This beam was fixed at one end and loaded in the radial direction, using a micrometer, at the other end. Two double-exposure specklegrams were recorded, one with the optical axis at the fixed end and the other with the optical axis at the center of the curved beam. The resulting specklegrams were filtered using both whole-field and pointwise techniques. Figure 7.17 shows the whole-field isothetics from the two specklegrams. These isothetics represent the

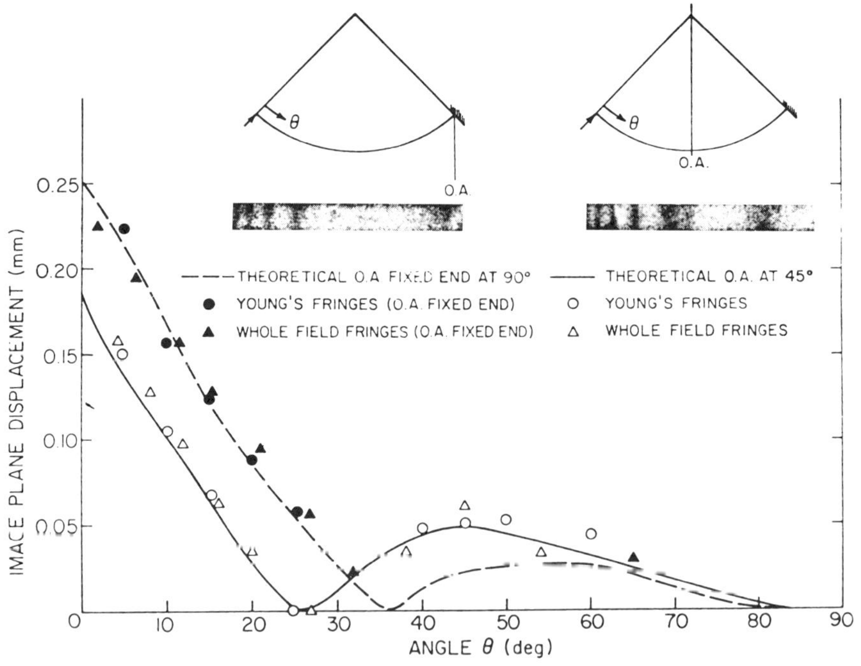

Figure 7.17 Whole-field fringes representing deformation of a curved cantilever beam as obtained from two different positions of optical axes of recording camera.

combined in-plane and out-of-plane displacements. They agree fairly well with the corresponding theoretical displacements obtained from the theory of elasticity. These two curves can then be used for separating the individual displacement components as discussed in Section 7.5.

As a further demonstration, the rubber ball of the first experiment was reused. In this case deformation was induced by inflating the rubber ball between exposures. Two photographic plates in tandem were used to record the double-exposure pattern. The pointwise-filtered fringes along three latitudes are shown in Fig. 7.18. The 0° and 15° patterns were obtained from the first plate and the 40° pattern from the second. The whole-field isothetics are shown in Fig. 7.19, with the ones on the left obtained from the first plate and the other two from the second. The outer dark rings in the set of pictures on the left and the central dark disks in the set on the right are the areas where the speckles are defocused. No attempt was made to compare the results with the actual values, as neither the applied pressure nor the material properties of the rubber were known. However, it is evident that a quantitative analysis can be made from these fringe patterns

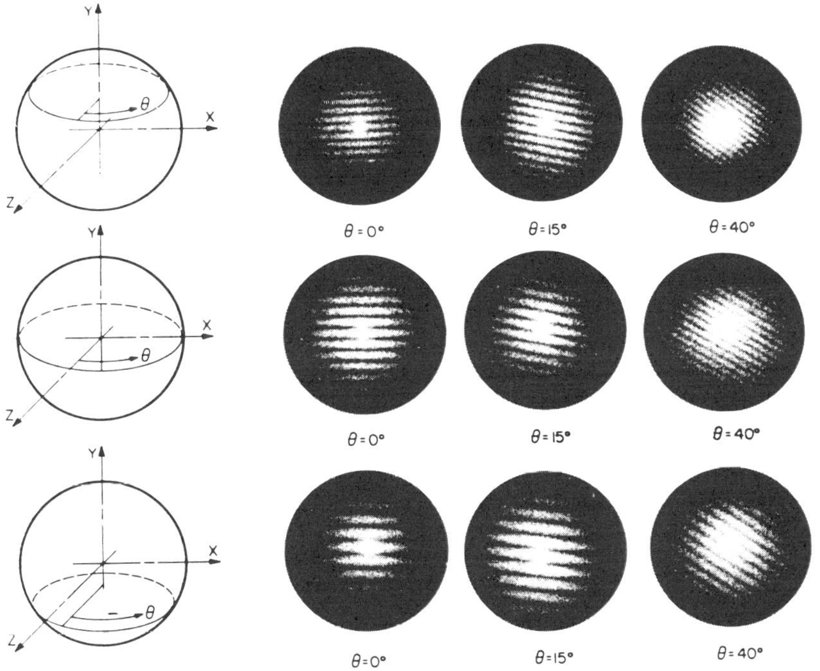

Figure 7.18 Pointwise-filtered fringes on an inflated spherical ball along three latitudes.

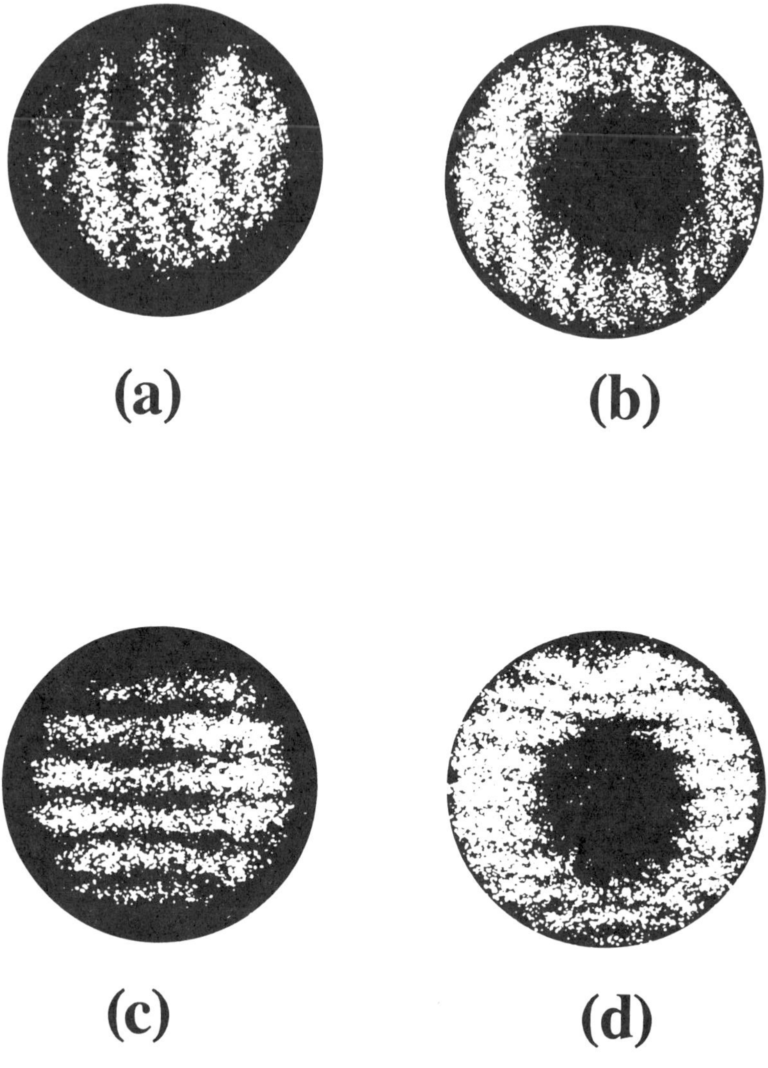

Figure 7.19 Whole-field (a, b) u and (c, d) v isothetics of the spherical ball under internal pressure using tandem plates.

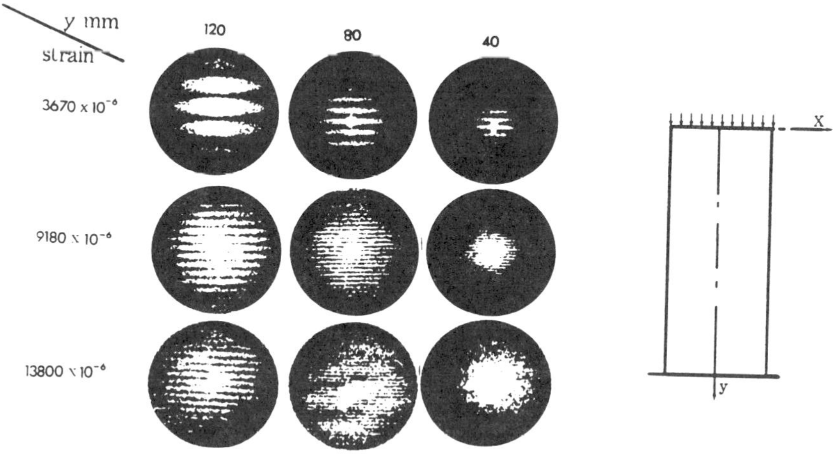

(a)

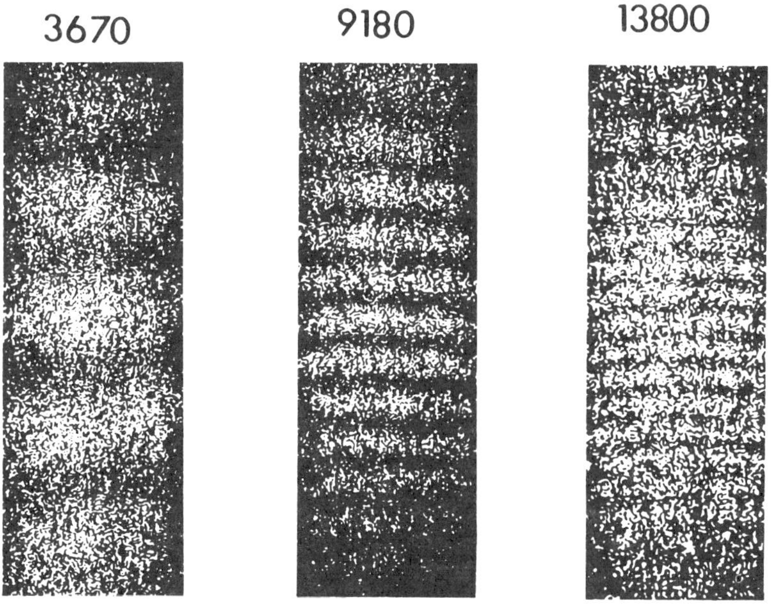

(b)

Figure 7.20 (a) Pointwise and (b) whole-field fringes on a rubber cylinder under various levels of strain.

7.8 LARGE-DEFORMATION MEASUREMENT

The size of the largest deformation that can be measured using the white light speckle method is determined solely by the ability to resolve pointwise-filtered fringes in the presence of secondary speckle noise. Decorrelation due to changing speckled structure or tilts in the presence of large deformation are not a problem as is the case with the laser speckle method.

The technique is similar to other techniques elaborated earlier except that the deformation between exposures is large. The feasibility of using the white light speckle method to determine large displacements is demonstrated with a urethane rubber cylinder whose surface was coated with retroreflective paint [17]. The cylinder was uniformly compressed by increasing levels of strain, and a double-exposure specklegram was recorded. Figure 7.20a shows the pointwise-filtered fringes for three applied strain levels at different y coordinates of the specimen. The origin is at the loaded end, and the other end is resting on a rigid table. The fringe density thus increases as the loaded end is approached. The filtered fringes are still

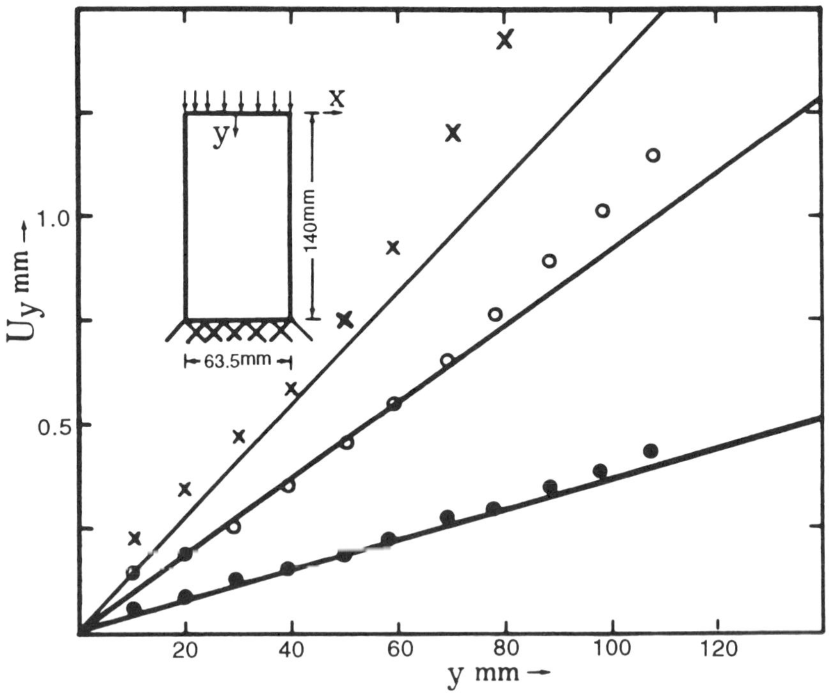

Figure 7.21 Displacement distribution for different applied strains indicating deviation from linearity under high strain.

discernible even at high strain levels. Fringe visibility degrades as the fringe spacing approaches the size of secondary speckles. Figure 7.20b shows the whole-field fringes' contouring displacement component in the *y* direction. The fringe constant is 0.08 mm/fringe on the object plane, and fringe contrast is good. Figure 7.21 plots the data obtained experimentally for the three strain levels. For small strains, the displacement distribution is linear, agreeing well with the applied constant-strain value. However, for higher deformations, the nonlinear terms in strain–displacement equations become significant, and deviations from linearity are observed. The higher the applied strain, the greater is the deviation from linearity.

7.9 OTHER APPLICATIONS

The range of applications of the white light speckle method has been demonstrated in earlier sections. In this section a few other extensions of the white light speckle method are described.

7.9.1 Objective White Light Speckle Method

Objective speckle recording, similar in definition to laser speckle recording, requires no imaging lens. Since in the white light speckle method the speckles are present only on the surface, to faithfully copy these onto film, close contact between the specimen and the recording film is essential. Illumination is provided by a white light source, which can either come through the model onto the film provided the specimen is transparent or come through the film [15]. The experiments were conducted on specimens made of transparent materials. One of the faces of each specimen was roughened by a special process. Analysis of the specklegrams yields results in good agreement with theory. Fringes of poor contrast are obtained when there is not proper contact between the object and the recording medium over the entire surface of the object.

7.9.2 Contouring Using the Defocus Effect

The effect of defocus in the white light speckle method is to cut off the higher frequency components in a recorded speckle pattern. A curved three-dimensional object that is imaged would have speckles recorded with varying contrast as shown in Fig. 7.15. Pointwise filtering at various locations along the recorded image yields halos of various sizes. The radius of the dark rings in conjunction with Eq. (7.9) thus provides information on the distance of the object from the plane of focus. To make this focus plane as narrow as possible, ideally one should use the camera lens at a low *F* number.

Whole-field filtering gives a more striking visual effect [43]. A spherical ball was chosen as an object, and a single-exposure specklegram was recorded. This specklegram was replaced in the whole-field filtering setup, the filtering aperture on the Fourier plane was moved progressively away from the optical axis, and the appearance of the object was monitored. When the aperture is near the optic axis, allowing only low-frequency components to enter, the entire object appears bright. This implies that this band of frequencies is present everywhere. As a higher band of frequencies is selected, the outer regions of the sphere, which were defocused the most, turn dark; that is, this higher band of frequencies is not present in the specklegram. Moving further away, the dark region moves progressively inward. It was assumed in this experiment that the speckle frequencies are radially symmetric and thus the filtering aperture could be placed along any radial direction. It would be preferable to have a filter consisting of two concentric circles (a ring aperture) centered on the optical axis. Such a filter would be sensitive to radial variations in defect of focus.

7.9.3 Vibration Analysis

With the appropriate choice of experimental setup, the white light speckle method can be used to obtain nodal and antinodal lines as well as amplitudes of in-plane and out-of-plane vibration of plates. The choice of experimental setup is guided by the similarity of this method to the moiré method. As with the moiré method, and to some extent the laser speckle method, time-averaged images of speckles on or as seen through the vibrating plate are recorded. Familiar Bessel-type fringes characteristic of time-average vibrations are formed. For in-plane vibration measurement the speckles have to be created on the surface of the object, while for out-of-plane vibration analysis speckles are reflected off or projected onto the vibrating plate depending on whether antinodal or nodal lines are required.

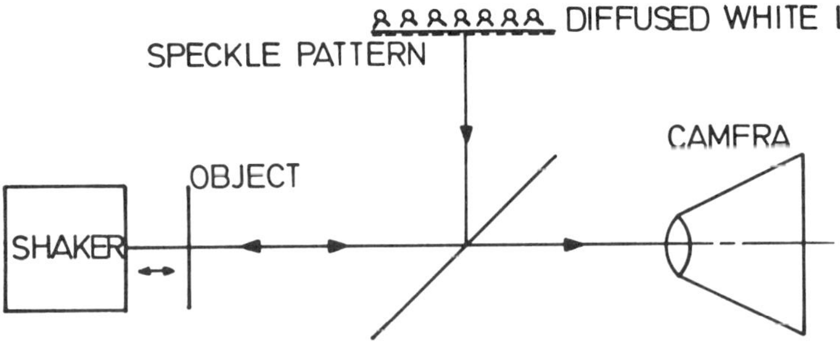

Figure 7.22 Reflection speckle setup.

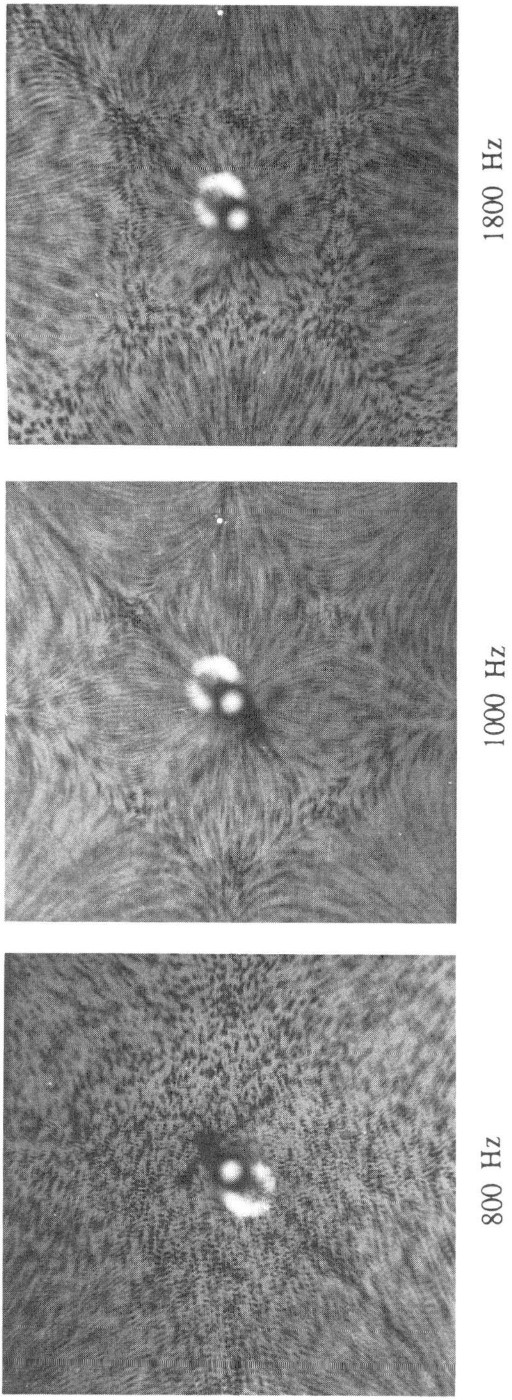

Figure 7.23 Time-averaged reflection speckle patterns of a vibrating square plate.

In this section only the reflection speckle method is demonstrated along with its similarities and differences with the reflection moiré method. The setup for a reflection speckle method shown schematically in Fig. 7.22 is identical to the setup for a reflection moiré method except that a speckle pattern replaces the grating. The requirement for this method is a specular object surface. Perspex does quite well, as it is fairly specular itself. A time-average record is made of the reflected image of the speckle pattern as seen in the mirrorlike specimen, with the plate vibrating at one of its resonant frequencies. Figure 7.23 shows the resulting photographs obtained

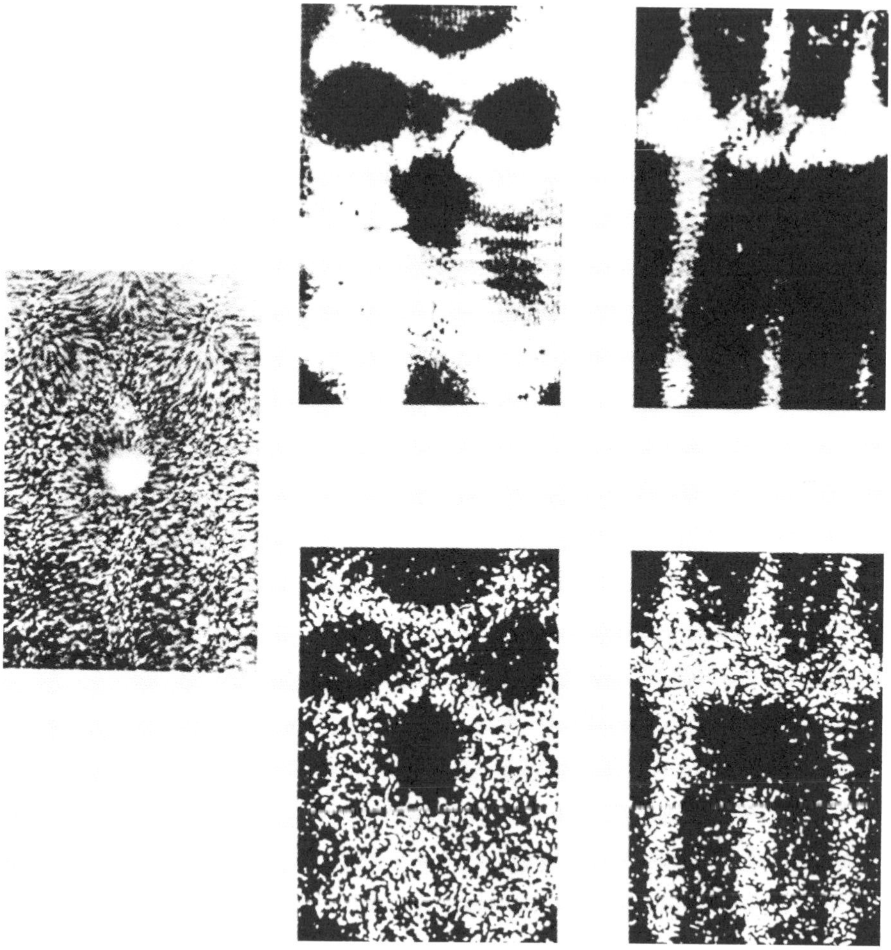

Figure 7.24 Similarity of reflection moiré and filtered reflection speckles for display of partial antinodal patterns.

for three different frequencies of vibration of a square plate. The patterns show regions where the speckles are sharp and regions where the speckles are blurred. The regions where the speckles are sharp correspond to points where the slopes of the vibrating plate are zero, i.e., antinodal regions. Unlike the reflection moiré method, wherein use of line gratings produces only partial antinodal lines, the speckle method provides the complete antinodal pattern. As a comparison, Fig. 7.24 shows the partial antinodal patterns on a rectangular plate obtained using the time-average reflection moiré method [44]. As shown in the same figure, similar patterns can be obtained by filtering the time-averaged speckle pattern along the horizontal and vertical directions. Incidentally, the moiré patterns were filtered to enhance contrast.

7.10 CONCLUSION

The versatility and ease of use of the white light speckle method is demonstrated in this chapter. Fringe interpretation and analysis with this method is much simpler than with its laser counterpart because speckle movements faithfully follow object movements. The method thus bears a remarkable resemblance to conventional moiré methods. Indeed, it has been referred to as a random grating method in view of its postrecording sensitivity selection capability. Thus all moiré methods can be (and have been) addressed by the white light speckle method. An added convenience is that it is not necessary to print gratings on the surface, which is especially cumbersome on curved objects. Principally, like the moiré method, the white light speckle method is best suited for large-deformation measurement on large objects. Nevertheless, its versatility enables it to be adapted to small deformations on small objects, thus covering the entire spectrum of deformation.

REFERENCES

1. L. Glass, Moiré effect from random dots, *Nature*, *223*, 578–580 (1969).
2. E. Bernabeu, J. C. Amar'e, and M. P. Arroyo, White light speckle method of measurement of flow velocity distribution, *Appl. Opt.*, *21*, 2583–2586 (1982).
3. A. Asundi and F. P. Chiang, Vibration analysis using the white light speckle method, Tech. Rep. 399, College of Engineering and Applied Science, SUNY at Stony Brook, N.Y., 1982.
4. J. M. Burch and C. Forno, A high sensitivity moiré grid technique for studying deformation in large objects, *Opt. Eng.*, *15*, 178–185 (1975).
5. C. Forno, White light speckle photography for measuring deformation, strain and shape, *Opt. Laser Technol.*, *16*, 217–221 (1975).

6. P. Boone and L. C. De Backer, Speckle methods using photography and reconstruction in incoherent light, *Optik*, *44*, 343–355 (1976).
7. F. P. Chiang and A. Asundi, White light speckle method of experimental strain analysis, *Appl. Opt.*, *18*, 409–411 (1979).
8. P. Jacquot and P. K. Rastogi, Influence of out-of-plane deformation and its elimination in white light speckle photography, *Opt. Lasers Eng.*, *2*, 33–55 (1981).
9. A. Asundi, White light speckle method for strain analysis, Ph.D. dissertation, SUNY at Stony Brook, N.Y., 1981.
10. F. P. Chiang and A. Asundi, Interior displacement and strain measurement using white light speckles, *Appl. Opt.*, *19*, 2252–2256 (1980).
11. F. P. Chiang and A. Asundi, White light speckle method with tandem plates for 3-D displacement and deformation measurement on curved surfaces. *Appl. Opt.*, *20*, 2167–2169 (1981).
12. F. P. Chiang and A. Asundi, White light speckle method applied to the determination of stress intensity factor and displacement field around a crack tip, *Eng. Fracture Mech.*, *15*, 115–121 (1981).
13. A. Asundi and F. P. Chiang, Defocussed white light speckle effect, *Appl. Opt.*, *21*, 1887–1888 (1982).
14. A. Asundi and F. P. Chiang, Perspective effect in white light speckle method, *Appl. Opt.*, *21*, 1708–1710 (1982).
15. A. Asundi and F. P. Chiang, Objective white light speckles and their application to stress intensity factor determination, *Opt. Lett.*, *7*, 378–379 (1982).
16. A. Asundi and F. P. Chiang, Theory and application of the white light speckle method for strain analysis, *Opt. Eng.*, *21*, 570–580 (1982).
17. A. Asundi and F. P. Chiang, Measurement of large deformation using the white light speckle method, *Mech. Res. Commun.*, *9*, 325–330 (1982).
18. A. Asundi, Application of the white light speckle method to fracture studies, *Adv. Fracture Res.*, *5*, 3329–3335 (1984).
19. M. Jonsson and L. Benckert, Measurement of residual deformations in butt-welded plates by use of the white light speckle technique, *J. Strain Anal.*, *21*, 243–246 (1986).
20. A. Asundi and F. P. Chiang, Applications of the white light speckle method to interior displacement measurement, *J. Strain Anal.*, *18*, 23–26 (1983).
21. F. P. Chiang and A. Asundi, White light speckle method of experimental strain analysis, *Appl. Opt.*, *18*, 409–411 (1979).
22. G. H. Kaufmann, Double pulse white light speckle photography, *Appl. Opt.*, *23*, 194–196 (1984).
23. G. Cloud, R. Falco, R. Radke, and J. Peiffer, Non-coherent light photography for measurements of fluid velocity fields, *Proc. SPIE*, *243*, 150 (1980).
24. E. Conley and G. Cloud, Practical applications of double exposure non-coherent light speckle photography, *Appl. Opt.*, *25*, 2246–2248 (1986).
25. E. Conley and G. L. Cloud, Enhancing sensitivity of the white light speckle method in terrestrial applications, *Exp. Mech.*, *29*, 138–143 (1989).
26. E. Conley and G. Cloud, Resolution experiment using white light speckle method, *Appl. Opt.*, *30*, 795–800 (1991).

27. F. P. Chiang, B. C. Liu, and S. T. Lin, Multi-aperture white light speckle method applied to the strain analysis of cylinders with holes under compression, *Opt. Laser Technol.*, *2*, 151 (1981).
28. N. Krishna Mohan, C. Joenathan, D. K. Sharma, and R. S. Sirohi, Applications of multi-aperture white light speckle photography, *Optik*, *76*, 7–11 (1987).
29. N. Krishna Mohan, C. Joenathan, and R. S. Sirohi, Investigations on multi-aperture white light speckle photography, *Opt. Commun.*, *67*, 27–33 (1988).
30. F. P. Chiang and D. W. Li, Random (speckle) patterns for displacement and strain measurement: some recent advances, *Opt. Eng.*, *24*, 936–945 (1985).
31. E. Archbold, A. E. Ennos, and M. S. Virdee, Speckle photography for strain measurement—a critical review, *Proc. SPIE*, *136*, 258–264 (1977).
32. K. A. Stetson, Miscellaneous topics in speckle metrology, in *Speckle Metrology*, R. K. Erf, Ed., Academic, New York, 1978, pp. 295–320.
33. T. C. Chu, W. F. Ransom, M. A. Sutton, and W. H. Peters, Applications of digital image-correlation techniques to experimental mechanics, *Exp. Mech.*, *25*, 232–244 (1985).
34. F. P. Chiang, J. Adachi, R. Anastasi, and J. Beatty, Subjective laser speckle method and its application to solid mechanics problems, *Opt. Eng.*, *21*, 379–390 (1982).
35. X. M. Hu, S. J. P. Palmer, and J. E. Field, The application of high speed photography and white light speckle to the study of dynamic fracture, *Opt. Laser Technol.*, *16*, 303–306 (1984).
36. N. Krishna Mohan, D. K. Sharma, and R. S. Sirohi, Determination of longitudinal displacement using white light speckle photography, *Opt. Commun.*, *78*, 222–226 (1990).
37. C. Joenathan, C. S. Narayana Murthy, and R. S. Sirohi, Localisation of fringes in speckle photography due to axial motion of the diffuse object, *J. Opt. Soc. Am. A*, *5*, 1035–1040 (1988).
38. C. S. Vikram and K. Vedam, Complete 3-D deformation analysis in the white light speckle method, *Appl. Opt.*, *22*, 213–214 (1983).
39. A. Asundi and F. P. Chiang, Separation of 3-D displacement components in the white light speckle method, *Opt. Laser Technol.*, *15*, 41–45 (1983).
40. L. Benckert, M. Jonsson, and N. E. Molin, Measuring true in-plane displacements of a surface by stereoscopic white light speckle photography, *Opt. Eng.*, *26*, 167–169 (1987).
41. F. P. Chiang, A new family of 2D and 3D experimental stress analysis techniques using laser speckles, *Solid Mech. Arch.*, *3*, 1–32 (1978).
42. F. P. Chiang, A new three dimensional strain analysis technique by scattered light speckle interferometry, in *Proceedings of the Conference on the Engineering Uses of Coherent Optics*, E. R. Robertson, Ed., Cambridge Univ. Press, New York, 1976, pp. 249–262.
43. A. Asundi, Defocused white light speckle method for object contouring, *Proc. SPIE*, *1385*, 239–245 (1991).
44. F. P. Chiang and A. Asundi, White light speckle method for vibration of plates, *Proc. SPIE*, *814*, 149–152 (1987).

8

Surface Roughness Evaluation

J. D. Briers

Kingston University
Kingston upon Thames, England

8.1 INTRODUCTION

With the invention of the laser in 1960, the phenomenon of laser speckle started to make its presence felt. Its initial impact was detrimental and annoying, as it limited the resolution that could be achieved when laser illumination was used. Soon, however, applications of this "noise" phenomenon began to appear in the literature. One of these applications was the measurement of surface roughness.

This account is in the nature of a review, so a large number of articles are referenced and listed at the end of the chapter. Certain key references are, however, repeatedly cited. These include definitive works on laser speckle and on surface roughness and also previous reviews of the application of laser speckle to the measurement of surface roughness.

The definitive text on laser speckle and its applications is the book *Laser Speckle and Related Phenomena* edited by J. C. Dainty and published by Springer-Verlag. First published in 1975, a second edition appeared in 1984 [1]. As far as the specific references made in this chapter are concerned, there is very little difference between the two editions.

The chapters of Dainty's book that are particularly relevant to surface roughness measurement, and that are extensively referred to in the present account, are J. W. Goodman's contribution on the statistical properties of laser speckle patterns [2] and G. Parry's chapter on speckle patterns in partially coherent light [3].

The main source text on the nature and measurement (by non-speckle means) of surface roughness has been *Introduction to Surface Roughness and Scattering* by J. M. Bennett and L. Mattsson, published in 1989 by the Optical Society of America [4].

Three previous reviews of the application of laser speckle to the measurement of surface roughness have been used extensively in the preparation of this chapter. These are W. T. Welford's 1977 review article in *Optical and Quantum Electronics* [5]; T. Asakura's contribution on surface roughness measurement in *Speckle Metrology*, edited by R. K. Erf and published in 1978 by Academic Press [6]; and Chapter 9, "The Study of Surface Roughness," in M. Françon's book *Laser Speckle and Applications in Optics*, translated by H. H. Arsenault and published by Academic Press in 1979 [7]. There is also a brief summary of the field in the chapter on recent developments in the second (1984) edition of Dainty's book [1].

Notation

For ease of reference, the following is a complete list of the notation adopted in this chapter.

C_A	autocovariance of complex amplitude
C_R	autocovariance of surface roughness (variation in surface height)
c	normalized autocovariance of intensity
c_A	normalized autocovariance of complex amplitude
c_R	normalized autocovariance of surface roughness
$\mathcal{C}$	speckle contrast (defined as $= \sigma/\langle I \rangle$)
D	distance from the scattering surface to the plane of observation in the far field
G_R	autocorrelation function of surface roughness
I	intensity
i	incremental counter
J_1	first-order Bessel function
k	$= 2\pi/\lambda$
N	number of scatterers contributing to the speckle pattern
P	probability density function
R_a	average roughness (CLA variation of surface height)
r	distance from center of illuminated area on the scattering surface
s	radius of the scattering area, when considered circular
u, v	spatial frequencies in the ξ and η directions in the speckle pattern (or in the x and y directions on the scattering surface)

V	rms difference between two speckle intensity distributions
W_R	Wiener spectrum (power spectrum) of the surface roughness
x, y	coordinate directions in the plane of the scattering surface
$\delta x, \delta y$	increments in x and y
z	surface height in the plane of the scattering surface
δz	variation in surface height
α	angular distance from the optical axis in the speckle pattern
γ	fringe contrast
θ	angle of incidence
λ	wavelength of the light
Λ	effective wavelength in two-wavelength techniques
ξ, η	coordinate directions in the far-field speckle pattern
$\delta\xi, \delta\eta$	increments in ξ and η
ρ	ratio of background intensity to total intensity
σ^2	variance of intensity
σ_ϕ^2	variance of phase distribution
σ_k	rms spectral deviation
σ_R	rms surface roughness (variation of surface height)
σ_α	rms angular elongation of polychromatic speckles
ϕ	resultant phase of scattered light
$\langle\ \rangle$	ensemble average

8.2 WHAT IS SURFACE ROUGHNESS?

8.2.1 The Importance of Surface Roughness

For many applications, especially in today's high-technology world, surfaces need to be as smooth as possible. In the optics industry this has always been the case, as glass components can only transmit images when the surfaces are highly polished. Recent advances in the electronics and precision mechanics industries now demand very smooth surfaces in many other materials. Even in the optics field, the requirements of new developments such as ring laser gyros have stretched existing manufacturing and measuring capabilities.

There is thus a need to be able to specify in some detail the surface characteristics of various materials. In the extreme, this can be achieved only by mapping the surface with high precision. This can be done, for example, by means of the optical profilometers that are discussed briefly in Section 8.10. However, this approach would result in every single surface being unique, and comparisons and judgments about "fitness for use" would be extremely difficult. What is needed is some simple way to characterize the nature of a surface such that it can be compared with other similar surfaces and, more important, with a reference standard or with a specification.

8.2.2 Aspects of Surface Quality

Even so, it is necessary to differentiate in a broad way between different aspects of surface quality. In the optics field, a clear distinction has traditionally been assumed to exist between surface "form" or "figure," surface polish, and surface defects or flaws.

Surface form is the departure of the general surface shape from that of the ideal shape, usually a plane or a sphere, and is universally measured by means of optical interferometry [8]. This technique has excellent depth resolution, down to nanometers if required, but is very limited in lateral resolution. This, however, is usually not important, because the interest is in the gradual departure from the ideal shape across the surface.

Surface polish (or lack of roughness!) and localized defects ("scratch and dig") are usually assessed visually and qualitatively by trained inspectors without recourse to sophisticated measurement techniques. In the last few years, this relatively simple differentiation between large-scale departures (form) and small-scale departures (polish and flaws) from the ideal surface has become complicated by the growing use of machining techniques in the optics industry [9]. This has introduced the new dimension—long known in the mechanics industry—of small-scale periodic errors usually referred to as "waviness" and of directionality or "lay." The periodicities involved are long compared with the scale of surface roughness but much shorter than those usually encountered as form errors. This blurring of the boundaries between surface form and small-scale irregularities, together with the need for higher precision and better quality optics, has prompted a fresh look at the entire question of surface finish and its assessment. This in turn has led to a significant dialogue between the optics, electronics, and mechanics industries.

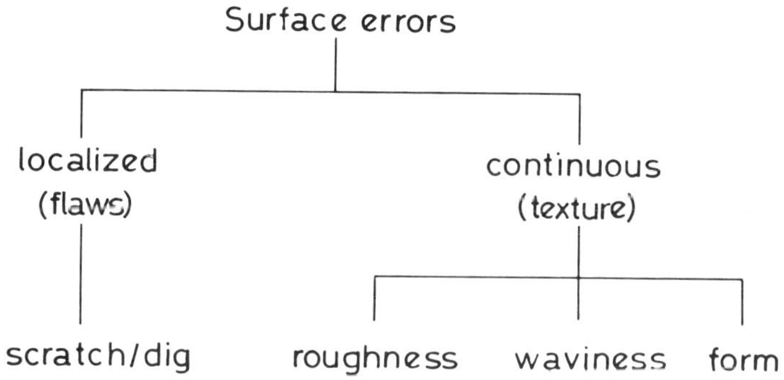

Figure 8.1 Classification of surface errors.

The various aspects of surface finish can be brought together in the "family tree" of Fig. 8.1. Note, however, that the boundaries between form, waviness, and roughness are arbitrary, the three forming a continuum of surface height variation. As a rough rule, the "waviness" regime can be taken to be spatial wavelengths (roughness correlation lengths, see Section 8.2.5) between 1 and 10 mm. There is a shortage of measuring instruments in this range, although attempts are being made to fill the gap.

8.2.3 Defining and Characterizing Surface Roughness

The factors described in the preceding paragraphs have resulted in a need to move away from the traditional subjective assessment of small-scale surface finish toward more reliable measurement methods. This is the trend even in the case of scratch-and-dig measurements, and a recent revision of the relevant British standard now recommends the use of comparator-type instruments rather than the former reliance on human visual judgment [10]. However, in the case of roughness, the question arises as to what exactly should be measured. The problem is that the actual variations in surface height embodied in the concept of roughness can be very varied. For example, they can be periodic or random, isotropic or directional. To some extent, the nature of the roughness will dictate the measurement technique. Since the purpose of this chapter is to investigate the use of laser speckle as a measurement technique, and as it will be shown later that laser speckle is a random phenomenon, only random, isotropic (non-directional) roughness will be considered. Even then, there is an infinite variety of distributions of surface height variations that might be encountered. This has led to the use of a very large number of parameters to characterize surface roughness, different ones being of more or less relevance to different industries and processes. Although most of these parameters are of particular value only in specific applications, there are still several that find wider use. The most important of these are described briefly below. For more details, the reader is referred to the literature [4].

8.2.4 First-Order Statistics of Surface Roughness

Surface roughness can be regarded as the variation of surface height. Assuming this variation to be random and isotropic, the roughness can be characterized in a statistical manner. The most obvious parameter is then the average departure of surface height from its mean value, usually called the *average roughness* or the *centerline average* (CLA) *roughness*. This is the parameter most frequently used for machined surfaces [11] and is usually designated by the symbol R_a.

Mathematically, it is very simply defined in terms of the surface height variations δz_i measured from the mean surface level:

$$R_a = \frac{1}{N} \sum_{i=1}^{N} |\delta z_i| \tag{8.1}$$

An alternative measure of the average variation in surface height is the *root-mean-square (rms) roughness*. This is the quantity commonly used in the optics industry. It is defined as

$$\sigma_R = \left[\frac{1}{N} \sum_{i=1}^{N} \delta z_i^2 \right]^{1/2} \tag{8.2}$$

There will obviously be some definite relationship between R_a and σ_R. However, this will depend on the particular statistical distribution of surface height present in any given case. In the machining industry, the relationship usually assumed is [7]

$$\sigma_R = 1.11 R_a \tag{8.3}$$

The multiplying factor will, however, be larger for surfaces with abnormally high numbers of large deviations from the mean surface height.

In the optics industry, surface form has traditionally used a *peak-to-valley* measure for deviation of the surface from the desired shape. This can be extended to surface roughness measurements but is usually of limited value. An alternative is the so-called *ten-point height* used in the machining industry, in which the average of the five deepest valleys is subtracted from the average of the five highest peaks.

For some applications, slope variations are more important than actual height variations, and a *root-mean-square (rms) slope* can be defined similarly to the rms roughness.

Other parameters are used in the machining industry to indicate further details of the surface height distribution. *Skewness* indicates whether the distribution of surface heights is biased (skewed) toward positive or negative height deviations compared with, say, a Gaussian distribution. A high positive skewness would indicate a generally smooth surface with occasional but relatively large "hills"; a high negative skewness would imply a surface with pits or holes. Skewness is usually defined in terms of a mean-cubed roughness:

$$\text{Skewness} = \frac{1}{\sigma_R^3} \left(\frac{1}{N} \right) \sum_{i=1}^{N} \delta z_i^3 \tag{8.4}$$

Kurtosis is defined in a similar way to skewness, but in terms of the fourth power of the surface height deviations:

$$\text{Kurtosis} = \frac{1}{\sigma_R^4}\left(\frac{1}{N}\right)\sum_{i=1}^{N} \delta z_i^4 \qquad (8.5)$$

A Gaussian distribution has a kurtosis of 3. A higher value indicates that more of the surface height deviations than usual are in the higher ranges (positive and negative); a lower value, that the surface has fewer highs and lows than the corresponding Gaussian profile.

Further details of these and other parameters of the surface height distribution can be found in the literature and in the relevant standards [4, 11, 12].

8.2.5 Second-Order Statistics of Surface Roughness

All of the foregoing discussion has been concerned with the first-order statistics of the surface height distribution. In other words, only the actual values of surface height have been considered, and not how rapidly the variations occur as one moves over the surface. To analyze the lateral scale of the surface roughness, the typical "spatial frequencies" or "spatial wavelengths" involved, it is necessary to invoke second-order statistics.

The two most important second-order statistics are the autocovariance (or its close relative, the autocorrelation function) and the power spectrum. *Autocovariance* is a measure of how well the surface height at one point can be predicted from a knowledge of the height at a second point. Obviously, if the two points are very close together this prediction can be made with very high confidence: the heights at the two points are highly correlated. As the distance between the two points (the *lag*) is increased, the correlation reduces until, after a certain distance, there is no correlation between the two surface heights at all. The distance over which the decorrelation occurs is called the *correlation length* and is a measure of the average lateral scale of the surface height variations. Physically, the correlation process can be visualized as the sliding of the surface over itself and an assessment of the similarity between the surface and its displaced self for each value of lag. Mathematically, the *autocorrelation function* $G_R(\delta x, \delta y)$ is defined by multiplying the surface height at each point on the surface by the height at the point displaced from it by the lag distance (components δx and δy), averaging over the whole surface, and repeating for different values of lag:

$$G_R(\delta x, \delta y) = \langle z(x, y) \cdot z(x + \delta x, y + \delta y) \rangle \qquad (8.6)$$

where $\langle\ \rangle$ indicates the ensemble average.

The *autocovariance* $C_R(\delta x, \delta y)$ is similarly defined, except that the mean value of z is subtracted from each z value before the multiplications are carried out:

$$C_R(\delta x, \delta y) = \langle[z(x, y) - \langle z\rangle][z(x + \delta x, y + \delta y) - \langle z\rangle]\rangle \tag{8.7}$$

As will be seen later, when the statistics of speckle patterns are considered, it is also convenient to define a *normalized autocovariance*:

$$c_R(\delta x, \delta y) = C_R(\delta x, \delta y)/\sigma_R^2 \tag{8.8}$$

The advantage of the normalized version of the autocovariance is that it has a maximum value of 1 (at zero lag) and tends to zero at infinite lag.

The *power spectrum*, or *Wiener spectrum*, of the surface height variations, like its temporal counterpart in electronics, gives the distribution of the spatial frequencies in the roughness. Mathematically, it is given by the Fourier transform of the autocovariance:

$$W_R(u, v) = \iint_{-\infty}^{+\infty} C_R(\delta x, \delta y) \exp[-2\pi i(\delta x \cdot u + \delta y \cdot v)]\, d(\delta x)\, d(\delta y) \tag{8.9}$$

The emphasis in the past has been on the measurement of the first-order statistics of surface roughness, in particular the rms roughness σ_R or the CLA roughness R_a. Recently, however, there has been increasing interest in the second-order statistics of spatial wavelength (or correlation length) and power spectrum.

8.2.6 Measurement of Surface Roughness

The traditional method of measuring surface roughness is to pass a stylus probe across the surface and measure its movement as it follows the surface profile. This technique has been developed to a very sophisticated level and can achieve surprisingly high precision [4]. However, there are limits set by the finite radius of the tip of the probe; for example, peaks that are narrower than the probe tip will still be followed by the probe but will be recorded as being broader than they really are, while the probe will be physically incapable of reaching the bottom of narrow crevices. In other words, some of the surface height information is being filtered out. Results also seem to depend on the actual shape of the stylus and on its loading. In addition, many applications are unsuitable for contact techniques, which could damage the surface. Finally, the mechanical probe technique measures the roughness along a line on the surface, and a large number of such lines must be used if the entire surface is to be sampled. It is for these reasons that much effort has been put into the development of noncontact, usually optical, techniques. Diffraction, scattering, and interferometry have all been used in different approaches, and some of these will be discussed briefly later. The main purpose of this chapter, however, is to report on the technique of using laser speckle patterns.

8.2.7 Laser Speckle As a Technique for Measuring Surface Roughness

As will be seen in Section 8.4, laser speckle patterns are random and can be described only in statistical terms. It therefore seems logical that they could be used only, if at all, to measure random roughnesses. In general this turns out to be the case, and for the rest of this chapter only randomly rough surfaces will be considered. In fact, the assumption of a Gaussian distribution of surface height will generally be assumed, and no attempt will be made to measure such parameters as skewness and kurtosis. However, if *comparative* measurements on surfaces known to have similar statistics are to be made, rather than absolute measurements, the techniques to be described will, in fact, have a wider applicability than has been assumed.

8.3 THE ORIGIN OF THE LASER SPECKLE EFFECT

8.3.1 The Speckle Phenomenon

When a diffusely reflecting object is illuminated with laser light, a high-contrast "speckle pattern" is produced [13]. In the early days of lasers, the effect was regarded purely as a nuisance because it severely affected resolution when laser light was used, for example in holography, and much effort was directed to reducing speckle in images formed in laser light [14]. However, it was not long before scientists started to study speckle for its own sake, and practical applications of the phenomenon began to be reported in the literature. Since speckle was produced only by rough surfaces—it was not observed in specular reflection from mirrorlike surfaces—one of the first questions that was addressed was whether the speckle pattern could be used to measure the roughness of surfaces.

Even an uninitiated person seeing a speckle pattern for the first time realizes that there is something odd about it. He gets the impression that he cannot focus his eyes properly on the surface exhibiting the speckle pattern, and he notices that the speckles move as he moves his head. If he were to try a few simple experiments he would discover even stranger effects. For example, if he were to observe the speckle pattern through a pinhole held in front of his eye, he would find that the speckles appeared larger. If he were to look at the light scattered by the object and falling on a screen, he would find that this, too, produced a speckle pattern and that in this case the speckle size depended on the area of the object surface illuminated: the smaller this area, the larger the speckles. This latter type of speckle pattern will be referred to in this chapter as *far-field speckle*, while the speckle observed on an illuminated surface will be called *image speckle*. (In Gabor's terminology [15] they are referred to as "objective" and "subjective" speckle, respectively.)

The fact that speckle patterns only came into prominence with the invention of the laser suggests that the cause of the phenomenon is the high degree of coherence of the laser. Further investigation shows that this is, indeed, the case, although it is, in fact, possible to observe speckle patterns with other, less coherent, sources such as mercury lamps and even, in certain circumstances, with sunlight.

8.3.2 A Qualitative Explanation of the Speckle Effect

Speckle can be regarded as an interference pattern produced by coherent light reflected from different parts of the illuminated surface. If the surface is rough compared with the wavelength of the light, rays reflected from different parts of the surface within a resolution cell (the size of which is determined by the limiting aperture of the observing system) will traverse different optical path lengths in reaching the eye of the observer (or the observing screen in the case of far-field speckle), and the resulting intensity at a given point on the object (or on the observing screen) will be determined by the coherent addition of the complex amplitudes associated with each of these rays. If the resultant amplitude is zero, a dark "speckle" will be seen at the point, whereas if all the rays arrive at the point in phase, an intensity maximum will be observed.

8.3.3 Far-Field Speckle

In the case of far-field speckle, illustrated in Fig. 8.2, the whole of the illuminated area of the object contributes to each speckle observed on the screen. The intensity at any point on the screen is given by the squared modulus of the resultant amplitude obtained by the coherent addition of complex amplitudes of rays arriving at the point from all parts of the object. It is intuitively reasonable to deduce that the larger the surface contributing to the pattern, the more rapid will be the variation in resultant amplitude across the far-field speckle pattern and hence the smaller will be each individual speckle. In fact, it can be shown that the average speckle size is equal to that of the Airy disk produced in the far field by diffraction from an aperture the size of the illuminated area.

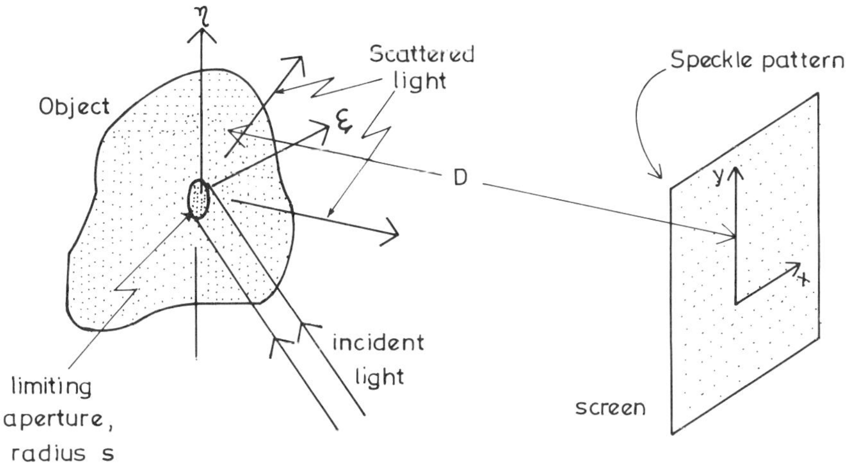

Figure 8.2 Experimental arrangement for observing far-field speckle patterns.

8.3.4 Image Speckle

In image speckle, the size of a speckle is also determined by an aperture, but in this case it is the limiting aperture of the observing system that is important. This is illustrated in Fig. 8.3. When an object is viewed with the unaided eye, the limiting aperture is the pupil of the eye itself. The

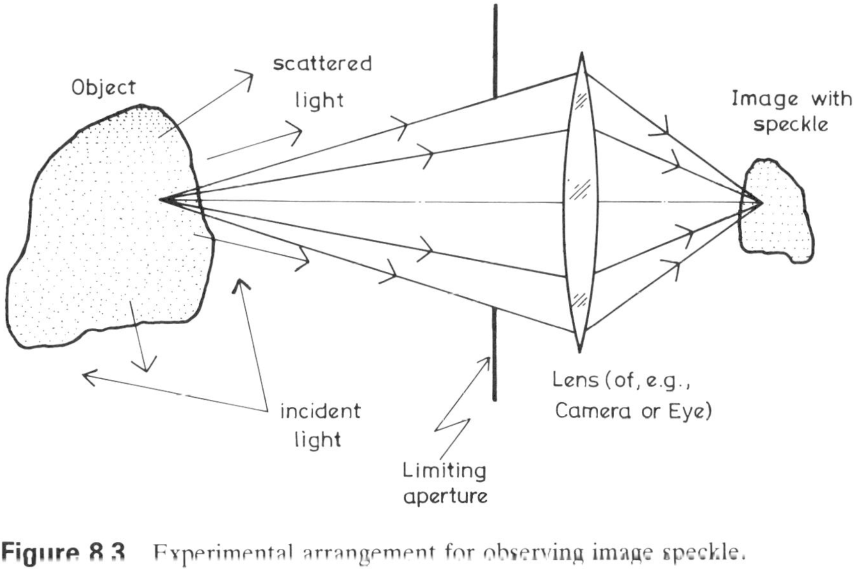

Figure 8.3 Experimental arrangement for observing image speckle.

speckle size in the case of image speckle is equal to the resolution limit of the optical system used, and hence once again the speckle size increases as the aperture of the system is reduced. In this case, of course, the light rays contributing to the intensity of a particular speckle can only come from that same resolution cell; this is an important difference from far-field speckle, in which light from the entire illuminated surface contributes to every speckle. However, it can be shown that the same basic mathematical theory governs the properties of both types of speckles, providing a large number of scattering centers contribute to each speckle. In the case of far-field speckle, this condition is met simply by arranging for the illuminated area to be very large compared with the roughness of the surface.

8.4 STATISTICS OF LASER SPECKLE PATTERNS

The statistics of laser speckle patterns were first derived in detail by Goodman [2, 16, 17]. Since speckle statistics play a major role in the use of speckle techniques for measuring surface roughness, the most important mathematical properties of speckle patterns are stated below, but without proof. The reader is referred to the papers cited above for formal derivations of the formulas.

The case of far-field speckle is specifically considered. Similar results are obtained for image speckle, providing the optical arrangement ensures that the assumptions regarding Gaussian statistics are valid. (In practice, the optics often do not guarantee this, and the formulation of image-plane speckle statistics can become very complex [5].) It is also assumed, for now, that the surface is rough compared with the wavelength of light and that both the coherence length of the light and the size of the scattering area are much greater than the path differences caused by the roughness of the surface.

8.4.1 First-Order Statistics

As in the case of surface roughness statistics, considered in Section 8.2, the first-order statistics describe the properties of a speckle pattern point by point. They are not concerned with the relationship between intensities at different points in the speckle pattern.

The *probability density function* of intensity is found to follow a negative exponential law:

$$P(I) = \frac{1}{\langle I \rangle} e^{-I/\langle I \rangle} \tag{8.10}$$

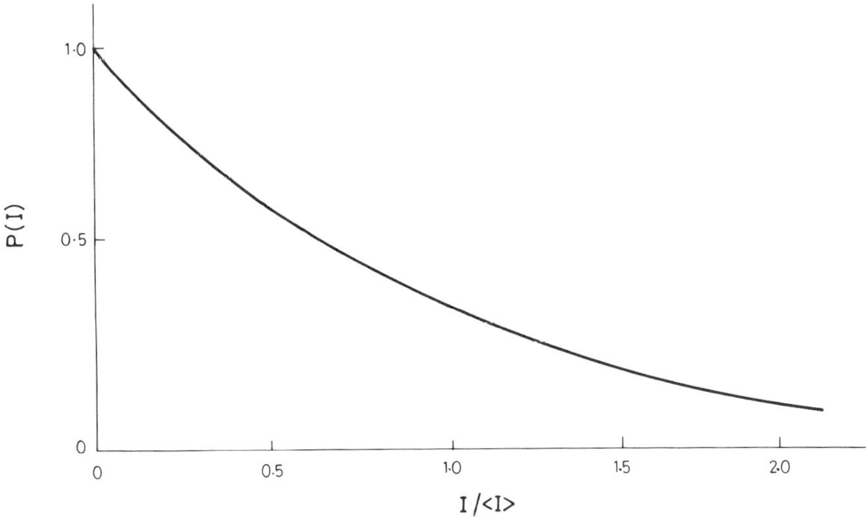

Figure 8.4 Probability density function of intensity in a speckle pattern.

This equation, plotted in Fig. 8.4, gives the probability that the intensity at a point in the speckle pattern will have the value I and leads to two apparently paradoxical results:

1. The most probable intensity at any point is zero.
2. It is possible to have an infinite intensity at some points.

The first of these properties is, in fact, true and is what causes the very high contrast observed in a speckle pattern. It has been confirmed experimentally [18–20]. The second cannot, of course, be true, and the intensity at any point in the speckle pattern is limited absolutely by the total amount of energy scattered by the surface and by the fact that this energy must be spread over at least one speckle, whose minimum size is limited by diffraction theory. However, it has been shown by Parry [21] that the intensity at a point in the far-field speckle pattern can exceed the mean intensity by a factor of several thousand.

The other important first-order property of the speckle pattern is its *standard deviation* σ (or its square, the *variance*), which describes the spread of values of intensity about the mean and hence is a measure of the contrast of the speckle pattern. The standard deviation of the intensity in a speckle pattern is, in fact, equal to the mean intensity:

$$\sigma = \langle I \rangle \tag{8.11}$$

Note that the angular brackets ⟨ ⟩ indicate the *ensemble average*; for the purposes of the present discussion this can be taken to be the spatial average over the speckle pattern.

The simplicity of this relationship leads to a very useful parameter for describing the contrast of speckle patterns, the ratio $\sigma/\langle I\rangle$. This will be discussed in more detail in Section 8.4.4.

8.4.2 Second-Order Statistics

The second-order statistics of a speckle pattern describe how rapidly the intensity varies from point to point in the pattern. They thus give an indication of the size of the speckles and the distribution of speckle size in the pattern. (Again this is exactly analogous to the surface roughness statistics considered in Section 8.2.)

The most common second-order statistic used in the study of speckle patterns has been the *normalized autocovariance* of intensity (called by some authors, but not strictly correctly, the *autocorrelation function*). This is defined as follows:

$$c^{(2)}(\delta\xi, \delta\eta) = \frac{\langle [I(\xi, \eta) - \langle I\rangle][I(\xi + \delta\xi, \eta + \delta\eta) - \langle I\rangle]\rangle}{\sigma^2} \tag{8.12}$$

Again the angular brackets ⟨ ⟩, strictly speaking, represent the ensemble average but can be taken to indicate the spatial average over the whole speckle pattern. Following the practice of workers in the field of photon counting and intensity fluctuation spectroscopy [22], the superscript (2) is used to indicate that this is the *second-order* normalized autocovariance, i.e., of intensity rather than amplitude.

As defined, the quantity $c^{(2)}$ has a maximum value of 1 at $\delta\xi = \delta\eta = 0$ (i.e., at zero lag) and tends to zero at infinite lag. The value of lag ($\delta\xi$ or $\delta\eta$) at which $c^{(2)}$ falls to a given (arbitrary) value, say 1/2, or $1/e$, or $1/e^2$, or even zero if there is a zero crossing, can then be used as a measure of the average speckle size.

Another important second-order statistic is the *power spectrum* or *Wiener spectrum*, which describes the distribution of speckle sizes in the pattern. As in the case of the surface roughness statistics considered earlier, this is given by the Fourier transform of the autocovariance.

It can also be shown that the autocovariance of intensity in the far field can be expressed in terms of the Fourier transform of the intensity distribution of the light scattered by the surface. For the case of a uniformly illuminated circular scattering area of radius s (see Fig. 8.2), this takes the form of a first-order Bessel function:

$$c^{(2)}(\delta\xi, \delta\eta) = \frac{\langle I\rangle^2}{\sigma^2}\left[\frac{2J_1\{(2\pi s/\lambda D)\sqrt{(\delta\xi)^2 + (\delta\eta)^2}\}}{(2\pi s/\lambda D)\sqrt{(\delta\xi)^2 + (\delta\eta)^2}}\right]^2 \tag{8.13}$$

(Note the link with the Airy disk, which can also be expressed as a first-order Bessel function.)

Thus the autocovariance (along with all second-order statistics) is determined entirely by the size of the illuminated area on the scattering surface.

8.4.3 Effect of a Finite Aperture

It should be emphasized that the above account of speckle statistics is purely theoretical and is based on the concept of intensity measurements being taken at a "point" in the speckle pattern. When practical measurements of speckle patterns are carried out, the intensity measurements must be made through finite apertures and are thus integrated over an area, albeit a very small area, of space. This has the effect of smoothing out the speckle pattern and modifying the statistics. This effect has been investigated by several authors, and appropriate corrections have been derived [17, 18, 20, 23, 24]. However, provided *relative* rather than absolute measurements of speckle statistics are carried out, using the same detector setup, these modifications do not usually need to be considered.

8.4.4 "Partially Developed" Speckle Patterns

The first- and second-order statistics discussed above assume that the incident light is fully coherent and that the surface is rough compared with the wavelength of light. The resulting speckle patterns, in the terminology of Pedersen [25], are "fully developed." As shown by Eq. (8.11), the standard deviation of the intensity variations is equal to the mean intensity. In practice, speckle patterns often have a standard deviation that is *less* than the mean intensity, and this is observed as a reduction in the contrast of the speckle pattern. In fact, it is usual to define the speckle contrast as the ratio of the standard deviation to the mean intensity:

$$\text{Speckle contrast } \mathscr{C} = \sigma/\langle I \rangle \tag{8.14}$$

Thus for a fully developed speckle pattern, the contrast is defined to be unity:

$$\mathscr{C} = \sigma/\langle I \rangle = 1 \tag{8.15}$$

"Partially developed" speckle patterns (sometimes called "reduced" speckle patterns [14]) exhibit lower speckle contrast, with $\mathscr{C} < 1$.

The reduction in speckle contrast can be due to a number of causes. The most obvious are a reduction in the coherence of the light source or a reduction in the roughness of the surface. (It will be recalled that the initial assumptions when considering the statistics of speckle were that the coherence length of the source was long compared with the surface roughness, which was in turn greater than the wavelength of the light.) Other causes might be the addition to the speckle pattern of a uniform background of light, which may be either coherent with the speckle pattern or completely incoherent. It can be shown [26, 27] that if the ratio of the background intensity to the average total intensity of the speckle pattern plus background is ρ, then the speckle contrast is given in the coherent case by

$$\mathcal{C} = \sigma/\langle I \rangle = \sqrt{1 - \rho^2} \tag{8.16}$$

and in the incoherent case by

$$\mathcal{C} = \sigma/\langle I \rangle = 1 - \rho \tag{8.17}$$

8.5 SURFACE ROUGHNESS FROM SPECKLE CONTRAST MEASUREMENTS

8.5.1 Speckle Statistics and Surface Roughness

As there is no doubt that speckle patterns are caused by the roughness of a surface, it is reasonable to assume that the statistics of the speckle patterns produced might be used to provide a measure of this roughness. However, a glance at the formulas for both the first- and second-order statistics of far-field speckle patterns [Eqs. (8.10)–(8.14)] shows that in neither case does any parameter appear that is related to the surface roughness. In fact, the first-order statistics (speckle contrast) are the same for *all* rough surfaces (rough compared with the wavelength of light), and the second-order statistics depend only on the size of the illuminated area. *Thus far-field laser speckle patterns contain no information about the roughness of scattering surfaces that are rough compared with the wavelength of the light used* [5].

On the other hand, it is clear that a very smooth surface will not produce a speckle pattern at all: the incident light will be specularly reflected. It is reasonable to ask: What happens as the roughness of the surface is gradually increased?

8.5.2 Speckle Contrast from "Weak Scatterers"

The answer, of course, is that as the surface roughness is increased, the speckle pattern gradually appears, being at first only "partially developed," with speckle contrast $\mathcal{C}$ less than 1 (i.e., $\sigma < \langle I \rangle$). The speckle contrast will then steadily increase, approaching unity as the surface roughness approaches the wavelength of the light. (Note that this is not in contradiction to the formulas quoted in Section 8.2, because in the derivation of these formulas it was assumed that the surface roughness was large compared with the wavelength.)

In principle, then, it should be possible to measure surface roughness (variation of surface height) by measuring the contrast of the speckle pattern produced by it, providing the roughness does not exceed the wavelength of the light used. Such surfaces are usually referred to as "weak scatterers" [5].

8.5.3 A Model for Weak Scatterers

In practice, there is an additional complication in that the speckle contrast varies with scattering angle. Physically, this can be regarded as being due to the fact that some of the light is still being specularly reflected (or transmitted without scattering in the case of a transparent scatterer), and this light is providing a (coherent) background to the speckle pattern, thus reducing its contrast. (This is similar to a model that was proposed to explain the angular variation in speckle fluctuations when biological specimens are illuminated with laser light [28].) Equation (8.16) therefore applies:

$$e = \sigma/\langle I \rangle = \sqrt{1 - \rho^2}$$

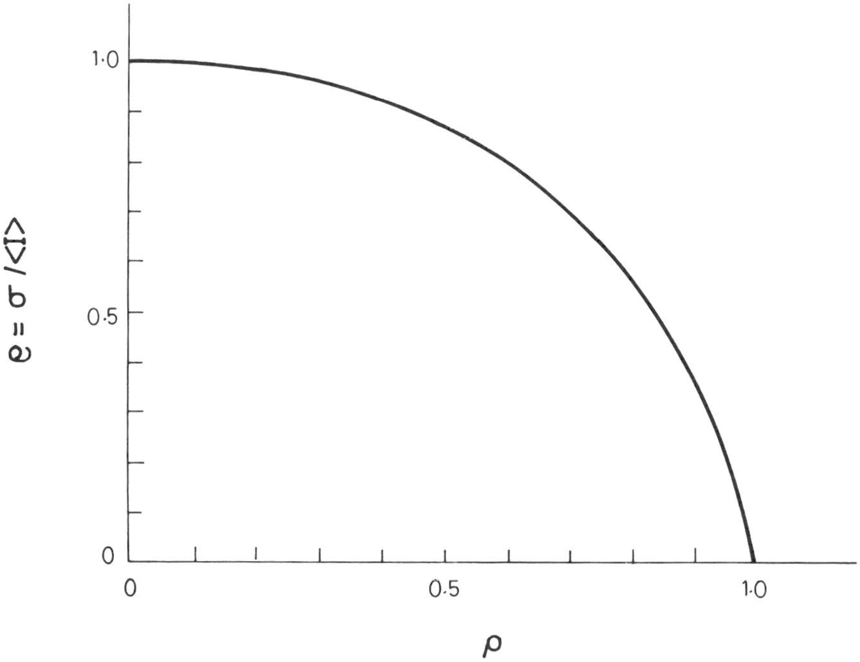

Figure 8.5 Variation of speckle contrast with coherent background.

In this equation ρ is the ratio of the intensity of the specular beam to the total intensity (specular plus scattered). A plot of $\mathscr{C}$ versus ρ is given in Fig. 8.5.

8.5.4 The Relationship Between Speckle Contrast and Roughness

There are two major problems in applying the foregoing theory to the measurement of surface roughness. First, the relationship between the parameter ρ and the surface roughness has to be established. This involves applying scattering theory (or, more fundamentally, solving Maxwell's equations for the particular geometry and polarization conditions used) and proves to be possible only if we make several simplifying assumptions. When this is done, as has been shown by Welford [5, 29, 30] and will be discussed in Section 8.9, the result points the way to a method of measuring surface roughness that does not actually involve the speckle contrast at all! However, the speckle effect itself *can* be used, and Pedersen has carried out a detailed analysis of the relationship between the surface roughness of weak scatterers and the speckle contrast, using the above approach of speckle pattern plus coherent background [25, 31]. This theory has been further developed, for both far-field and image speckle, by Asakura's group [32–38], and a detailed review of this work is given by Asakura [6].

8.5.5 Effect of Second-Order Statistics of Roughness

Asakura and his coworkers also showed, first in computer simulations of speckle patterns from weak scatterers of different roughnesses and later in experimental investigations, that the correlation length of the surface height variations—i.e., the spatial scale of the roughness—can also be deduced from speckle contrast measurements. This is achieved by measuring the speckle contrast as a function of the aperture of the imaging lens (in the image speckle case) or of the area illuminated (in the far-field speckle case). It is found that as this parameter is increased, the speckle contrast at first increases, reaches a maximum value, and then falls again. It is the *maximum* value reached by the contrast that is used in the calculation of the rms surface roughness, both theory and experiment showing an approximately linear relationship between these two quantities for rms roughnesses σ_R between 0.02 and 0.30 wavelength [6]. This maximum speckle contrast is found to occur when the diameter of the Airy disk (resolution cell) of the imaging lens in the image speckle case, or the effective diameter of the illuminated area in the far-field speckle case, is approximately twice the correlation length of the roughness. More practically, it is found that the speckle contrast when the diameter of this effective aperture is equal to the roughness correlation length is equal to very nearly 0.80 times the maximum speckle contrast.

8.5.6 Practical Realization of the Speckle Contrast Technique

Thus the technique is to find the effective aperture that gives the maximum contrast, use this value of contrast to give the rms roughness, then reduce the effective aperture until the speckle contrast is reduced to 0.80 times this maximum and take the diameter of this effective aperture as the correlation length of the roughness.

The second problem referred to above is the practical one of measuring speckle contrast. Speckle contrast is a form of "texture," and texture measurements are notoriously difficult to carry out. One method, which was adopted in a technique designed to measure blood flow by the reduction in speckle contrast caused by the flow [39], is to use spatial filtering techniques to convert the contrast variations to intensity fluctuations [40]. Real time techniques that involve modulating or scanning the speckle and measuring the depth of modulation are preferable. This was achieved in

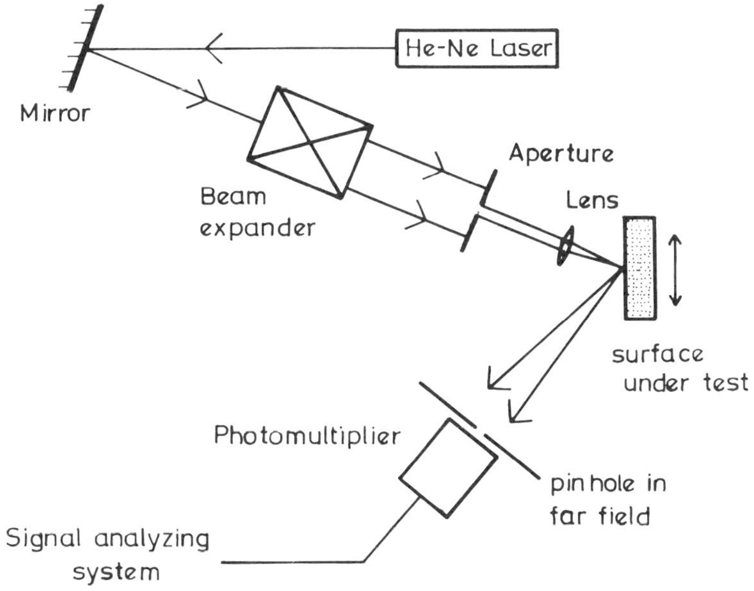

Figure 8.6 Schematic diagram of far-field speckle technique for surface roughness measurement. (After Asakura [6].)

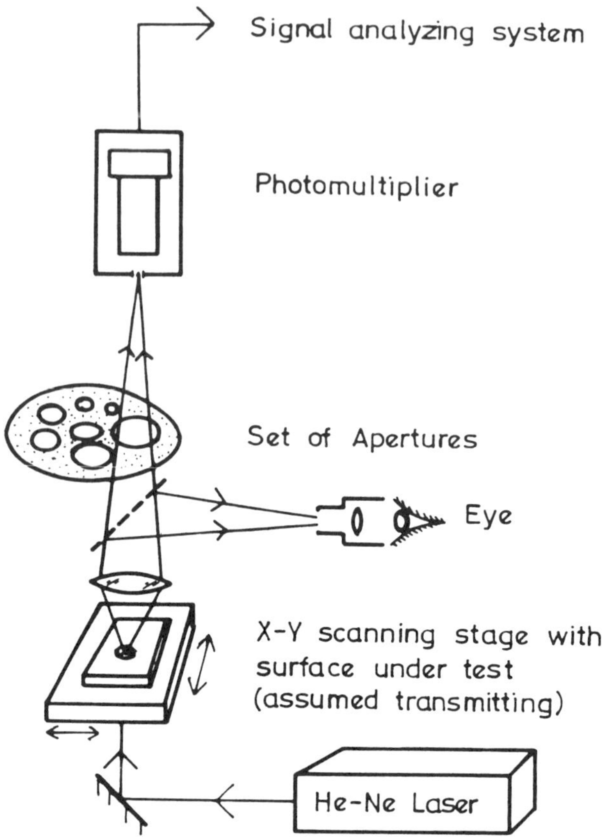

Figure 8.7 Schematic diagram of image-plane speckle technique for surface roughness measurement. (After Asakura [6].)

experiments carried out by Asakura's group by moving the test surface laterally, and laser speckle roughness meters, illustrated schematically in Figs. 8.6 and 8.7, were designed for both far-field and image speckle techniques [6]. The image speckle method gave the more convincing results, and an experimental curve of the maximum average speckle contrast as a function of rms surface roughness is reproduced in Fig. 8.8.

8.5.7 Dynamic Range of the Technique

As mentioned above, the range of rms roughnesses over which the speckle contrast is observed to vary is from about 0.02 to about 0.3 wavelength of

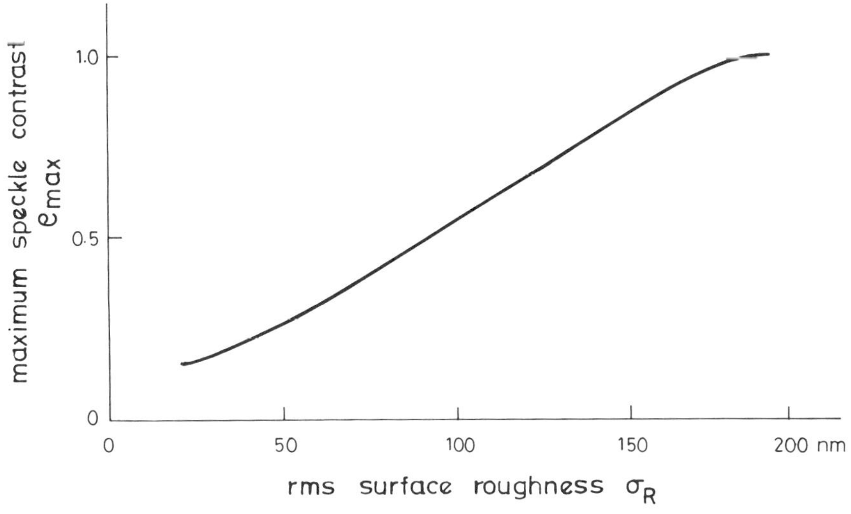

Figure 8.8 Experimental curve of maximum average image-plane speckle contrast against rms surface roughness. (After Asakura [6].)

the light used. If a helium-neon laser (wavelength 633 nm) is used, this implies values of σ_R between about 15 and 200 nm. The range can be extended downwards, to less than 10 nm, by using blue laser light instead of red, and even lower by using ultraviolet lasers. Similarly, the use of infrared wavelengths would allow the range to be extended upwards.

The range can also be extended upwards, and quite considerably, if oblique illumination is used. It is well known that even rough surfaces become specularly reflecting at large angles of incidence, and this fact has been used to extend the range of surfaces for which classical interferometry can be used [41–43]. In the same way, a cosine obliquity factor comes into play in the speckle technique for surface roughness measurement. Thus the upper limit on surface roughness is raised by a factor of 2 at an angle of incidence of 60° and by a factor of 4 at an angle of incidence of 75°. The rule of thumb for choosing an appropriate angle of incidence is that it must be sufficiently large for a specular component to be visible in the scattered light; this could be very simply assessed visually before any measurements are carried out. The only limitation is the practical difficulty of using these large angles of incidence.

Varying the wavelength and/or the angle of incidence, as described above, effectively introduces another degree of freedom into the technique. This, in fact, is the key to extending the range of surface roughnesses for which speckle techniques can be used. It can be argued that this is what

Asakura and coworkers did when they used a variable aperture in the method described above. In fact, as Welford argues [5], the use of a limited aperture means that they were operating in the non-Gaussian region. As discussed in Section 8.9.1, this means that the usual assumptions regarding Gaussian statistics are invalid, and a full mathematical treatment becomes extremely difficult. In practice, the technique devised by Asakura's group would require calibration on surfaces of known roughness. It is, however, the fact that they were in the non-Gaussian region that enabled them to include the measurement of the roughness correlation length in their technique.

Another way to introduce an additional degree of freedom is to relax the coherence requirements of the light source. This is the basis of the techniques to be described in the next section.

8.6 REDUCED-COHERENCE TECHNIQUES

8.6.1 Theory of Partially Coherent Speckle Patterns

As mentioned in the previous section, in order to use speckle statistics to measure surface roughness greater than the wavelength of the light used, it is necessary to introduce another degree of freedom. One technique is to use partially coherent light. Either the spatial or temporal coherence of the light source may be reduced, and both approaches have been used.

The mathematical formulation of the statistics of speckle patterns presented in Section 8.3 assumed that the illumination was perfectly coherent. The theory must be modified for the more general case of partial coherence, and several workers have contributed to this development [16, 19, 44–60]. A detailed review of this theory, including a general treatment of speckle patterns produced by light of arbitrary coherence, is given by Parry [3]. The effects of partial coherence are also mentioned in the reviews by Welford [5], Asakura [6], and Françon [7].

The statistics of partially coherent speckle patterns turn out to be rather complicated. However, the equations become more manageable if some simplifying assumptions are made. For example, if it is assumed that the light used is spatially coherent but that the frequency spectrum is of Gaussian shape with a finite spectral bandwidth σ_k [rms spectral deviation from the central value k $(= 2\pi/\lambda)$], that the light is incident normally on the surface, that the far-field speckle pattern is viewed in the specular direction, and that the surface height distribution is Gaussian, then the speckle contrast is given by [3, 47–49]

$$\mathcal{C} = \sigma_I/\langle I\rangle = 1/[1 + (4\sigma_k\cdot\sigma_R)^2]^{1/4} \qquad (8.18)$$

This function is plotted in Fig. 8.9.

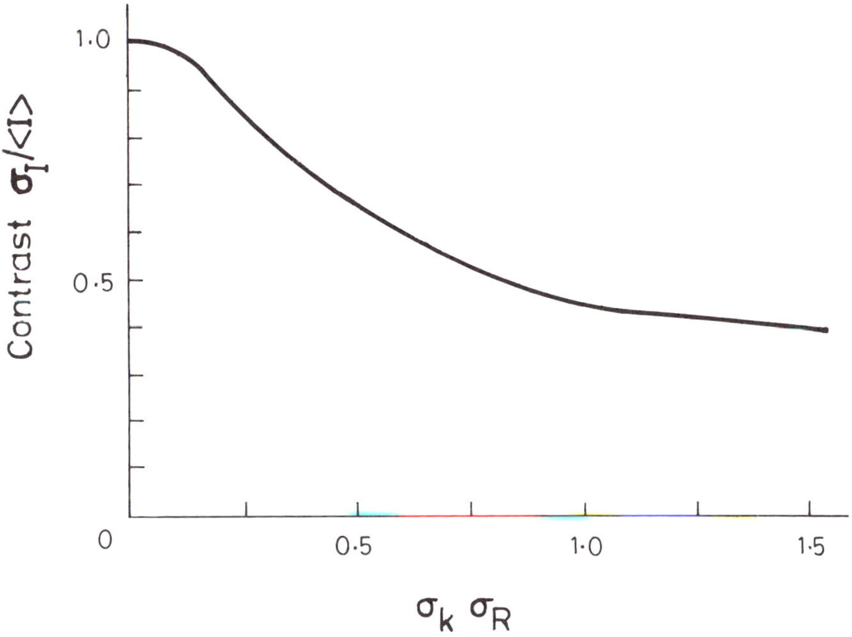

Figure 8.9 Speckle contrast as a function of surface roughness σ_R and spectral bandwidth σ_k: theoretical curve. (After Parry [3].)

Note that if the spectral bandwidth is *reduced*, the range of surface roughnesses over which contrast variations are recorded is *increased*. In other words, a narrower bandwidth gives a larger dynamic range for the technique (but with less precision of measurement). This apparently contradictory result will be discussed later, in Section 8.6.3, when dichromatic speckle patterns are considered.

The analysis can be extended to nonspecular directions if the further assumption of a Gaussian aperture function is made, the expression becoming slightly more complex.

Making similar assumptions, the second-order statistics can also be analyzed [3]. The results indicate that the speckles become elongated radially from the specular direction, giving rise to a "fibrous" appearance [61]. The degree of elongation depends on the distance from the center of the pattern (i.e., the specular direction), the spectral bandwidth of the light source, and the roughness of the surface. The rms angular elongation σ_α is given by [3, 48]

$$\sigma_\alpha = \{2\alpha^2\sigma_k^2/k^2[1 + (8\sigma_k\sigma_R)^2]\}^{1/2} \tag{8.19}$$

It is worth noting, as pointed out by Parry [3], that these radial structures were observed long before the advent of the laser—in the quasi-monochromatic diffraction patterns produced by particles and random apertures [62–68]. Also, they are observed only in the far-field speckle patterns and not in image speckles [52].

If the mutual coherence function of the illuminating light is included in the derivation of the speckle statistics, it is found that the statistics do depend on this function [3, 6, 31, 48–51, 53, 59, 60, 69–73] and that the use of partially coherent light affects the dependence of the speckle contrast on the surface roughness. Experimentally, however, the use of partial coherence is by no means easy [5], and practical applications have not, in general, been pursued.

8.6.2 Polychromatic Speckle

The first reports of the use of partially coherent speckle patterns to measure surface roughness are usually attributed to Sprague [44] and to Elbaum et

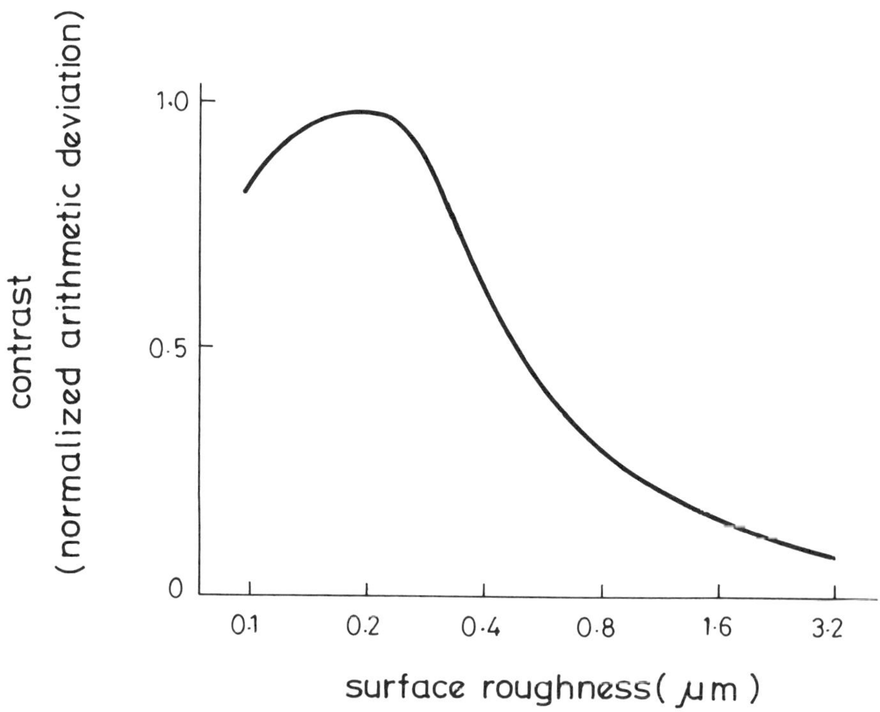

Figure 8.10 Contrast of polychromatic speckle as a function of surface roughness. (After Sprague [44].)

al. [45], although the ideas are based on the work of Goodman [16]. Sprague used a tungsten-zirconium arc as the source, not a laser, and was thus using quite broadband light with a coherence length of about 1.5 μm. His results are shown schematically in Fig. 8.10. Note that he defined contrast as the normalized *arithmetic* average deviation of the intensity, normalized with respect to the mean intensity, rather than the more usual rms definition. The results were independent of the method of preparation of the rough surface and suggest that the technique could be used for roughnesses between about 0.2 and 3 μm.

Instead of using light of a finite bandwidth, it is also possible to use a mixture of several discrete laser lines. This approach was adopted by Parry [3, 51], who proposed the use of a multimode argon laser, which has an output consisting of six dominant lines in the blue and green.

The theoretical relationship between the speckle contrast and the surface roughness was calculated by Parry and is reproduced schematically in Fig. 8.11. This suggests that the multiwavelength technique could be used for roughnesses between about 0.5 and 2.0 μm, at which point a saturation effect is found.

The above theoretical and empirical results indicate that the first-order statistics (contrast) of polychromatic speckle can be used to measure rms

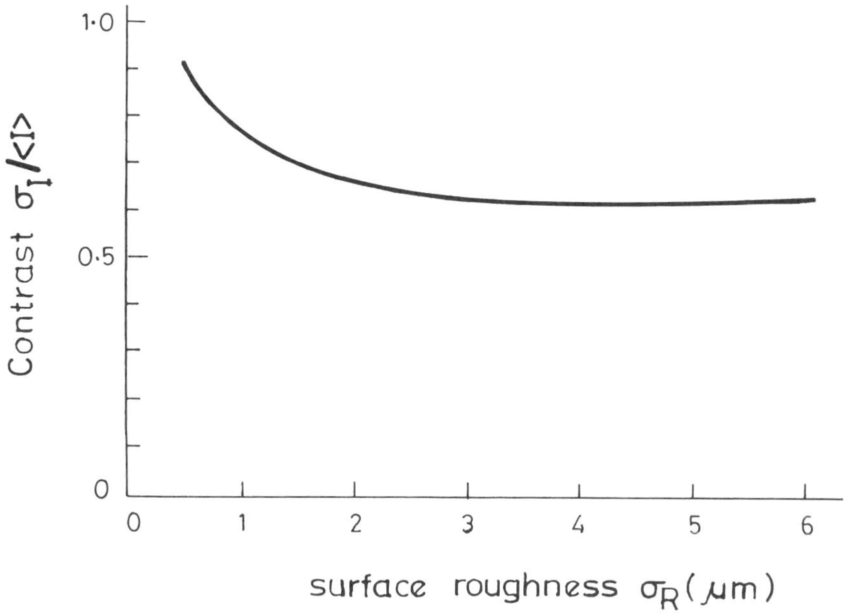

Figure 8.11 Variation of speckle contrast with surface roughness using a multimode argon laser: theoretical curve. (After Parry [3].)

surface roughnesses up to 2 or 3 μm. (This restriction, however, arises only because broadband light was used by Sprague and because *all* the argon laser lines were used by Parry. As was mentioned in Section 8.6.1, and as will be discussed in Section 8.6.3, the range of roughnesses that can be measured can be extended, virtually at will, by *limiting* the range of wavelengths used.)

The fibrous radial structure observed in far-field polychromatic speckle is easy to observe but less easy to measure. A simpler approach is to use just two wavelengths and measure the correlation between the two speckle patterns formed. These correlation techniques are described in Section 8.7.

8.6.3 Dichromatic Speckle Patterns

Instead of using light with a finite bandwidth, or the multiwavelength output of a multimode laser, it is possible to use just two discrete wavelengths. Fujii and Lit [74] defined an rms difference between the two normalized far-field speckle intensity distributions produced by the two spectral components:

$$V(\xi_1, \xi_2; k_1, k_2) = \left\langle \left[\frac{I(\xi_1, k_1)}{\langle I(\xi_1, k_1)\rangle} - \frac{I(\xi_2, k_2)}{\langle I(\xi_2, k_2)\rangle} \right]^2 \right\rangle^{1/2} \tag{8.20}$$

(The square of this function is, in fact, the so-called structure function often used to describe the magnitude of fluctuations in random media [75].) Considering only the specular direction, i.e., putting $\xi_1 = \xi_2 = 0$, Fujii and Lit then showed that this function displayed a simple dependency on the rms surface roughness σ_R and on the spectral difference between the two illuminating wavelengths, represented by $|k_1 - k_2|$:

$$V(0, 0,; k_1, k_2) = \{2[1 - \exp(-\sigma_R^2(k_1 - k_2)^2/4)]\}^{1/2} \tag{8.21}$$

[Note that a printing error led to the factor of 2 in the above expression being omitted in Eq. (14) of the paper by Fujii and Lit [74]. Also, the factor of 4 that appears in Eq. (8.21) is absent from the version given by Fujii and Lit; this is because the rms optical path difference used by Fujii and Lit has been replaced by the rms surface roughness.] The function is plotted in Fig. 8.12.

Subsequent experimental results agreed reasonably well with this theoretical curve. The function is seen to be approximately linear over the range $\sigma_R|k_1 - k_2| = 0$ to $\sigma_R|k_1 - k_2| = 0.5$. This limit of $\sigma_R|k_1 - k_2| = 0.5$ can be taken to be the upper limit for which the technique can be used. If an argon ion laser is used, the two extreme wavelengths of 458 and 514 nm give the value of $|k_1 - k_2|$ as

$$|k_1 - k_2| = 2\pi[1/\lambda_1 - 1/\lambda_2] = 1.5\ \mu\text{m}^{-1} \tag{8.22}$$

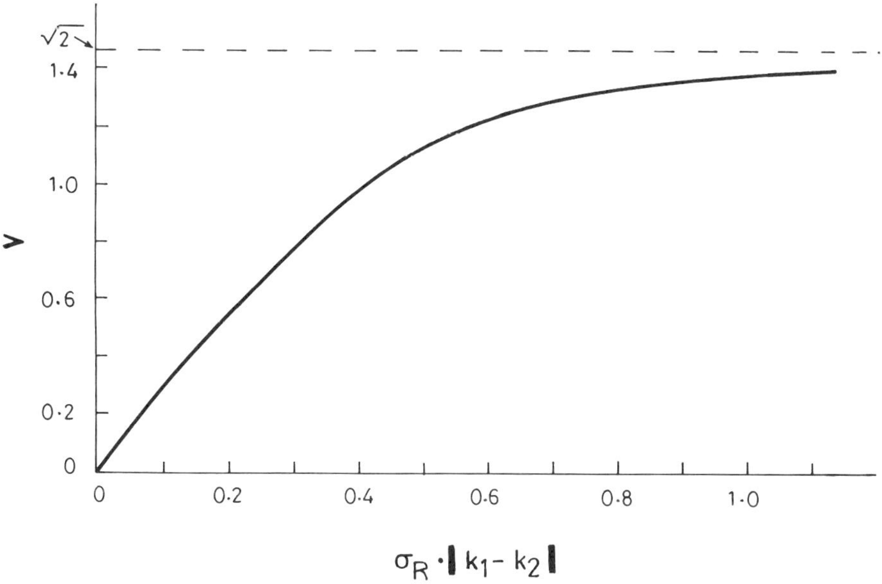

Figure 8.12 Theoretical curve of the function V versus the product of rms roughness σ_R and spectral difference $|k_1 - k_2|$. (After Fujii and Lit [74].)

and the limiting value of σ_R is then $0.5/1.5 = 0.33$ μm. Taking the two closest wavelengths of 497 and 502 nm, on the other hand, leads to a value for $|k_1 - k_2|$ of 0.13 μm^{-1} and a corresponding limit for σ_R of 3.8 μm. Thus to measure rougher surfaces, the two wavelengths should be chosen to be as close together as possible. This is the same trend that was encountered in the case of polychromatic speckle discussed in Sections 8.6.1 and 8.6.2. Although at first sight unexpected, it is, in fact, correct and also arises in the use of two-wavelength interferometry [76] or holography [77] for measuring the shape (form or figure) of steep aspheric surfaces or of rough surfaces [78–83]. In this case the effective wavelength, Λ, is given by

$$\Lambda = \lambda_1\lambda_2/|\lambda_1 - \lambda_2| \tag{8.23}$$

and again the technique is most "desensitized" (for measuring aspherics) or can handle rougher surfaces when the two wavelengths are closest together. For the dichromatic speckle technique, then, as well as for the polychromatic speckle techniques, the range of surface roughnesses that can be measured can be extended at will by restricting the bandwidth of the light used. However, the precision with which the roughness can be measured is correspondingly reduced.

8.6.4 Speckle Patterns in Partially Spatially Coherent Light

The use of partially spatially coherent light certainly affects the contrast of speckle patterns [60, 69, 72, 73], but varying the degree of coherence has only a limited effect on the range of surface roughnesses for which the contrast varies. Thus there is no real advantage in using spatial coherence as a variable for the measurement of surface roughness.

8.7 CORRELATION TECHNIQUES

8.7.1 Use of Two Wavelengths

As was mentioned in Section 8.6.2, the fibrous radial patterns observed in polychromatic speckle patterns are difficult to analyze directly. A better technique is to restrict the illumination to just two wavelengths and measure the correlation between the two speckle patterns formed. It is found that the rate at which the speckle patterns decorrelate as the two wavelengths are moved farther apart depends on the surface roughness. The decorrelation of speckle patterns plays a role in several other applications of laser speckle [84], and an elegant method of measuring the correlation of two speckle patterns is to displace the two patterns when they are recorded (on photographic film, for example) and carry out a Fourier transform on the combined pattern (for example, by observing its far-field diffraction pattern). If the two speckle patterns are fully correlated, high-contrast fringes will be observed; if they are completely decorrelated, no fringes will be produced; and in general the fringe contrast will be a measure of the degree of correlation between the two speckle patterns [84–86]. This approach was used by Tribillon [87], who showed that complete decorrelation implies that

$$\sigma_R > \lambda^2/2\,\Delta\lambda \tag{8.24}$$

The technique, then, is to increase the value of $\Delta\lambda$, the difference between the two wavelengths, until the fringes in the far-field diffraction pattern produced by the two superimposed speckle patterns just disappear, and then insert this value of $\Delta\lambda$ into Eq. (8.24) to give σ_R. If a dye laser is used as the light source, $\Delta\lambda$ can be varied gradually and over a reasonable range, allowing roughnesses to be measured over a range from about 0.5 μm upwards. A modification by Tribillon [7, 87] of the technique suggested by Mendez and Roblin [88] uses two fixed wavelengths and deduces the value of σ_R from the contrast of the fringes. A related technique, but one that uses the contrast of fringes formed on two-wavelength holographic interferograms, was suggested by Ribbens [89]. This was an extension of an earlier proposal of his [90], suitable for weak scatterers, that used single-exposure holographic interferograms.

The main disadvantage of these two-wavelength correlation techniques is that they are necessarily two-stage processes and therefore do not allow real-time measurements to be made.

8.7.2 Use of Two Angles of Incidence

In this method, proposed by Léger et al. [7, 91], two speckle patterns obtained with two different angles of incidence of the laser beam are recorded on the same photographic plate. The two speckle patterns will be displaced with respect to each other, and the far-field diffraction pattern of the combined pattern will again show the characteristic Young's fringes. However, once again the two speckle patterns will decorrelate if the surface is sufficiently rough, and the fringes will lose contrast. It can be shown that if θ and $\theta + \Delta\theta$ are the two angles of incidence, then the fringe contrast γ is given by

$$\gamma = \exp\left[-\left(\frac{2\pi\sigma_R}{\lambda}\sin\theta\,\Delta\theta\right)^2\right] \tag{8.25}$$

Hence σ_R can be deduced by measuring the fringe contrast γ.

This method has been used by Dalmases et al. [92] to measure the roughness of skin—or rather of an impression or replica taken from the skin, as the microscopic movements of the skin itself rendered direct measurements impossible. Their version of the experimental arrangement is illustrated in Fig. 8.13, and their setup for measuring the fringe contrast in the Fourier plane of the resulting double-exposure photograph of the two speckle patterns is illustrated in Fig. 8.14. They found that the lower limit of rms roughness for the technique was about 4 μm, and they claimed a precision of around 10%. Root-mean-square roughnesses of 10–15 μm were measured for the skin of the hand, and it was suggested that this noncontact technique could be of value in dermatology.

The technique suffers from the same disadvantage as the two-wavelength method discussed in Section 8.7.1: real-time measurement is not possible. However, real-time operation *is* possible if the intensity correlation is replaced by an amplitude correlation. This can be done [7, 93] by illuminating the surface simultaneously with two beams from the same laser but at slightly different angles of incidence; this can be achieved by means of an interferometer such as a Michelson. The two speckle patterns will normally be displaced but may be brought back into superposition by using a second identical interferometer in the diffracted beams. The arrangement used by Léger [93] is illustrated in Fig. 8.15.

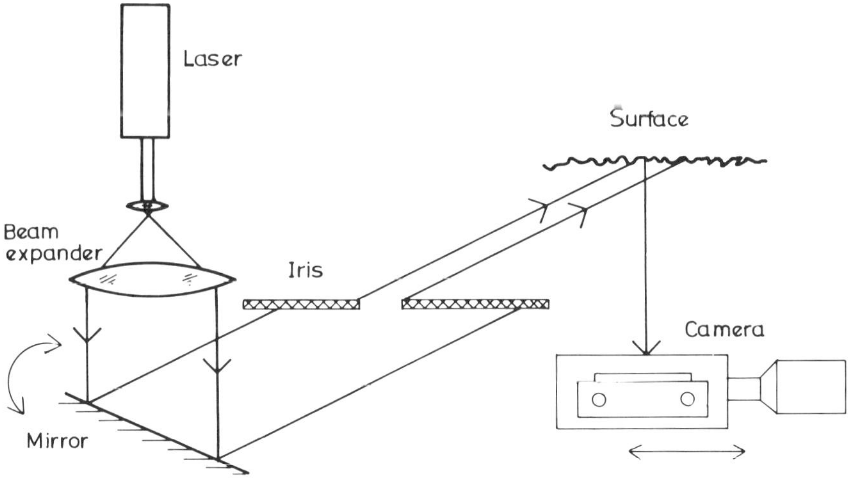

Figure 8.13 Arrangement for recording speckle patterns at two angles of incident illumination. (After Dalmases et al. [92].)

If the second interferometer is not adjusted for a zero path difference, the two speckle patterns will be displaced along the beams, and circular interference fringes will be observed if the beams are focused with a lens. The contrast of these fringes, γ, is related to the rms surface roughness σ_R as follows:

$$\gamma = \frac{1}{2} \exp \left[-\left(\frac{2\sqrt{2}\pi\sigma_R}{\lambda} \sin \Delta\theta \right)^2 \right] \tag{8.26}$$

A more practical method of measuring the fringe contrast is to readjust the second interferometer for a zero path difference, in which case the circular fringes in the focal plane of the lens are replaced by a uniform intensity—effectively, a single fringe fills the field of view. If a detector is now placed in this plane and the path difference between the two interfering beams is modulated (for example, by means of a piezoelectric transducer attached to one of the mirrors of the interferometer), then the intensity recorded by the detector is modulated as the path difference varies, and the depth of modulation of this signal is proportional to the fringe contrast and can be calibrated to yield the same information.

The main disadvantage of this technique is the difficulty of keeping the Michelson interferometers in adjustment. If the restriction of always using the same angle between the two illuminating beams can be tolerated, then one way to overcome this problem is to use Wollaston prisms instead of the Michelson interferometers [7]. As these devices are quite robust, their

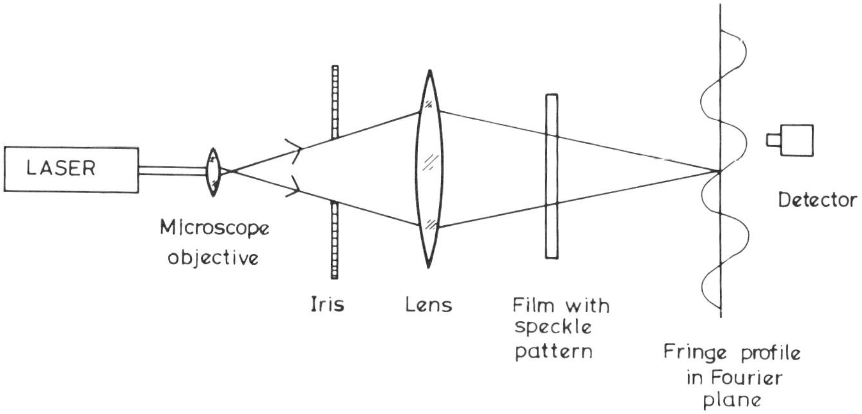

Figure 8.14 Arrangement for measuring the degree of correlation between two speckle patterns. (After Dalmases et al. [92].)

use leads to a very stable and practical technique. In addition to the restriction of fixing the angle between the two beams, however, there is also the problem that some surfaces will depolarize the light and render the method inapplicable. In practice, however, this does not appear to be a major problem.

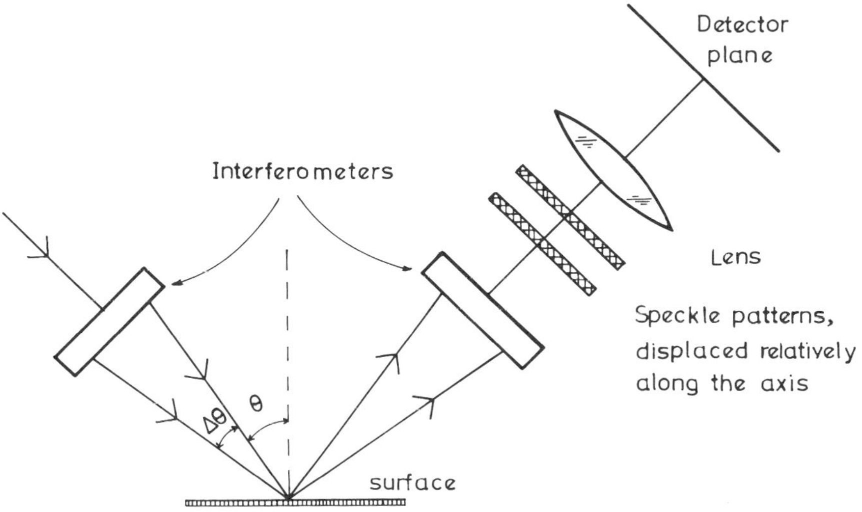

Figure 8.15 Arrangement for speckle amplitude correlation technique using two angles of incidence. (After Léger [93].)

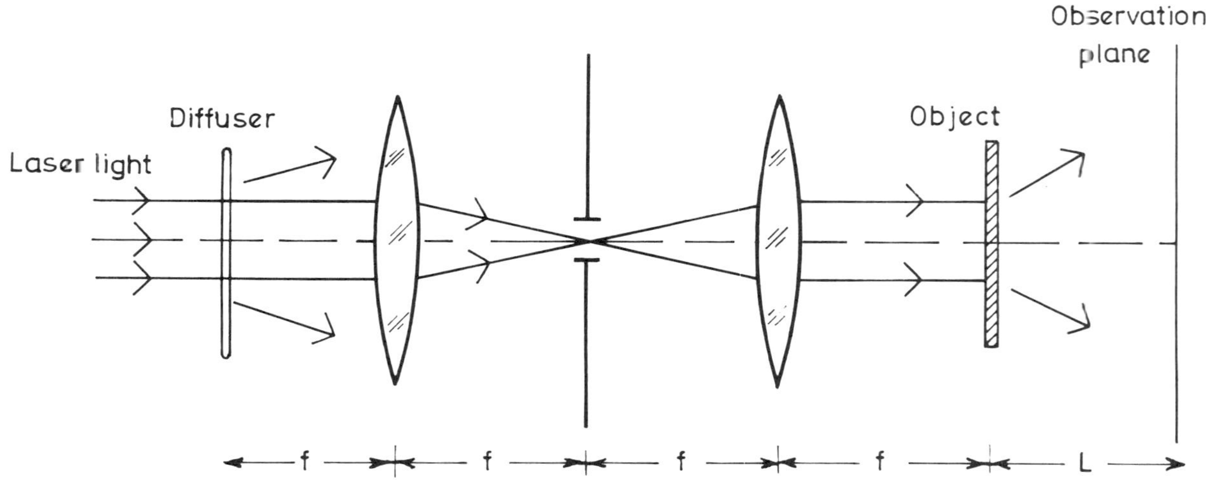

Figure 8.16 Speckle pattern illumination for roughness measurement. (After Yoshimura et al. [96].)

8.7.3 Use of Speckle Pattern Illumination

A rather different approach to the correlation technique was adopted by Yoshimura et al. [94–96]. They illuminated the surface under examination with a speckle pattern produced by illuminating a diffuser with laser light. A variable aperture in the optical system allowed the speckle size in the illuminating beam to be varied, thus introducing the additional degree of freedom necessary. Their experimental arrangement is illustrated in Fig. 8.16. As presented, the method is applicable to transmitting diffusers such as ground glass, but presumably it could also be applied in reflection to opaque surfaces.

A mathematical relationship was derived that connects the correlation function of the far-field speckle pattern produced in the observing plane with the speckle size of the illuminating beam and of both components (specular and diffuse) of the light scattered by the surface under observation, and with the rms surface roughness and the correlation length of the roughness. The conclusion was that the speckle contrast in the final speckle pattern is always equal to 1 and is independent of the surface roughness, but that the speckle size does depend on the surface roughness, the speckles becoming smaller as the surface becomes rougher. The dependence is, however, also affected by the correlation length of the roughness.

A related technique was proposed by Dzialowski and May [97], who showed that the distance a surface can be displaced axially before correlation of the far-field speckle pattern is lost is related to the correlation length of the scatterers on the surface.

8.8 OTHER ASPECTS OF SPECKLE

8.8.1 Non-Gaussian Speckle

The theory of the statistics of speckle patterns presented in Section 8.4 assumes that a sufficient number of scatterers contribute to the speckle pattern for the central limit theorem [98] to be valid [2]. If this is not the case, and there are only a few scatterers in the illuminated area, then the theory of speckle statistics becomes more complicated. This is the case of so-called non-Gaussian speckle, which has been studied by Jakeman and Pusey [99–102]. The use of non-Gaussian speckle extends the range of surface roughnesses for which the speckle contrast is a function of roughness, and, as was explained by Welford [5], this was effectively what Asakura's group was doing when the illuminated area was controlled by aperture stops [6]. However, as pointed out by Dainty [1, second edition only], the speckle contrast in the non-Gaussian case is also very sensitive to the form of the surface roughness correlation length, and this tends to complicate any method that uses non-Gaussian statistics.

8.8.2 More Rigorous Scattering Theories

The scalar diffraction theory used to derive the formulas of scattered light and in particular of laser speckle statistics is, of course, an approximation. More rigorous formulations have been tried, although in general they do not lead to any improvement in practical techniques for the measurement of surface roughness [5]. Examples include a vector theory of reflective scattering from surfaces with finite conductivity [103], which reduces to Beckmann's formula as a special case, and a scattering theory that takes into account evanescent waves (surface plasmons) [104–107].

8.9 NON-SPECKLE LIGHT-SCATTERING TECHNIQUES

8.9.1 A Model for Scattering from Relatively Smooth Surfaces

In their investigation of the possibilities of the direct use of speckle for measuring surface roughness, Chandley and Welford [5, 30, 108–111] reformulated the classical Beckmann theory of scattering from rough surfaces [112, 113]. They used a "thin phase screen" model [114] to represent surfaces whose roughness was less than the wavelength of the light. As discussed in Section 8.5.3, the scattered light from such a "weak scatterer" can be regarded as being made up of two components, one scattered and one specular (or unscattered), which add coherently to form the observed speckle pattern. One result of this is an angular variation in speckle contrast across the pattern. As Welford points out, it is then relatively easy to separate out the components; one way to achieve this is shown in Fig. 8.17.

The specular component is focused by the second lens and produces an intensity peak A on the screen. The scattered component gives rise to a background speckle pattern. It is easy to see that the distribution of phase in the light scattered by the surface, $\phi(x, y)$, is related to the distribution of surface height on the surface, $z(x, y)$, as follows:

$$\phi(x, y) = (4\pi/\lambda)z(x, y)\cos\theta \tag{8.27}$$

where θ is the angle of incidence.

8.9.2 Surface Roughness Variance from Intensity Measurements

Kirchhoff diffraction theory can now be used to show that in the far field (on the screen in Fig. 8.17) the ratio of the intensity of the specular com-

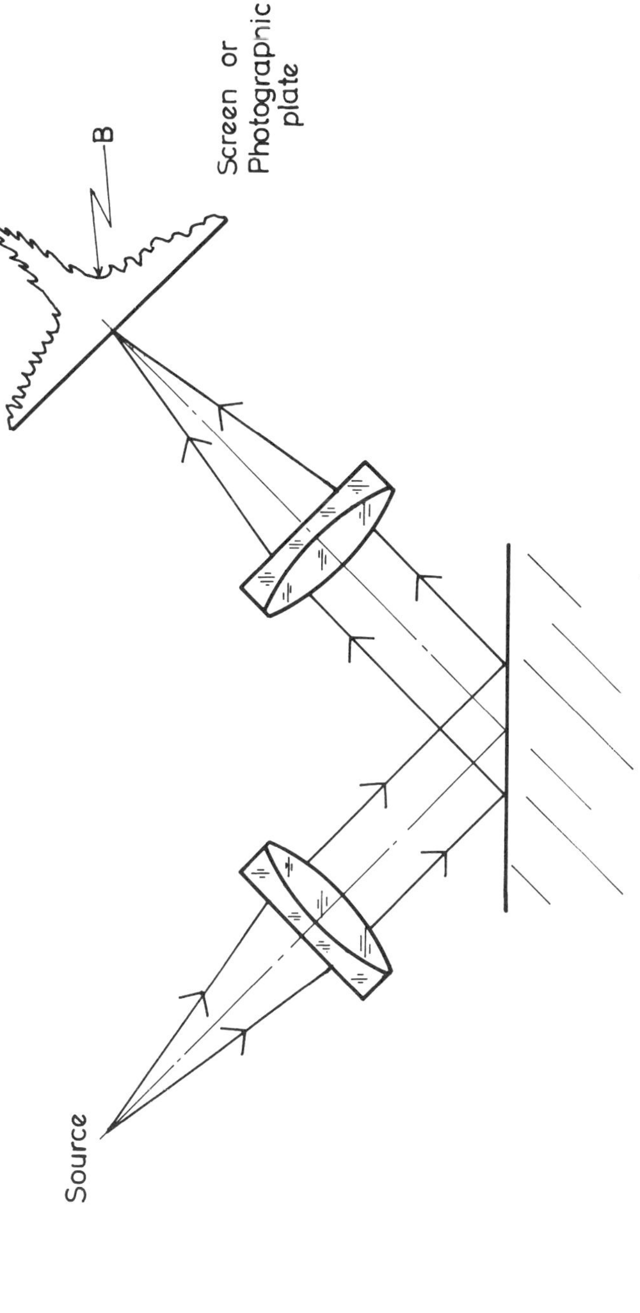

Figure 8.17 Measurement of specular and scattered light.

ponent I to the intensity of the specular reflection from a perfectly smooth (mirror) surface, I_0, is given by

$$I/I_0 = \exp(-\sigma_\phi^2) \tag{8.28}$$

where σ_ϕ^2 is the variance of the phase distribution in the scattered light.

Equations (8.27) and (8.28) can now be used to obtain the variance of surface height, which is the most commonly used measure of surface roughness.

In applying this technique to measure surface roughness, it must be remembered that I is the intensity of the specular peak *above the background speckle*. Hence the intensity near the specular peak—at point B, say, in Fig. 8.17—must be measured, averaged (since the background is a speckle pattern), and subtracted from the intensity at A to give the required value of I. For convenience, this technique will be referred to as the "peak/background" technique.

A practical application of a technique using the above concept was reported recently by Theocharous [115, 116]. An online noncontact sensor was developed to monitor the surface roughness of silicon carbide monofilaments of diameter 100 μm being produced by a CVD process.

8.9.3 Extension to Rougher Surfaces

As in the case of the direct use of speckle contrast discussed in Section 8.5, the peak/background technique described in Section 8.9.2 is in principle restricted to variations in surface height (roughness) less than the wavelength of the light used. However, the techniques described in Section 8.5.7 can be used to "desensitize" the method and extend its range of applicability. Again there is a cosine obliquity factor [see Eq. (8.27)] involving the angle of incidence θ, and by using high angles of incidence (approaching grazing incidence) the surface roughness can effectively be scaled down. The use of shorter wavelengths is again a possibility, and an additional method for *transmitting* diffusers, that of immersing the surface to be measured in a liquid of near-matching refractive index, was suggested by Welford [5, 30].

8.9.4 The Limited Role of "Speckle"

It will be noted that although the peak/background technique and its theory have been developed using the ideas of speckle, the only quantities that have to be measured in order to apply the technique are intensities, and the speckle structure is not actually involved at all! Hence there is no need to use the highly coherent light of a laser, and thermal sources can be used. In fact, there is even an advantage in using light of less coherence, as the absence of fully developed speckle in the background makes the measurement of the background intensity (at point B in Fig. 8.17) easier (there is no need to carry out the averaging). [Some restriction on the spectral bandwidth of the source must, however, be maintained, because the effect is wavelength-dependent, as can be seen from Eq. (8.27).]

8.9.5 Measuring the Second-Order Statistics of Surface Roughness

The peak/background technique gives the variance of the surface height variations, which is a first-order statistic of the surface roughness. It is also possible to extend the method to measure the second-order statistics of the roughness, that is, to determine how rapidly the surface height varies from point to point on the surface. A measure of this is the autocovariance of surface height, which is defined as follows [Eq. (8.7)]:

$$\begin{aligned} C_R(\delta x, \delta y) &= \langle [z(x, y) - \langle z \rangle][z(x + \delta x, y + \delta y) - \langle z \rangle] \rangle \\ &= \langle z(x, y) z(x + \delta x, y + \delta y) \rangle - \langle z \rangle^2 \end{aligned}$$

Now the autocovariance of intensity in the far-field speckle pattern is given by the Fourier transform of the intensity distribution of the scattered light (see Section 8.4.2), and it seems reasonable that there will be a connection between the latter, expressed, say, as the autocovariance of the complex amplitude of the scattered light leaving the surface, and the autovariance of the surface height. This is not as simple to prove as the relationship between the variances. However, with the aid of several simplifying assumptions, most notably that the speckle pattern is symmetrical about its center and that the distribution of surface heights is Gaussian, it can be shown [30, 110] that the autocovariance of the surface height is given in terms of the normalized autocovariance of the complex amplitude of the scattered light and the variance of surface height:

$$C_R(\delta r) = 1 + \frac{\ln \{c_A(\delta r)[1 - \exp(-\sigma_R^2)] + \exp(-\sigma_R^2)\}}{\sigma_R^2} \tag{8.29}$$

where $\delta r^2 = \delta z^2 + \delta y^2$; $c_A(\delta r)$, the normalized autocovariance of complex amplitude, is equal to $C_A(\delta r)/C_A(0)$; and σ_R^2 is the variance of the surface height distribution.

Now the autocovariance of the complex amplitude in the scattered light, of course, merely gives an indication of the width of the central peak in the far-field speckle pattern. Hence, once again, the structure of the actual speckle pattern is not involved, and there is no need to use highly coherent light.

8.9.6 Summary of the Peak/Background Scattered Light Technique

The conclusions can be stated as follows [30]:

1. When the surface height variation (the surface "roughness") is less than the wavelength of the light, there is a central peak of intensity in the light scattered into the far field.
2. The height of this central peak, compared with the background intensity close to the peak and with the intensity that would be produced by a specular surface of the same reflectivity, can be used to calculate the variance of the surface height distribution (i.e., the first-order statistics of the surface roughness).
3. The width of the central intensity peak can be used to calculate the autocovariance of the surface height distribution (i.e., the second-order statistics of the surface roughness).
4. As the structure of the speckle pattern is not actually used in either calculation, there is no need to produce a speckle pattern and no need to use coherent light, and the technique is, in fact, not really a speckle technique at all! (Though it has been analyzed in terms of the mathematics used to describe speckle statistics.)
5. There are the usual possibilities of extending the technique to surfaces whose roughness is greater than the wavelength of the light used—for example, by using oblique incidence illumination, shorter wavelengths, or immersion liquids.

8.9.7 Other Scattering Techniques

Other light-scattering techniques for the measurement of surface roughness, which have no recourse to the concept of speckle, have also been developed. Although it is outside the brief of this book to describe them in detail, the most important are mentioned here for the sake of comparison. An excellent account of these and other methods of measuring surface roughness is given by Bennett and Mattsson [4].

The two most important developments have been the total integrated scattering (TIS) and angle-resolved scattering (ARS) techniques. Based on what is usually called Beckmann theory [112, 113], but developed independently by Bennett and Porteous [117, 118] from the earlier work of Davies [119], the TIS method involves measuring the ratio of the light scattered into a complete hemisphere to the total light (scattered plus specularly reflected) returning from the illuminated surface. The method has now been embodied in a U.S. standard for the measurement of surface roughness [120]. Traceability is achieved by calibrating the instrument with reference surfaces.

The ARS method [121–123] is considerably more complex, involving the measurement of the scattered light as a function of the scattering angle, and taking into consideration the polarization state of both the incident and scattered light. The method allows the measurement of the spatial structure of the surface roughness as well as the rms height variations. However, there are still some discrepancies between theory and experiment [124, 125], and more work needs to be done in this area. Several instruments have been designed to measure ARS, including the detector array for laser light angular scattering (DALLAS) developed at the U.S. National Institute of Standards and Technology [126, 127].

Another scattered light parameter that has gained some prominence in recent years is the bidirectional reflectance distribution function (BRDF) and its transmission counterpart, the BTDF [128–130]. These functions are identical to the ARS function, apart from an obliquity factor equal to the cosine of the scattering angle, as a result of the use of radiometric terms in their definition. BRDF and BTDF can therefore be regarded as variants of ARS.

Finally, an instrument developed for the objective measurement of scratch-and-dig surface defects also has the potential to be used for the measurement of surface roughness. Called the microscope image comparator (MIC) [131], it operates by measuring the amount of light that has been removed from the specular beam—by whatever process—by comparing the remaining intensity in the main beam with a reference.

There is still much work going on in both the theory and the practice of measuring surface roughness by means of scattered light measurements. Just a few examples are given here as an indication of this continuing activity. A major problem on the theoretical side is the effect of multiple scattering: the phenomenon of enhanced backscatter has been observed and explained [132–134], and attempts are being made to complete the theory of multiple scattering [135, 136]. Several investigations into the relationship between roughness parameters calculated from scattering techniques and those measured by stylus instruments have been carried out, a typical example being the detailed investigation by NIST into the DALLAS ARS instrument [137]. An interesting combination of interferometry and intensity measurements, performed on the light reflected and scattered at the specular angle, has been proposed by Angelsky and Maksimyak [138] as a means of characterizing "slightly rough" surfaces.

Meanwhile, so-called scatterometers, previously confined to standards laboratories because of their high cost and difficulty of use, are being miniaturized and made more rugged and are becoming commercially available. At least one manufacturer now offers a hand-held device capable of determining rms surface roughness from a single TIS or BRDF/BTDF measurement [139]. (These devices are much more sophisticated than the so-called glossmeters, which have been available for some time. The latter usually rely on the measurement of the light scattered in one or two fixed directions, one of these being normally the specular direction, and empirical calibration against stylus measurements. Several instruments of this type have been described by Westbury [140].)

Specific applications are also being reported, such as the determination of the surface roughness of dental fillings from measurements of the intensity of specular reflection described by Bouchard et al. [141].

8.10 OTHER METHODS OF MEASURING SURFACE ROUGHNESS

8.10.1 Other Optical Techniques

Laser speckle is not, of course, the only optical technique available for measuring surface roughness. As described in the previous section, the analysis of scattered light need not involve the speckle phenomenon. Indeed, the inherent advantage of noncontact methods for measuring surface roughness has led to the development of a variety of optical techniques. To put the speckle methods described in this chapter into context, some of these developments are outlined briefly here.

Closely related to the speckle and scattered light techniques discussed in Sections 8.4–8.9 are methods based on diffraction. There are many examples in the literature, and only two will be mentioned here. An instrument based on extracting features from the Fraunhofer diffraction pattern of the surface was described by Brodman et al. [142], and a more sophisticated approach, using a Bayesian classification of surfaces based on information extracted from the two-dimensional diffraction pattern and incorporating a learning technique to recognize predefined classes of surfaces, has been proposed by Milana et al. [143].

Optical interferometry, well established for measuring the surface form or figure of optical components [8, 144], is an obvious candidate for the noncontact measurement of surface roughness, but the problem for deterministic measurements lies in achieving the lateral resolution necessary to resolve the roughness structure. One solution is to combine interferometry with microscopy, and this has led to the development of the science of "optical profilometry," a noncontact analog of the mechanical stylus. Commercial instruments are now available, typically based on the Mirau, Michelson, and Linnik interferometers [145–147], on common-path interferometers [148–150], and on the Nomarski polarization microscope [151–153]. An excellent review of optical profilometry techniques, including details of the commercial instruments available at the time (1989], is included in the book by Bennett and Mattsson [4].

Some optical profilometers have been developed into mapping instruments, giving the profile of the surface over areas whose sides vary from a few micrometers to over 10 mm in length. They also compute surface height data and other surface statistics. Several of these are described by Bennett and Mattsson [4]. For greater resolution, even atomic resolution if this is required, topographical maps can also be produced by scanning tunneling microscopy (STM), atomic force microscopy (AFM), and electron microscopy (TEM, SEM, and STEM). Again, these techniques are reviewed by Bennett and Mattsson [4].

Several other optical techniques for measuring surface roughness have been reported in the literature, including the use of FECO fringes (fringes of equal chromatic order) [154], ellipsometry for metal surfaces [155, 156], and, borrowing an idea used in the pickup of optical disk players, the detection of astigmatic focus errors [157].

8.10.2 Stylus Techniques

The classical tool for measuring surface roughness is, of course, the mechanical stylus. The technique is well established, is well understood, and remains the standard against which all other methods are compared and judged. The resolution of stylus instruments is surprisingly high and easily competes with that of optical techniques. As mentioned in Section 8.2.6, there can be a problem in that its response is not symmetrical—the stylus tip has a finite radius and tends to exaggerate the width of sharp peaks in the surface as it rides over them, whereas it fails to find the bottom of similarly shaped valleys. Its main disadvantage, though, is that it is a contact device and there are many instances where it is not advisable, or even feasible, to contact the surface.

8.11 LASER SPECKLE COMPARED WITH OTHER TECHNIQUES

8.11.1 Summary of Speckle Techniques

To compare the various speckle techniques discussed in this chapter, an attempt will now be made to summarize them in terms of their dynamic range (rms surface roughness they can handle) when visible light is used. Where applicable, their ability to measure the roughness correlation length and their ease of use as a practical technique are also mentioned.

For strong scatterers, those with rms surface roughnesses exceeding half the wavelength of the light used, the statistics of fully developed far-field speckle patterns are independent of the surface roughness. For weak scatterers, on the other hand, the presence of a specular peak superimposed on the speckle pattern modifies the speckle statistics around the specular direction and allows roughnesses in this range—say 10–200 nm rms roughness—to be measured.

The use of image speckle, with the size of the scattering area determined by the aperture of the imaging system, or the deliberate use of apertures in the far-field technique leads to operation in the non-Gaussian region. This requires a much more complex theory and results, for all practical purposes, in techniques that must be calibrated against other, better understood methods. On the other hand, there are benefits in that the range of measurable roughnesses can be extended slightly and the roughness correlation length can also be measured.

Reduction of the temporal coherence of the light source leads to an upward extension of the range of surface roughnesses that can be measured by means of speckle. The use of broadband sources allows rms roughnesses between 200 nm and 3 μm to be measured. Using several discrete wavelengths simultaneously, such as those of an argon ion laser, allows the range 500 nm to 2 μm to be covered, and the two-wavelength technique (dichromatic speckle) extends this range to 300 nm to 4 μm. These multiwavelength techniques are, however, difficult to implement in practice.

Speckle correlation techniques, although even more difficult to implement, do allow the range of roughnesses that can be handled to be considerably extended. Both the two-wavelength and two-angle methods can deal with rms roughnesses from 500 nm upwards, almost without limit. A range of 4–15 μm, with a precision of 10%, was quoted for one particular application.

The technique of illuminating the surface to be measured with a speckle pattern was used by its originators to measure surface roughnesses in the "weak scatterer" regime, with rms roughnesses between 300 and 800 nm. This method does, however, also allow the measurement of the roughness correlation length.

Finally, all these techniques can have their dynamic ranges moved upwards by the use of various devices. Thus the use of oblique incidence can achieve factors of up to 4 or more in the surface height variations that can be measured, and the use of immersion liquids (when practicable) can extend this range almost without limit for transmitting diffusers. Also, of course, since the dynamic range is wavelength-dependent, it can be extended in either direction by moving out of the visible spectrum; in particular, the range can be extended upwards by using infrared wavelengths such as the 10.6-μm line of the CO_2 laser.

8.11.2 Summary of Other Noncontact Techniques

The comparison of specular and scattered intensities is limited to weak scatterers, i.e., to rms surface roughnesses less than half the wavelength of the light used. Hence the dynamic range is the same as that of the direct speckle method, say 10–200 nm. However, a measure of the roughness correlation length is also possible. The technique is relatively simple to carry out and could easily be embodied in a roughness meter.

The theory of the TIS, ARS, and BRDF/BTDF techniques and diffraction methods assumes that the surface roughness is much less than the wavelength of the light used. For visible light, then, the dynamic range is typically 0.1–35 nm rms roughness. Many of the practical realizations of these techniques have the option of a 10.6-μm source, thus extending the range to about 10–600 nm. In principle, roughness correlation lengths can be calculated from TIS and ARS measurements; the problem is, however, an inverse one and is subject to the difficulties of all such calculations. Commercial instruments are available for most of these techniques.

Although no figures are available, the dynamic range of the microscope image comparator (MIC) is also likely to be limited to the low-roughness end of the spectrum. An ability to measure roughness correlation lengths seems unlikely. The instrument is commercially available but is designed primarily for scratch-and-dig measurement. Work is proceeding on extending its applicability to surface roughness.

Optical profilometers, both interferometric and those using techniques such as Nomarski imaging, can have very large dynamic ranges, typically from 0.1 nm to 5 μm rms roughness. It should be pointed out, however, that coverage of this large range requires changes of components such as microscope objectives and the use of additional modifications such as two-wavelength operation. Since actual profiles are measured, roughness correlation lengths can also be measured. Several very sophisticated optical profilometers are already on the market and have established themselves as very practical and useful instruments.

The new microscopies based on tunneling and associated phenomena are mainly concerned with very high resolution and are thus more suited to the measurement of very smooth surfaces. For example, the dynamic range of the STM is typically 0.1–50 nm rms roughness. Commercial instruments are already on the market, but the field is changing rapidly as new techniques are developed.

8.11.3 Stylus Techniques

The classical tool for measuring surface roughness is undoubtedly the mechanical stylus. Despite its disadvantages—mainly that it is a contact device and that measurements sometimes depend on the stylus shape and loading—it remains the standard against which all other techniques must be measured. Its dynamic range is from 0.1 nm to 50 μm, though the entire range is not, of course, covered by any single instrument. Since it produces a plot of the actual profile, measurement of the roughness correlation length is also possible. And, of course, several stylus instruments are available commercially.

8.11.4 Comparing the Various Techniques

As concluded by Bennett and Mattsson [4], surface roughness is such a complex matter that there is no such thing as a "best" method for its measurement. Much depends on the reason for requiring a knowledge of the roughness and which roughness parameters are important for the particular application. Cost, of course, must also be a factor: Some of the techniques mentioned in this chapter—STM, for example—are very expensive, and there will usually be a trade-off between cost and sophistication of the measurement. On the other hand, the range of roughnesses to be measured will certainly help to determine which techniques can be used. Supersmooth surfaces, for example, from which only very low scattering can be permitted, will dictate the use of techniques such as TIS or ARS or even, in the extreme, the new microscopies such as STM. (It is worth noting that the normal integrating sphere version of TIS can suffer from severe noise problems when very smooth surfaces are being measured, and other variations of the technique are preferred.)

However, for some applications there will be several possible techniques, or even commercial instruments, to choose from, and some sort of comparative exercise is desirable. This has been done by several authors, perhaps the most comprehensive being that carried out by Stedman [158]. The performance of each instrument is summarized on a so-called amplitude/surface-wavelength (correlation length) map. Separate analyses have been carried out for contact stylus instruments [159] and for noncontact optical profilometers [160]. Several specific intercomparisons of techniques have been carried out, and many of these are surveyed by Bennett and Mattsson [4]. Some show good agreement between techniques. For example, a comparison of an optical and a stylus profilometer showed good agreement over a surface roughness range of 0.1–4 nm rms. However, a study on five aluminized roughness standards by Bennett herself [161], with roughnesses varying from 1 to 15 nm rms, showed rather more divergence, with TIS giving higher values of roughness than profilometers and a contact stylus instrument giving higher values than either of the two optical profilometers used. Some of these discrepancies can be attributed to the averaging that occurs with the optical profilometers. There is also the possibility of noise in TIS measurements, caused by Mie or Rayleigh scattering from dust particles, especially in the infrared. Bennett and Mattsson [4] concluded that optical and mechanical profilometers should give comparable roughness values for smoother (i.e., optical) surfaces, provided there are not large numbers of scratches, etc., which the mechanical instruments would detect but the optical profilometers would average over. TIS and ARS techniques will also give comparable results for "normal" surfaces. However, for diamond-machined surfaces, or other surfaces with longer correlation lengths (larger structure), the roughnesses measured by the different techniques can turn out to be quite different. In general, the divergence between mechanical and optical profilometers starts to become significant when the correlation length exceeds about 3 μm.

A notable absentee from all of the above comparative exercises is any speckle technique! The reason, of course, is that the comparisons were, in general, being made of available *instruments* for measuring surface roughness, and, so far as is known, no speckle instruments are on the market. It is necessary, therefore, to address the potential role of speckle techniques in the study of surface roughness.

8.11.5 The Role of Laser Speckle in Surface Roughness Measurement

It is significant that in the excellent book by Bennett and Mattsson on surface roughness and its measurement [4], speckle techniques are virtually absent, warranting only the briefest of casual mentions. One reason, as stated above, is that speckle techniques have not, in general, been converted into practical instruments. (An exception is the laser speckle roughness meter, an image-plane speckle device described by Asakura [6], but it is not clear whether a commercial version even of this instrument was ever made available.) The underlying cause is probably the fact that the speckle techniques have, in general, been developed by researchers working in the speckle field: Speckle, like holography—and even like the laser itself, in its early days—has been very much a "solution in search of a problem." Engineers to whom surface roughness is a problem, however, tend to look for "black-box" instruments to allow them to measure the surfaces, or, failing that, to devise, or at least to instigate, their own solutions to the problem. Thus the mechanical stylus instrument was developed as the obvious answer to a specific measurement problem. Engineers also tend to prefer deterministic measurements such as those given by stylus instruments and optical profilometers; speckle techniques (and other scattering methods such as TIS) give surface statistics without the reassuring "picture" of the surface profile. Finally, the dynamic range of surface roughness height variation is well covered by the non-speckle techniques.

However, the fact that there is still so much work going on into surface roughness and its measurement, with new papers still appearing regularly and entire scientific meetings being devoted to the topic [162], suggests that the problems are still far from solved and that more work on the subject is needed. Many of the instruments mentioned in this review are very expensive, and it may well be that speckle techniques do have a role to play in the measurement of surface roughness: as a cheaper alternative to the more esoteric techniques, when the nondeterministic nature of its measurements can be accepted.

REFERENCES*

1. J. C. Dainty, Ed., *Laser Speckle and Related Topics*, Vol. 9 in Series *Topics in Applied Physics*, Springer-Verlag, New York, 2nd ed., 1984 (or 1st ed., 1975).
2. J. W. Goodman, Statistical properties of laser speckle patterns, in Ref. 1, pp. 9–75.
3. G. Parry, Speckle patterns in partially coherent light, in Ref. 1, pp. 77–122.
4. J. M. Bennett and L. Mattsson, *Introduction to Surface Roughness and Scattering*, Optical Society of America, Washington, 1989.
5. W. T. Welford, Optical estimation of statistics of surface roughness from light scattering measurements, *Opt. Quant. Electron.*, *9*, 269–287 (1977).
6. T. Asakura, Surface roughness measurement, in *Speckle Metrology*, R. K. Erf, Ed., Academic, New York, 1978, pp. 11–49.
7. M. Françon, *Laser Speckle and Applications in Optics*, translated by H. H. Arsenault, Academic, New York, 1979, pp. 121–131.
8. J. D. Briers, Interferometric testing of optical systems and components: a review, *Opt. Laser Technol.*, *4*, 28–41 (1972).
9. J. D. Briers and A. Devlin, Aspheric optics by diamond machining, *Photonics Spectra*, *20*(12), 89–92 (1986).
10. British Standard BS-4301 (1991), Recommendations for Preparation of Drawings for Optical Elements and Systems, British Standards Institution, London, 1991.
11. Standard ANSI/ASME B46.1-1985, Surface Texture (Surface Roughness, Waviness and Lay), The American Society of Mechanical Engineers, New York, 1985.
12. K. J. Stout, C. Obray, and J. Jungles, Specification and control of surface finish: empiricism versus dogmatism, *Opt. Eng.*, *24*, 414–418 (1985).
13. J. D. Rigden and E. I. Gordon, The granularity of scattered optical maser light, *Proc. IRE*, *50*, 2367–2368 (1962).
14. T. S. McKechnie, Speckle reduction, in Ref. 1, pp. 123–170.
15. D. Gabor, Laser speckle and its elimination, *IBM J. Res. Dev.*, *14*, 509–514 (1970).
16. J. W. Goodman, Statistical properties of laser sparkle patterns, Tech. Rep. SEL-63-140 (TR2303-1), Stanford Univ. Electronics Labs, 1963.

*References 1–7 are regarded as key references.

17. J. W. Goodman, Some effects of target-induced scintillation on optical radar performance, *Proc. IEEE*, *53*, 1688–1700 (1965).
18. M. A. Condie, An experimental investigation of the statistics of diffusely reflected coherent light, Ph.D. Thesis, Stanford Univ., Stanford, Calif., 1966.
19. J. C. Dainty, Some statistical properties of random speckle patterns in coherent and partially coherent illumination, *Opt. Acta*, *17*, 761–772 (1970).
20. T. S. McKechnie, Measurement of some second order statistical properties of speckle, *Optik*, *39*, 258–267 (1974).
21. G. Parry, Speckle patterns in partially coherent light, Ph.D. Thesis, Univ. London, 1975.
22. H. Z. Cummins and E. R. Pike, Eds., *Photon Correlation and Light Beating Spectroscopy*, NATO Adv. Study Inst. Ser. B: Physics, Vol. 3, Plenum, New York, 1974.
23. J. C. Dainty, Detection of images immersed in speckle noise, *Opt. Acta*, *18*, 327–339 (1971).
24. R. Barakat, First-order probability densities of laser speckle patterns observed through finite-size scanning apertures, *Opt. Acta*, *20*, 729–740 (1973).
25. H. M. Pedersen, The roughness dependence of partially developed, monochromatic speckle patterns, *Opt. Commun.*, *12*, 156–159 (1974).
26. J. D. Briers, Some applications of holographic interferometry and speckle correlation techniques to the study of plant growth and physiology, Ph.D. Thesis, Univ. London, 1975.
27. J. D. Briers, A note on the statistics of laser speckle patterns added to coherent and incoherent uniform background fields, and a possible application for the case of incoherent addition, *Opt. Quant. Electron.*, *7*, 422–424 (1975).
28. J. D. Briers, Wavelength dependence of intensity fluctuations in laser speckle patterns from biological specimens, *Opt. Commun.*, *13*, 324–326 (1975).
29. W. T. Welford, First order statistics of speckle produced by weak scattering media, *Opt. Quant. Electron.*, *7*, 413–416 (1975).
30. W. T. Welford, Laser speckle and surface roughness, *Contemp. Phys.*, *21*, 401–412 (1980).
31. H. M. Pedersen, Theory of speckle dependence on surface roughness, *J. Opt. Soc. Am.*, *66*, 1204–1210 (1976).
32. H. Fujii, T. Asakura, and Y. Shindo, Measurement of surface roughness properties by means of laser speckle techniques, *Opt. Commun.*, *16*, 68–72 (1976).
33. J. Ohtsubo and T. Asakura, Statistical properties of speckle patterns produced by coherent light at the image and defocus planes, *Optik*, *45*, 65–72 (1976).
34. H. Fujii, T. Asakura, and Y. Shindo, Measurement of surface roughness properties by using image speckle contrast, *J. Opt. Soc. Am.*, *66*, 1217–1222 (1976).
35. H. Fujii, J. Uozumi, and T. Asakura, Computer simulation study of image speckle patterns with relation to object surface profile, *J. Opt. Soc. Am.*, *66*, 1222–1236 (1976).

36. J. Uozumi, H. Fujii, and T. Asakura, Further computer simulation study of image speckle patterns with relation to object surface profile, *J. Opt. Soc. Am.*, *67*, 808–815 (1977).
37. I. Ohtsubo and T. Asakura, Statistical properties of laser speckle produced in the diffraction field, *Appl. Opt.*, *16*, 1742–1753 (1977).
38. H. Fujii and T. Asakura, Effect of the point spread function on the average contrast of image speckle patterns, *Opt. Commun.*, *21*, 80–84 (1977).
39. A. F. Fercher and J. D. Briers, Retinal blood-flow visualization by means of laser speckle photography, *Invest. Ophthalmol. Vis. Sci.*, *22*, 255–259 (1982).
40. J. D. Briers, Optical filtering techniques to enhance speckle contrast variations in single-exposure laser speckle photography, *Optik*, *63*, 265–276 (1983).
41. N. Abramson, The "Interferoscope," a new type of interferometer with variable fringe separation, *Optik*, *30*, 56–71 (1969).
42. J. D. Briers, Interferometric flatness testing of nonoptical surfaces, *Appl. Opt.*, *10*, 519–524 (1971).
43. K. G. Birch, Oblique incidence interferometry applied to non-optical surfaces, *J. Phys. E: Sci. Instrum.*, *6*, 1045–1048 (1973).
44. R. A. Sprague, Surface roughness measurement using white light speckle, *Appl. Opt.*, *11*, 2811–2816 (1972).
45. M. Elbaum, M. Greenebaum, and M. King, A wavelength diversity technique for reduction of speckle size, *Opt. Commun.*, *5*, 171–174 (1972).
46. R. Dändliker and F. M. Mottier, Determination of coherence length from speckle contrast on a rough surface, *Z. Angew. Math. Phys.*, *22*, 369–381 (1971).
47. H. M. Pedersen, On the contrast of polychromatic speckle patterns and its dependence on surface roughness, *Opt. Acta*, *22*, 15–24 (1975).
48. H. M. Pedersen, Second-order statistics of light diffracted from Gaussian, rough surfaces with applications to the roughness dependence of speckles, *Opt. Acta*, *22*, 523–535 (1975).
49. G. Parry, Some effects of temporal coherence on the first order statistics of speckle, *Opt. Acta*, *21*, 763–772 (1974).
50. G. Parry, Some effects of surface roughness on the appearance of speckle in polychromatic light, *Opt. Commun.*, *12*, 75–78 (1974).
51. G. Parry, The scattering of polychromatic light from rough surfaces: first order statistics, *Opt. Quant. Elect.*, *7*, 311–318 (1975).
52. N. George and A. Jain, Speckle in microscopy, *Opt. Commun.*, *6*, 253–257 (1972).
53. N. George and A. Jain, Speckle reduction using multiple tones of illumination, *Appl. Opt.*, *12*, 1202–1212 (1973).
54. H. Fujii and T. Asakura, Coherence measurement of quasi-monochromatic thermal light using speckle patterns (II), *Optik*, *39*, 284–302 (1974).
55. G. Ross, Light scattering in amorphous media: coherence of the radiation field and its effects, *Opt. Acta*, *16*, 611–628 (1969).

56. I. Yamaguchi, Granulation im Beugungs- und Bildfeld von rauhen Objekten. Ableitungen der statistischen Funktionen für lineare Systeme, Teil I, *Optik*, *35*, 591–607 (1972).
57. I. Yamaguchi, Granulation im Beugungs- und Bildfeld von rauhen Objekten. Ableitungen der statistischen Funktionen für lineare Systeme, Teil II, *Optik*, *36*, 173–184 (1972).
58. I. Yamaguchi, Granulation im Bildfeld von Objekten idealer Rauheit, *Optik*, *37*, 141–163 (1973).
59. H. Fujii and T. Asakura, A constant variation of image speckle intensity under illumination of partially coherent light, *Opt. Commun.*, *12*, 32–38 (1974).
60. H. Fujii and T. Asakura, Statistical properties of image speckle pattern in partially coherent light, *Nouv. Rev. Opt.*, *6*, 5–14 (1975).
61. P. Hariharan, Speckle patterns: a historical retrospect, *Opt. Acta*, *19*, 791–793 (1972).
62. K. Exner, Über die Fraunhofer'schen Ringe, die Quetelet'schen Streifen und verwandte Erscheinungen, *Sitzungsber. Kaiserl. Akad. Wiss. (Wien)*, *76*, 522–550 (1877) plus two diagrams.
63. K. Exner, Über die Newton'schen Staubringe, *Wiedemanns Ann. Phys. Chem.*, *9*, 239–260 (1880).
64. M. von Laue, Mathematische Betrachtungen über die Beugungserscheinungen an vielen unregelmässig verstreuten Teilchen, *Mitt. Physik. Ges. (Zurich)*, *18*, 90–102 (1916).
65. M. von Laue, Ein Versagen der klassischen Optik, *Verhandl. Deut. Phys. Ges.*, *19*, 19–21 (1917).
66. E. Buchwald, Ein Versagen der klassischen Optik?, *Verhandl. Deut. Phys. Ges.*, *21*, 492–494 (1919).
67. W. J. de Haas, On the diffraction phenomenon caused by a great number of irregularly distributed apertures or opaque particles, *Proc. Sect. Sci. Kon. Akad. Wet. Amsterdam*, *20*, 1278–1288 (1918).
68. W. J. de Haas, Die Beugungserscheinungen, welche an einer grossen Anzahl unregelmässig verstreuter Öffnungen oder undurchlässigen Teilchen auftreten, *Ann. Phys. IV*, *57*, 568–580 (1918).
69. E. Jakeman, E. R. Pike, G. Parry, and B. Saleh, Speckle patterns in polychromatic light, *Opt. Commun.*, *19*, 359–364 (1976).
70. N. George and A. Jain, Space and wavelength dependence of speckle intensity, *Appl. Phys.*, *4*, 201–212 (1974).
71. N. George, A. Jain, and R. D. S. Melville, Experiments on the space and wavelength dependence of speckle, *Appl. Phys.*, *7*, 157–169 (1975).
72. J. Ohtsubo and T. Asakura, Statistical properties of speckle intensity variations in the diffraction field under illumination of partially coherent light, *Nouv. Rev. Opt.*, *6*, 189–195 (1975).
73. J. Ohtsubo and T. Asakura, Statistical properties of speckle intensity variations in the diffraction field under illumination of coherent light, *Opt. Commun.*, *14*, 30–34 (1975).

74. H. Fujii and J. W. Y. Lit, Measurement of surface roughness using dichromatic speckle, *Opt. Commun.*, *22*, 231–235 (1977).
75. V. I. Tatarski, *Wave Propagation in a Turbulent Medium*, Dover, New York, 1961.
76. C. Polhemus, Two-wavelength interferometry, *Appl. Opt.*, *12*, 2071–2074 (1973).
77. J. C. Wyant, Testing aspherics using two-wavelength holography, *Appl. Opt.*, *10*, 2113–2118 (1971).
78. A. F. Fercher and H. Z. Hu, Two wavelength heterodyne interferometry, Proc. 6th Int. Congress Laser 83 Optoelektronik, 1984, pp. 142–146.
79. A. F. Fercher, Neue Interferometer zur Optikprüfung, *Laser Optoelektron*, *15*, 301–308 (1983).
80. Y.-Y. Cheng and J. C. Wyant, Two-wavelength phase-shifting interferometry, *Appl. Opt.*, *23*, 4539–4543 (1984).
81. A. F. Fercher, H. Z. Hu, and U. Vry, Rough surface interferometry with a two-wavelength heterodyne speckle interferometer, *Appl. Opt.*, *24*, 2181–2188 (1985).
82. A. F. Fercher and U. Vry, Two-wavelength speckle interferometric technique for rough surface contour measurement, *Opt. Eng.*, *25*, 623–626 (1986).
83. A. F. Fercher, U. Vry, and W. Werner, Two-wavelength speckle interferometry on rough surfaces using a mode hopping diode laser, *Opt. Lasers Eng.*, *11*, 271–279 (1989).
84. A. E. Ennos, Speckle interferometry, in Ref. 1, pp. 203–253.
85. E. Archbold, J. M. Burch, and A. E. Ennos, Recording of in-plane surface displacement by double exposure speckle photography, *Opt. Acta*, *17*, 883–898 (1970).
86. J. N. Butters and J. A. Leendertz, A double exposure technique for speckle pattern interferometry, *J. Phys. E: Sci. Instrum.*, *4*, 277–279 (1971).
87. G. Tribillon, Corrélation entre deux speckles obtenus avec deux longueurs d'onde: application à la mesure de la rugosité moyenne, *Opt. Commun.*, *11*, 172–174 (1974).
88. J. A. Mendez and M. L. Roblin, Utilisation des franges d'interférence en lumière diffuse pour l'étude de l'état de surface d'un diffuseur, *Opt. Commun.*, *13*, 142–147 (1975).
89. W. B. Ribbens, Surface roughness measurement by two wavelength holographic interferometry, *Appl. Opt.*, *13*, 1085–1088 (1974).
90. W. B. Ribbens, Surface roughness measurement by holographic interferometry, *Appl. Opt.*, *11*, 807–810 (1972).
91. D. Léger, E. Mathieu, and J. C. Perrin, Optical surface roughness determination using speckle correlation technique, *Appl. Opt.*, *14*, 872–877 (1975).
92. F. Dalmases, R. Cibrián, M. Buendía, C. Romero, R. Salvador, and J. Montilla, Speckle correlation technique to determine roughness in the dermotologic interval, *Phys. Med. Biol.*, *33*, 913–922 (1988).
93. D. Léger, Deux méthodes de mesure de rugosité par corrélation de speckle, Doctoral Thesis, Univ. Orsay, France, 1976.

94. T. Yoshimura and K. Nakagawa, Properties of light scattered by a diffuse object under dynamic speckle illumination and their application to the roughness distribution measurement, *Opt. Commun.*, *60*, 139–144 (1986).
95. T. Yoshimura, K. Kato, N. Morino, and K. Nakagawa, Measurements of two roughness parameters under speckle pattern illumination, *Opt. Commun.*, *75*, 208–214 (1990).
96. T. Yoshimura, K. Kato, and K. Nakagawa, Surface-roughness dependence of the intensity correlation function under speckle-pattern illumination, *J. Opt. Soc. Am. A*, *7*, 2254–2259 (1990).
97. Y. Dzialowski and M. May, Correlation of speckle pattern generated by laser point source illuminated diffusers, *Opt. Commun.*, *16*, 334–339 (1976).
98. D. Middleton, *Introduction to Statistical Communication Theory*, McGraw-Hill, New York, 1960.
99. E. Jakeman and P. N. Pusey, The statistics of light scattered by a random phase screen, *J. Phys. A*, *6*, L88–L92 (1973).
100. E. Jakeman and P. N. Pusey, Light scattering from electrohydrodynamic turbulence in liquid crystals, *Phys. Lett.*, *44A*, 456–458 (1973).
101. E. Jakeman and P. N. Pusey, Non-Gaussian fluctuations in electromagnetic radiation scattered by a random phase screen. I. Theory, *J. Phys. A*, *8*, 369–391 (1975).
102. P. N. Pusey and E. Jakeman, Non-Gaussian fluctuations in electromagnetic radiation scattered by a random phase screen. II. Application to dynamic scattering in a liquid crystal, *J. Phys. A*, *8*, 392–410 (1975).
103. J. C. Leader, Bidirectional scattering of electromagnetic waves from rough surfaces, *J. Appl. Phys.*, *42*, 4808–4816 (1971).
104. E. Kretschmann, Die Bestimmung der Oberflächenrauhigkeit dünner Schichten durch Messung der Winkelabhängigkeit der Streustrahlung von Oberflächenplasmaschwingungen, *Opt. Commun.*, *10*, 353–356 (1974).
105. E. Kretschmann and E. Kröger, Reflection and transmission of light by a rough surface, including results for surface-plasmon effects, *J. Opt. Soc. Am.*, *65*, 150–154 (1975).
106. J. Bodesheim and A. Otto, On the quantitative measurement of the roughness spectrum of silver films, *Surf. Sci.*, *45*, 441–456 (1974).
107. G. S. Agarwal, Scattering from rough surfaces, *Opt. Commun.*, *14*, 161–166 (1975).
108. P. J. Chandley and W. T. Welford, A re-formulation of some results of P Beckmann for scattering from rough surfaces, *Opt. Quant. Electron.*, *7*, 393–397 (1975).
109. P. J. Chandley, Surface roughness measurements from coherent light scattering, *Opt. Quant. Electron.*, *8*, 323–327 (1976).
110. P. J. Chandley, Determination of the autocorrelation function of height on a rough surface from coherent light scattering, *Opt. Quant. Electron.*, *8*, 329–333 (1976).
111. P. J. Chandley, Determination of the probability density function of height on a rough surface from far-field scattering, *Opt. Quant. Electron.*, *11*, 413–418 (1979).

112. P. Beckmann and A. Spizzichino, *The Scattering of Electromagnetic Waves from Rough Surfaces*, Pergamon, Oxford, 1963.
113. P. Beckmann, Scattering of light by rough surfaces, *Prog. Opt.*, *6*, 55–69 (1967)
114. J. A. Ratcliffe, Some aspects of diffraction theory and their application to the ionosphere, *Rep. Prog. Phys.*, *19*, 188–267 (1956).
115. E. Theocharous, On-line optical surface roughness monitor, presented at Optics 90, Nottingham, U.K., 1990.
116. E. Theocharous, private communication, 1992.
117. H. E. Bennett and J. O. Porteous, Relation between surface roughness and specular reflection at normal incidence, *J. Opt. Soc. Am.*, *51*, 123–129 (1961).
118. J. O. Porteous, Relation between the height distribution of a rough surface and the reflectance at normal incidence, *J. Opt. Soc. Am.*, *53*, 1394–1402 (1963).
119. H. Davies, The reflection of electromagnetic waves from a rough surface, *Proc. Inst. Elec. Eng. IV*, *101*, 209–214 (1954).
120. Standard ASTM Doc. F1048-87, Standard test method for measuring the effective surface roughness of optical components by total integrated scattering, American Society for Testing and Materials, New York, 1987.
121. J. M. Elson, H. E. Bennett, and J. M. Bennett, Scattering from optical surfaces, in *Applied Optics and Optical Engineering*, Vol. 7, R. R. Shannon and J. C. Wyant, Eds., Academic, New York, 1979, pp. 191–244.
122. J. M. Elson and J. M. Bennett, Vector scattering theory, *Opt. Eng.*, *18*, 116–124 (1979).
123. J. M. Elson and J. M. Bennett, Relation between the angular dependence of scattering and the statistical properties of optical surfaces, *J. Opt. Soc. Am.*, *69*, 31–47 (1979).
124. J. M. Bennett, H. H. Hurt, J. P. Rahn, J. M. Elson, K. H. Guenther, M. Rasigni, and F. Varnier, Relation between optical scattering, microstructure and topography of thin silver films. 1. Optical scattering and topography, *Appl. Opt.*, *24*, 2701–2711 (1985).
125. J. M. Bennett, J. M. Elson, and J. P. Rahn, Angle-resolved scattering: comparison of theory and experiment, *Proc. SPIE*, *401*, 234–246 (1983).
126. T. V. Vorburger, E. C. Teague, F. E. Scire, M. J. McLay, and D. E. Gilsinn, Surface roughness studies with DALLAS—Detector Array for Laser Light Angular Scattering, *J. Res. Natl. Bur. Stand.*, *89*, 3–16 (1984).
127. L.-X. Cao, T. V. Vorburger, A. G. Lieberman, and T. R. Lettieri, Light-scattering measurement of the rms slopes of rough surfaces, *Appl. Opt.*, *30*, 3221–3227 (1991).
128. F. E. Nicodemus, Reflectance nomenclature and directional reflectance and emissivity, *Appl. Opt.*, *9*, 1474–1475 (1970).
129. F. O. Bartell, E. L. Dereniak, and W. L. Wolfe, The theory and measurement of bidirectional reflectance distribution function (BRDF) and bidirectional transmittance distribution function (BTDF), *Proc. SPIE*, *257*, 154–160 (1980).
130. E. L. Dereniak, L. G. Brod, and J. E. Hubbs, Bidirectional transmittance

distribution function measurement on ZnSe, *Appl. Opt.*, *21*, 4421–4425 (1982).

131. L. R. Baker, Microscope image comparator, *Opt. Acta*, *31*, 611–614 (1984).
132. E. R. Mendez and K. A. O'Donnell, Observation of depolarization and backscattering enhancement in light scattering from Gaussian random surfaces, *Opt. Commun.*, *61*, 91–95 (1987).
133. K. A. O'Donnell and E. R. Mendez, Experimental study of scattering from characterized random surfaces, *J. Opt. Soc. Am. A*, *4*, 1194–1205 (1987).
134. M.-J. Kim, J. C. Dainty, A. T. Friberg, and A. J. Sant, Experimental study of enhanced backscattering from one- and two-dimensional random rough surfaces, *J. Opt. Soc. Am. A*, *7*, 569–577 (1990).
135. N. C. Bruce and J. C. Dainty, Multiple scattering from random rough surfaces using the Kirchhoff approximation, *J. Mod. Opt.*, *38*, 579–590 (1991).
136. N. C. Bruce and J. C. Dainty, Multiple scattering from rough dielectric and metal surfaces using the Kirchhoff approximation, *J. Mod. Opt.*, *38*, 1471–1481 (1991).
137. E. Marx and T. V. Vorburger, Direct and inverse problems for light scattered by rough surfaces, *Appl. Opt.*, *29*, 3613–3626 (1990).
138. O. V. Angelsky and P. P. Maksimyak, Optical diagnostics of slightly rough surfaces, *Appl. Opt.*, *31*, 140–143 (1992).
139. J. G. Valliant, Scatterometry on the shop floor, *Photonics Spectra*, *26*(1), 118–119 (1992).
140. J. Westbury, Development of objective methods for judging the quality of ground and polished surfaces in production, *Proc. Inst. Mech. Eng.*, *182*, Pt. 3K, 260–273 (1968).
141. M. Bouchard, M. Doucet, and G. April, Surface roughness measurements of dental fillings by optical methods, *Proc. SPIE*, *1647*, 137–147 (1992).
142. R. Brodman, O. Gerstorfer, and G. Thurn, Optical roughness measuring instrument for fine-machined surfaces, *Opt. Eng.*, *24*, 408–413 (1985).
143. E. Milana, C. Morandi, C. Truzzi, and F. Ziprani, Bayesian classification of machined surface finishing by two-dimensional analysis of diffraction patterns, *Opt. Eng.*, *30*, 1461–1467 (1991).
144. D. Malacara, Ed. *Optical Shop Testing*, Wiley, New York, 1978.
145. B. Bhushan, J. C. Wyant, and C. L. Koliopoulos, Measurement of surface topography of magnetic tapes by Mirau interferometry, *Appl. Opt.*, *24*, 1489–1497 (1985).
146. E. R. Cochran and J. C. Wyant, Longscan surface profile measurements using a phase-modulated Mirau interferometer, *Proc. SPIE*, *680*, 112–117 (1986).
147. K. Creath and J. C. Wyant, Absolute measurement of surface roughness, *Appl. Opt.*, *29*, 3823–3827 (1990).
148. G. E. Sommargren, Optical heterodyne profilometry, *Appl. Opt.*, *20*, 610–618 (1981).
149. M. J. Downs, W. H. McGivern, and H. J. Ferguson, Optical system for measuring the profiles of super-smooth surfaces, *Precis. Eng.*, *7*, 211–215 (1985).

150. M. J. Downs, N. M. Mason, and J. C. C. Nelson, Measurement of the profiles of super smooth surfaces using optical interferometry, *Proc. SPIE*, *1009*, 14–17 (1989).
151. T. C. Bristow and K. Arackellian, Surface roughness measurements using a Nomarski type scanning instrument, *Proc. SPIE*, *749*, 114–118 (1987).
152. T. C. Bristow and D. Lindquist, Surface measurements with a non-contact Nomarski-profiling instrument, *Proc. SPIE*, *816*, 106–110 (1987).
153. T. C. Bristow, A. Bouzid, and J. Bietry, Surface measurements and applications for manufactured parts using a noncontact profilometer, *Proc. SPIE*, *954*, 217–225 (1988).
154. J. M. Bennett, Measurement of the rms roughness, autocovariance function, and other statistical properties of optical surfaces using a FECO scanning interferometer, *Appl. Opt.*, *15*, 2705–2721 (1976).
155. O. Keller and P. Sønderkaer, A description of scattering of light from a rough surface and measurements of the surface topography of a metal, *Proc. SPIE*, *680*, 150–159 (1986).
156. J. R. Blanco and P. J. McMarr, Roughness measurements of Si and Al by variable angle spectroscopic ellipsometry, *Appl. Opt.*, *30*, 3210–3220 (1991).
157. K. Mitsui, M. Sakai, and Y. Kizuka, Development of a high resolution sensor for surface roughness, *Opt. Eng.*, *27*, 498–502 (1988).
158. M. Stedman, Basis for comparing the performance of surface-measuring machines, *Precis. Eng.*, *9*, 149–152 (1987).
159. M. Stedman and K. Lindsey, Limits of surface measurement by stylus instruments, *Proc. SPIE*, *1009*, 56–61 (1989).
160. M. Stedman, Limits of surface measurement by optical probes, *Proc. SPIE*, *1009*, 62–67 (1989).
161. J. M. Bennett, Comparison of techniques for measuring the roughness of optical surfaces, *Opt. Eng.*, *24*, 380–387 (1985).
162. Surface Roughness and Scattering, Optical Society of America meeting, June 1992.

9

Automatic Fringe Analysis Procedures in Speckle Metrology

G. H. Kaufmann

Rosario Institute of Physics
Rosario, Argentina

9.1 INTRODUCTION

During the last decade, the wide availability of electro-optical devices has provided a practical approach to the development of automatic systems for data evaluation in the area of speckle metrology. More recently, the widespread introduction of fast and inexpensive personal computers and powerful framegrabbers has also had an important impact in this area, stimulating the development of accurate processing algorithms capable of achieving good noise immunity. Nowadays, it can be affirmed that the use of automatic fringe analysis procedures in speckle metrology has increased the range of application of these techniques and the effectiveness of speckle metrology as a practical measuring tool.

The present chapter reviews the most important developments that have taken place in this area during the last decade. Its aim is to present a description of the several optical systems and processing algorithms that were developed and provide an introduction to the literature on the subject. The limitations, accuracy, and reliability of the various approaches are also discussed.

9.2 ANALYSIS OF FRINGES GENERATED FROM DOUBLE-EXPOSURE SPECKLE PHOTOGRAPHS

9.2.1 Semiautomated Procedures for the Measurement of Displacement Fields

As noted from previous chapters, speckle photography is now a well-established technique for deformation and strain measurement in solid mechanics and for velocity measurement in fluid mechanics. Analysis of both laser speckle and particle image velocimetry photographs involves mapping the displacement of a random pattern. When a double-exposure technique is used, this analysis is normally done by probing the specklegram point by point with a narrow laser beam. The far-field diffraction pattern consists of a speckled diffraction halo modulated by cosine-squared interference fringes, called Young's fringes, with spacing inversely proportional to the magnitude and orientation of fringes perpendicular to the direction of the required displacement. In the case of in-plane displacement, its normalized two-dimensional irradiance can be expressed as

$$I(x, y) = I_0(x, y)\{1 + V \cos [2\pi(k_x x + k_y y + c)]\} \qquad (9.1)$$

where x and y are the coordinates over the diffraction plane, $I_0(x, y)$ is the diffraction halo corresponding to a single-exposure speckle distribution; V is the fringe visibility; k_x and k_y are the number of fringes cutting the x and y axes, respectively, and are proportional to the displacement components u_x and u_y; and c is a constant.

Since several hundred displacement measurements must be performed to obtain a detailed understanding of the deformation or velocity fields, manual evaluation of the Young's fringes is a time-consuming, tedious, and impractical procedure. For this reason, considerable effort has been put into developing optical systems with digital signal processng to automate the analysis procedure. However, as fringe patterns generated from both laser speckle photography and particle image velocimetry are heavily modulated by noise, this is not a simple task.

The first systems for automating the analysis of the Young's fringes generated in speckle photography were developed around the end of the 1970s [1–4]. These were semiautomated systems in which the fringes were manually aligned in a direction parallel to the columns of the imaging device. As the direction of the Young's fringes was known, the determination of the fringe spacing was reduced to a one-dimensional problem. Specklegrams were positioned in the reading laser beam by means of a motor-driven XY stage, thus allowing a host computer to analyze the fringes from any point on them. To reduce the amount of data to be processed and the speckle noise that modulates the diffraction halo, these systems usually included the integration of the fringes along its direction before the spacing evaluation.

Maddux et al. [1] used a long photodiode mounted on a motor-driven stage for scanning the fringes mechanically. In Ref. 1, they pointed out for the first time that a small shift in fringe maxima is introduced due to the varying irradiance of the diffraction halo. For this reason, they presented a simple routine to compute the average fringe spacing from the location of fringe minima. From a calibration test, the system accuracy was determined to be within 2% when a sampling of 128 points was used.

Ineichen et al. [2] used a TV camera to acquire irradiance data and a Dove prism to manually align the Young's fringes parallel to the TV lines. To reduce the speckle noise, the video signal was integrated along each TV line by means of a fast analog integrator. This integrated irradiance, sampled at 256 points, was transferred to the computer memory. The influence of the diffraction halo was minimized by recording its average irradiance and by subtracting it from the irradiance stored in the computer. Further increase in the signal-to-noise ratio was obtained by calculating the autocorrelation of the bias-free fringe function. Fringe spacing was finally determined by calculating the average distance between the peaks of the autocorrelation function. By testing an object subjected to a pure rotation, it was shown that this system was capable of measuring displacements over a range of fringe densities of 2–20 fringes per pattern with an accuracy better than 1% for the magnitude of 1° for the direction.

A further step in the development of a semiautomated fringe analysis system was given by Kaufmann et al. [3] with the design of a compact optical setup that projects without distortion the one-dimensional Fourier transform of a small area of a specklegram onto a linear photodiode array. This optical system was composed of two lenses, a high-quality camera lens and a cylindrical lens, which focused the fringes into a line whose width was smaller than the depth of the individual diodes of the array. Fringe patterns were detected with an array of 512 elements, which gave an output signal similar to the one shown in Fig. 9.1.

For each point on the specklegram, two functions were recorded: the cross section across the fringe pattern and the average irradiance across the diffraction halo. The expected cosine-squared function that describes the Young's fringes was obtained by dividing the first function by the second. In this way, the influence of the diffraction halo in the fringe spacing determination was minimized. The fringe spacing was evaluated by computing the one-dimensional Fourier transform of the quotient function. As the Fourier transform did not provide enough frequency resolution, transforms were calculated for nonintegral frequencies using a z transform method, and the position of its maximum gave the required fringe spacing.

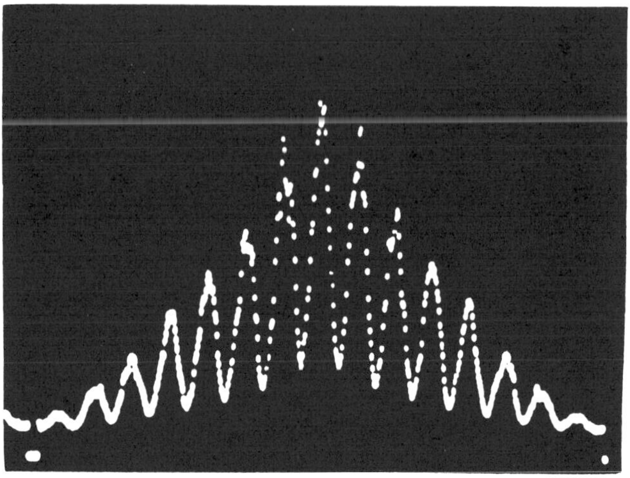

Figure 9.1 Typical output signal from the linear photodiode array used in the fringe analysis system of Kaufmann et al. [3].

The performance of the system was extensively tested by using an experimental method to record specklegrams that stored a simulated uniform radial strain field. This was done by imaging a flat surface at slightly differing magnifications. These tests proved that the standard deviation in any displacement measurement was 0.1 μm, which was about 1/30 of the average speckle size, and that the system was capable of measuring to this accuracy for densities of 2–50 fringes. To assess its accuracy in practical situations, the system was also tested under conditions of reduced fringe contrast. These specklegrams were recorded with a two-beam illumination, with one of the beams including a glass plate that was rotated slightly between exposures. Processing of experimental data showed that the system could deal with fringe contrast as low as 20%, though with reduced accuracy.

More recently, Chitsaz and Moslehy [5] used a 1024-pixel charge-coupled device to acquire irradiance data from scan lines perpendicular to the fringes. They used data from eight scans, four to the right and four to the left of the center of the fringe pattern. To convert digitized data in a time signal, the data were fed into a spectrum analyzer and then Fourier transformed. Fringe spacing was calculated by locating the peak of the average of the eight transforms.

As pointed previously, fringe maxima do not occur at the maxima of the cosine-squared fringe pattern due to the varying irradiance of the diffraction halo. When fringe visibility is near unity, displacements can be evaluated from fringe minima that do not shift. However, in practice, fringe visibility falls because of speckle decorrelation effects and lens aberrations. So errors are also introduced even if fringe minima are taken for spacing evaluation. The influence of removing the diffraction halo before fringe analysis, a quite important problem when numerical processing is used, was investigated for the first time by Kaufmann [6]. In this work, the errors caused by the diffraction halo-induced shift on fringe maxima and minima positions were calculated considering the halo irradiance as the square of the autocorrelation of the aperture function. Later, it was pointed out by Meynart [7] that this irradiance distribution is correct only for incoherent speckle patterns. The study performed by Kaufmann [6], which was done for a square aperture, showed that considerable errors may arise if the diffraction halo is not removed before the fringe analysis. It was also found that relative errors in displacements increased as fringe density and visibility decreased.

As a continuation of this work, the diffraction halo effect was extensively studied by other authors. Vikram and Vedam [8] calculated analytically the fringe shifts for the case of sinusoidal vibrations. Later, several approaches for error elimination in specklegrams recorded through different apertures were developed by Vikram [9–12] and Vikram and Vedam [13, 14]. More recently, new aspects such as position-dependent visibility were studied by Georgieva [15, 16], and the production of a constant irradiance diffraction halo through an appropriate aperture function by Joenathan and Sirohi [17]. A generalized theory of the diffraction halo effects, including position-dependent visibility and fringe phase, for particle image velocimetry and for coherent and incoherent speckle photography has been presented by Hinsch [18].

Earlier works on the investigation of the diffraction halo effect promoted new studies in this area. Using a one-dimensional computer simulation of the diffraction halo for a square aperture, Kaufmann [19] explored the influence of not removing the halo irradiance when Young's fringes were processed via discrete Fourier transform. It was shown that the Fourier method relaxed the influence of the diffraction halo and introduced relative errors much lower than those calculated from fringe minima. For densities larger than six fringes, errors lower than 0.5% were obtained for most of the visibility values considered. Similar results were obtained using experimentally recorded data [20].

9.2.2 Automated Procedures for the Measurement of Displacement Fields

In the 1980s, the development of new electro-optical components and more powerful digital image processing systems made possible the analysis of complete two-dimensional speckle fringe patterns in reasonable time. The first fully automated processing algorithm to be used in these systems was developed by Robinson [21]. As it is currently done, Young's fringes were acquired by a TV camera and its output was transferred to the host computer as an image of 512 × 512 pixels with 256 gray levels. On the assumption that the center of the field of view was always a bright fringe, an approximate direction of the fringes was found by adding irradiance values along a series of radial vectors passing through the pattern center. By means of an iterative approach, as shown in Fig. 9.2, the process was repeated using a smaller range of angular scan and a longer radial vector. The vector giving the largest summed value was taken as the fringe direction. Once this direction was evaluated, speckle noise was reduced by averaging the pattern along a direction parallel to the fringes. In this way, the two-dimensional pattern was reduced to a one-dimensional pattern whose period was equal to the fringe spacing.

An algorithm based on the phase detection technique [22], which is well known in the field of communication for detecting the presence of periodic

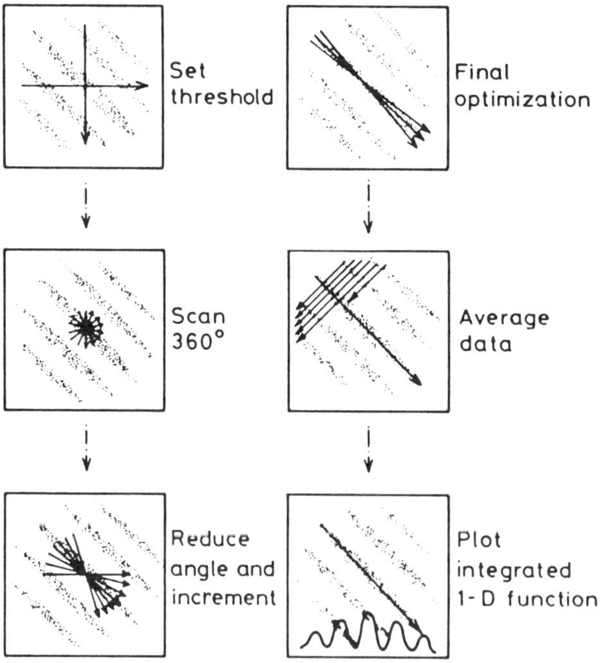

Figure 9.2 Analysis sequence in the algorithm developed by Robinson. (From Ref. 21.)

signals in noise, was developed by Toyooka et al. [23]. Briefly, this approach begins by assuming an approximate value f' for the spatial frequency of the Young's fringe pattern, so that irradiance data can be multiplied by two reference signals, sin $(2\pi f'y)$ and cos $(2\pi f'y)$. This approximate value is estimated by calculating the Fourier spectrum for one scanning line of the detector. From the ratio of the two products, the phase term of the periodic irradiance distribution can be computed along any line of the detector. The spacing and direction of the fringe pattern are obtained by determining an average phase plane from all phase values for 100 detector lines.

Huntley [24] presented an alternative approach based on the two-dimensional Fourier transformation of the Young's fringes. The Fourier transform of a noisy cosine-squared signal consists of just three major peaks, as shown in Fig. 9.3a, and noise is spread fairly uniformly over the entire frequency domain. However, this noise immunity is achieved at a cost of about $2N^2 \log N$ multiplications and additions per transform, a procedure that is computationally demanding. From Eq. (9.1), it is seen that the $N \times N$ array containing the digitized image has the form

$$I(m, n) = I_0(m, n)\{1 + V \cos [2\pi(k_m m + k_n n + c)/N]\} \tag{9.2}$$

The Fourier transform of $I(m, n)$ consists of a delta function of unit amplitude at $k_x = 0$, $k_y = 0$ and delta functions of amplitude $V/2$ at $k_x = \pm k_m$ and $k_y = \pm k_n$ convolved with the Fourier transform of $I_0(m, n)$. The finite width of the zero-order peak has the effect of shifting slightly the position of nearby peaks, causing errors in the patterns of low fringe density. A similar problem was previously shown by Kaufmann for the one-dimensional case [19]. For this reason, the zero-order peak due to the diffraction halo was removed, calculating the array $x(m, n)$ as

$$x(m, n) = I(m, n) - (s_1/s_2)I_o(m, n) \tag{9.3}$$

where

$$s_1 = \sum_{m=0}^{N-1} \sum_{n=0}^{N-1} I(m, n)$$

$$s_2 = \sum_{m=0}^{N-1} \sum_{n=0}^{N-1} I_0(m, n) \tag{9.4}$$

so that the mean of $x(m, n)$ was equal to zero. $I_0(m, n)$ was estimated by averaging 16 diffraction halos from different points on a single-exposure specklegram. The modulus of the Fourier transform of $x(m, n)$ is shown in Fig. 9.3b.

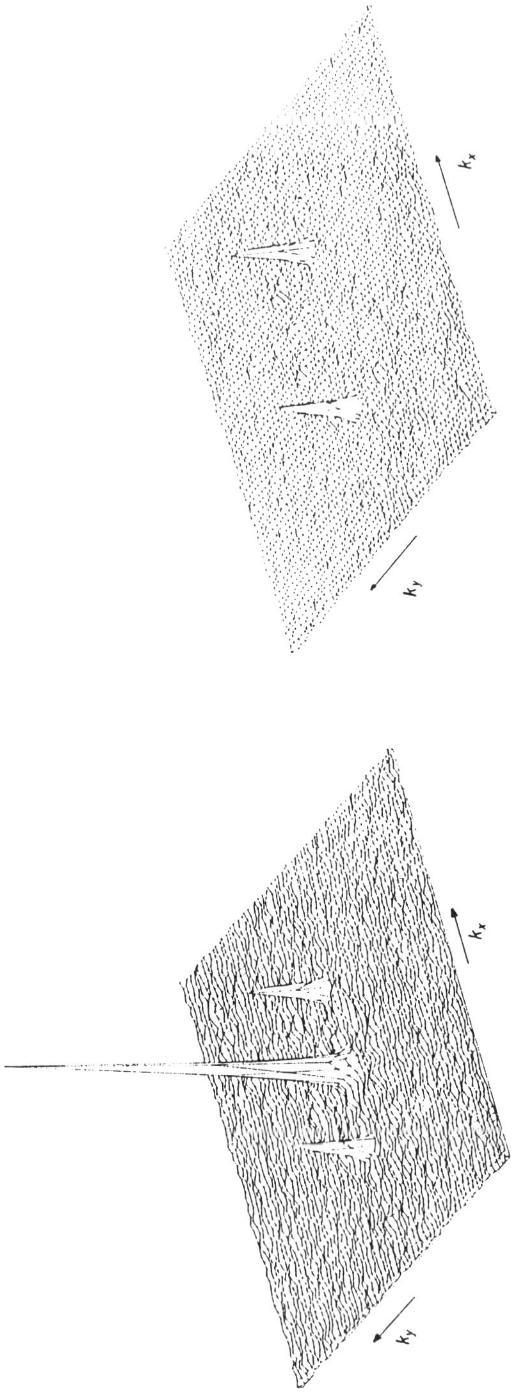

(a) (b)

Figure 9.3 Modulus of the discrete Fourier transform of (a) Young's fringe pattern; (b) $x(m, n)$. (From Ref. 24.)

In general, k_m and k_n will be nonintegral, so the peak of the transform will lie between sample points. For this reason, first the two-dimensional discrete Fourier transform was computed and the position of its maximum was taken as an initial estimate of (k_m, k_n). This operation was performed efficiently using a fast Fourier transform algorithm. The initial estimate was then refined by maximizing the modulus of the Fourier transform using a bounded Newton–Raphson algorithm. Due to the separable nature of the two-dimensinal Fourier transform, the problem was reduced to a series of one-dimensional optimizations along lines of constant k_x and k_y.

The reliability and accuracy of this algorithm were experimentally tested using specklegrams that generated fringes of reduced contrast. These specklegrams were made by recording the two exposures for different durations. For the relatively low image resolution of 64×64, results showed that the algorithm could deal with fringe visibilities of under 10%, giving a standard deviation in displacement magnitude of 0.05 μm and an angle of 0.1°. A later work by Huntley et al. [25] illustrates the application of this system to the solution of practical engineering problems: the study of strain distributions in plastic-bonded explosive disks and calculation of stress intensity factors from the evaluation of displacement fields around the tips of stationary and moving cracks.

To decrease the computation time, in two subsequent works Huntley [26, 27] investigated the use of the Walsh transform [28]. The two-dimensional discrete Walsh transform can be expressed in terms of the Walsh functions, which are a complete orthogonal set taking only two amplitude values, +1 and −1. As the evaluation of this transform requires additions and subtractions but no multiplications, it is up to an order of magnitude faster than the Fourier transform. Although the Walsh spectrum is more complex than the Fourier spectrum, it can be interpreted unambiguously by further cross-correlation of the fringes with square waves. Comparison of this method with the Fourier transformation showed that on average the Walsh method gave marginally larger errors. While the Walsh method started to fail at fringe visibilities below 12% compared with about 6% for the Fourier transformation, it was proved to be at least three times faster.

Ansari and Ciurpita [29] presented a simple approach to provide fully automated evaluation of the fringe direction, a procedure that was developed mainly to avoid the use of an expensive image recognition system with large memory capacity. In their system, the diffraction halo was scanned by a photodetector that follows a circular path around the center of the pattern, and its output was transferred and stored in a microcomputer as a vector $i(n)$. To reduce speckle noise, this irradiance waveform was smoothed using median filtering. Figure 9.4a shows a typical vector $i(n)$ after the averaging procedure. In the circular path, the zero-order fringe

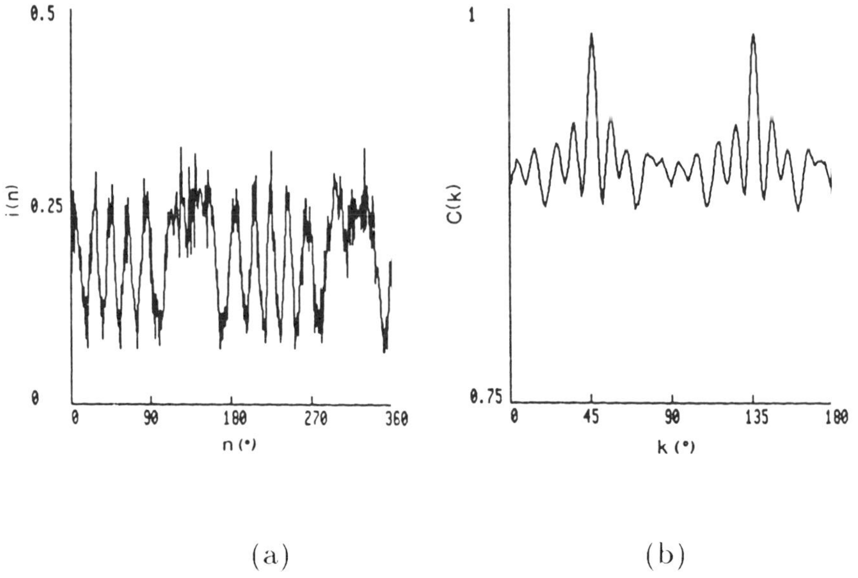

Figure 9.4 (a) Typical averaged irradiance waveform measured along a circular path: (b) cross-correlation coefficient. (From Ref. 53.)

was intercepted at two positions 180° apart, so it bisected the pattern symmetrically. A cross-correlation technique was used to measure the similarity between the two halves of the vector $i(n)$. The cross-correlation coefficient $C(k)$ was defined as

$$C(k) = \frac{\sum_{n=0}^{N} i(k-n)i(k+n)}{\left(\sum_{n=0}^{N} i^2(k-n) \sum_{n=0}^{N} i^2(k+n)\right)^{1/2}} \tag{9.5}$$

where k is the point on the waveform around which the cross-correlation was taken and N is half the number of waveform points minus one.

The cross-correlation coefficient $C(k)$ calculated from the waveform irradiance of Fig. 9.4a is shown in Fig. 9.4b, and its peak determines the fringe direction. However, as a line drawn along the direction perpendicular to the zero-order fringe also bisects the diffraction halo, a second peak appears in Fig. 9.4b. In practice, it was found that this second peak was usually lower than the first one. Fringe spacing can be calculated from the angular distance between the selected seals and either of the two adjacent peaks. The performance of this system was tested from the analysis of specklegrams with known rigid-body displacements, and calculated displacements agreed reasonably well with micrometer readings.

To simplify the reading process of the specklegram, Arnold and Hinsch [30] developed a parallel optical evaluation scheme. Instead of scanning the specklegram, a holographic optical element was used to produce a two-dimensional array of narrow interrogating light beams. These beams formed an array of tiny Young's fringe systems as shown in Fig. 9.5a. The subsequent evaluation of the fringe systems was also done in parallel. The array of fringe patterns was recorded on a high-resolution film and introduced into the array of reading beams from the holographic optical element, with the position carefully adjusted so that the fringe patterns coincided with the beams. This setup produced an array of nonoverlapping autocorrelation functions as shown in Fig. 9.5b. It is seen that each function consists of a symmetrical pair of two side peaks around a central peak. The distance between the peaks is directly proportional to the magnitude of the displacements, and the direction of displacement is given by the line connecting the two peaks.

Höfling [31] described a new analytical method based on a fast image-fitting comparison of the Young's fringes under analysis with a set of synthetic fringe patterns by means of a videoprocessor of pipeline architecture. The synthetic fringes are constructed as the product of a pure cosine fringe pattern with a diffraction halo distribution using the equations given in the

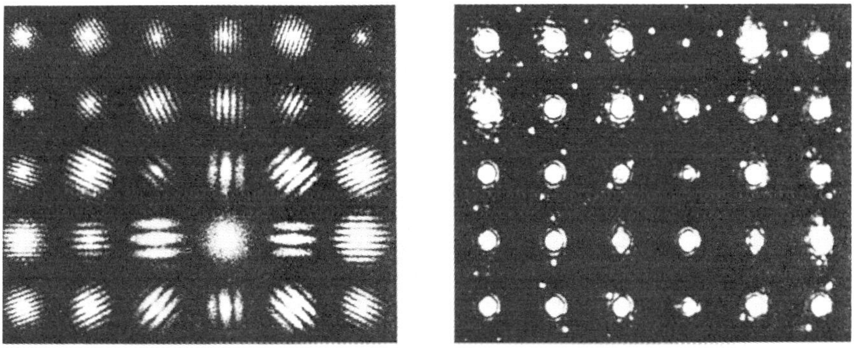

(a) (b)

Figure 9.5 Parallel optical evaluation of a specklegram of a rotated plate. (a) Array of Young's fringes; (b) array of autocorrelation functions. (From Ref. 30.)

literature [7]. The comparison of the fringe pattern $I(d_{01}, d_{02})$ with the set of synthetic fringes $I_s(d_1, d_2)$ is done by calculating the correlation as

$$C(d_1, d_2) = \frac{\sum_{ij} I_{ij}(d_{01}, d_{02}) I_{sij}(d_1, d_2)}{\left(\sum_{ij} I_{ij}^2(d_{01}, d_{02}) \sum_{ij} I_{sij}^2(d_1, d_2) \right)^{1/2}} \tag{9.6}$$

where (d_{01}, d_{02}) are the speckle displacement components, i and j denote a location in the discrete image, and the double sums are taken over all picture elements. The isolines of a typical correlation surface $C(d_1, d_2)$ together with the fringe pattern are shown in Fig. 9.6, where the peaks are found to be located at (0, 0) and $\pm(d_{01}, d_{02})$. Experimental verification was done using both experimental and simulated data, and standard deviations of about 0.1–0.2 μm were found for displacements of 1–25 μm. It was also proved that this method was well suited for the measurement of displacements in the subfringe range, where the diffraction halo was only slightly deformed.

Deng and Yamaguchi [32] present another algorithm for analyzing patterns of very low fringe density, from 4 to 0.5 fringes within the diffraction halo. It can also be used to analyze narrower fringes with a higher accuracy.

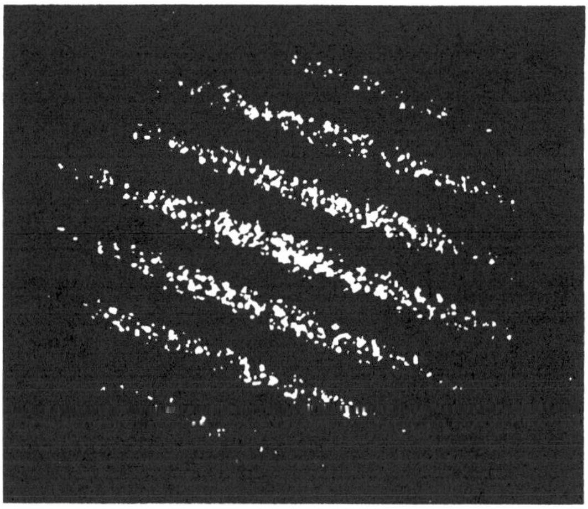

(a)

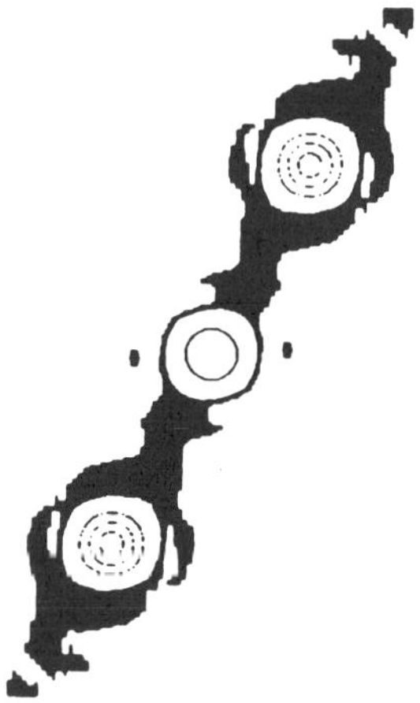

(b)

Figure 9.6 (a) Young's fringes; (b) lines of equal correlation $C(d_1, d_2)$. (From Ref. 31.)

The method begins by sampling the fringe pattern on a mesh of 128 × 128 pixels whose central point coincides with the zero-order spot. The pattern is divided by a smoothed diffraction halo to obtain uniform fringe amplitude. Fringe direction is found by adding the irradiance of the enhanced pattern over bands having various directions within the 0–180° range (Fig. 9.7a). Each irradiance is multiplied by a weighting function $W(k)$ defined by

$$W(k) = \begin{cases} W_0 - k, & -k_0 \le k \le k_0 \\ 0 & \text{otherwise} \end{cases} \tag{9.7}$$

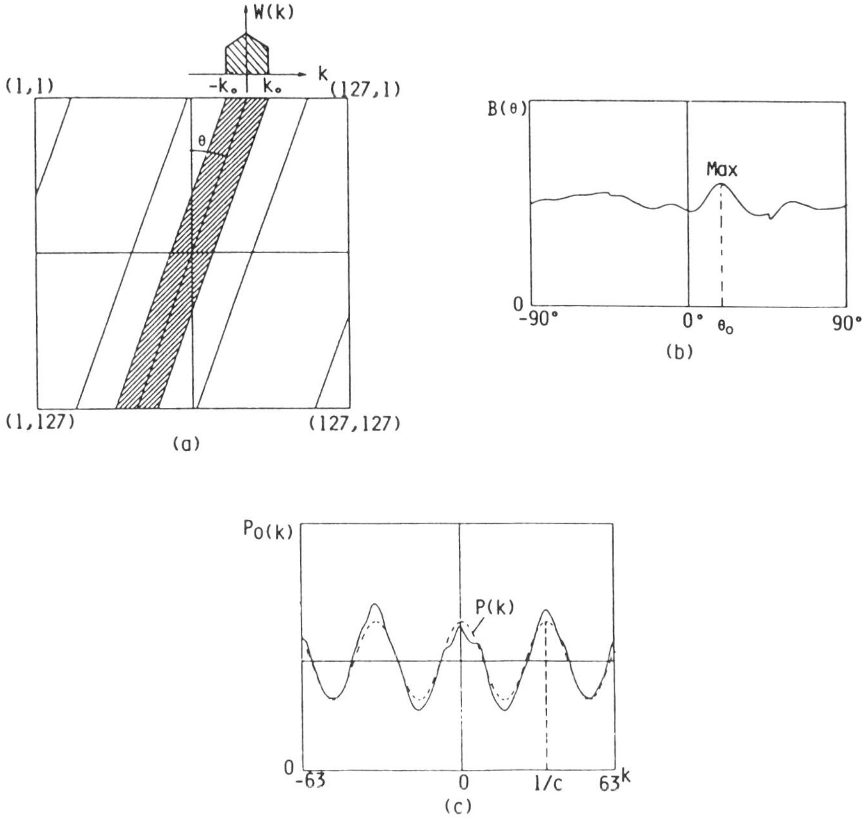

Figure 9.7 Schematics of the algorithm proposed in by Deng and Yamaguchi [32]. (a) Addition band and weighting function; (b) distribution of the sum along band orientation; (c) projection in the fringe direction and curve fitting.

where the width $2k_0 + 1$ of the band is determined by considering the expected range of fringe spacing analyzed and W_0 is chosen through experience. The direction θ_0 in which the sum reaches its maximum is regarded as the fringe orientation (Fig. 9.7b). To determine fringe spacing, the projection of the data irradiance along the detected direction θ_0 is taken. As this projection $P_0(k)$ resembles a cosine curve with a bias, it is fitted to follow a curve by a least squares method,

$$P(k) = a + b \cos (2\pi ck) \tag{9.8}$$

from which the spacing is determined as $1/c$ (Fig. 9.7c).

The reliability and accuracy of this algorithm were investigated using specklegrams with known in-plane translations. Results showed that patterns with less than one fringe within the diffraction halo can be evaluated, so an effective extension for the lower limit of analysis is obtained. The algorithm was also used to evaluate the strain field of a plate subjected to extension, and values derived from displacement components coincided with those measured by strain gauges attached to the sample.

Another numerical method was recently presented by Chen and Chiang [33]. In their approach, first a fast Fourier transform algorithm is applied to four one-dimensional arrays obtained from the original fringe pattern. These arrays are taken along the x and y axes and at 45° and −45° lines with respect to the x axis. In the next step, the fringe-shifting effect resulting from the diffraction halo distribution is eliminated using a division procedure. Then a fringe-enhancement process is applied by convolving four Hanning windows oriented along the same directions to the two-dimensional fringe pattern. In dealing with different fringe densities, rather than changing the size of the Hanning windows, a resolution compression technique is used before the enhancement. For three to seven fringes, the image is compressed to a resolution of 64 × 64. For eight to 18 fringes, it is compressed to 128 × 128, and for more than 19 fringes, it retains the original resolution (256 × 256). The detection of the fringe peaks along the four directions becomes quite simple because good quality fringe patterns are achieved in the process. Experimental tests showed that the standard deviation of a displacement measurement was 0.04 μm in magnitude and 0.12° in orientation and that a 100% success rate was obtained for fringe visibilities higher than 10%.

One final system based on hybrid analog and digital electronic processing techniques was described by Zhang et al. [34]. In this approach, the video signal is fed to an electronic unit where it is processed as shown in Fig. 9.8. There, a gating circuit produces a window that cuts out the TV sync pulses. A specially designed tracking filter separates the fringe signal from

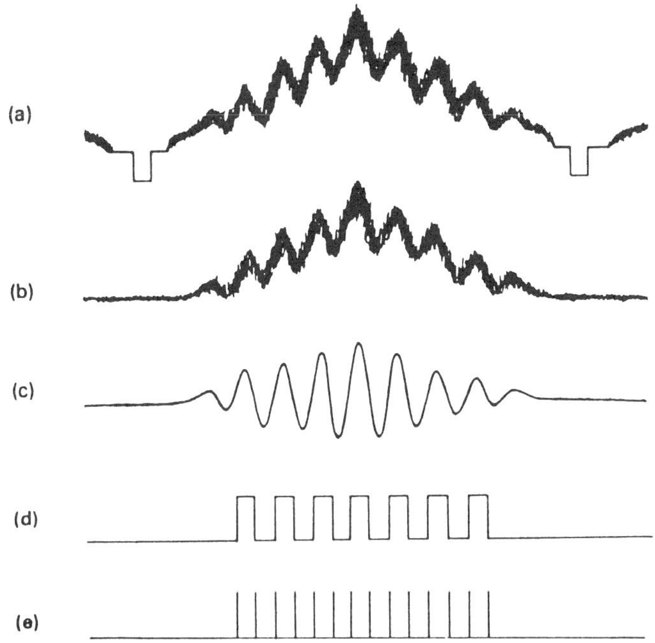

Figure 9.8 Processing of the video signal. (a) Video signal of one TV line; (b) gated video signal; (c) filtered fringe signal; (d) shaped fringe signal; (e) fringe edge pulses. (From Ref. 34.)

the speckle noise and the background. Then the filter signal is converted into a square-wave sequence by a shaper, and a localizer detects the rising and falling edges of every square wave, producing a series of pulses. The localizer gives a coordinate with respect to the TV sync pulse, and pulses are counted continuously and stored in a buffer. A microcomputer reads out fringe coordinates and then calculates fringe spacings along the x direction between adjacent fringes in sequence. To reduce error, a number of lines are taken and results are averaged. The same procedure was carried out with the TV camera scanning vertically, and fringe spacings along the y direction were calculated. Experiments showed that the system gives an accuracy of 1 μm in the range of 20–100 μm.

9.2.3 Measurement of Fluid Velocity Fields

As previously shown, particle image velocimetry (PIV) is a technique derived from laser speckle photography for the measurement of two-dimensional velocity vector fields from a plane of interest within a fluid flow. In this case, double-exposure speckle photography is used to record the position of seeding particles within a thin sheet of light illuminating a section of the flow. The displacement of the particle images recorded in a small region of the photograph can be related to the local fluid velocity provided the time interval between exposures and the magnification of the recording optics are known. However, Young's fringe patterns generated by PIV are heavily modulated by noise, so automated measurement of fringe spacing and orientation is more problematic than in solid mechanics.

The first works on digital analysis of Young's fringes recorded in PIV were reported by Meynart [35–37], who used two complementary methods: one-dimensional averaging and averaged autocorrelation. In the first method, the aim is to obtain a periodic signal by averaging over the lines of the image. However, it is necessary to measure the fringe angle independently, which is accomplished by superimposing a computer-generated line on the digitized image of $N \times N$ pixels and adjusting its direction until it is parallel to the fringes. If $I(m, n)$ is the digitized irradiance and α the angle of the fringes along the n axis, then the frequency

$$f(m) = \sum_{n=0}^{N-1} I\left[m + \left(n - \frac{N}{2} - 1\right)\tan\alpha, n\right], \qquad 0 \leq m \leq N - 1 \tag{9.9}$$

is proportional to the component along the m axis. It is noted that this procedure eliminates the need for mechanical rotation of the fringes. Fourier transformation of $f(m)$ yields the value of its periodicity. To minimize the diffraction halo influence, a constant-amplitude function must be recovered from $f(m)$. However, the diffraction halo in fluid mechanics applications is not a smooth curve because of the non-Gaussian speckle statistics. For this reason, the logarithm of $f(m)$ is Fourier transformed. This method, based on the cepstrum technique used in acoustics [38], generates a frequency spectrum of the diffraction halo well localized about the origin. In this way, periodicity measurement is possible even for low fringe densities. Computations have shown that this technique is very fast. However, it does not always yield clearly identifiable peaks.

As alignment of the fringes is a drawback of the preceding method, an alternative method was proposed by Meynart. Averaging over the autocorrelations of each line gave a periodic signal whose frequency was proportional to the projection of the velocity onto the lines of the acquisition system. The velocity component along the m axis was obtained by computing

$$g(u) = \sum_{n} \left(\frac{\sum_{m=0}^{N-1} I(m, n)I(m + u, n)}{\sum_{m=0}^{N-1} I^2(m, n)} \right), \qquad -N + 1 \leq u \leq N + 1 \tag{9.10}$$

The individual autocorrelations in $g(u)$ were normalized to give an equivalent weight to each line, and lines where the periodic signal was too weak were discarded from the sum. The averaged autocorrelation exhibited an important peak around $u = 0$, which was due mainly to the speckle present on the diffraction halo. The height of this peak was decreased by computing the function

$$g'(u) = \frac{g(u)}{1 + \exp(-u^2/p^2)} \tag{9.11}$$

where p is an adjustable parameter. The n component was evaluated by computing the averaged autocorrelation of the columns of $I(m, n)$.

Compared with the one-dimensional averaging method, this method required substantially more computation time. Another shortcoming was the difficulty of measuring velocities when the fringe density became too low. Both methods were applied to the measurement of very low velocities generated by a Rayleigh–Bénard flow in a cylindrical cell [36]. An application to velocities exceeding 1 m/s, the evolution of unsteady axisymmetrical rolls appearing near the nozzle of an unexcited round jet, was reported in a subsequent work using a double-pulsed ruby laser [37]. Accuracy was found to be about 1% for both methods. It was also concluded that accuracy was ultimately limited by the film resolution, a conclusion based on the supposition that the uncertainty in the recorded image position was on the order of the effective grain size.

Yao and Adrian [39] presented an optical system for the analysis of particle image data that improves the efficiency of data reduction in pulsed-laser velocimetry. This approach reduces the general $N \times N$ pixel two-dimensional problem to two N-element one-dimensional problems by compressing the information in orthogonal directions using integration. In this system, the photographic record is interrogated by a laser beam and the image is passed through a high spatial filter, after which the image beam is divided using a beamsplitter. The images are compressed by cylindrical lenses located perpendicular to each other and are then projected onto linear photodiode arrays oriented in the x and y directions.

Two improvements in the optical setup of a conventional readout system were introduced by Erbeck [40]. The additional speckle noise caused by light scattered from the rough surfaces of the screen used to project the Young's fringes is suppressed by using a rotating ground glass. Thereby, the interference conditions for forming the additional speckles vary with high speed, and as a result the fluctuating small-scale speckle noise disappear. Because, in this scheme, the irradiance signal is digitized in only two shades of gray (0 and 1), the Young's fringes are preprocessed before acquisition. This is done by a special compensation plate with decreasing transmission toward its center, which is placed in the light path behind the photographic record. Figure 9.9 shows the effect of the various preprocessing steps for a digitized Young's fringe pattern. Fringe spacing and orientation are calculated by computing the correlation function along three directions (horizontal, vertical, and a diagonal). Since only two shades of gray are used, computing speed is quite high.

Reynolds et al. [41] applied an autocorrelation method similar to that suggested by Meynart to analyze aerodynamic data. Because nonuniformities in the flow tend to concentrate the seeding particles in some areas,

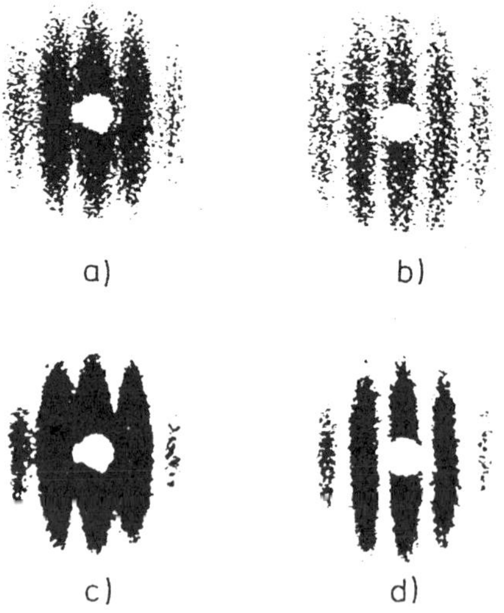

Figure 9.9 Examples of compensated Young's fringe pattern. (a) Without preprocessing; (b) with compensator; (c) with rotating ground glass; (d) with rotating ground glass and compensator. (From Ref. 40.)

leaving others with relatively sparse seeding, it is not practical to assume data on a regular grid pattern. For this reason, a searching scheme is implemented to avoid areas in which no fringes can be recognized. The technique was applied to the flow field about a delta wing at a high angle of attack to generate an unsteady vortex flow. Experimental results demonstrated the feasibility of analyzing turbulent flows in wind tunnel applications.

Pickering and Halliwell [42] developed a theoretical model to compute the fringe pattern irradiance in PIV and to examine the one-dimensional integration and two-dimensional autocorrelation methods. For the case of a Gaussian diffraction halo, errors in the measurement of fringe spacing were evaluated as a function of fringe density and visibility. It was shown that measurement errors lower than 1% can be achieved by dividing or subtracting the diffraction halo from the fringe pattern. Regarding accuracy, both diffraction halo-removal schemes were demonstrated to be essentially equivalent.

Kirita et al. [43, 44] proposed a new method of data reduction that uses the cross-correlation between two separated lines of fringe data. In this way, the analysis of a fringe pattern can be performed fully automatically with a significant decrease in computation time compared to existing techniques. If the digitized fringe data along the two lines labeled M and $M + L$ in Fig. 9.10 are considered, the peak corresponding to the center of a bright fringe in M will be shifted with respect to the equivalent peak in $M + L$ by an amount equal to $L \tan \theta$, where θ is the fringe orientation. By measuring this shift, it is possible to determine fringe orientation di-

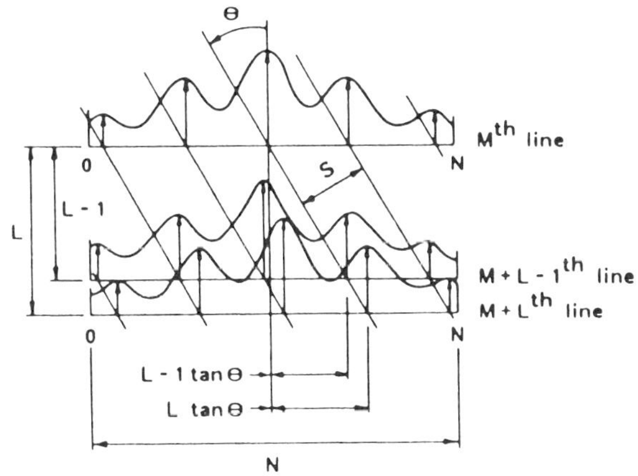

Figure 9.10 Digitized data in the cross-correlation method. (From Ref. 43.)

rectly. Given an accurate measurement of the orientation, fringe spacings are evaluated by locating the coincidence of peaks in two normalized correlation functions along lines L and $L - 1$. Provided the distances L and $L - 1$ are noninteger multiples of each other, only one pair of coincidence peaks will be detected.

Recently, Cho [45] analyzed analytically the multiple-exposure images recorded in PIV for three types of seed particles: elements of infinitesimal size, circular elements of a finite size, and elements of various sizes and shapes. It was demonstrated that the Fourier transform of the multiple-exposure image fields can be expressed as the product of two factors. One factor is merely the Fourier transform of the initial single-exposure image field, which in conjunction with the finite size of image elements reduces the fringe visibility. The other factor is the fringe function, which depends only on the number of exposures and the image displacement, being independent of the size and shape of the image elements and the distribution of the seed particles. This last function provides all the information necessary for the determination of the velocity field. Cho proposed a new technique, digital particle image velocimetry, in which a time sequence of images is captured using a high-speed video camera with each frame containing the image from a single exposure. Then a finite number of single-exposure images with a prescribed separation in time are linearly superimposed to construct the image, which is digitally Fourier transformed to obtain the velocity field. Figure 9.11 shows a triple-exposure image of

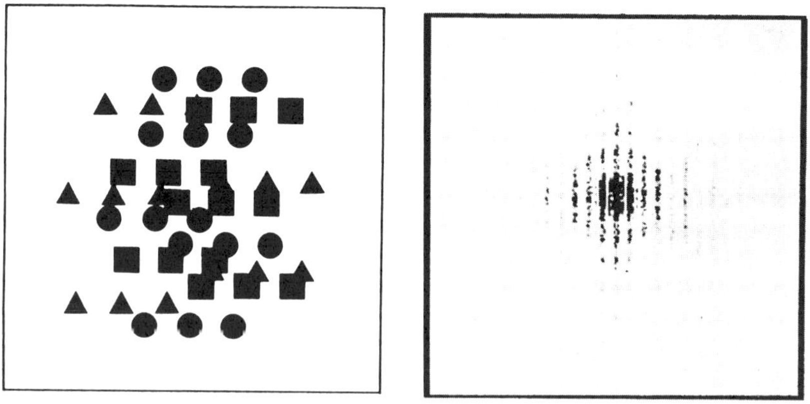

(a) (b)

Figure 9.11 (a) Triple-exposure image of various image elements with uniform displacement; (b) modulus of the Fourier-transformed image. (From Ref. 45.)

various elements with a uniform displacement and the modulus of its Fourier transform. It was shown that this technique greatly enhances the dynamic range of the velocity measurement and unequivocally determines the direction of the velocity vector. As the whole process can be carried out electronically, it can be used as a real-time technique.

More recently, Rösgen et al. [46] discussed the use of a two-dimensional fringe analysis algorithm based on the fast Fourier transform to obtain the power spectrum of the original image. The influence of the diffraction halo, which affects the reliability of the required peak detection, is reduced by subtracting a replica of the power spectrum, rotated by 90°, from the original one. To improve peak detection and reduce the influence of noise, a Hanning window is multiplied onto the raw data before the Fourier transform is computed.

9.2.4 Evaluation and Comparison of Different Processing Methods

In many of the data evaluation systems mentioned previously, information about displacement errors does not seem to be available. Furthermore, some of these systems have not been tested under conditions of reduced fringe contrast to assess their reliability in practical situations. For these reasons, the evaluation of the accuracy of different algorithms and comparison of their performance have recently attracted great interest.

Isacson and Kaufmann [47] used computer-generated fringe patterns to analyze the two-dimensional irradiance data acquired by an image processing system. The main advantage of using simulated fringes instead of real ones is that they are well characterized, so that the speckle displacement is known precisely with no systematic drifts or instrumental errors. The two-dimensional irradiance $I(x, y)$ of a Young's fringe pattern, for convenience parallel to the y axis, was generated using an expression similar to Eq. (9.1). The diffraction halo $I_0(x, y)$ was taken for the case of an incoherent speckle pattern recorded through a circular imaging aperture [7],

$$I_0(x, y) = (2/\pi)^2(\cos^{-1}\rho - \rho(1 - \rho^2)^{1/2})^2 \qquad (9.12)$$

where $\rho = (x^2 + y^2)^{1/2}/\rho_{max}$ is the normalized position in the halo, ρ_{max} being its radius. From $I(x, y)$, a discrete matrix $I(m, n)$ was calculated for different patterns with a resolution of 256 points along both axes. To simulate experimental data, each image was digitized to 8 bits. Then irradiance data were averaged for each two-dimensional fringe pattern along a direction parallel to the fringes and a one-dimensional function was obtained. For comparison, the irradiance along a direction perpendicular to the fringes was also computed. Fringe shifts that appeared if the diffraction halo was not removed prior to fringe spacing evaluation were calculated for different noise-free patterns as a function of fringe density and visibility. Fringe spacings were evaluated by means of two methods: a least squares curve-fitting algorithm and the Fourier transform method described by Kaufmann [19]. It was concluded that even in the case of spacings evaluated from minima, for low-density patterns with less than six fringes and visibility $V \leq 0.75$, the diffraction halo should be divided from the fringe pattern if measurements were determined with relative errors lower than 1%.

Real fringe patterns are contaminated by noise because the diffraction halo itself is modulated by a speckle pattern. Besides this optical noise, other noise still exists due to certain intrinsic sources localized at the detector and the electronic circuits of the image processing system. Common examples include shot and thermal noise. For this reason, Isacson and Kaufmann [48] extended the previous work for the case of noisy fringes using a model proposed by Froehlich et al. [49]. In this model, each pixel of the discrete irradiance $I(m, n)$ was contaminated with additive noise, so the two-dimensional noisy irradiance $I'(m, n)$ was written as

$$I'(m, n) = I(m, n) + kI(m, n)n_1 + n_2 \tag{9.13}$$

where n_1 and n_2 are signal-independent random processes and k is a scalar constant. The second and third terms on the right give the signal-dependent and signal-independent noise contributions, respectively. It was assumed that the noise term n_1 is a zero-mean Gaussian random variable with a standard deviation σ and that the signal-independent term n_2 is uniformly distributed between zero and a value of 255τ. Using a random number subroutine, a set of 256 numbers was generated for a given seed, and 256 sets were used to contaminate each two-dimensional irradiance pattern. Figure 9.12a shows the noise-free irradiance across the two-dimensional diffraction halo along a direction perpendicular to the fringes. This irradiance after contamination by signal-dependent ($k\sigma = 0.3$) and signal-independent ($\tau = 0.01$) noise is shown in Fig. 9.12b. The one-dimensional averaged irradiance that follows from the integration of the two-dimensional noisy fringe pattern is plotted in Fig. 9.12c, which effectively illustrates the improvement in the signal-to-noise ratio. Fringe patterns of different density and visibility were analyzed for several noise regimes, and results indicated that an extra frequency shift due to noise appeared in addition to that originated by the varying irradiance of the diffraction halo.

The directions of the fringe patterns were assumed to be known in both previous works. For this reason, in a subsequent paper, Navone and Kaufmann [50] evaluated the accuracy of the algorithm developed by Robinson

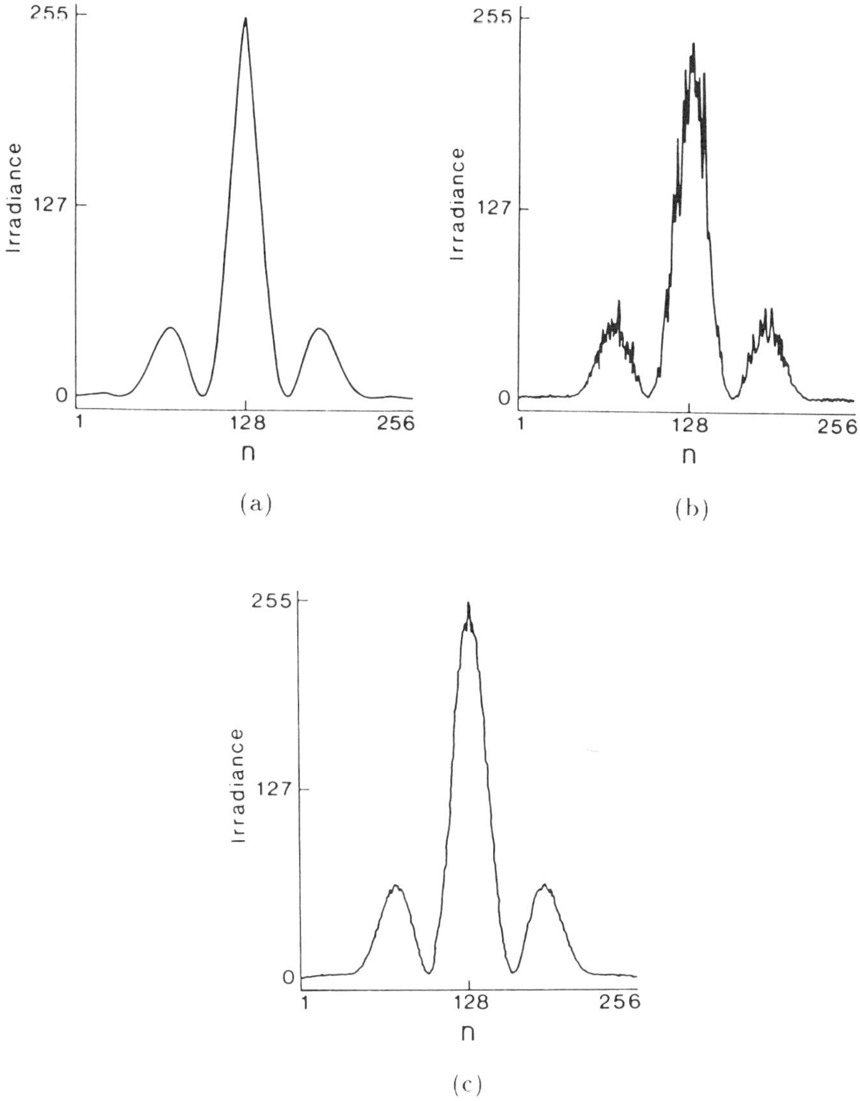

Figure 9.12 Computer simulation of a Young's fringe pattern. (a) Noise-free irradiance across the diffraction halo; (b) irradiance degraded by noise; (c) averaged irradiance. (From Ref. 48.)

[21] for the determination of the fringe direction. Using the described numerical model, computer-simulated patterns were generated for a known direction of the Young's fringes. This direction was numerically evaluated by the processing algorithm for different fringe densities, visibilities, and noise regimes. Numerical results indicated that even in the extreme cases of patterns with very low fringe densities and visibilities contaminated with high noise levels, fringe direction can be determined with a fair degree of accuracy (0.5°). The reliability of this algorithm was also found to be very high.

Diez et al. [51] tested various fringe reduction methods through the analysis of a known case of rigid-body displacement. The tested methods included gray scan along the fringes, autocorrelation of the fringe pattern, discrete Fourier transform, and least squares regression. For all these methods, Diez et al. reported an uncertainty of about three to four orders of magnitude less than that of the measured displacement. They concluded that the Fourier transform used with a geometric reconstruction technique provided the most automated and elegant method for fringe reduction.

Using the simulation model presented in [48], Navone and Kaufmann [52] analyzed the performance of two processing algorithms to compare accuracy and computer run time. One of the algorithms was the one developed by Robinson [21], and the other was a two-dimensional Fourier analysis method based on the calculation of the fringe frequency along both axes by one-dimensional Fourier transformation. Each algorithm was extensively tested for different fringe densities, visibilities, and noise levels, and average relative errors in displacement components were also evaluated. Results showed that the Fourier analysis algorithm is faster and more accurate than Robinson's. A reduction by a factor of 2 was obtained in the average relative error (0.4% for a pattern of 10 fringes and a visibility $V = 0.7$ using a resolution of 128×128 pixels) and by a factor of 4 for the computer run time.

Using the same computer simulation, Navone and Kaufmann [53] later tested the performance of the cross-correlation method developed by Ansari and Ciurpita [29] to evaluate the fringe direction. Patterns of different fringe densities were analyzed for the most unfavorable case of low-visibility fringes contaminated with high-level noise. Errors in the evaluation of fringe direction were calculated and compared with those computed by Robinson's method. It was shown that the cross-correlation algorithm was faster and more accurate than the other method. From the calculation of fringe spacing, displacement components were also determined. Accuracy and computer run time were compared with those obtained using Robinson's method and two-dimensional Fourier analysis. It was shown that the Fourier analysis was the most accurate method for calculating displacement components and also gave a significant time saving when a moderate resolution of 128 × 128 pixels was used.

Recently, Huntley [54] developed a computer model to simulate the entire process of recording and pointwise filtering a specklegram. He considered that the speckle distribution over plane 1 of the lens used to image the object onto the photographic plate (plane 2) consisted of a large number of independent scatterers with the phases uniformly distributed over $(-\pi, \pi)$. Relating the complex amplitude over planes 1 and 2 by the Huygens–Fresnel principle expressed in the Fresnel approximation and considering the filtering operation of the imaging lens, a speckle pattern was generated over the specklegram plane. Figure 9.13a shows the pattern generated by this simulation procedure.

To create a double-exposure specklegram, two speckle patterns were added, with the second one displaced by (k_m, k_n) pixels. Partial decorrelation effects that appear in practice were simulated by adding an independent speckle pattern multiplied by a decorrelation parameter δ that ranged between 0 and 1. The development of the photographic plate was simulated by considering a linear film response characterized by two constants. Then the specklegram placed at plane 3 was probed with a Gaussian laser beam to generate the required Young's fringes at the diffraction plane, plane 4. The irradiance distribution at this last plane was obtained from

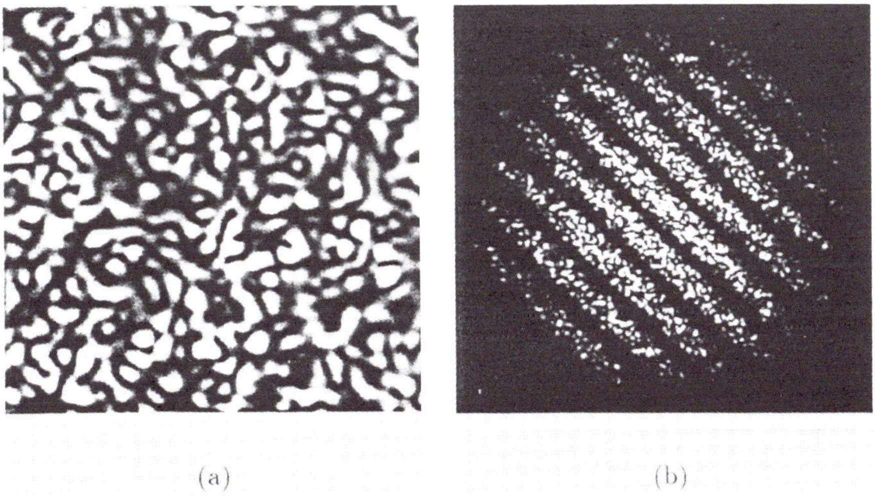

Figure 9.13 (a) Computer-generated speckle pattern. (b) Computer-generated Young's fringe pattern. (From Ref. 54.)

the complex amplitude at plane 3 by discrete Fourier transformation. Figure 9.13b shows a typical fringe pattern.

The computer model was used to generate 10 data sets, which were used to investigate the effects of displacement angle, displacement magnitude, film characteristics, laser probe radius, and speckle decorrelation. Four fringe analysis algorithms were compared: the one-dimensional integration method, which requires the fringe angle to be known; the one-dimensional autocorrelation method, which involves analysis by columns and rows; and the two-dimensional Fourier and Walsh transforms. Interesting observations emerged from the results. It was found that fringe-shift errors due to the diffraction halo were negligible when the halo was simply subtracted from the fringe pattern. The two-dimensional Fourier transform method showed the lowest random errors, and its accuracy was not significantly affected by either displacement angle or magnitude. Also, values of the film transmission parameters were found to be of no importance. By contrast, the fringe visibility and laser probe radius influenced the accuracy considerably. It was found empirically that the standard deviation in each displacement component was proportional to the square of the subjective laser speckle size and inversely proportional to the laser probe radius and the visibility factor. The two-dimensional Walsh transform algorithm runs faster but is the most complex algorithm to code. A final conclusion was that the sampling area of each element is as important as the number of pixels in the digital image. It was shown that aliasing errors can be reduced if a square element of side equal to the interpixel distance is used, so that standard deviation values can still be valid despite any undersampling.

9.3 ANALYSIS OF FRINGES GENERATED IN DUAL-PLATE SPECKLE PHOTOGRAPHY

9.3.1 Heterodyne Speckle Photogrammetry

The heterodyne analysis of interferograms has become a useful tool to automate the evaluation of fringe data [55]. In general, the optical frequencies of the two beams in the interferometer are Doppler-shifted so that the fringes sweep across the output field. There, a photodetector receives sinusoidally modulated illumination, the phase of which is related to fringe order. The detector output signal is analyzed by an electronic meter to measure the phase to 1/1000 of a fringe order.

The application of the heterodyne readout to the analysis of dual-plate speckle photography fringes was introduced by Smith and Stetson [56].

The heterodyne photogrammetry technique, as it was called, is based on the recording of two images corresponding to two states of deformation of the specimen on separate photographic plates. At the analysis stage, the two diffraction halos originating by pointwise filtering from the same point on both plates are combined in a heterodyne interferometer, which results in moving Young's fringes. In principle, this technique offers several advantages over digital fringe analysis techniques. As different exposure plates are recorded, it easily permits the analysis of consecutive loading steps in irreversible processes. Since the heterodyne approach is based on direct local phase measurements instead of irradiance variations, it is much less susceptible to the fringe shifts due to the diffraction halo. Moreover, it does not necessitate the presence of fringe minima within the diffraction halo. This allows the lower limit of measurable displacement to reach the limit set by the noise and the system stability, or well into the subspeckle size displacement regime. Also, the ambiguity of the assignment of the displacement sign is removed.

Figure 9.14 shows a practical system for performing heterodyne photogrammetry. Two Doppler-shifted collinear beams of vertical and horizontal polarizations are created by a rotating half-wave plate followed by a quarter-wave plate and are brought to focus at the output plane by a lens. The mirrors and the compensated beamsplitter are located so that specklegrams are at the same distance from the output beamsplitter. Both specklegrams are mounted on a common translation stage that moves parallel to their surfaces. Photodetectors in the transform plane convert the scan-

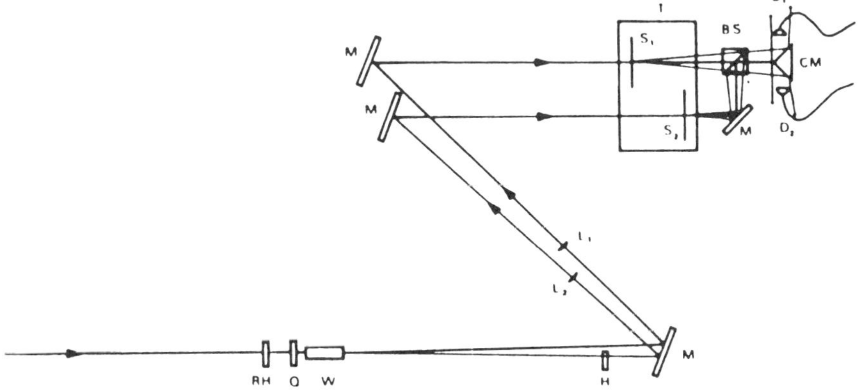

Figure 9.14 Readout system for heterodyne photogrammetry. (From Ref. 56.)

ning fringes into sinusoidal signals, which are filtered to remove noise and fed to an electronic phase meter. The measured strain ϵ is obtained from

$$\epsilon = \Delta\theta \, R\lambda/(2\pi d \, \Delta x) \tag{9.14}$$

where $\Delta\theta$ is the phase change in radians, R is the distance from the specklegrams to the transform plane, λ is the wavelength of the light, d is the separation of the detectors, and Δx is the distance by which both specklegrams are translated.

The accuracy with which phase measurements can be made is limited by the effect of scintillation noise and speckle grain noise. These effects were studied by Stetson [57, 58] and Kim et al. [59, 60]. The scintillation noise results from imperfect correlation of the two diffraction halos. The interfering fields not only give the desired signal but also provide a background of speckles scintillating at the same frequency with random phases and amplitudes. As the specklegrams are scanned, scintillation randomly perturbs the output signal and creates errors in strain measurement, thus setting a practical limit on accuracy. Speckle grain noise is an effect that persists even when the two diffraction halos are perfectly correlated. It results from the fact that, in general, the phase of the halo interference does vary across the detector and is composed of interfering speckles. The brightest speckles make the greatest contribution to the output signal and therefore emphasize the phase corresponding to their position on the fringe pattern. Because speckles change randomly as the illumination beam is scanned, the measured phase is perturbed and strain errors will appear. It was shown that both of these effects can be reduced by using a large number of detectors and properly processing their signals.

9.3.2 Phase Shifting of Whole-Field Fringes

The heterodyne approach unfortunately rests on a sequential pointwise analysis technique, and the most severe requirement is that the scanning of both plates should be made such that they remain parallel to themselves throughout the entire procedure. From tests, it was proved that tilts due to wobbles in the driving stage or due to different temperature changes between the two plate holders can give rise to appreciable fictitious displacements. Also, the angular instability of the probing beam is another serious problem affecting the system accuracy. To overcome these limitations, Kaufmann and Jacquot [61] recently developed a new readout scheme based on the application of a phase-shifting technique, so that it is not necessary to scan both plates. The basic idea is to combine the far-field diffraction patterns generated by the two specklegrams in a Mach–Zehnder interferometer. An arrangement of the first prototype is shown in Fig. 9.15a.

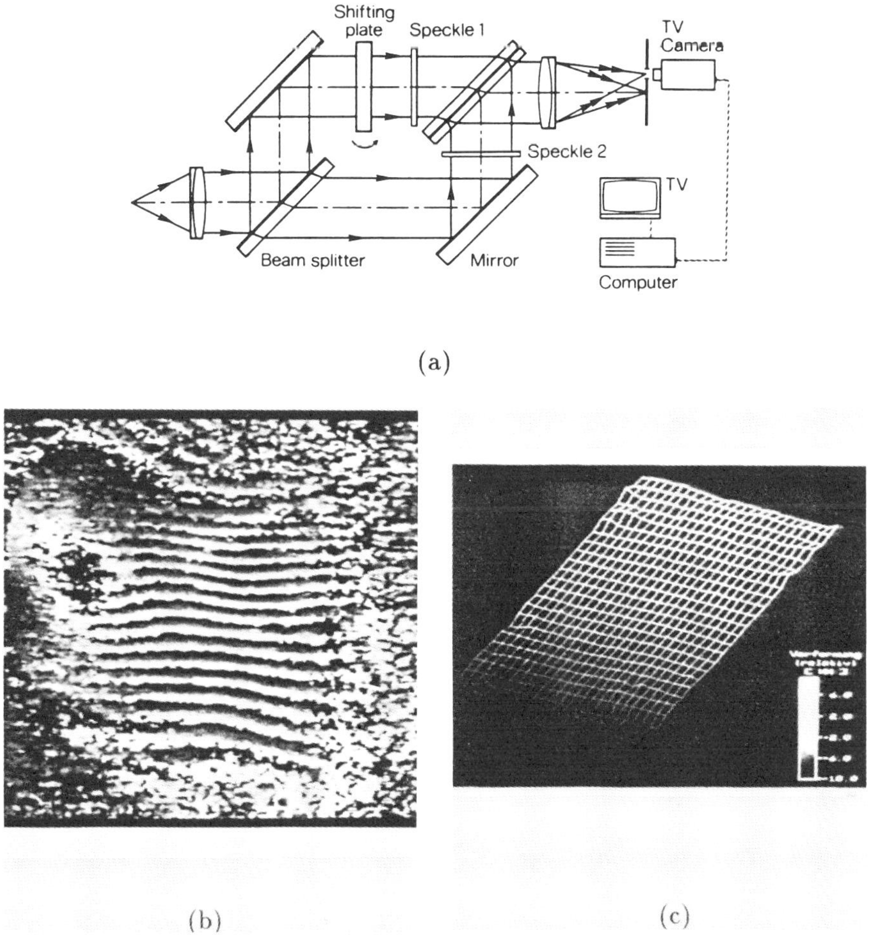

Figure 9.15 (a) Phase-shifting readout system. (b) Phase computed for an in-plane rotation. (c) Isometric plot of computed phase. (From Ref. 61.)

The diffraction patterns of both specklegrams are superimposed by means of a compensated beamsplitter and are collected by a transform lens. Whole-field fringes are obtained by optically filtering both combined diffraction halos using the offset aperture of a TV camera on the focal plane of the transform lens. If α is the angle between the optical axis and the line joining the center of the transform lens and the aperture location, the fringe pattern satisfies

$$d_\alpha \sin \alpha = n\lambda \tag{9.15}$$

where d_α is the component of the in-plane displacement along the aperture direction, n is the fringe order, and λ is the laser wavelength.

A tilted glass plate placed in one of the interferometer arms was used to shift the resulting fringes. In addition to $I_1(x, y)$, two sets of irradiance data, $I_2(x, y)$ and $I_3(x, y)$, were recorded as the glass plate was twice tilted at an angle that gave two successive 120° phase shifts. The output of the TV camera was digitized and stored in a personal computer, where the phase angle $\phi(x, y)$ was calculated from [62]

$$\phi(x, y) = \tan^{-1}\left(\sqrt{3}\,\frac{I_3(x, y) - I_2(x, y)}{2I_1(x, y) - I_2(x, y) - I_3(x, y)}\right) \tag{9.16}$$

The surface displacement component d_α can be calculated as

$$d_\alpha = \phi(x, y)\lambda/(2\pi \sin \alpha) \tag{9.17}$$

Figure 9.15b shows the computed phase for an in-plane rotation of one of the specklegrams along a direction parallel to one of its sides. This gray-level representation was obtained before removing the modulo 2π uncertainties. An isometric plot of the calculated phase is shown in Fig. 9.15c, where it is seen that the phase is constant along a line parallel to one side of the plate and increases linearly in a perpendicular direction.

A simpler on-line readout system, which offers several advantages over the interferometric approach previously mentioned, was recently presented by Kaufmann and Jacquot [63]. In this scheme, the pair formed by specklegrams P_1 and P_2 is illuminated by a parallel laser beam at normal incidence, and the two superimposed halos are collected by a Fourier transform lens as shown in Fig. 9.16. When the pair is seen through an offset aperture located along the y axis at the focal plane, whole-field fringes that are contours of the equal in-plane displacement component d_y are visualized. These fringes are phase-shifted by giving to the specklegram P_2 a translation Δy along the y direction. For this purpose, a piezoelectric transducer with submicrometer resolution is used. In response to the translation of specklegram P_2, a constant phase increment $\Delta\phi$ is added at each point of the interferogram and the fringes shift accordingly as a whole.

As in the previous interferometric system, three successive frames of irradiance data, I_1, I_2, and I_3, are stored in the personal computer as the phase is changed stepwise with a step of 120°. A high-voltage amplifier commanded by the personal computer provides a linear signal to drive the phase shifter. The phase $\phi(x, y)$ of the resulting fringe pattern is calculated using Eq. (9.16), and then the displacement component d_y is computed from Eq. (9.17).

Preliminary results obtained from in-plane rotations of one of the specklegrams are encouraging. Compared with the earlier interferometric

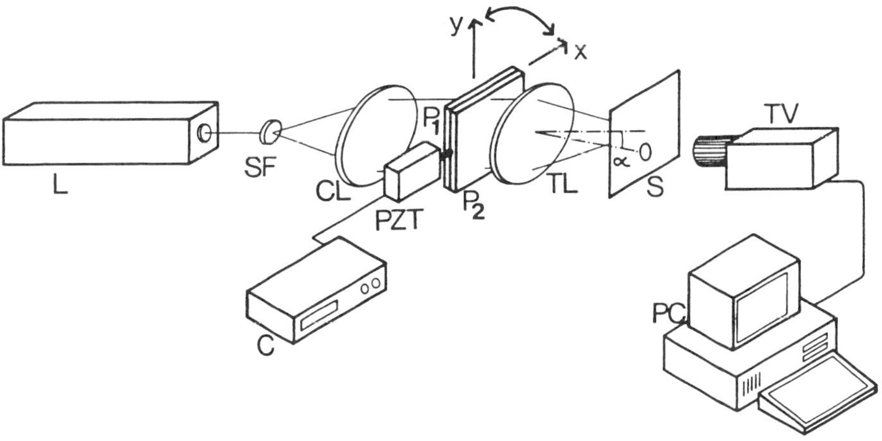

Figure 9.16 On-line phase-shifting readout system. (From Ref. 63.)

scheme, the new readout system presents several advantages. The optical setup is quite simple to align and does not require the use of high-quality optical components. Stability requirements for maintaining good fringe visibility are less severe. Finally, the other in-plane displacement component d_x can be simply determined by turning both the screen aperture and the TV camera by 90° and translating the specklegram P_2 in the x direction.

9.3.3 Circular Fringes

A major limitation of double-exposure speckle photography is the effect of rigid-body movements of the object. When these movements become too large, no fringe modulation is seen in the diffraction halo. It is well known that dual-plate speckle photography provides a technique for canceling these displacements by recording the images corresponding to the two deformation states on separate photographic plates. However, as plate emulsions are not infinitely thin, speckle fringes in the diffraction halo are found to be curved. This effect, which occurs even when the reading is made using a sandwich setup in which the two emulsions are placed against each other, complicates the automatic evaluation of data. Some years ago, Beranek and Bruinsma [64] proposed a new approach to extract displacement data in dual-plate speckle photography, in which both plates are placed in a holder so that the glass base of one of them is between the emulsions. Figure 9.17a shows both specklegrams, each of width t, separated by a distance s. If A_2 is the point corresponding to A_1 after a small radial displacement u_r over the xy plane is introduced between exposures,

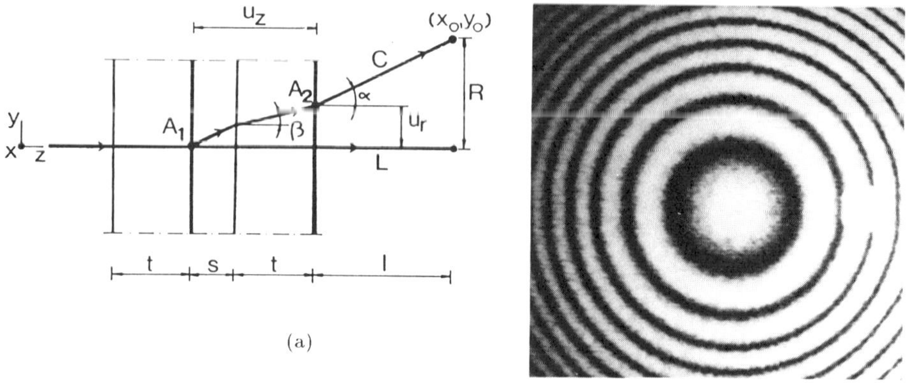

(b)

Figure 9.17 (a) Generation of circular fringes in dual-plate speckle photography. (b) Simulated circular fringe pattern. (From Ref. 65.)

these points will behave as coherent light sources when a laser beam in the direction of the z axis illuminates both specklegrams. From interference theory, it can be shown that a fringe pattern of concentric circles centered at the intersection of the line connecting A_1 and A_2 is formed over the plane $z = l$. These fringes are described by

$$(x - x_0)^2 + (y - y_0)^2 = \frac{(u_z + m\lambda)^2 - u_z^2}{(u_z + m\lambda)^2} l^2 \tag{9.18}$$

where x_0 and y_0 are the coordinates of the fringe pattern center, u_z is the distance between A_1 and A_2 along the z axis, λ is the laser wavelength, $m = 1, 2, \ldots$ for constructive interference (bright fringes), and $m = 0.5, 1.5, \ldots$ for destructive interference (dark fringes). From Fig. 9.17a, it is seen that the radial displacement u_r can be calculated from

$$u_r = (s + t/n)(R/l) \tag{9.19}$$

where n is the refractive index of the glass base of the photographic plates and R is the distance between the center of the circular fringe pattern and the undiffracted laser beam. Thus, measuring the coordinates of the fringe pattern center for points on a square mesh lattice over the specklegrams, the complete displacement field can be evaluated. In this way, the circular feature of the fringes is gainfully used and the readout process is also simplified. Moreover, the intrinsic 180° ambiguity in the displacement direction is easily eliminated.

Recently, Kaufmann and Galizzi [65] evaluated the performance of a processing algorithm to automate the analysis of the circular fringes obtained in dual-plate speckle photography. The assessment of the algorithm was made by means of computer-simulated fringe patterns. After the parameters u_z, λ, and l were chosen, the simulation was started by computing the array $C(m, n)$ from the circular fringe pattern given by Eq. (9.18) centered at the pixel (m_0, n_0). To include the effect of partial decorrelation of the two speckle patterns, this image was multiplied by a visibility parameter V that ranged between 0 and 1. The second step consisted in constructing the diffraction halo $I_0(m, n)$ generated by the specklegram pair. The corresponding array $I_0(m, n)$ was calculated from the theoretical diffraction halo generated by a single speckle pattern recorded through a circular imaging aperture [7]. The third step was to introduce the diffraction halo modulation by multiplying $C(m, n)$ by $I_0(m, n)$. The noise-free fringe pattern was modified in the fourth step by adding the Gaussian profile of the undiffracted laser beam. The last step was to introduce the optical noise, which modulates the fringe pattern by means of the same model used in [48]. Figure 9.17b shows a typical computer-simulated fringe pattern.

The center of curvature of the circular fringe pattern was determined by means of an algorithm similar to the one described by Robinson [21], originally developed to analyze Newton's ring type fringes. The analysis proceeds as follows. Straight vectors chosen empirically from the pattern are read, and the bright fringe peak with maximum irradiance is located. As the success of the algorithm depends on the correct location of a bright fringe peak, checks are made to eliminate false peak detection. The approximate tangent to the fringe is determined by adding irradiance values along a series of radial straight vectors passing through the determined peak. The vector giving the largest summed value is taken as the approximate tangent to the fringe. To evaluate the center of curvature of the fringe, irradiance values along a series of circular vectors passing through the peak are summed. The circular vector giving the maximum summed value is taken as the best fit to the fringe. To determine the center of the circular vector previously determined, this center is propagated along a circular path. The best fit to the circular fringe and its center are found by iteratively repeating the last two steps.

The computer model was used to generate several fringe patterns, which allowed an investigation of the effect of displacement magnitude, orientation, and fringe visibility. Relative errors in displacements were also evaluated. The algorithm reliability was proved to be quite good. Numerical results showed that the algorithm is capable of accurately determining in-plane displacement components in dual-plate speckle photography, but it needs a rather long computer time.

More recently, Kaufmann and Galizzi [66] developed a new processing algorithm for the automatic analysis of circular fringe patterns that significantly reduces the computer run time. For the case of simulated data, the undiffracted laser beam is firstly centered at the pixel (255, 255). To increase fringe contrast, the small area containing the undiffracted laser beam is set to zero, and to obtain a pattern with nearly uniform amplitude, the image is divided by an averaged diffraction halo. In the case of real data, these steps are carried out by appropriately locating the TV camera used to record the fringe pattern and blocking off the laser beam with a beam stop. For each line of the resultant pattern, bright fringe peaks are located. Then the distances between consecutive peaks are determined and the greatest one is identified. The line giving the greatest distance is taken as an approximate diameter of the first circular fringe, and its center is determined. To optimize the location of the center, irradiance values along a circular path are added. This step is iteratively repeated by propagating the center in the neighborhood of the pixel determined previously and also by varying the radius of the first circular fringe. As the success of the algorithm depends on the correct location of bright fringe peaks, a special subroutine was developed for this purpose. False peak detection is eliminated by comparing the peak irradiance with those of several neighbors. Also, the number of neighbors used in each comparison grows as the distance between consecutive peaks increases. Finally, the modulation of a fringe must be greater than a given value.

Numerical results show that the algorithm is capable of analyzing circular fringe patterns with good accuracy. In fact, absolute errors in the distance R were less than 1 pixel in nearly all patterns analyzed. The accuracy is not significantly affected by the displacement magnitude, angle, or fringe visibility, and the algorithm can analyze patterns with a visibility factor $V = 0.4$ giving relative errors less than 1%. It was also shown that this algorithm gives both lower relative errors and a significant time saving.

9.4 IMAGE PROCESSING USING PHOTOREFRACTIVE MATERIALS

In the schemes described so far, photographic films are used to record the necessary speckle patterns, thus requiring troublesome wet processing procedures. This drawback can be overcome by replacing the photographic film with new electro-optical materials, which permit the recording of fringe patterns in situ. For this purpose, several correlator designs have been recently published [67]. Collicott and Hesselink [68] developed a system that uses a BGO photorefractive crystal in the transverse electro-optic configuration as recording medium. These authors showed that double-exposure speckle patterns produced by a solid body in motion illuminated by two *Q*-switched laser pulses can be recorded into the crystal with no need for a reference beam as in holographic recording. They also demonstrated that speckle patterns can be immediately interrogated with an optical processor similar to the conventional one used for measuring speckle displacements recorded on photographic film. However, it was discovered that the diffraction halo generated from the device was not circularly symmetric, so the maximum number of observable fringes was directionally dependent and fringe contrast was quite low.

The visibility of the Young's fringes was investigated by Nakawaga and Minemoto [69] using a PROM (Pockel's readout optical modulator) device. This device has the advantage that a large electro-optical effect can be generated in a thin BSO crystal in the longitudinal mode. It was shown that the specklegram recorded by underexposure can be transformed to a higher contrast specklegram by the optimum reduction of the applied voltage to the PROM device in the readout operation and that the visibility of the generated Young's fringes improved also.

Recently, Coupland and Halliwell [70] developed an automated optical method that uses a photorefractive crystal to perform full two-dimensional Fourier transformation of Young's fringe patterns. This method, which is equivalent to calculating the two-dimensional spatial autocorrelation of the

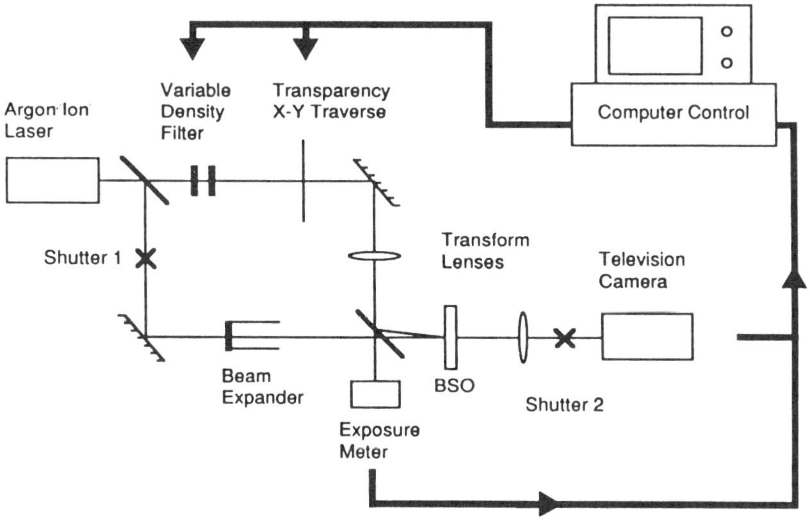

Figure 9.18 Automated readout system with photorefractive recording. (From Ref. 70.)

amplitude transmitted by each illuminated region, is incorporated into an automated transparency analysis system as shown in Fig. 9.18. The optical autocorrelator is a new design of the system reported earlier [71]. In this scheme, first a Vander Lugt filter is recorded as a Fourier hologram of the light amplitude immediately behind the interrogated region. The reference beam is then shuttered off, and the autocorrelation function of this region is subsequently formed in the output plane. A simple algorithm is used to locate and measure the position of the relevant peaks in each autocorrelation. This system was successfully used to analyze a specklegram recording two-dimensional flow past a step. Also, it is worth noting that processing time was quite low, since about 4 s was required to make a single displacement measurement.

9.5 FRINGE ANALYSIS IN ESPI

In Chapter 4, it is seen that electronic speckle pattern interferometry (ESPI) relies on both electronic and optical phenomena to generate a correlation fringe pattern. Since speckle fields are used as information carriers, a high level of noise is inherent to the technique, and fringe analysis is quite a difficult problem. For this reason, until recently ESPI has provided only qualitative data in the form of fringe patterns, requiring a skilled operator to interpret them. Since 1980, major advancements have been made with the application of digital image processing techniques. Using this approach, Nakadate et al. [72, 73] and Höfling and Osten [74] removed electronic noise by frame averaging and suppressed speckle noise by means of special filtering operations. By extracting fringe skeletons, quantitative measurements were performed.

During the last few years, numerous research groups have investigated different approaches to automate the analysis of ESPI data, most of them based on the phase-shifting method. Because of its importance, this method will be discussed next in detail. Basically, the optical setup of a phase-shifting ESPI system is similar to that of a conventional speckle interferometer but includes a PZT-actuated mirror in either the reference or object beam. This mirror is used to change the optical path difference between the two beams by applying a high voltage to the PZT transducer. To find the phase of a deformation, irradiance data are taken for the object before it is stressed, and the interference pattern is calculated using standard phase-measurement techniques [62]. A common algorithm for phase calculation is the four-frame technique, in which four irradiance sets $I_i(x, y)$ are recorded,

$$I_i(x, y) = I_0\{1 + V \cos [\phi(x, y) + \alpha_i]\} \tag{9.20}$$

where I_0 is the average irradiance, V is the modulation of the interference term, $\phi(x, y)$ is the phase of the interference between the reference and the speckle pattern scattered by the object, and $\alpha_i = 0, \pi/2, \pi, 3\pi/2$. The phase at each point is calculated from

$$\phi(x, y) = \tan^{-1}\left(\frac{I_4 - I_2}{I_1 - I_3}\right) \tag{9.21}$$

This equation is sufficient for only a modulo π calculation. To determine the phase modulo 2π, the signs of the numerator and denominator of Eq. (9.21) must be examined.

After the object is deformed, a second set of irradiance data are taken, and the phase values $\phi'(x, y)$ for this state are also calculated. Then, the phase of the deformation $\Delta\phi\ (x, y)$ is computed from

$$\Delta\phi\ (x, y) = \phi'(x, y) - \phi(x, y) \tag{9.22}$$

Due to the arctan function, phase-shifting ESPI yields a calculated phase lying in the range from $-\pi$ to π. Therefore, phase unwrapping must be carried out to restore the unknown multiple of 2π to each pixel. In principle, this procedure can be achieved by comparing the phase difference between adjacent pixels. When the phase difference is greater than π, 2π is either added or subtracted from the remaining pixels in the row, and the process is then repeated along the columns.

Phase-shifting ESPI systems for measurement of out-of-plane and in-plane deformations were described by Nakadate and Saito [75]. Both systems used a reference beam generated by a rough surface and median filtering to reduce speckle noise. Figure 9.19 shows the phase distribution in nine states of in-plane deformation of a brass plate with a hole.

Systems for measuring static deformations were reported by Creath [76], Robinson and Williams [77], Kerr and Tyrer [78], and Kerr et al. [79], which included different approaches to filter the raw phase to smooth speckle noise content while retaining the 2π discontinuities. Creath and Slettemoen [80] and Vikhagen [81] applied phase-shifting ESPI to the measurement of vibration amplitudes. Kerr et al. [82] developed an alternative method that requires only two fringe patterns at 0° and 90° phase difference. This method in conjunction with a pulsed laser was applied by Mendoza Santoyo et al. [83] to study resonant in-plane vibrations of a metal plate. The modulo 2π phase map for a plate vibrating at 31.5 kHz is shown in Fig. 9.20.

By using three different illumination geometries around a single imaging system, Shellabear and Tyrer [84] developed a system to measure the orthogonal components of vibration amplitude independently. Figure 9.21

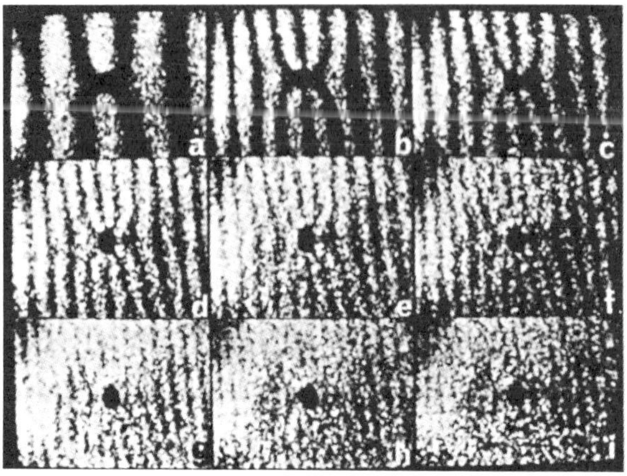

Figure 9.19 Phase distribution in nine states of in-plane deformation of a brass plate with a hole. (From Ref. 75.)

shows the case of volume vibrations at 24 kHz of a thick annular cylinder, where the amplitude vector contains significant x, y, and z components.

Vikhagen [85] developed a phase-shifting technique that does not need to fully stabilize the object during recording, so it can be used with slowly changing specimens such as those loaded by heating. This technique was combined with phase gradient calculation to evaluate the actual phase gradient on each point on the image, thus giving the same information as shearography. Figure 9.22 shows a triangular object between plies in a polymer laminate heated between exposures.

The development of new algorithms to automatically unwrap the phase of complex ESPI fringe patterns is an area of current interest. This is

Figure 9.20 Phase map for a plate vibrating at 31.5 kHz. (From Ref. 83.)

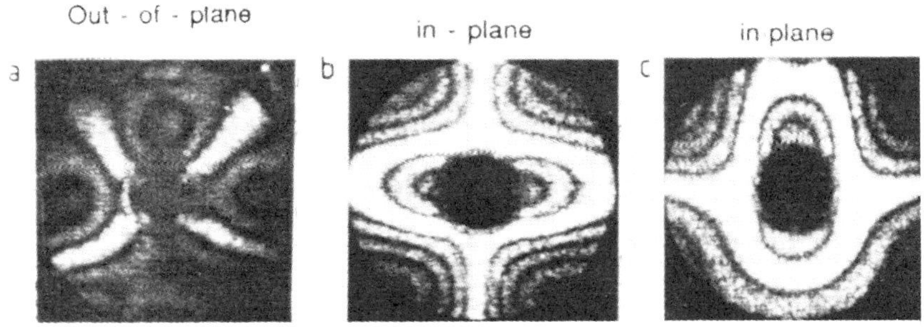

Figure 9.21 Vibration analysis of a thick annular cylinder at 24 kHz. (From Ref. 84.)

because conventional unwrapping algorithms, which are computationally efficient, do not give reliable results in many cases, for the following reasons. Electronic and speckle noise, the most familiar error sources, have a time-dependent character, so the noise configuration is not identical for the various phase-shifted frames. Hence, computed principal value phases are corrupted by noise to such an extent that localized inconsistent regions appear. Discontinuities on the object such as holes, shaded regions, or irregular boundaries can cause abrupt phase changes. Consequently, they are another source of incorrect unwrapping. Finally, discontinuous and displaced fringes such as those due to cracks also give rise to dislocations in the phase distribution. This unsteady behavior results in inconsistent areas that must be avoided to obtain a correct unwrapping. To overcome these problems, strategies have recently been developed that allow the automated analysis of ESPI fringes.

Huntley [91] proposed a new unwrapping algorithm, combining both noise immunity and computational efficiency, based on the simple requirement that the unwrapped map should be independent of the route by which unwrapping takes place. One of the main advantages of this algorithm is its ability to deal with regions of an image that contain no fringe information.

Towers et al. [86] developed an approach based on the segmentation of the fringe field into small rectangular areas called tiles. The strength of this approach stems from the local information extracted from the individual tiles, and the rejection of tiles containing bad data further improves confidence in the solution. Tiles are connected using a minimum weight spanning tree that produces an optimal solution for the whole-phase field.

Spik and Robinson [87] used an alternative strategy to improve the slow speed of some unwrapping methods. This is based on a fully path-inde-

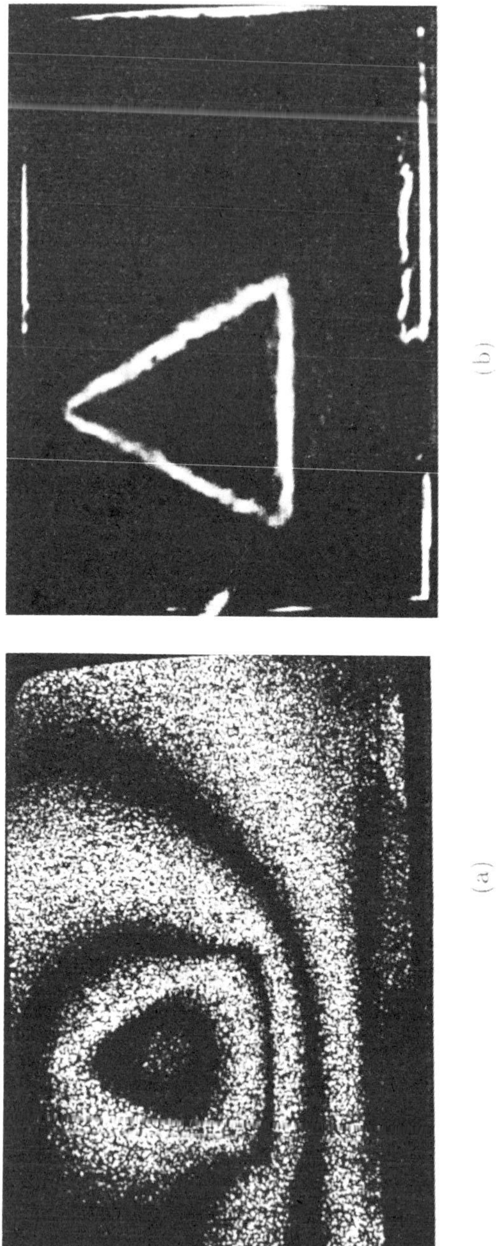

Figure 9.22 Triangular object between plies in a laminate. (a) Ordinary ESPI fringes; (b) deformation phase gradients. (From Ref. 85.)

pendent algorithm, cellular automata, which effectively propagates the unwrapping process in all directions at once. As the cellular automata method has natural parallelism, it was implemented in an array processor, thereby decreasing the computing time.

A major disadvantage of ESPI is that after the interference fringe data are processed, terms proportional to the phase of the speckle are produced by the interference between the reference and object beams, causing noisy fringes. By processing the data differently, Stetson and Brohinsky [88] and Stetson et al. [89] showed that the speckle term can be removed. This method, called electro-optic or electronic holography, uses the same optical setup as ESPI but provides real-time images that are comparable to those obtained from photographic holograms. This technique uses four frames of data with 90° relative phase shifts between the object and reference beams, which are stored in an array processor. The eight frames of data, four taken before deformation and four after, are combined in a pipeline image processing system to produce fringes proportional to the change due to the object displacement.

To conclude, it must be pointed out that single-mode fiber optics allow the design of compact systems with little to adjust or go out of alignment. Additionally, these fiber systems permit the laser to be located remotely from the experimental area, and stroboscopic illumination can be achieved using fiber switches, removing the need for a Bragg or Pockel cell. Phase shifts can be introduced by wrapping a length of fiber around a piezoelectric transducer, and phase feedback can be used to eliminate phase modulator errors and to compensate for phase drifts caused by environmental disturbances [90]. It is hoped that these and new developments to come will enable ESPI techniques to be flexibly used to solve a number of industrial problems.

9.6 CONCLUDING REMARKS

In summary, automatic fringe pattern analysis procedures are having a major impact in the field of speckle metrology, both by extending the range of applications and by improving its accuracy. Nowadays, new techniques are being actively pursued. It may be expected that future developments and decreasing costs of parallel computers will also stimulate new advances in this area. Finally, I believe that different automatic procedures possess great potential for being developed into a practical tool suitable for nondestructive inspection in industrial environments.

ACKNOWLEDGMENTS

I wish to express my gratitude to A. E. Ennos and J. M. Burch, pioneers in the field of speckle metrology, who introduced me to the area of automatic fringe pattern analysis during my 1978–1979 stay at the Division of Mechanical and Optical Metrology of The National Physical Laboratory, Teddington, U.K. The names of my collaborators are appropriately cited in the text; my debt of appreciation to them is very great. This research has been generously supported by the National Council of Scientific and Technological Research of Argentina.

REFERENCES

1. G. E. Maddux, S. L. Morman, and R. C. Corwin, A programmable data-retrieval system for in-plane displacements for speckle photographs, U.S. Air Force Tech. Rep. AFFDL-TM-78-109-FBE (1978).
2. B. Ineichen, P. Eglin, and R. Dändliker, Hybrid optical and electronic image processing for strain measurements by speckle photography, Proceedings of the 11th Congress of the International Commission for Optics, Madrid, 1978, p. 563; *Appl. Opt.*, *12*, 2191 (1980).
3. G. H. Kaufmann, A. E. Ennos, B. Gale, and D. J. Pugh, An electro-optical read-out system for analysis of speckle photographs, *J. Phys. E*, *13*, 579 (1980).
4. H. Kreitlow and T. Kreis, Automatic evaluation of Young's fringes related to the study of in-plane deformations by speckle techniques, *Proc. SPIE*, *210*, 18 (1980).
5. B. Chitsaz and F. A. Moslehy, A computerized displacement data analysis in laser speckle photography, *Exp. Tech.*, *13*, 21 (1989).
6. G. H. Kaufmann, On the numerical processing of speckle photograph fringes, *Opt. Laser Technol.*, *12*, 207 (1980).
7. R. Meynart, Diffraction halo in speckle photography, *Appl. Opt.*, *23*, 2235 (1984).
8. C. S. Vikram and K. Vedam, Speckle photography of lateral sinusoidal vibrations: error due to varying halo intensity, *Appl. Opt.*, *20*, 3388 (1981).
9. C. S. Vikram, Error in speckle photography of lateral sinusoidal vibrations: a simple analytical solution, *Appl. Opt.*, *21*, 1710 (1982).
10. C. S. Vikram, Simple approach to process speckle-photography data, *Opt. Lett.*, *7*, 374 (1982).
11. C. S. Vikram, Analysis of Young's fringes in speckle photography: generalized square imaging aperture, *Appl. Phys. B*, *31*, 221 (1983).
12. C. S. Vikram, Interpretation of Young's fringes in speckle photography for lateral vibration analysis: imaging with circular aperture, *Optik*, *65*, 263 (1983).
13. C. S. Vikram and K. Vedam, Processing of speckle photography data: circular imaging aperture, *Appl. Opt.*, *22*, 653 (1983).
14. C. S. Vikram and K. Vedam, Selective counting path of Young's fringes in speckle photography for eliminating diffraction halo effects, *Appl. Opt.*, *22*, 2242 (1983).
15. J. Georgieva, Diffraction halo effect in speckle photography, *Appl. Opt.*, *25*, 3970 (1986).

16. J. Georgieva, Removing the diffraction halo effect in speckle photography, *Appl. Opt.*, *28*, 21 (1989).
17. C. Joenathan and R. S. Sirohi, Elimination of error in speckle photography, *Appl. Opt.*, *25*, 1791 (1986).
18. K. Hinsch, Fringe positions in double-exposure speckle photography, *Appl. Opt.*, *28*, 5298 (1989).
19. G. H. Kaufmann, Numerical processing of speckle photography data by Fourier transform, *Appl. Opt.*, *20*, 4277 (1981).
20. G. H. Kaufmann, Digital analysis of speckle photography fringes: processing of experimental data, *Appl. Opt.*, *21*, 3411 (1982).
21. D. W. Robinson, Automatic fringe analysis with a computer image-processing system, *Appl. Opt.*, *22*, 2169 (1983).
22. S. Toyooka and M. Tominaga, Spatial fringe scanning for optical phase measurement, *Opt. Commun.*, *51*, 68 (1984).
23. S. Toyooka, Y. Iwaasa, M. Kawahashi, K. Hosoi, and M. Suzuki, Automatic processing of Young's fringes in speckle photography, *Opt. Lasers Eng.*, *6*, 203 (1985).
24. J. M. Huntley, An image processing system for the analysis of speckle photographs, *J. Phys. E*, *19*, 43 (1986).
25. J. M. Huntley, S. J. P. Palmer, and J. E. Field, Automatic speckle photography fringe analysis: application to the crack propagation and strength measurement, *Proc. SPIE*, *814*, 276 (1988).
26. J. M. Huntley, Speckle photography fringe analysis by the Walsh transform, *Appl. Opt.*, *25*, 382 (1986).
27. J. M. Huntley, Fast transforms for speckle photography fringe analysis, *Opt. Lasers Eng.*, *7*, 149 (1986/87).
28. K. G. Beauchamp, *Walsh Functions and Their Applications*, Academic, New York, 1975.
29. F. Ansari and G. Ciurpita, Automated fringe measurement in speckle photography, *Appl. Opt.*, *26*, 1688 (1987).
30. W. Arnold and K. D. Hinsch, Parallel optical evaluation of double-exposure records in optical metrology, *Appl. Opt.*, *28*, 726 (1989).
31. R. Höfling, Fringe analysis by automatic image fitting and its application to speckle pattern photography, *Opt. Lasers Eng.*, *11*, 49 (1989).
32. N. Deng and I. Yamaguchi, Automatic analysis of speckle photographs with extended range and improved accuracy, *Appl. Opt.*, *28*, 296 (1990).
33. D. J. Chen and F. P. Chiang, Digital processing of Young's fringes in speckle photography, *Opt. Eng.*, *29*, 1413 (1990).
34. J. Zhang, P. Sun, J. Chen, and B. Ding, Pointwise automatic analysis of speckle photographs, *Opt. Eng.*, *30*, 382 (1991).
35. R. Meynart, Digital image processing for speckle flow velocimetry, *Rev. Sci. Instrum.*, *53*, 110 (1982).
36. R. Meynart, Convective flow field measurement by speckle velocimetry, *Rev. Phys. Apl.*, *17*, 301 (1982).
37. R. Meynart, Instantaneous velocity field measurements in unsteady gas flow by speckle velocimetry, *Appl. Opt.*, *22*, 535 (1983).

38. R. Schafer and A. Oppenheim, *Digital Signal Processing*, Prentice-Hall, Englewood Cliffs, N.J., 1975.
39. C. S. Yao and R. J. Adrian, Orthogonal compression and 1-D analysis technique for measurement of 2-D particle displacements in pulsed laser velocimetry, *Appl. Opt.*, *23*, 1687 (1984).
40. R. Erbeck, Fast image processing with a microcomputer applied to speckle photography, *Appl. Opt.*, *24*, 3838 (1985).
41. G. A. Reynolds, M. Short, and M. C. Whiffen, Automated reduction of instantaneous flow field images, *Opt. Eng.*, *24*, 475 (1985).
42. C. J. D. Pickering and N. A. Halliwell, Automatic analysis of Young's fringes in speckle photography and particle image velocimetry, *Opt. Lasers Eng.*, *7*, 227 (1986/87).
43. A. Kirita, C. J. D. Pickering, and N. A. Halliwell, Particle image velocimetry: a new approach to fringe analysis, *Opt. Lasers Eng.*, *7*, 137 (1986/87).
44. A. Kirita, C. J. D. Pickering, and N. A. Halliwell, Particle image velocimetry: automatic fringe analysis by cross-correlation, *Opt. Eng.*, *27*, 188 (1988).
45. Y. C. Cho, Digital image velocimetry, *Appl. Opt.*, *28*, 740 (1989).
46. T. Rösgen, K. Wozniak, and G. Wozniak, Image processing for laser speckle velocimetry using 2-D fast Fourier transform, *Appl. Opt.*, *29*, 5298 (1990).
47. S. A. Isacson and G. H. Kaufmann, 2-D digital processing of speckle photography fringes. 1. Diffraction halo influence for the noise-free case, *Appl. Opt.*, *24*, 189 (1985).
48. S. A. Isacson and G. H. Kaufmann, 2-D digital processing of speckle photography fringes. 2. Diffraction halo influence for the noisy case, *Appl. Opt.*, *24*, 1444 (1985).
49. G. K. Froehlich, J. F. Walkup, and R. B. Asher, Optical estimation in signal dependent noise, *J. Opt. Soc. Am.*, *68*, 1665 (1978).
50. H. D. Navone and G. H. Kaufmann, 2-D digital processing of speckle photography fringes. 3. Accuracy in angular determination, *Appl. Opt.*, *26*, 154 (1987).
51. E. Diez, D. Chambless, W. Swinson, and J. Turner, Image-processing techniques in laser-speckle photography with application to hybrid stress analysis, *Exp. Mech.*, *26*, 230 (1986).
52. H. D. Navone and G. H. Kaufmann, Automatic digital processing in speckle photography: comparison of two algorithms, *Appl. Opt.*, *28*, 350 (1989).
53. H. D. Navone and G. H. Kaufmann, Evaluation of an automated fringe measurement method for analysis of speckle photography data, *Opt. Eng.*, *28*, 1009 (1989).
54. J. M. Huntley, Speckle photography fringe analysis: assessment of current algorithms, *Appl. Opt.*, *28*, 4316 (1989).
55. R. Dändliker, Heterodyne holographic interferometry, *Prog. Opt.*, *17*, 1–84 (1980).
56. G. B. Smith and K. A. Stetson, Heterodyne readout of specklegram halo interference fringes, *Appl. Opt.*, *19*, 3031 (1980).
57. K. A. Stetson, Effect of scintillation noise in heterodyne speckle photogrammetry, *Appl. Opt.*, *23*, 920 (1984).

58. K. A. Stetson, The theory and applications of heterodyne speckle photogrammetry, in *Holography and Speckle Phenomena and Their Industrial Applications*, R. S. Sirohi, Ed., World Scientific, Singapore, 1990, p. 313.
59. J. S. Kim, M. Giglio, U. Perini, and S. Musazzi, Speckle grain noise in heterodyne techniques for holographic interferometry and speckle photogrammetry, *Appl. Opt.*, *26*, 4653 (1987).
60. J. S. Kim, S. Musazzi, U. Perini, and M. Giglio, Heterodyne readout system for dual plate speckle photography: analysis of error sources and performance evaluation, *Appl. Opt.*, *28*, 1862 (1989).
61. G. H. Kaufmann and P. Jacquot, Phase shifting of whole field speckle photography fringes, *Appl. Opt.*, *29*, 3570 (1990).
62. K. Creath, Phase-measurement interferometry, *Prog. Opt.*, *25*, 349–393 (1988).
63. G. H. Kaufmann and P. Jacquot, Phase shifting of whole field fringes in dual plate speckle photography, *Appl. Opt.*, *31*, 3201 (1992).
64. W. J. Beranek and A. J. A. Bruinsma, Sandwich speckle interferometry, Proceedings, VIth Int. Conf. on Experimental Stress Analysis, Munich, 1978, p. 417.
65. G. H. Kaufmann and G. E. Galizzi, Digital processing of dual plate speckle photography data, *Appl. Opt.*, *31*, 1537 (1992).
66. G. H. Kaufmann and G. E. Galizzi, Dual plate speckle photography data processing algorithm, *Opt. Lasers Eng.* (under review).
67. L. Pichon and J. P. Huignard, Dynamic joint-Fourier transform correlator by Bragg diffraction in photorefractive $Bi_{12}SiO_{20}$ crystals, *Opt. Commun.*, *36*, 277 (1981).
68. S. H. Collicott and L. Hesselink, Real-time photorefractive recording and optical processing for speckle velocimetry, *Opt. Lett.*, *13*, 348 (1988).
69. K. Nakagawa and T. Minemoto, Improvement of Young's fringes visibility in speckle photography by the use of BSO-PROM device, *Opt. Commun.*, *70*, 288 (1989).
70. J. M. Coupland and N. A. Halliwell, Automated optical analysis of Young's fringes—optical autocorrelation, *Opt. Lasers Eng.*, *14*, 351 (1991).
71. J. M. Coupland and N. A. Halliwell, Particle image velocimetry: rapid transparency analysis using optical correlation, *Appl. Opt.*, *27*, 1919 (1988).
72. S. Nakadate, T. Yatagai, and H. Saito, Electronic speckle pattern interferometry using digital image processing techniques, *Appl. Opt.*, *19*, 1879 (1980).
73. S. Nakadate, T. Yatagai, and H. Saito, Computer-aided speckle pattern interferometry, *Appl. Opt.*, *22*, 237 (1983).
74. R. Höfling and W. Osten, Displacement measurement by image-processed speckle patterns, *J. Mod. Opt.*, *34*, 607 (1987).
75. S. Nakadate and H. Saito, Fringe scanning speckle-pattern interferometry, *Appl. Opt.*, *24*, 2172 (1985).
76. K. Creath, Phase shifting speckle interferometry, *Appl. Opt.*, *24*, 3053 (1985).
77. D. W. Robinson and D. C. Williams, Digital phase stepping speckle interferometry, *Opt. Commun.*, *57*, 26 (1986).

78. D. Kerr and J. Tyrer, Use of high resolution real-time image processing techniques in generation and analysis of ESPI fringe patterns, *Opt. Lasers Eng.*, *8*, 109 (1988).
79. D. Kerr, F. Mendoza Santoyo, and J. R. Tyrer, Manipulation of the Fourier components of speckle fringe patterns as part of an interferometric analysis process, *J. Mod. Opt.*, *36*, 195 (1989).
80. K. Creath and G. Slettemoen, Vibration-observation techniques for digital speckle-pattern interferometry, *J. Opt. Soc. Am. A*, *2*, 1629 (1985).
81. E. Vikhagen, Vibration measurement using phase shifting TV-holography and digital image processing, *Opt. Commun.*, *69*, 214 (1989).
82. D. Kerr, F. Mendoza Santoyo, and J. R. Tyrer, Extraction of phase data from electronic speckle pattern interferometric fringes using a single-phase step method: a novel approach, *J. Opt. Soc. Am. A*, *7*, 820 (1990).
83. F. Mendoza Santoyo, M. C. Shellabear, and J. R. Tyrer, Whole field in-plane vibration analysis using pulsed-phase stepped ESPI, *Appl. Opt.*, *30*, 717 (1991).
84. M. C. Shellabear and J. R. Tyrer, Application of ESPI to three-dimensional vibration measurements, *Opt. Lasers Eng.*, *15*, 43 (1991).
85. E. Vikhagen, Nondestructive testing by use of TV holography and deformation phase gradient calculation, *Appl. Opt.*, *29*, 137 (1990).
86. D. P. Towers, T. R. Judge, and P. J. Bryanston-Cross, Automatic interferogram analysis techniques applied to quasi-heterodyne holography and ESPI, *Opt. Lasers Eng.*, *14*, 239 (1991).
87. A. Spik and D. W. Robinson, Investigation of the cellular automata method for phase unwrapping and its implementation on an array processor, *Opt. Lasers Eng.*, *15*, 14 (1991).
88. K. A. Stetson and W. R. Brohinsky, Electro-optics holography system for vibration analysis and nondestructive testing, *Opt. Eng.*, *26*, 1234 (1987).
89. K. A. Stetson, W. R. Brohinsky, J. Wahid, and T. Bushman, An electro-optic holography system with real time arithmetic processing, *J. Nondestruct. Eval.*, *8*, 69 (1989).
90. C. M. Mercer and G. Beheim, Fiber optic phase stepping system for interferometry, *Appl. Opt.*, *30*, 729 (1991).
91. J. M. Huntley, Noise-immune phase unwrapping algorithm, *Appl. Opt.*, *28*, 3268 (1989).

10

Speckle Metrology Using Hololenses

Chandra Shakher

Indian Institute of Technology
New Delhi, India

10.1 INTRODUCTION

Holography is becoming a widespread technique for fabricating inexpensive, lightweight optical components (gratings, beamsplitters, mirrors, lenses, lens arrays, etc.). A hologram of a point source can be generated by interference between two spherical waves or between a spherical wave and a plane wave, and such a hologram can be regarded as a holographic lens. A hologram of a point source can be used to transfer an optical wave front in much the same manner as a lens [1]. With a holographic lens instead of a glass lens, the image of an object can be generated in the usual way.

When the holographic lens (hololens) is used to form the image of an object of finite size, only one point on the object will give rise to the wave front that was recorded. Other points on the object will be less accurately focused, and aberrations will occur corresponding to those found in conventional lenses. Consequently the hololens can be used over large apertures but, like conventional aspheric components, will tolerate only small field angles. Although classical lenses perform their required functions through refraction at the optical interface to change the direction of the ray, hololenses work on the phenomenon of diffraction. Gabor and others [2–4] explored the correspondence between on-line holograms and zone plates. Horman and Chau [5] showed that zone plates are special types of holograms. In this analogy, the recording of holographic fringe patterns forms an off-axis Gabor zone plate. Since the imaging characteristics of zone plates have been known and studied for many years, the interpretation of a hologram as a superposition of zone plates provides a direct and useful understanding of imaging properties of holograms.

The performance features of holographic lenses can best be shown through a simple conjugate distance equation. A two-point hololens has a focal length defined by

$$\frac{1}{f} = \frac{\lambda_c n}{\lambda_0}\left[\frac{1}{R_1} + \frac{1}{R_2}\right] \tag{10.1}$$

where R_1 is the distance from the point source to the hololens vertex, R_2 the distance from the second point source to the hololens vertex, n the diffracted order number, λ_0 the construction wavelength, and λ_c the wavelength used for reconstruction. From Eq. (10.1) it can be seen that the focal length of a hololens varies linearly with wavelength. Axial chromatic aberration of a single hololens for visible wavelengths is more than an order of magnitude larger than that of a conventional lens. The strong dispersive characteristics of hololenses limit the use of a single hololens in wideband systems except, of course, where the designer desires chromatic dispersion. Hololenses can play a key role in many systems that make use of narrowband sources such as lasers. A combination of hololenses can be designed to achieve very well corrected imagery over the extended object field [6]. This type of system can be used advantageously in speckle metrology and optical data processing [7–11].

10.2 DUAL HOLOLENS IMAGING SYSTEM FOR SPECKLE PHOTOGRAPHY

10.2.1 Recording and Playback Geometry

To make a typical dual hololens imaging system for speckle metrology, the lens recording geometry is given in Fig. 10.1. The lenses used in the experiment described in Section 10.3 were recorded using a diverging spherical wave and a plane wave. To prevent the formation of a spurious grating, the holograms were recorded under index-matching conditions. The angle between the two beams (45°) was optimized to achieve high-efficiency operation at all points of the lens aperture. A typical system used for imaging and conducting experiments with speckles is shown in Fig. 10.2. The first half of this system consists of source S, object O, and hololens HL_1, and the second half consists of an aperture stop, hololens HL_2, and image plane. The entire system, therefore, consists of two spectrum analyzers in cascade. When the imaging properties of the lenses are fully taken

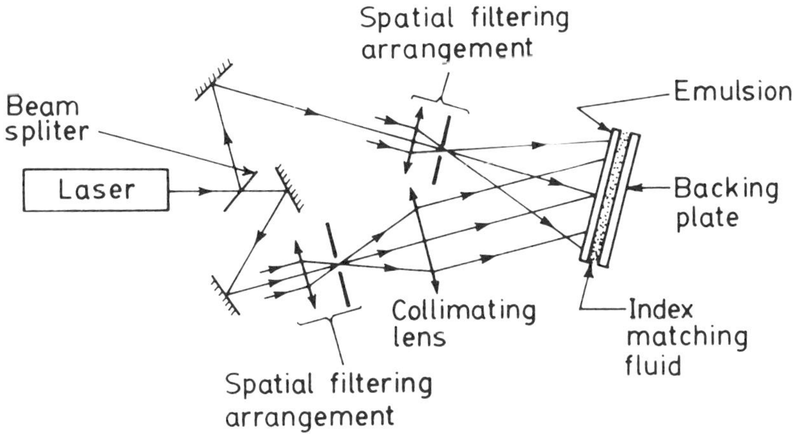

Figure 10.1 Schematic of the geometry for recording hololenses.

into account [12] to achieve accurate detail in the image with respect to phase and amplitude, the distance between HL_1 and HL_2 is equal to $f_1 + f_2$. An aperture stop placed between the two lenses restricts the spread of the ray cones that form each point in the image. This allows clarity and overall brightness in the image.

The lenses were played back to obtain a plane wave front from each object point and a real point focus from a plane wave. Hololens HL_1 is used in the imaging setup after being rotated through 180° with respect to the axis perpendicular to the plane of Fig. 10.1 and illuminated by a di-

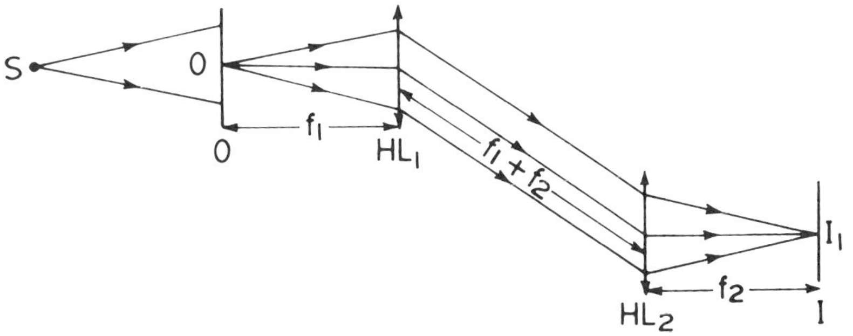

Figure 10.2 Schematic of typical imaging system consisting of two hololenses used for coherent imaging. S, monochromatic point source; O, diffused object; HL_1, HL_2, hololenses 1 and 2; I, image plane; $f_1 = f_2 = 178$ mm.

verging beam. The radius of curvature of the wave propagated from HL_1 is given by

$$\frac{1}{R_{I_1}} = \frac{1}{R_{C_1}} \quad \mu \left(\frac{1}{R_{r_1}} - \frac{1}{R_{o_1}}\right) \tag{10.2}$$

where $\mu(= \lambda_c/\lambda_r = 1)$ is the ratio of the wavelengths used for reconstructing and recording the holograms and

$$R_{C_1} = R_{r_1}, \qquad R_{I_1} = R_{o_1} = \infty$$

where R_{o_1}, R_{r_1}, R_{C_1}, and R_{I_1} are the distances of object, reference, reconstruction, and image points for lens HL_1.

Hololens HL_2 is used in the imaging system after being rotated through 180° with respect to the axis perpendicular to the plane of Fig. 10.1 and illuminated by a plane wave generated by HL_1. The radius of curvature of the wave propagated from HL_2 is given by

$$\frac{1}{R_{I_2}} = \frac{1}{R_{C_2}} - \mu \left(\frac{1}{R_{r_2}} - \frac{1}{R_{o_2}}\right) \tag{10.3}$$

for $R_{C_2} = R_{r_2} = \infty$ and $R_{I_2} = R_{o_2}$, where R_{o_2}, R_{r_2}, R_{C_2}, and R_{I_2} are the distances of object, reference, reconstruction, and image points for lens HL_2. This shows that the radius of curvature of the wave exiting lens HL_2 is the same as was recorded.

10.2.2 Imaging Characteristics

To assess the aberrations introduced by HL_1 and HL_2 for the conditions in which they are fabricated and used in the imaging system, we use the expressions of Champagne [13]. The coefficients of spherical aberration (S), coma (C_x, C_y), astigmatism (A_x, A_y, A_{xy}), curvature of field (F), and distortion (D_x, D_y) can be written as

$$S = \frac{1}{R_C^3} - \left(\frac{1}{R_o^3} - \frac{1}{R_r^3}\right) - \frac{1}{R_I^3}$$

$$C_x = \frac{x_C}{R_C^3} - \left(\frac{x_o}{R_o^3} - \frac{x_r}{R_r^3}\right) - \frac{x_I}{R_I^3}$$

$$C_y = \frac{y_C}{R_C^3} - \left(\frac{y_o}{R_o^3} - \frac{y_r}{R_r^3}\right) - \frac{y_I}{R_I^3}$$

$$A_x = \frac{x_C^2}{R_C^3} - \left(\frac{x_o^2}{R_o^3} - \frac{x_r^2}{R_r^3}\right) - \frac{x_I^2}{R_I^3}$$

$$A_y = \frac{y_C^2}{R_C^3} - \left(\frac{y_o^2}{R_o^3} - \frac{y_r^2}{R_r^3}\right) - \frac{y_I^2}{R_I^3}$$

$$A_{xy} = \frac{x_C y_C}{R_C^3} - \left(\frac{x_o y_o}{R_o^3} - \frac{x_r y_r}{R_r^3}\right) - \frac{x_I y_I}{R_I^3}$$

$$F = \frac{x_C^2 + y_C^2}{R_C^3} - \frac{x_o^2 + y_o^2}{R_o^3} + \frac{x_r^2 + y_r^2}{R_r^3} - \frac{x_I^2 + y_I^2}{R_I^3}$$

$$D_x = \frac{x_C^3 + x_C y_C^2}{R_C^3} - \frac{x_o^3 + x_o y_o^2}{R_o^3} + \frac{x_r^3 + x_r y_r^2}{R_r^3} - \frac{x_I^3 + x_I y_I^2}{R_I^3}$$

$$D_y = \frac{y_C^3 + y_C x_C^2}{R_C^3} - \frac{y_o^3 + y_o x_o^2}{R_o^3} + \frac{y_r^3 + y_r x_r^2}{R_r^3} - \frac{y_I^3 + y_I x_I^2}{R_I^3}$$

where (x_C, y_C), (x_o, y_o), (x_r, y_r), and (x_I, y_I) are coordinates of the points giving constructing, object, reference, and reconstructed beams, respectively.

We see that a plane wave propagates between two hololenses, i.e., $R_{I_1} = \infty$ for the first lens. Lens HL_1 is constructed under the condition $R_{o_1} = \infty$ and played back after being rotated through 180°. Under conditions $x_{C_1} = x_{r_1}$, $y_{C_1} = y_{r_1}$, $R_{C_1} = -R_{r_1}$, the coefficients of spherical aberration (S), coma (C_x, C_y), astigmatism (A_x, A_y, A_{xy}), field curvature (F), and distortion (D_x, D_y) become zero.

The reconstructed beam from the first lens, HL_1, acts as the reconstructing beam for the second lens, HL_2, under the conditions

$$R_{C_2} = R_{r_2} = \infty, \qquad R_{o_2} = -R_{I_2}$$
$$x_{o_2} = x_{I_2}, \qquad y_{o_2} = y_{I_2}$$

The coefficients of spherical aberration, coma, astigmatism, field curvature, and distortion for HL_2 also become zero.

The above analysis shows that through the imaging system shown in Fig. 10.2, a point object can be imaged without any conventional aberration. To analyze aberrations over an extended object field one has to resort to ray tracing based on diffraction calculations.

Ray-tracing equations for holographic optical elements can be derived by considering the fringe spacing and orientation of the local diffraction grating at the points where rays strike the hologram. The type of recording material and whether the hologram is thin or thick are not considered, as these factors determine the diffraction efficiency and hence the brightness of the reconstructed image but do not affect the geometrical optics. The full derivation of the ray-tracing formula based on vectors is given by Welford [14].

10.2.3 Finite Ray-Tracing Equations

Let (x_C, y_C, z_C) be the coordinates of a reconstruction point with respect to an origin of coordinates in the plane of the first hologram. At (x_1, y_1) we need to know the direction cosines of the two original rays used in making the first holographic lens, HL_1. The direction cosines $(L_{r_1}, M_{r_1}, N_{r_1})$ for the reference wave and $(L_{o_1}, M_{o_1}, N_{o_1})$ for the object wave are given by

$$L_{r_1} = -x_1/R_{r_1} \tag{10.4a}$$
$$M_{r_1} = -y_1/R_{r_1} \tag{10.4b}$$
$$N_{r_1} = z_{H_1}/R_{r_1} \tag{10.4c}$$

where

$$R_{r_1} = z_{H_1}\sqrt{1 + \frac{x_1^2 + y_1^2}{z_{H_1}^2}} \qquad \text{and } L_{o_1} = 0 \tag{10.5a}$$
$$M_{o_1} = \sin(\text{ang}_{H_1}) \tag{10.5b}$$
$$N_{o_1} = \cos(\text{ang}_{H_1}) \tag{10.5c}$$

For the reconstruction wave at the first hologram, the distance R_{C_1} is

$$R_{C_1} = z_C\left(\frac{1 + (x_C - x_1)^2 + (y_C - y_1)^2}{z_C^2}\right)^{1/2}$$

The direction cosines $(L_{C_1}, M_{C_1}, N_{C_1})$ are given by

$$L_{C_1} = (x_C - x_1)/R_{C_1} \tag{10.6a}$$
$$M_{C_1} = (y_C - y_1)/R_{C_1} \tag{10.6b}$$

and

$$N_{C_1} = (z_C - z_1)/R_{C_1} \tag{10.6c}$$

The diffracted waves after the first hologram have directions given by

$$L = L_{C_1} + m\mu(L_{o_1} - L_{r_1}) \tag{10.7a}$$
$$M = M_{C_1} + m\mu(M_{o_1} - M_{r_1}) \tag{10.7b}$$
$$N = \sqrt{(1 - (L^2 + M^2))} \tag{10.7c}$$

where m is the order of diffraction and μ is the ratio of the playback wavelength to the formation wavelength.

Let (x_2, y_2) and (x_s, y_s) be the coordinates in the plane of the second hologram and stop, respectively. If the center of the second hologram is at $(0, y_{\text{center H}_2}, z)$, then the diffracted rays strike the second hologram at

$$x_2 = x_1 + \frac{L}{N} z \tag{10.8a}$$

$$y_2 = y_1 + \frac{M}{N} z - y_{\text{center H}_2} \tag{10.8b}$$

and the stop with center at $(y_{\text{pupil}}, z_{\text{pupil}})$ at

$$x_s = x_1 + \frac{L}{N} z_{\text{pupil}} \tag{10.9a}$$

$$y_s = y_1 + \frac{M}{N} z_{\text{pupil}} - y_{\text{pupil}} \tag{10.9b}$$

At (x_2, y_2) the direction cosines for object and reference rays are

$$L_{o_2} = -x_2/R_{o_2} \tag{10.10a}$$

$$M_{o_2} = -y_2/R_{o_2} \tag{10.10b}$$

$$N_{o_2} = z_{\text{H}_2}/R_{o_2} \tag{10.10c}$$

where

$$R_{o_2} = z_{\text{H}_2}[1 + (x_2^2 + y_2^2)/z_{\text{H}_2}^2]^{1/2}$$

and

$$L_{r_2} = 0 \tag{10.11a}$$

$$M_{r_2} = \sin(\text{ang}_{\text{H}_2}) \tag{10.11b}$$

$$N_{r_2} = \cos(\text{ang}_{\text{H}_2}) \tag{10.11c}$$

After diffraction the final ray direction (L_2, M_2, N_2) is given by

$$L_2 = L + m\mu(L_{o_2} - L_{r_2}) \tag{10.12a}$$

$$M_2 = M + m\mu(M_{o_2} - M_{r_2}) \tag{10.12b}$$

$$N_2 = \sqrt{1 - (L_1^2 + M_1^2)} \tag{10.12c}$$

and the ray strikes the image plane at a distance $z_{\text{play H}_2}$ from the second hologram, at

$$x_3 = x_2 + \frac{L_2}{N_2} z_{\text{play H}_2} \tag{10.13a}$$

$$y_3 = y_2 + \frac{L_2}{N_2} z_{\text{play H}_2} \tag{10.13b}$$

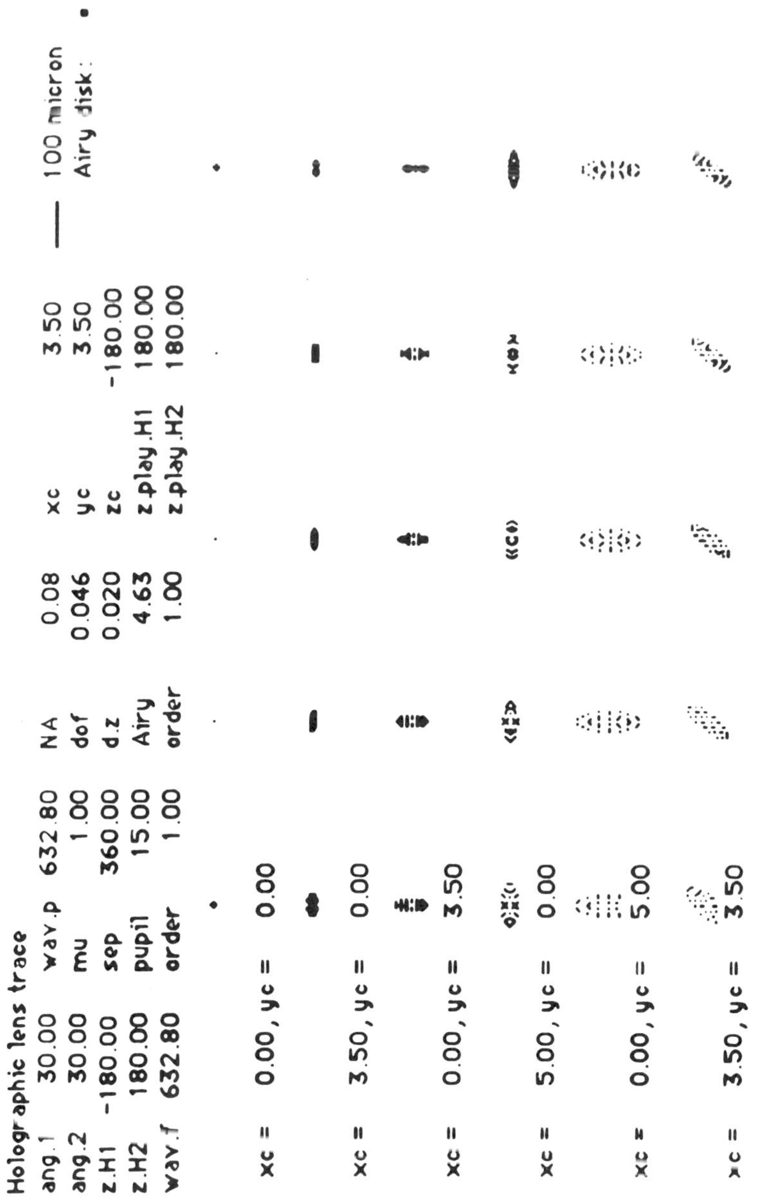

Figure 10.3 Size and shape of the image spot under the conditions shown on the figure.

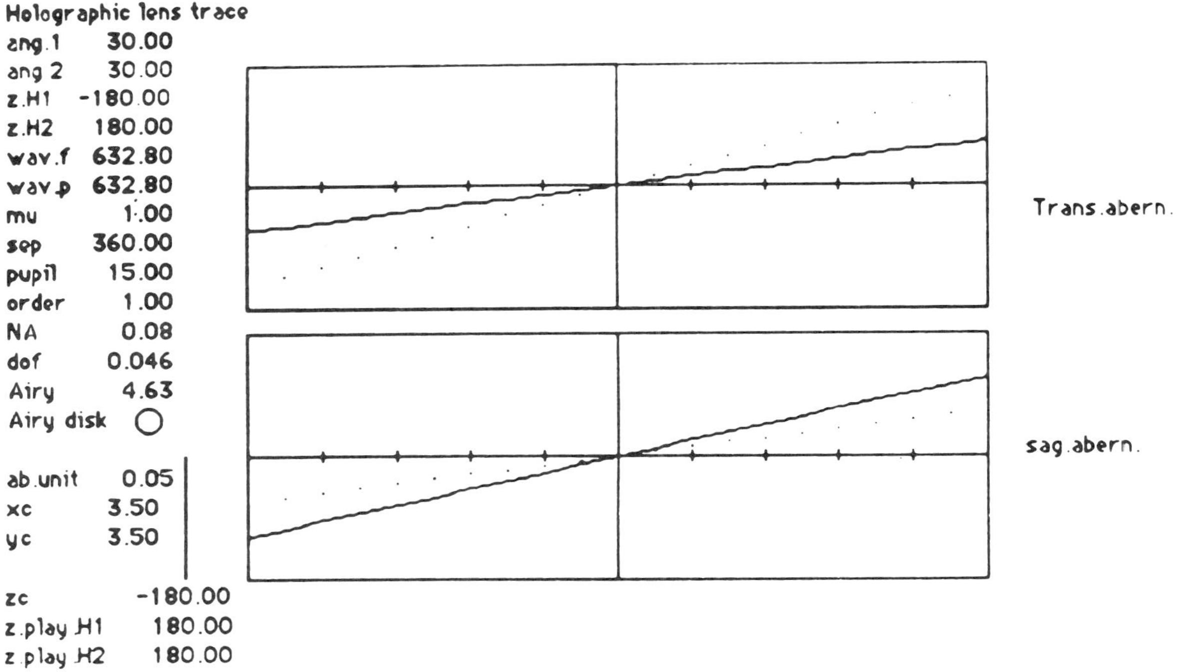

Figure 10.4 Transverse ray aberration curves under the conditions shown on the figure.

Tangential Rays

If p refers to the principal ray that strikes the first hologram at $(0, y_{1_p}, 0)$, the transverse ray aberrations for rays incident at $(0, y_1, 0)$ are

$$x_{\text{trans}\,t} = x_3 - x_{3_p} \tag{10.14a}$$

$$y_{\text{trans}\,t} = y_3 - y_{3_p} \tag{10.14b}$$

Sagittal Rays

Similarly, for the sagittal ray striking the first hologram at $(x_1, 0, 0)$, the transverse ray aberrations are

$$x_{\text{trans}\,s} = x_3 - x_{3_s} \tag{10.15a}$$

$$y_{\text{trans}\,s} = y_3 - y_{3_s} \tag{10.15b}$$

Figures 10.3 and 10.4 show the size and shape of the image spot and transverse ray aberrations in the image of the two-hololens imaging system under the conditions shown on the figures.

The transverse ray aberration plots in Fig. 10.4 show astigmatism and focal position error. The best compromise focal position will be that where all the curves have the same slope. By moving to this best compromise focal plane, the net total transverse error reduces to approximately 0.0125 mm, i.e., a blur with a diameter of 0.0125 mm. This corresponds to the resolution limit of approximately 160 lines/mm.

The average size of subjective speckles produced by the lens/imaging system of f number F is approximately [15]

$$\sigma = 1.2\lambda F \tag{10.16}$$

where λ is the wavelength of the light in use.

10.3 LASER SPECKLE PHOTOGRAPHY USING A DUAL-HOLOLENS IMAGING SYSTEM

Laser speckle photographic techniques employing conventional optical elements have been used for measurement of in-plane translation [16], in-plane rotation [17], in-plane vibration amplitudes, out-of-plane tilt [18], and slopes [19–21] for the study of displacements around crack tips [22, 23] and nondestructive testing [24–26], depending on the mode of recording. In this section I discuss the implementation of a few speckle photographic techniques using a dual-hololens imaging system.

The speckle pattern in the image of a coherently illuminated scattering surface will have a characteristic periodicity that depends on the numerical aperture of the imaging lens. For a lens of f number F, the speckle size is given approximately by Eq. (10.16). This corresponds to a speckle size on the object surface of

$$\sigma_o = 1.2M\lambda F \tag{10.17}$$

where M is the demagnification factor of the imaging system.

A double-exposure recording in which the speckle pattern as a whole is shifted sideways by a distance d between the two exposures can be examined either on a point-by-point basis or by using a whole-field filtering technique [17].

10.3.1 Measurement of In-Plane Translations and In-Plane Rotations

A diffuse object illuminated with an expanded laser beam was imaged with the help of a dual-hololens imaging system as shown in Fig. 10.2. The two hololenses were identical, having a diameter of 25 mm and a focal length of 178 mm.

A large number of experiments were conducted using the imaging system shown in Fig. 10.2 to determine the lower and upper limits of measurements for in-plane translation and in-plane rotation that can be measured using speckle photography. A dial gauge (Mitutoya, Japan) having a least count of 1 μm was used for monitoring the displacement given to the diffuse object. Double-exposure speckle photographs were recorded on Agfa Scientia 10E75 plates. The object was given in-plane translation and/or in-plane rotation between the two exposures. A measurable fringe pattern could be recorded corresponding to in-plane translations of 5.5–100 μm. The minimum in-plane displacement of 5.5 μm could be measured as compared to the theoretical limit of 5.4 μm. Figure 10.5 shows the Young's fringes for minimum and maximum values of rigid-body in-plane translation. In the case of rigid-body in-plane rotation the speckle displacement increases linearly with radial distance from the axis of rotation. This limits the range of observable Young's fringes, which in turn limits the maximum angle of rotation that can be measured using speckle photography. The minimum and maximum values of in-plane rotation measured were 0.6 and 9.8 mrad, respectively. Figure 10.6 shows the Young's fringes obtained from different regions of double-exposure specklegrams, where the object has undergone in-plane rotation between exposures. Table 10.1 gives the measured values of in-plane displacements using double-exposure speckle photography implemented with the dual-hololens imaging system.

The values of in-plane displacements were calculated by analyzing the double-exposure specklegrams using a point-by-point filtering technique.

(a) (b)

Figure 10.5 Photographs of the Young's fringes for minimum and maximum values of in-plane translation. (a) Displacement = 6 μm; (b) displacement = 100 μm.

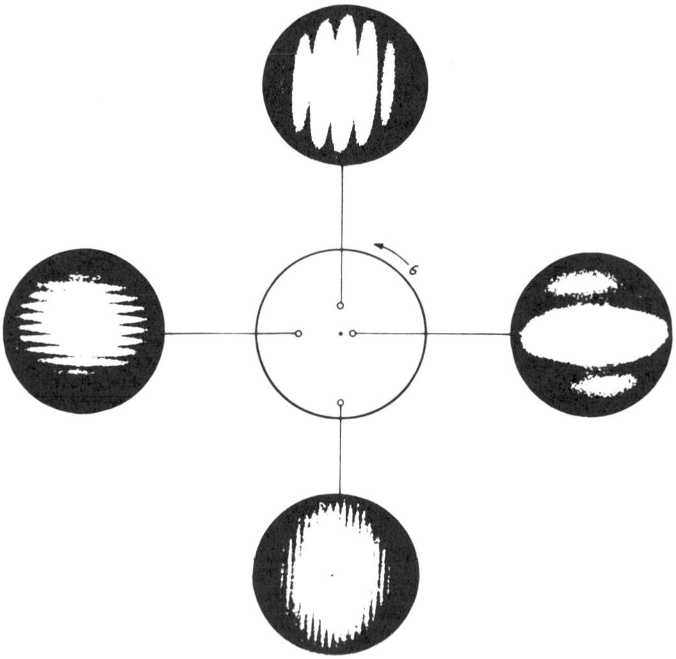

Figure 10.6 Photographs of the Young's fringes obtained by viewing through different regions of a double-exposure specklegram. The object was rotated in its own plane between exposures.

Table 10.1 Measured Values of In-Plane Displacements with Corrections for the Fringe Spacing Due to Diffraction Halo Corresponding to the Different Values of the Displacements (Recorded with a Dial Gauge) Given to the Object Between Exposures

In-plane displacement recorded with a dial gauge, d_a (μm)	Distance between screen and specklegram, Z (mm)	Average fringe spacing measured, β (mm)	Correction for the measured value of fringe spacing, Δβ (mm)	Corrected value of fringe spacing, β_c (mm)	Corrected value of displacement d_c (μm)	% Error of d_c' with respect to d_a (%)
5	500	68.34	3.10	65.24	4.85	3.0
10	500	34.24	1.66	32.58	9.71	2.9
15	500	22.75	1.05	21.70	14.58	2.8
25	500	13.67	0.62	13.05	24.25	3.0
40	500	8.55	0.40	8.15	38.84	2.9
70	500	4.88	0.23	4.65	68.11	2.7
100	500	3.41	0.16	3.25	97.30	2.7

For the imaging geometry under consideration, where the magnification of the system is unity, the magnitude of the displacement vector **d** on the object surface is determined by the relation [17]

$$|\mathbf{d}| = \lambda Z/\beta \tag{10.18}$$

where Z is the distance between screen and specklegram, λ is the wavelength of light used, and β is the average spacing of the Young's fringes. Due to the effect of the variable intensity distribution in the diffraction halo, the fringe maxima come closer to the center and the minima go away from the center. This phenomenon affects the accuracy of the measurement of the fringe spacing β. To get rid of the error due to this phenomenon in the measurement, the appropriate corrections in the fringe spacing were implemented as suggested by Vikram and Vedam [27]. In Table 10.1, β_c and d_c are corrected values of fringe spacing and displacement, respectively.

10.3.2 Measurement of Displacements Around a Crack Tip

A crack is a singularity, and hence any experimental data that are gathered to arrive at crack characteristics must be acquired from the region where the effect of the singularity is present. This region is generally not so good from the point of view of acquiring experimental stress/strain data [28].

The laser speckle photographic technique has been used to measure crack mouth opening displacement (CMOD) and crack opening displacement (COD), at any particular point lying between the crack tip and the crack mouth, in an aluminum beam specimen having a central edge crack subjected to three-point bending. Using the principle of similarity of triangles, the crack tip opening displacements (CTODs) were calculated with the help of measured values of CMODs and CODs for different values of load. From the calculated values of CTODs the other important fracture parameters such as K_I (stress intensity factor for mode I opening), r_p^*, and r were determined using empirical formulas. The comparison of experimentally achieved values of CMOD, CTOD, K_I, r_p^*, and r with theoretically calculated values shows very good agreement.

Experiments were conducted on cracked aluminum beam specimens having a central edge crack as shown in Fig. 10.7 subjected to three-point bending using focused plane speckle photography. The specimens were prepared according to ASTM standards so that experimental results could be compared with theoretically predicted values calculated from the empirical formulas. The thickness t of the specimen is kept between $0.25W$ and W (where W is the width of the specimen) to ensure the plane strain condition prevailing at the crack tip [29]. The dimensions of the specimen

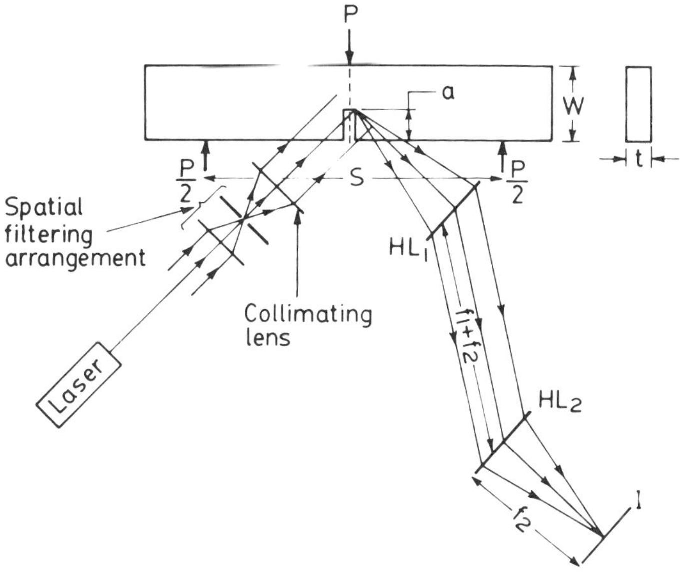

Figure 10.7 Schematic diagram of experimental setup for the measurement of crack opening displacement. $a = 8$ mm; $W = 20$ mm; $t = 6.2$ mm; $S = 80$ mm. $f = 178$ mm and diameter $= 25$ mm for both HL_1 and HL_2.

are shown in Fig. 10.8. The cracked area of the aluminum beam specimen was illuminated with laser light. An area nearly 100 mm square around the crack tip was imaged through a dual-hololens imaging system. The double-exposure specklegrams, with the specimen loaded between the two exposures, were recorded on Agfa Scientia 10E75 plates. The specklegrams were developed and fixed following standard procedures.

Measurement of Crack Mouth Opening Displacement

The recorded specklegrams were examined by using the pointwise filtering technique. The magnitude of the displacement vector **d** at the crack mouth opening is determined by analyzing the crack mouth with an unexpanded laser beam. For the present imaging geometry, where the magnification is unity, the magnitude of the displacement vector **d** on the object surface is determined by relation (10.18), i.e.,

$$|\mathbf{d}| = \lambda Z/\beta$$

Knowing the fringe spacing and orientation of the fringes with respect to the x or y axis, the displacement vector at a particular point can be calculated. Figure 10.8 shows the Young's fringes at the crack mouth opening

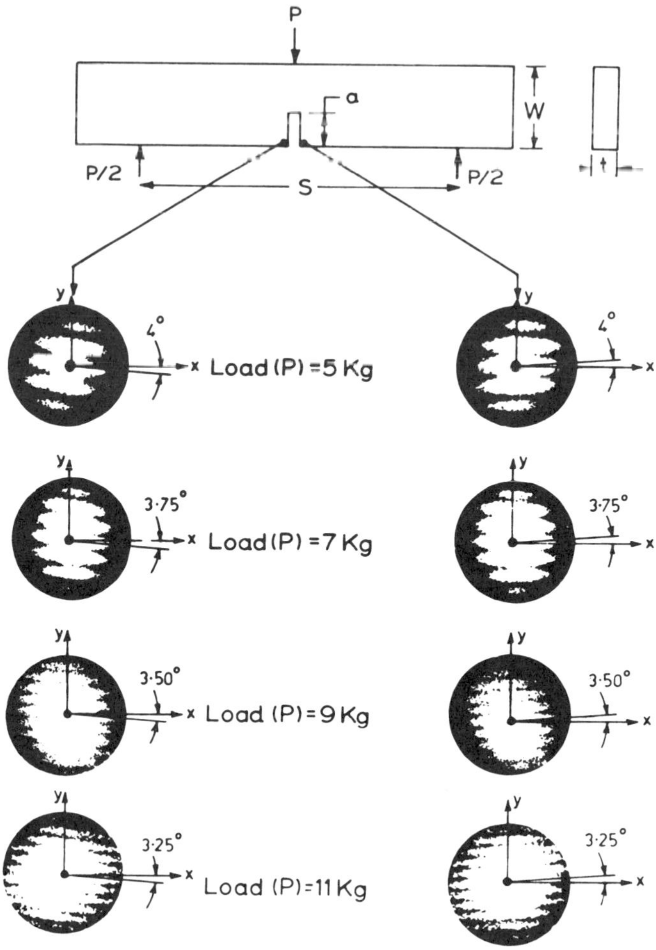

Figure 10.8 Photographs of the Young's fringes at the crack mouth opening at different loads.

at different loads. For the symmetrical loading geometry under consideration, the orientation θ of the fringes with respect to the x axis at either side of the crack mouth will be the same. In such a case, CMOD is given by [29]

$$\text{CMOD} = 2|\mathbf{d}| \sin \theta \tag{10.19}$$

A large number of experiments were conducted for different loading conditions. The experimental values of CMOD calculated by the data extracted from the double-exposure specklegrams are compared with theoretical CMOD values calculated by the empirical formula [30]

$$\text{CMOD} = \frac{6PSa}{E'tW} V\left(\frac{a}{W}\right) \tag{10.20}$$

where E' is the plane strain modulus $[= E/(1 - \gamma^2)]$, E is the Young's modulus of the specimen material, γ is the Poisson ratio, S is the span of the specimen, a is the crack length, t is the thickness of the specimen, W is the width of the specimen, P is the load applied on the specimen, and

$$\begin{aligned} V(a/W) = {} & 1.45 - 2.18(a/W) + 13.71(a/W)^2 \\ & - 5.96(a/W)^3 - 36.9(a/W)^4 + 70.7(a/W)^5 \end{aligned} \tag{10.21}$$

The above empirical relation is valid within $\pm 1\%$ of accuracy for $a/W \le 0.7$.

The aluminum specimen had the following specifications:

$E = 7.2 \times 10^3\ \text{kg/mm}^2, \quad \gamma = 0.33, \quad E' = 8.08 \times 10^3\ \text{kg/mm}^2$

$a = 8\ \text{mm}, \quad W = 20\ \text{mm}, \quad t = 6.2\ \text{mm}$

$S = 80\ \text{mm}$ and $V(a/W) = 2.1695$

A comparison of the experimental and theoretical values of CMOD at different loads is given in Table 10.2. Corrections are made to take care of diffraction halo intensity variations while calculating the displacements from Young's fringes [27].

Measurement of Crack Tip Opening Displacement

In the case of elastic deformations, the CTOD is negligibly small. In such cases better results on the measurement of CTOD can be had by measuring the crack opening displacements at the crack mouth and at any other point between the crack mouth and crack tip.

$$\text{CTOD} = \frac{Y \cdot \text{CMOD} - [\text{CMOD} - (\text{COD}_Y)]a}{Y} \tag{10.22}$$

where CMOD is the crack mouth opening displacement and $(\text{COD})_Y$ denotes the crack opening displacement at a distance Y above the crack mouth. Figure 10.9 shows the Young's fringes at distance $Y = 4$ mm above the crack mouth for loads of 5, 7, 9, and 11 kg. The theoretical values of CTOD are obtained with the equation [28]

$$\text{CTOD} = (4/\pi)(K_I^2/E\sigma_{ys}) \tag{10.23}$$

where K_I is the stress intensity factor and σ_{ys} is the yield strength of the material.

Table 10.2 Experimental and Theoretical Values of Crack Mouth Opening Displacement (CMOD) at Various Loads for the Aluminum Beam Specimen with a Central Edge Crack Subjected to Three-Point Bending[a]

Load (kg)	Distance between screen and specklegram, Z (mm)	Average fringe spacing measured, β (mm)	Magnitude of displacement calculated, $\lvert\mathbf{d}\rvert$ (μm)	Average fringe orientation (θ) w.r.t. x axis (deg)	CMOD, exptl. (μm)	CMOD, theoret. (μm)	Error in measurement of CMOD (%)
5	350	16.30	13.59	4.00	1.90	2.08	8.7
7	350	10.90	20.32	3.75	2.66	2.90	8.3
9	350	7.90	28.04	3.50	3.42	3.74	8.6
11	350	6.00	36.91	3.25	4.19	4.57	8.3

[a]See Fig. 10.7.

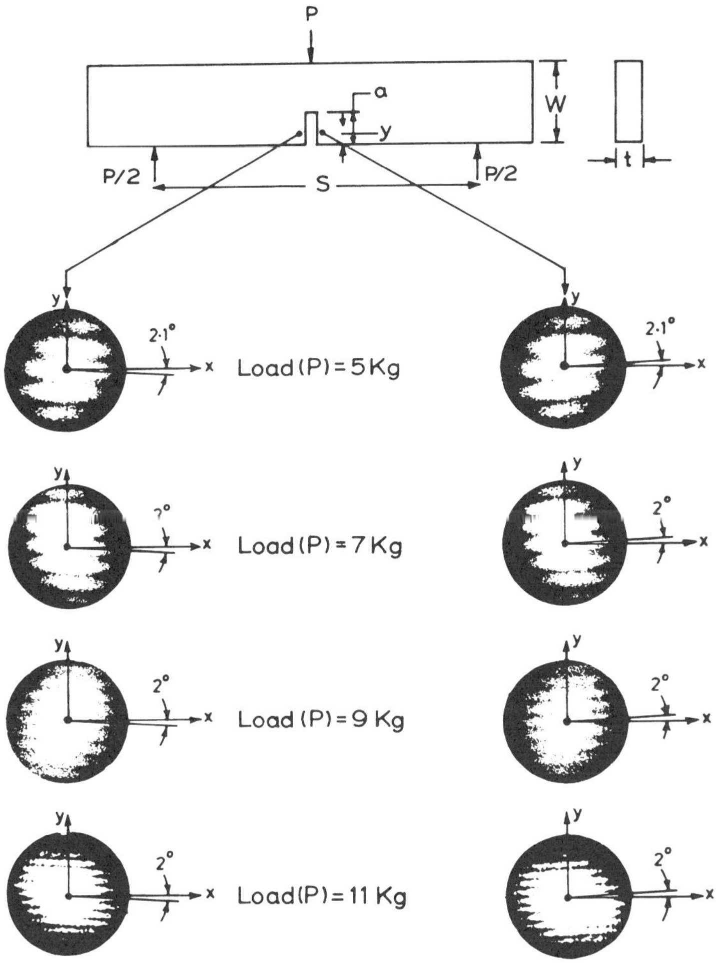

Figure 10.9 Photographs of the Young's fringes at a distance $y = 4$ mm above the crack mouth opening at different loads.

A comparison of theoretical and experimental values for loads of 5, 7, 9, and 11 kg is given in Table 10.3. Corrections are made to take care of diffraction halo intensity variations while calculating the displacements from Young's fringes.

Determination of Stress Intensity Factor

The stress intensity factor K_I determines the stress concentration around the crack tip for the mode I opening. For the three-point bend specimen

Table 10.3 Experimental and Theoretical Values of Crack Tip Opening Displacement (CTOD) at Various Loads for Aluminum Beam Specimen with Central Edge Crack Subjected to Three-Point Bending[a]

	Experimental values of COD at a point 4 mm above crack mouth opening – $(COD)_{Y=4\,mm}$								
Load (kg)	Distance between screen and specklegram, Z (mm)	Average fringe spacing measured, β (mm)	Magnitude of displacement calculated, $\lvert\mathbf{d}\rvert$ (mm)	Average fringe orientation, θ, with respect to x axis (deg)	$(COD)_{y=4\,mm}$ (μm)	$(CMOD)_{exptl}$ μm	$(CTOD)_{exptl}$ μm	$(CTOD)_{theoret}$ μm	Error in CTOD measurement (%)
5	350	16.60	13.34	2.1	0.978	1.90	0.056	0.065	13.8
7	350	11.15	19.86	2.0	1.386	2.66	0.112	0.128	12.5
9	350	8.58	25.81	2.0	1.802	3.42	0.184	0.212	13.2
11	350	6.92	32.01	2.0	2.234	4.19	0.278	0.320	13.1

[a] See Fig. 10.7.

under consideration, the CTOD is related to stress intensity factor K_I by Eq. (10.23),

$$\text{CTOD} = (4/\pi)(K_I^2/E\sigma_{ys})$$

or by

$$K_I = (\pi E\sigma_{ys}\cdot\text{CTOD}/4)^{1/2} \tag{10.24}$$

where σ_{ys} is the yield strength of aluminum, $\sigma_{ys} = 17.35\ \text{kg/mm}^2$.

Theoretical values of the stress intensity factor K_I are computed from the relation [31].

$$K_I = \frac{3Sp}{2tW^2}\sqrt{a\pi}F_1\left(\frac{a}{W}\right) \tag{10.25}$$

where

$$F_1(\alpha) = \frac{1.99 - \alpha(1 - \alpha)[2.15 - 3.93\alpha + 2.7\alpha^2]}{(1 + 2\alpha)(1 - \alpha)^{3/2}} \tag{10.26}$$

Equation (10.26) is valid within $\pm 0.5\%$ of accuracy for $a/W \leq 1.0$. The value of $F_1(a/W)$ for the specimen geometry under consideration comes out to be 2.0879. A comparison of experimental and theoretical values of K_I for loads of 5, 7, 9, and 11 kg is given in Table 10.4.

Determination of the Size of the Crack Tip Plastic Zone r_p^* and Rotational Factor r

The plastic zone size r_p^* at the crack tip is an important parameter in fracture mechanics analysis. If the plastic zone is small compared with crack size (a), linear elastic fracture mechanics can be successfully applied. If, however, the plastic zone size r_p^* is large compared with crack size, the elastic plastic fracture mechanics must be used.

The deformation caused by loading in a three-point bending specimen is assumed to occur by a hinge mechanism about the center of rotation at a distance $r(W - a)$ above the crack tip as shown in Fig. 10.10, which determines the position of the center of rotation with respect to the crack tip. The plastic zone size r_p^* and rotational factor r are related to experimentally determined parameters CMOD and CTOD by the relations [28]

$$r_p^* = \frac{\text{CTOD}\cdot E}{8\sigma_{ys}} \tag{10.27}$$

and

$$r = \frac{\text{CTOD}\cdot a}{(W - a)(\text{CMOD} - \text{CTOD})} \tag{10.28}$$

Table 10.4 Experimental and Theoretical Values of Stress Intensity Factor K_I, Crack-Tip Plastic Zone Size r_p^* and Rotational Factor r for Aluminum Beam Specimen with a Central Edge Crack Subjected to Three-Point Bending[a]

Load	Experimental/theoretical values of K_I			Experimental/theoretical values of r_p^*			Experimental/theoretical values of r		
P (kg)	$(K_I)_{expt}$ (kg/mm$^{3/2}$)	$(K_I)_{theo}$ (kg/mm$^{3/2}$)	Error (%)	$(r_p^*)_{expt}$ (μm)	$(r_p^*)_{theo}$ (μm)	Error (%)	r_{expt}	r_{theo}	Error (%)
5	2.35	2.53	7.1	2.90	3.39	14.5	0.0203	0.0215	5.6
7	3.32	3.54	6.2	5.81	6.63	14.1	0.0293	0.0308	4.9
9	4.25	4.56	6.8	9.54	10.99	13.2	0.0379	0.0401	5.5
11	5.23	5.57	6.1	14.42	16.40	12.1	0.0474	0.0502	5.6

[a]See Fig. 10.7.

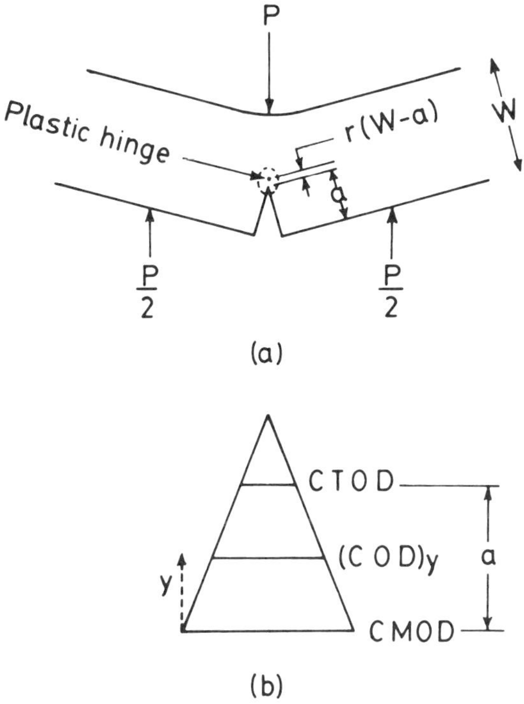

Figure 10.10 Schematic diagram of the crack opening mode. (a) Crack mouth opening mode and (b) crack opening displacements.

Putting values of CTOD and CMOD into Eqs. (10.27) and (10.28), we get values of r_p^* and r at various loads. The values of r_p^* and r are tabulated in Table 10.4.

Tables 10.2–10.4 show that all fracture parameters such as CTOD, K_I, r_p^*, and r can be measured quite accurately using laser speckle photography and a hololens imaging system.

10.4 DOUBLE-APERTURE SPECKLE INTERFEROMETER USING HOLOLENSES

Leendertz [16] suggested two-beam illumination speckle interferometry for measurement of the in-plane displacement component. The problems associated with the two-beam illumination method regarding splitting the laser beam and maintaining equal path lengths were overcome by Duffy [32, 33]. In Duffy's arrangement, instead of using two-beam illumination, the surface can be effectively viewed in two directions by means of twin apertures placed symmetrically over the imaging lens [32]. The sensitivity of the displacement measurement in this case depends on the angle θ, where θ is the half-angle subtended by the apertures on the image plane. The greater the value of θ, the higher will be the sensitivity of measurement. Unfortunately, θ cannot be as large as that of the two-beam method because of the limitation imposed by the finite size of conventional glass lenses. Duffy pointed out that for a double-aperture speckle interferometer the lens need not be diffraction-limited over the entire lens aperture but only over the two small areas encompassed by the double aperture.

In this case, the two subjective patterns are formed in the plane of the image, and these combine coherently to give each speckle a grating-like modulation that can be recorded on high-resolution recording material. Detection of this grating-like structure is the method used to generate fringe contours. Displacement of the surface in the line of sight z or in the y direction will affect the phase of light accepted by each aperture in a similar manner, and the speckle pattern modulation will remain unchanged. For a displacement in the x direction, the resultant phase changes will cause the periodic pattern to be displaced parallel to itself by one grating interval for every $\lambda/(2 \sin \theta)$ of surface movement.

The limitations on the sensitivity of measurement in Duffy's arrangement imposed by the finite size of the conventional glass lenses can be overcome, and almost diffraction-limited performance over the two apertures can be achieved by the use of suitably designed holographic lenses. In this section I describe a four-hololens imaging system that can be used as an alternative to Duffy's two-aperture speckle interferometer. In this interferometer, θ can be increased considerably using suitably designed hololenses [34–36].

10.4.1 Lens Recording and Playback Geometry

Four identical hololenses, each of diameter 30 mm and focal length 190 mm, were recorded using a plane wave for the object beam and a point source for the reference beam and vice versa. A schematic diagram of the recording geometry for recording hololenses to make a four-lens imaging system is shown in Fig. 10.11. The plane wave is made to fall normally on the hololenses recording plane. The angle between the two beams used for the recording hololenses was 18°. To prevent the formation of a spurious grating, the lenses were recorded under index-matching conditions [6].

Figure 10.12 shows the coherent imaging system with four hololenses in the setup shown in Fig. 10.11. Lenses HL_1 and HL_2 were used in the system after being rotated through 180° with respect to the axis perpen-

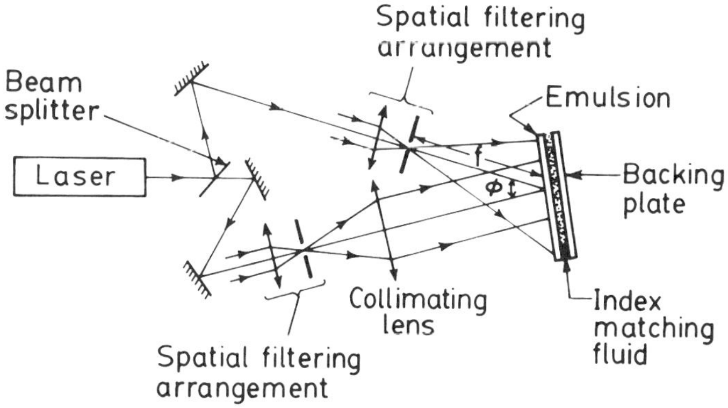

Figure 10.11 Schematic of the recording geometry for recording hololenses to generate four-hololens imaging configuration. $\phi = 18°$; $f = 190$ mm.

dicular to the plane of Fig. 10.11 to obtain a plane wave for each object point and a real point focus from the plane wave. The arm bb′ is the mirror image of the arm aa′. As discussed in Section 10.2, almost diffraction limited imaging can be achieved through both apertures for an object of diameter 7 mm. The imaging system shown in Fig. 10.12 behaves as an on-axis lens of focal length 180.7 mm. The effective focal length of the system can be calculated using the relation

$$f_{\text{eff}} = f \cos \phi = f \cos 18° \tag{10.29}$$

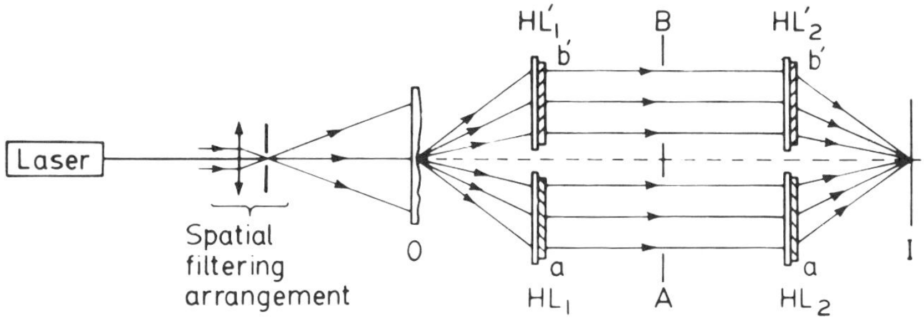

Figure 10.12 Schematic of the experimental set up used for the study of in plane displacement using low f number imaging configuration. O, object; A–B, aperture; HL_1, HL_1', HL_2, HL_2', identical hololenses, 30 mm diam, $f = 190$ mm; I, image plane.

10.4.2 Experimental

A diffuse object illuminated with an expanded laser beam was imaged through the imaging system just described. Figure 10.12 is a schematic diagram of the experimental setup used for measurement of the in-plane displacement component using a four-hololens imaging system. A large number of experiments were conducted to determine the lower and upper limits of measurement for in-plane translation along the x axis. Different values of in-plane displacements given to the object between the two exposures were monitored with a dial gauge having a least count of 1 μm. A dial gauge was used to measure linear displacements in the case of in-plane translations and the arc during in-plane rotations. Experimentally the displacements were calculated by analyzing the double-exposure specklegrams recorded on Agfa Scientia 10E75 plates by a pointwise analysis method, in both zero-order and first-order diffraction patterns. The unexpanded laser beam is made to fall perpendicular to the plane of the specklegram and the screen is held perpendicular to the first-order diffracted beam. The displacement measured by analyzing the Young's fringes contained in the first-order beam is very close to the displacement recorded by the dial gauge. Due to the high contrast of the Young's fringes contained in the first-order diffracted beam, the measurement of average fringe spacing is achieved with better accuracy, and hence error in the measurement is less. Figure 10.13 shows Young's fringes obtained from a double-exposure im-

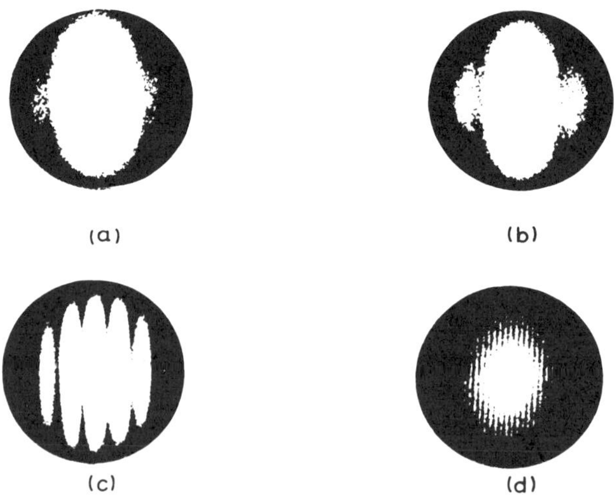

Figure 10.13 Photographs of the Young's fringes for in-plane displacements of (a) 2 μm; (b) 3 μm; (c) 15 μm; (d) 98 μm.

age-plane specklegram where the object has undergone in-plane translation between exposures.

Using the hololens imaging system of Fig. 10.12 and one-to-one imaging geometry, we could achieve a measurable fringe pattern from 2 to 98 μm for in-plane translation [36].

10.5 LASER SPECKLE PHOTOGRAPHY AND DOUBLE-APERTURE SPECKLE INTERFEROMETRY USING A COMPACT HOLOLENS IMAGING SYSTEM

While studying hololens system imaging in speckle photography and speckle interferometry as discussed in Sections 10.3 and 10.4, it was realized that if a compact hololens imaging system could be developed it would be more useful in speckle metrology. In a compact hololens imaging system, the problems encountered due to the bulkiness of the system and alignment of lenses could be handled more effectively. With these problems in mind, the typical two-hololens compact imaging shown in Fig. 10.14 was analyzed and studied. Figures 10.15 and 10.16 provide data on the size and shape of the image spot and the transverse ray aberrations in the image plane, respectively. A comparison of Figs. 10.4 and 10.16 shows that the compact system has less aberration, which may be due to the reduction of propagation error between the lenses. Using the compact hololens imaging systems shown in Figs. 10.14 and 10.17, preliminary experiments were conducted in speckle photography and speckle interferometry. Figures 10.18 and 10.19 show the Young's fringes obtained from a double-exposure im-

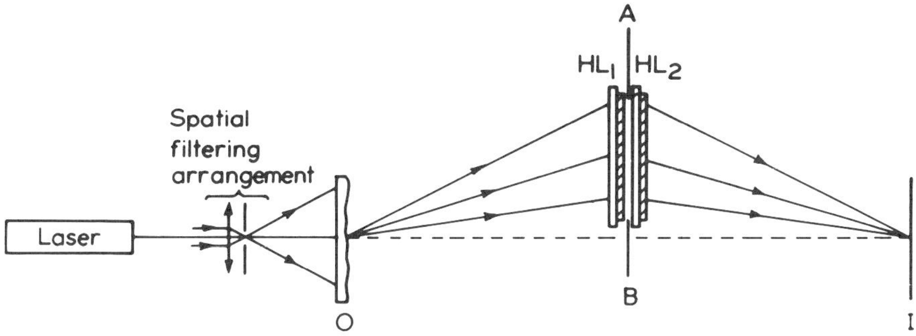

Figure 10.14 Schematic of the experimental setup used for the measurement of in-plane displacement and rotation about an axis parallel to the line of sight. O, diffused object; HL_1, HL_2, identical hololenses 30 mm diameter, $f = 190$ mm, $\phi = 18°$; I, image plane; A–B, aperture.

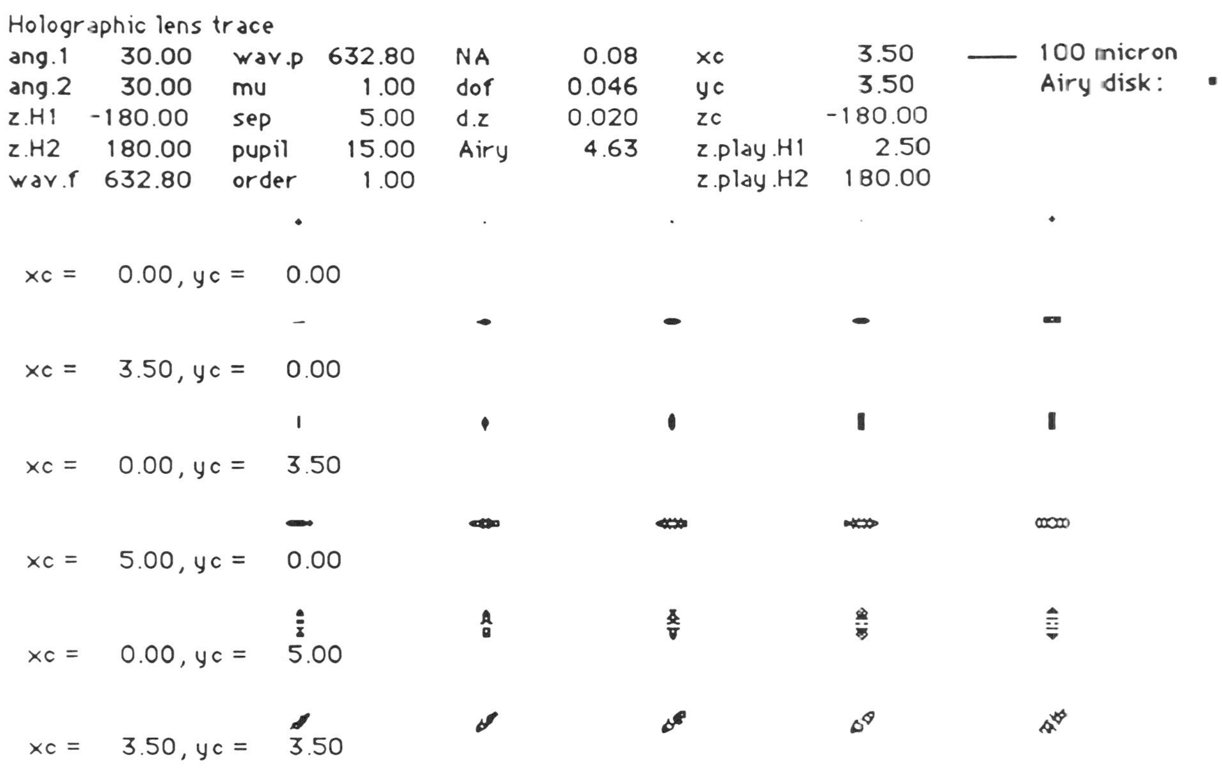

Figure 10.15 Size and shape of the image spot under the conditions shown on the figure.

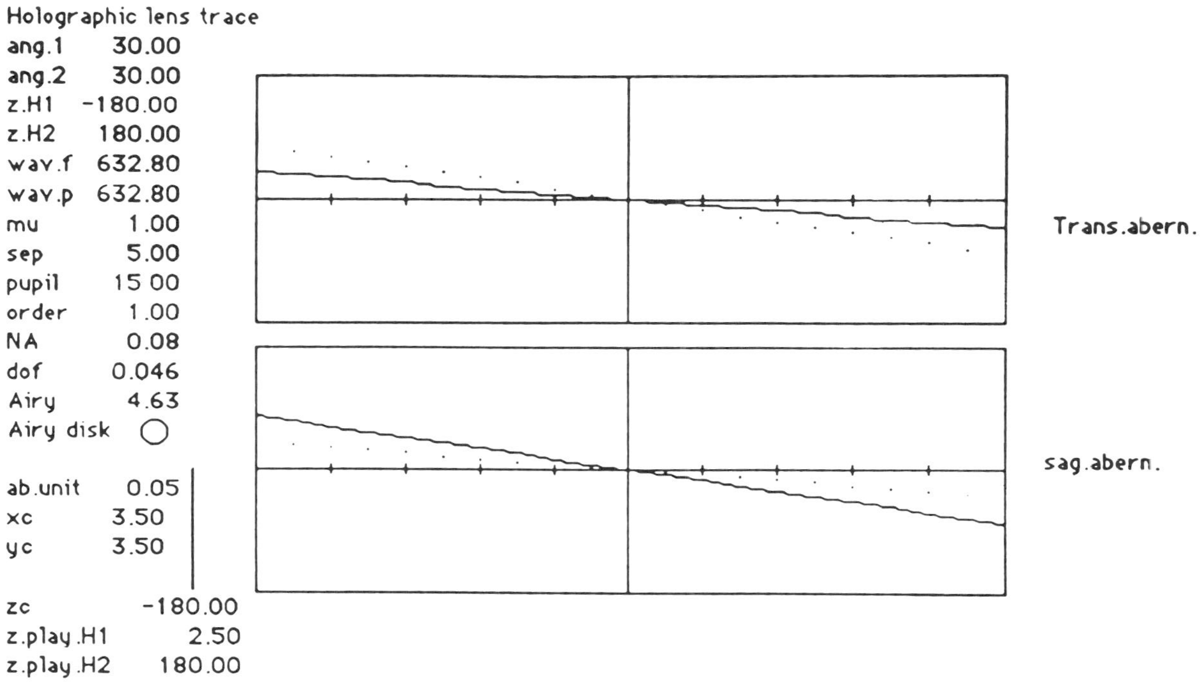

Figure 10.16 Transverse ray aberration curves under conditions shown on figure.

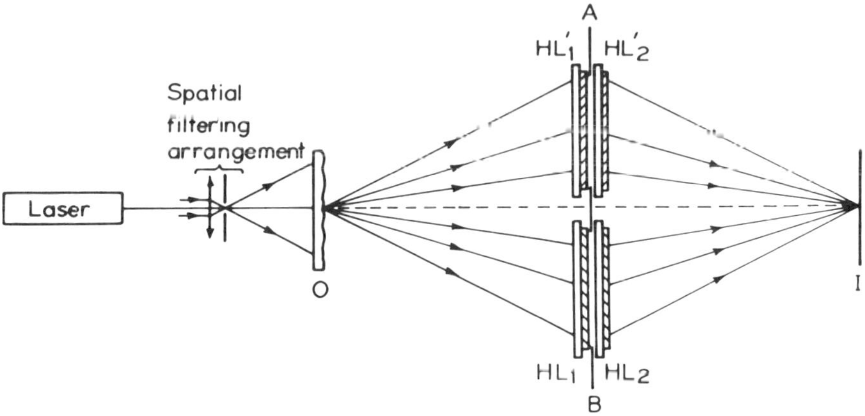

Figure 10.17 Schematic diagram of experimental setup for the study of in-plane displacement and rotation using four-hololens image configuration. HL_1, HL_2, HL_1', HL_2', identical hololenses of diameter 30 mm, $f = 190$ mm, $\phi = 18°$.

age-plane specklegram where the object has undergone in-plane translation and in-plane rotation between exposures. Using the compact hololens imaging systems for speckle photography and two-aperture speckle interferometry and one-to-one imaging, we could achieve measurable Young's fringes of 5–100 μm and 2–103 μm, respectively, for in-plane translation.

10.6 DISCUSSIONS AND CONCLUSIONS

Archbold and Ennos [17] pointed out that in speckle photography aberrations of the imaging system limit the number of observable Young's

(a) (b)

Figure 10.18 Photographs of Young's fringes using two-hololens compact imaging system for displacement of (a) 5 μm (b) 103 μm.

Figure 10.19 Photographs of Young's fringes using compact double-aperture speckle interferometer for displacement of (a) 2 μm (b) 100 μm.

fringes, thereby reducing the upper and lower limits of measurements. Further investigations carried out by Stetson [37] reveal that the spherical aberrations of the imaging system may lead not only to the loss of halo fringes but also to curvature of the halo fringes. When coma and astigmatism are present, the skewness of the rays may totally destroy some portions of the fringes. Stetson [38] concluded that adverse effects of lens aberrations in speckle photography can be eliminated by the use of properly designed lenses. It is quite improper to use normal camera lenses for 1:1 imaging as they are designed to record the demagnified image.

Theoretical and experimental results presented in this article show that properly recorded hololenses can be used to design the coherent imaging system for speckle photography for one-to-one imaging. It is shown that an almost complete solution to such a complicated problem as determining the displacements around a crack tip can be obtained using speckle photography and an imaging system made of hololenses. The results shown in Tables 10.1–10.4 are quite satisfactory.

Section 10.4 shows that properly designed hololenses can be advantageously used in speckle interferometry.

Recently Amitai and Goodman [39] gave a two-hololens configuration for achieving nearly diffraction limited imaging. The method is illustrated with holographic optical elements (HOEs) recorded at 488 nm and played back at 633 nm.

In conclusion, it appears that hololens imaging systems can be designed to achieve nearly diffraction limited imaging for an extended object field and these systems can be advantageously used in speckle metrology [6, 8] and some other specialized applications such as optical interconnects [40, 41]. To make the systems commercially viable, cemented two-hololens imaging systems and multielement compact hololens imaging systems for more extended object fields must be designed and realized.

ACKNOWLEDGMENTS

I am very grateful to Dr. Robin W. Smith, Head of the Applied Optics Group, Imperial College of Science and Technology, London, for his guidance in developing a computer program for calculating aberrations of hololens imaging systems. I am also grateful to Prof. R. S. Sirohi for his encouragement and many fruitful discussions. I thank Ms. A. J. Pramila Daniel for conducting speckle photographic and interferometric experiments with compact hololens imaging systems.

REFERENCES

1. M. J. R. Schwar, T. P. Pandya, and F. J. Weinberg, Point source holograms as optical elements, *Nature*, *215*, 239–241 (1967).
2. D. Gabor, A new microscope principle, *Nature*, *161*, 777–778 (1948).
3. G. L. Rogers, Experiments in diffraction microscopy, *Proc. Roy. Soc.* (*Edinburgh*), *63A*, 193–221 (1952).
4. E. N. Leith and J. Upatnieks, Reconstructed wavefronts and communication theory, *J. Opt. Soc. Am.*, *52*, 1123–1130 (1962).
5. M. H. Horman and H. H. M. Chau, Zone plate theory based on holography, *Appl. Opt.*, *6*, 317–322 (1967).
6. C. Shakher, Analysis and evaluation of a two hololens system imaging, *J. Opt.* (*Paris*), *19*, 27–31 (1988).
7. C. Shakher and G. V. Rao, Speckle metrology using hololenses, Part 2, *Appl. Opt.*, *26*, 654–657 (1987).
8. C. Shakher and G. V. Rao, Speckle metrology using hololenses, *Appl. Opt.*, *23*, 4592–4595 (1984).
9. C. Joenathan, R. K. Mohanty, and R. S. Sirohi, Hololenses in speckle and speckle shear interferometry, *Appl. Opt.*, *24*, 1294–1298 (1985).
10. A. Richter and F. P. Carlson, Holographically generated lens, *Appl. Opt.*, *13*, 2924–2930 (1974).
11. H. R. Manjunath and S. V. Pappu, Fourier spectrum of an object using holographic lens, *Appl. Opt.*, *14*, 2562 (1975).
12. J. D. Gaskill, *Linear Systems, Fourier Transforms and Optics*, Wiley, New York, 1978, p. 478.
13. E. B. Champagne, Non-paraxial imaging, magnification and aberration properties in holography, *J. Opt. Soc. Am.*, *57*, 51–55 (1967).
14. W. T. Welford, A vector ray tracing equation for hologram lenses of arbitrary shape, *Opt. Commun.*, *14*, 322–323 (1975).
15. A. E. Ennos, Speckle interferometry, in *Laser Speckle and Related Phenomena*, J. C. Dainty, Ed., Springer-Verlag, Berlin, 175, pp. 203–253.
16. J. A. Leendertz, Interferometric displacement measurement on scattering surfaces utilizing speckle effect, *J. Phys. E: Sci. Instrum.*, *3*, 214–218 (1970).
17. E. Archbold and A. E. Ennos, Displacement measurement from double exposure laser photographs, *Opt. Acta*, *19*, 253–271 (1972).

18. H. J. Tiziani, Application of speckling in in-plane vibration analysis, *Opt. Acta*, *18*, 891–902 (1971).
19. H. J. Tiziani, A study of use of laser speckle to measure small tilts, *Opt. Commun.*, *5*, 271–276 (1972).
20. D. A. Gregory, Basic physical principles of defocused speckle photography: a tilt topology inspection technique, *Opt. Laser Technol.*, *8*, 201–213 (1976).
21. R. P. Khetan and F. P. Chiang, Strain analysis by one beam laser speckle interferometry, single aperture method, *Appl. Opt.*, *15*, 2205–2215 (1976).
22. W. T. Evans and A. Luxmoore, Measurement of in-plane displacements around crack tips by laser speckle method, *Eng. Fracture Mech.*, *6*, 735–743 (1974).
23. E. J. Barbero, G. H. Kaufman, and S. R. Idelsohn, Fracture analysis of surface coated ceramic by speckle photography, *Opt. Laser Technol.*, *22*, 17–22 (1990).
24. J. N. Butters, Application of ESPI to NDT, *Opt. Laser Technol.*, *9*, 117–123 (1977).
25. C. Shakher, Laser speckle photography used for NDT of diaphragms, *Appl. Opt.*, *21*, 367–369 (1982).
26. R. Krishnamurthy, R. S. Sirohi, and M. P. Kothiyal, Detection of defects in plates and diaphragms using split lens shearing interferometry, *Non-Destruct. Testing*, *15*, 329–333 (1982).
27. C. S. Vikram and K. Vedam, Processing of speckle photographic data: circular imaging aperture, *Appl. Opt.*, *22*, 653–654 (1983).
28. D. Broek, *Elementary Engineering Fracture Mechanics*, Martinus Nijhoff, Dordrecht, Netherlands, 1986.
29. C. Shakher and H. L. Yadav, Use of holographic optical elements in speckle metrology, Part 3: Application to fracture mechanics, *Appl. Opt.*, *30*, 3607–3611 (1991).
30. H. Nisitani, H. Noguchi, and K. Mori, Analysis of single edge cracked specimen under three or four point bending by body force doublet method, *Trans. Jpn. Soc. Mech. Eng.*, *52*, 539 (1986).
31. J. E. Srawley, Wide range stress intensity factors expressions for ASTM E399 standard fracture toughness specimen, *Int. J. Fract. Mech.*, *12*, 475–476 (1976).
32. D. E. Duffy, Moiré gauging of in-plane displacement using double aperture imaging, *Appl. Opt.*, *11*, 1778–1781 (1972).
33. D. E. Duffy, Measurement of surface displacement normal to the line of sight, *Exp. Mech.*, *14*, 378–384 (1974).
34. C. Joenathan and R. S. Sirohi, Enhancement of sensitivity in in-plane displacement measurement in speckle photography, *Appl. Opt.*, *25*, 1380 (1986).
35. C. Shakher, H. L. Yadav, and A. K. Nirala, Design and analysis of low *f*-number imaging system using hololenses, *J. Opt.* (*Paris*), *20*, 259–262 (1989).
36. H. L. Yadav, Holographic lenses in speckle metrology and solar concentration, Ph.D. Thesis, IIT, Delhi, 1991.
37. K. A. Stetson, Vulnerability of speckle photography to lens aberrations, *J. Opt. Soc. Am.*, *67*, 1587–1590 (1977).

38. K. A. Stetson, Miscellaneous topics in speckle metrology, in *Speckle Metrology*, R. K. Erf, Ed., Academic, New York, 1978, pp. 295–320.
39. Y. Amitai and J. W. Goodman, Design of substrate-mode holographic interconnects with different recording and readout wavelengths, *Appl. Opt.*, *30*, 2376–2381 (1991).
40. J. W. Goodman, Fan-in and fan-out with optical interconnects, *Opt. Acta*, *32*, 1489–1496 (1985).
41. B. Robertson, M. R. Taghizadeh, J. Turunen, and A. Vasara, High-efficiency, high bandwidth optical interconnects in dichromated gelatin, *Opt. Lett.*, *15*, 694–696 (1990).

11

Correlation Speckle Interferometry in the Mechanics of Contact Interaction

Yuri I. Ostrovsky†

A. F. Ioffe Physical Technical Institute
St. Petersburg, Russia

V. P. Shchepinov

Physical Engineering Institute
Moscow, Russia

11.1 INTRODUCTION

The contrast (visibility) of interference fringes in speckle photography (one of the basic methods of speckle interferometry) depends on, among other factors, the degree to which the elements of the microrelief of the surface under study have changed. The change in surface microrelief causes decorrelation between the two speckle structures of the images of the object and, hence, a decrease in contrast of Young's fringes.

Although, on the one hand, these factors limit the applicability of speckle photography to measurements of the displacements of objects with diffusely reflecting surfaces, on the other hand, they permit the development of new techniques for studying the processes and phenomena that lead to microrelief changes of the surface under study by studying the variations in the fringe contrast. Let us call such methods correlation speckle interferometry or correlation speckle photography.

In this chapter we first derive expressions for the fringe contrast in correlation speckle photography for random variations of the surface microrelief. We evaluate the sensitivities of the two methods—correlation speckle photography and correlation holographic interferometry—and consider the various optical arrangements that permit a visualization of the surface areas where a variation in the microrelief takes place. These

---

†Deceased.

arrangements involve the use of carrier fringes or the optical image subtraction effect.

We then examine the practical applications of speckle photography. The most efficient way to solve the arising problems entails establishing an empirical correspondence among the physical, mechanical, and other parameters characterizing the process in question and the fringe contrast. Problems in the mechanics of contact interaction, such as the evaluation of the plastic component of the true contact surface, measurement of the contact contour surface, and determination of contact pressures are considered in this way. These parameters have to be known for calculating the friction and wear forces or evaluating contact strength and contact fatigue strength.

11.2 FRINGE CONTRAST IN SPECKLE PHOTOGRAPHY AS RELATED TO A CHANGE IN OBJECT SURFACE MICRORELIEF

11.2.1 Young's Fringe Contrast in Speckle Photography

The arrangement to record speckle photographs and the accepted notations for the vectors used in this subsection to analyze fringe contrast are shown in Fig. 11.1a [1]. The vector $\boldsymbol{\xi}$ lies in the plane of the object surface, and the vector $\mathbf{r}$ in the plane of the speckle photograph. The macrorelief of the surface area under study is assumed to be plane. Let us write the complex amplitude $u(\boldsymbol{\xi})$ of the light wave reflected from the object surface in the form

$$u(\boldsymbol{\xi}) = a(\boldsymbol{\xi}) \exp [i\varphi(\boldsymbol{\xi})]$$

where $a(\boldsymbol{\xi})$ and $\varphi(\boldsymbol{\xi})$ are, respectively, the amplitude and phase of the wave.

Let the speckle structure of the image of the initial surface state be recorded in the first exposure on a photographic plate. We denote the intensity distribution in the plane of the plate by $J_1(\mathbf{r})$. Now we assume that the process under study produces a random change in the microrelief on part of the object surface. After this action the complex amplitude $u_1(\boldsymbol{\xi})$ of the light wave can be written as

$$u_1(\boldsymbol{\xi}) = u(\boldsymbol{\xi}) \exp [i\psi(\boldsymbol{\xi})]$$

where $\psi(\boldsymbol{\xi})$ is a random function describing the phase change of the light wave. Note that the random functions $\varphi(\boldsymbol{\xi})$ and $\psi(\boldsymbol{\xi})$ are uncorrelated, i.e., $\langle\varphi(\boldsymbol{\xi})\psi(\boldsymbol{\xi})\rangle = 0$.

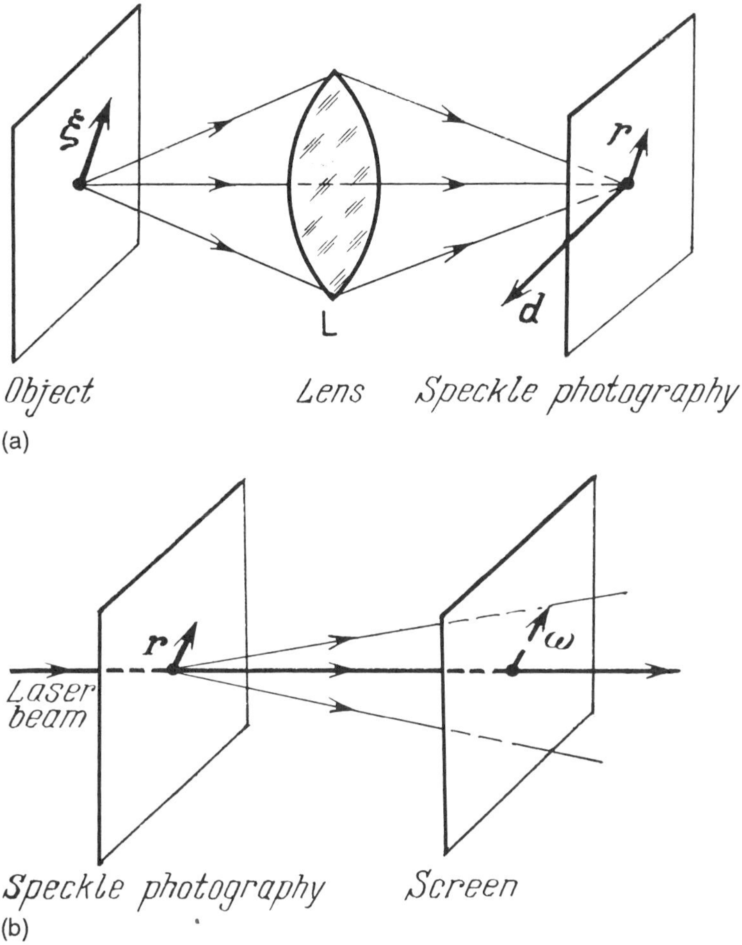

Figure 11.1 Notation accepted in the method of speckle photography. (a) Recording layout for a speckle photograph; (b) observation of Young's fringes.

The second exposure of the speckle photograph is made after the action on the surface under study and after the photographic plate has been shifted in its plane in the direction specified by the vector **d**. The intensity distribution recorded in the second exposure will be denoted by $J_2(\mathbf{r})$.

When the double exposure speckle photograph obtained in this way is scanned with an unexpanded laser beam (Fig. 11.1b), one will observe on the screen a Young's fringe pattern with the same fringe orientation and a period determined by the direction and magnitude of the displacement of the photographic plate. In the regions of the speckle photograph corresponding to the distorted areas of the surface, the fringe contrast will decrease.

The ensemble-averaged intensity $I_S(\omega)$ in the screen plane can be written in the form [2]

$$I_S(\boldsymbol{\omega}) = \langle|F(\boldsymbol{\omega})|^2\rangle = \left\langle\left|\int_{-\infty}^{\infty} [\exp(i\boldsymbol{\omega}\mathbf{r})][J_1(\mathbf{r}) + J_2(\mathbf{r})]\, d\mathbf{r}\right|^2\right\rangle \tag{11.1}$$

where $\boldsymbol{\omega}$ is a vector in the screen plane and $F(\boldsymbol{\omega})$ is the complex amplitude of the light wave in the screen plane. We rewrite Eq. (11.1) as

$$I_S(\boldsymbol{\omega}) = I_1(\boldsymbol{\omega}) + I_2(\boldsymbol{\omega}) + I_{12}(\boldsymbol{\omega}) + I_{21}(\boldsymbol{\omega}) \tag{11.2}$$

where

$$I_1(\boldsymbol{\omega}) = \int_{-\infty}^{\infty} \{\exp[i\boldsymbol{\omega}(\mathbf{r}_1 - \mathbf{r}_2)]\}\langle J_1(\mathbf{r}_1)J_1(\mathbf{r}_2)\rangle\, d\mathbf{r} \tag{11.3}$$

$$I_2(\boldsymbol{\omega}) = \int_{-\infty}^{\infty} \{\exp[i\boldsymbol{\omega}(\mathbf{r}_1 - \mathbf{r}_2)]\}\langle J_2(\mathbf{r}_1)J_2(\mathbf{r}_2)\rangle\, d\mathbf{r} \tag{11.4}$$

$$I_{12}(\boldsymbol{\omega}) = \int_{-\infty}^{\infty} \{\exp[i\boldsymbol{\omega}(\mathbf{r}_1 - \mathbf{r}_2)]\}\langle J_1(\mathbf{r}_1)J_2(\mathbf{r}_2)\rangle\, d\mathbf{r} \tag{11.5}$$

$$I_{21}(\boldsymbol{\omega}) = I_{12}^*(\boldsymbol{\omega}) \tag{11.6}$$

We assume that the optical system in Fig. 11.1a is a linear, spatially invariant system with an impulse response $h(\mathbf{r} - \boldsymbol{\xi})$. Then the expression for the intensity $J_1(\mathbf{r}_1)$ can be transformed to

$$J_1(\mathbf{r}_1) = \int_{-\infty}^{\infty} u(\xi_1)h(\mathbf{r}_1 - \xi_1)u^*(\xi_2)h^*(\mathbf{r}_1 - \xi_2)\, d\xi_1\, d\xi_2$$

For the autocorrelation function of the intensity, $J_1(\mathbf{r})$, we now have

$$\begin{aligned}\langle J_1(\mathbf{r}_1)J_1(\mathbf{r}_2)\rangle = \int_{-\infty}^{\infty} &\langle u(\xi_1)u^*(\xi_2)u(\xi_3)u^*(\xi_4)\rangle \\ &\cdot h(\mathbf{r}_1 - \xi_1)h^*(\mathbf{r}_1 - \xi_2)h(\mathbf{r}_2 - \xi_3)h^*(\mathbf{r}_2 - \xi_4) \\ &\times d\xi_1\, d\xi_2\, d\xi_3\, d\xi_4\end{aligned} \tag{11.7}$$

The fourth-order correlation moment of the complex amplitude of the light wave can be expressed in terms of the second-order correlation moments as follows [3]:

$$\begin{aligned}\langle u(\xi_1)u^*(\xi_2)u(\xi_3)u^*(\xi_4)\rangle = &\langle u(\xi_1)u^*(\xi_2)\rangle\langle u(\xi_3)u^*(\xi_4)\rangle \\ &+ \langle u(\xi_1)u^*(\xi_4)\rangle\langle u(\xi_3)u^*(\xi_2)\rangle\end{aligned} \tag{11.8}$$

For the δ-correlated complex amplitude $u(\boldsymbol{\xi})$, expression (11.8) can be rewritten as

$$\langle u(\boldsymbol{\xi}_1)u^*(\boldsymbol{\xi}_2)u(\boldsymbol{\xi}_3)u^*(\boldsymbol{\xi}_4)\rangle = \langle|u(\boldsymbol{\xi}_1)|^2\rangle\langle|u(\boldsymbol{\xi}_3)|^2\rangle\delta(\boldsymbol{\xi}_1 - \boldsymbol{\xi}_2)\delta(\boldsymbol{\xi}_3 - \boldsymbol{\xi}_4) + \langle|u(\boldsymbol{\xi}_1)|^2\rangle\langle|u(\boldsymbol{\xi}_3)|^2\rangle\delta(\boldsymbol{\xi}_1 - \boldsymbol{\xi}_4)\delta(\boldsymbol{\xi}_2 - \boldsymbol{\xi}_3) \quad (11.9)$$

where δ(·) is the Dirac delta function. Taking into account that the object is illuminated uniformly ($\langle|u(\boldsymbol{\xi})|^2\rangle = 1$) and substituting (11.9) into (11.7), we come to

$$\langle J_1(\mathbf{r}_1)J_1(\mathbf{r}_2)\rangle = \int_{-\infty}^{\infty} |h(\mathbf{r}_1 - \boldsymbol{\xi}_1)|^2\, d\boldsymbol{\xi}_1 \int_{-\infty}^{\infty} |h(\mathbf{r}_2 - \boldsymbol{\xi}_3)|^2\, d\boldsymbol{\xi}_3 + \int_{-\infty}^{\infty} h(\mathbf{r}_1 - \boldsymbol{\xi}_1)h^*(\mathbf{r}_2 - \boldsymbol{\xi}_1)\, d\boldsymbol{\xi}_1 \times \int_{-\infty}^{\infty} h^*(\mathbf{r}_1 - \boldsymbol{\xi}_3)h(\mathbf{r}_2 - \boldsymbol{\xi}_3)\, d\boldsymbol{\xi}_3 \quad (11.10)$$

Assuming that the impulse response is normalized to unity $[\int_{-\infty}^{\infty} h(\mathbf{r} - \boldsymbol{\xi})\, d\boldsymbol{\xi} = 1]$, Eq. (11.10) can be transformed to

$$\langle J_1(\mathbf{r}_1)J_1(\mathbf{r}_2)\rangle = 1 + |\Gamma(\mathbf{r}_1 - \mathbf{r}_2)|^2 \quad (11.11)$$

where

$$\Gamma(\mathbf{r}_1 - \mathbf{r}_2) = \int_{-\infty}^{\infty} h(\mathbf{r}_1 - \boldsymbol{\xi})h^*(\mathbf{r}_2 - \boldsymbol{\xi})\, d\boldsymbol{\xi}$$

Substituting Eq. (11.11) into (11.3) and making the change of variables $\mathbf{r}_1 - \mathbf{r}_2 = \boldsymbol{\eta}$ and $\mathbf{r}_1 = \mathbf{r}_1$, we obtain

$$I_1(\boldsymbol{\omega}) = \int_{-\infty}^{\infty} [\exp\,(i\boldsymbol{\omega}\mathbf{r}_1)][1 + |\Gamma(\boldsymbol{\eta})|^2]\, d\boldsymbol{\eta}\, d\mathbf{r}_1 \sim \delta(\boldsymbol{\omega}) + H(\boldsymbol{\omega}) \quad (11.12)$$

where $H(\boldsymbol{\omega})$ is the Fourier transform of the function $|\Gamma(\boldsymbol{\eta})|^2$.

Assuming that random changes in the microrelief of a surface do not affect its statistical properties, we obtain

$$\langle J_2(\mathbf{r}_1)J_2(\mathbf{r}_2)\rangle = \langle J_1(\mathbf{r}_1)J_1(\mathbf{r}_2)\rangle$$

Hence, Eq. (11.4) takes on the form

$$I_2(\boldsymbol{\omega}) \sim \delta(\boldsymbol{\omega}) + H(\boldsymbol{\omega}) \quad (11.13)$$

In a similar way we can derive an expression for the intensity cross-correlation function,

$$\langle J_1(\mathbf{r}_1)J_2(\mathbf{r}_2)\rangle = 1 + |\langle \exp(i\psi)\rangle|^2|\Gamma(\mathbf{r}_1 - \mathbf{r}_2 - \mathbf{d})|^2$$

After substituting this expression into Eq. (11.5) and changing the variables, $\mathbf{r}_1 - \mathbf{r}_2 = \boldsymbol{\eta}$ and $\mathbf{r}_1 = \mathbf{r}_1$, we come to

$$\begin{aligned} I_{12}(\boldsymbol{\omega}) &\sim \delta(\boldsymbol{\omega}) + |\langle \exp(i\psi)|^2[\exp(-i\boldsymbol{\omega}\mathbf{d})]H(\boldsymbol{\omega}) \\ &= I^*_{21}(\boldsymbol{\omega}) \end{aligned} \tag{11.14}$$

Substituting Eqs. (11.12)–(11.14) into Eq. (11.2) yields

$$I_S(\boldsymbol{\omega}) \sim 4\delta(\boldsymbol{\omega}) + 2H(\boldsymbol{\omega})[1 + |\langle \exp(i\psi)\rangle|^2 \cos(\boldsymbol{\omega}\mathbf{d})] \tag{11.15}$$

After filtration of the bright spot at the center of the screen, which is described by the delta function $\delta(\boldsymbol{\omega})$, Eq. (11.15) takes on the form

$$I_S(\boldsymbol{\omega}) \sim H(\boldsymbol{\omega})[1 + |\langle \exp(i\psi)\rangle|^2] \cos(\boldsymbol{\omega}\mathbf{d})$$

This equation can be used to derive an expression for the Young's fringe contrast γ_S,

$$\gamma_S = \frac{I_{S,\max} - I_{S,\min}}{I_{S,\max} + I_{S,\min}} = |\langle \exp(i\psi)\rangle|^2 \tag{11.16}$$

If $p(\psi)$ is the density function of random-phase variations, then

$$\gamma_S = \left\langle \left| \int_{-\infty}^{\infty} [\exp(i\psi)]p(\psi)\, d\psi \right|^2 \right\rangle = \langle \cos\psi\rangle^2 + \langle \sin\psi\rangle^2 \tag{11.17}$$

In the case of holographic interferometry, a change in the surface microrelief between two exposures results in the decorrelation of the reconstructed light waves, causing a reduction in the fringe contrast. The measuring method based on this phenomenon is called correlation holographic interferometry [4, 5].

Denote by u_1 and u_2 the complex amplitudes of the reconstructed light waves reflected from the surface under study before and after the change in its microrelief.

$$u_1 = a_1 \exp(-i\varphi_1), \qquad u_2 = a_2 \exp[-i(\varphi_1 + \psi + \varphi_0)] \tag{11.18}$$

Here a_1 and a_2 are the amplitudes of the reconstructed light waves, φ_1 is the phase of the object wave during the first exposure, φ_2 is the phase after the action on the surface, $\varphi_1 - \varphi_2 = \psi$ is the change in the object wave phase caused by the random change in the surface microrelief, and φ_0 is a regular phase variation introduced into the object wave to produce carrier interference fringes localized on the object surface.

The interference pattern observed under simultaneous reconstruction of object waves (11.18) recorded on a hologram is described by an intensity distribution I_H,

$$I_H = \langle |u_1 + u_2|^2 \rangle \tag{11.19}$$

where the brackets $\langle\ \rangle$ denote averaging over an area on the object surface corresponding to the limiting resolution of the optical system used to observe the object. We shall assume that the size of this area substantially exceeds the characteristic transverse size of the surface microinhomogeneities, i.e., the optical system is not capable of resolving individual details in the microrelief. If we also assume that the amplitudes a_1 and a_2 and phases φ_1 and φ_2 are statistically independent, Eq. (11.19) will transform after some straightforward algebra to

$$I_H = \langle a_1^2 \rangle + \langle a_2^2 \rangle + 2\langle a_1 a_2 \rangle \langle \cos (\varphi_0 + \psi) \rangle \tag{11.20}$$

Suppose also that the random nature of the change in surface microrelief affects only the phase of the reflected wave, i.e., $a_1 = a_2$. With this in mind, Eq. (11.20) becomes

$$I_H \sim 1 + \langle \cos (\varphi_0 + \psi) \rangle \tag{11.21}$$

Equation (11.21) can be transformed in the following way:

$$\begin{aligned} I_H &\sim 1 + \langle \cos \psi \rangle \cos \varphi_0 - \langle \sin \psi \rangle \sin \varphi_0 \\ &= 1 + \sqrt{\langle \cos \psi \rangle^2 + \langle \sin \psi \rangle^2} \cos \left[\varphi_0 + \tan^{-1} \left(\frac{\langle \sin \psi \rangle}{\langle \cos \psi \rangle} \right) \right] \end{aligned}$$

One can see that the contrast γ_H of the carrier holographic interference fringes is

$$\gamma_H = \frac{I_{H,\max} - I_{H,\min}}{I_{H,\max} + I_{H,\min}} = \sqrt{\langle \cos \psi \rangle^2 + \langle \sin \psi \rangle^2} \tag{11.22}$$

A comparison of Eqs. (11.17) and (11.22) yields

$$\gamma_S = \gamma_H^2 \tag{11.23}$$

Thus the method of correlation speckle photography has a higher sensitivity to changes in the surface microrelief than correlation holographic interferometry.

11.2.2 Fringe Contrast in Speckle Photography with Double Circular Aperture

The optical arrangement to record speckle photographs presented in Fig. 11.2a is distinguished from the one in Fig. 11.1a by an opaque mask placed in front of the lens L. This mask has two symmetric circular holes, denoted by a and b [6]. Write $u(\xi)$ for the complex amplitude of the light wave at

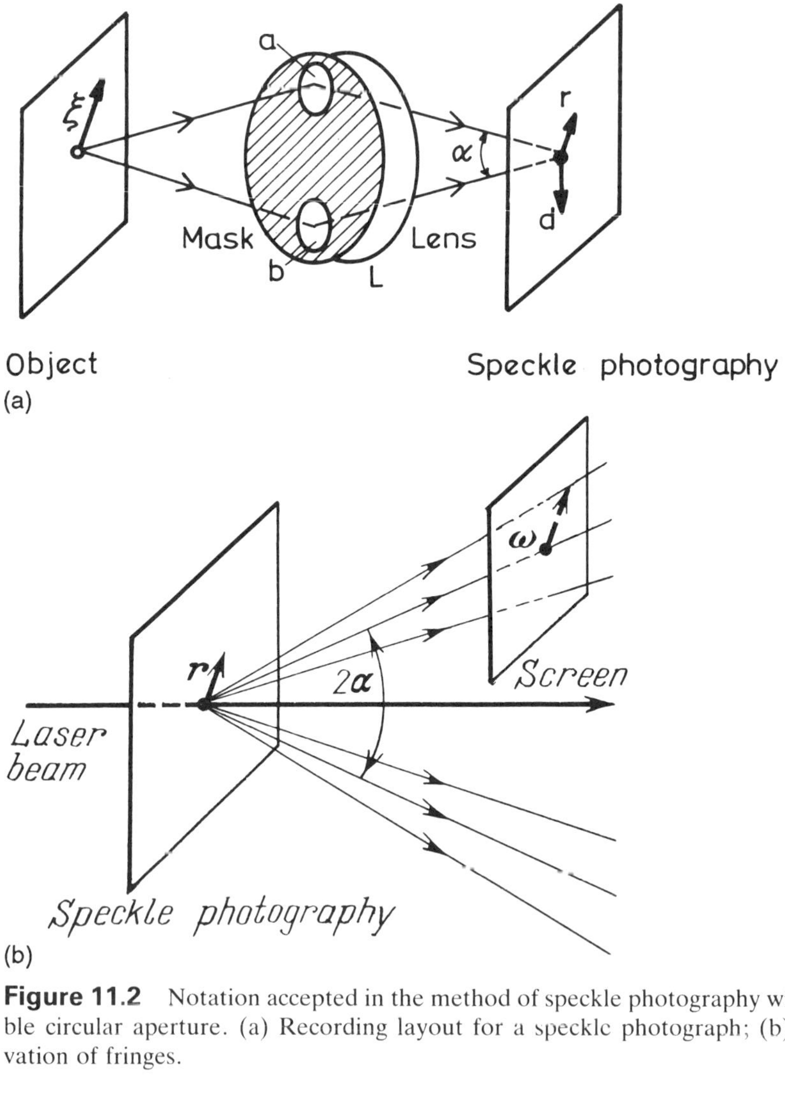

Figure 11.2 Notation accepted in the method of speckle photography with double circular aperture. (a) Recording layout for a speckle photograph; (b) observation of fringes.

the object surface. Then the expression for intensity distribution $J_1(\mathbf{r})$ in the speckle photograph plane during the first exposure can be written as

$$J_1(\mathbf{r}) = |u_{a1} + u_{b1}|^2 = |u_{a1}|^2 + |u_{b1}|^2 + u_{a1}u_{b1}^* + u_{a1}^*u_{b1} \tag{11.24}$$

The third term in Eq. (11.24) describes the regular interference pattern, which modulates each speckle.

For the intensity distribution $J_2(\mathbf{r})$ after the action on the object surface that has irreversibly modified its microrelief, we have

$$J_2(\mathbf{r}) = |u_{a2} + u_{b2}|^2 = |u_{a2}|^2 + |u_{b2}|^2 + u_{a2}u_{b2}^* + u_{a2}^*u_{b2} \tag{11.25}$$

When a double-exposure speckle photograph thus obtained is illuminated by an unexpanded laser beam (Fig. 11.2b) we obtain a reconstructed field in one of the lateral orders having a complex amplitude $v(\mathbf{r})$:

$$v(\mathbf{r}) = u_{a1}u_{b1}^* + u_{a2}u_{b2}^* \tag{11.26}$$

An expression for light intensity $I(\boldsymbol{\omega})$ on the screen plane (see Fig. 11.2b) can be written in the form

$$\begin{aligned} I(\boldsymbol{\omega}) &= \left\langle \left| \int_{-\infty}^{\infty} [\exp(i\boldsymbol{\omega}\mathbf{r})] v(\mathbf{r})\, dr \right|^2 \right\rangle \\ &= \int_{-\infty}^{\infty} \{\exp[i\boldsymbol{\omega}(\mathbf{r}_1 - \mathbf{r}_2)]\, \langle v(\mathbf{r}_1)v^*(\mathbf{r}_2)\rangle\, d\mathbf{r}_1\, d\mathbf{r}_2 \\ &= \int_{-\infty}^{\infty} \{\exp[i\boldsymbol{\omega}(\mathbf{r}_1 - \mathbf{r}_2)][\langle u_{a1}(\mathbf{r}_1)u_{a2}^*(\mathbf{r}_2)\rangle\langle u_{b1}^*(\mathbf{r}_1)u_{b1}(\mathbf{r}_2)\rangle \\ &\quad + \langle u_{a1}(\mathbf{r}_1)u_{a2}^*(\mathbf{r}_2)\rangle\langle u_{b1}^*(\mathbf{r}_1)u_{b2}(\mathbf{r}_2)\rangle \\ &\quad + \langle u_{a2}(\mathbf{r}_1)u_{a2}^*(\mathbf{r}_2)\rangle\langle u_{b2}^*(\mathbf{r}_1)u_{b2}(\mathbf{r}_2)\rangle \\ &\quad + \langle u_{a2}(\mathbf{r}_1)u_{a1}^*(\mathbf{r}_2)\rangle\langle u_{b2}^*(\mathbf{r}_1)u_{b1}(\mathbf{r}_2)\rangle]\, d\mathbf{r}_1\, d\mathbf{r}_2 \end{aligned} \tag{11.27}$$

If the holes in the mask are of equal diameters, then

$$\langle u_{a1}(\mathbf{r}_1)u_{a1}^*(\mathbf{r}_2)\rangle = \langle u_{a2}(\mathbf{r}_1)u_{a2}^*(\mathbf{r}_2)\rangle = \Gamma(\mathbf{r}_1 - \mathbf{r}_2) \tag{11.28}$$

$$\langle u_{b1}^*(\mathbf{r}_1)u_{b1}(\mathbf{r}_2)\rangle = \langle u_{b2}^*(\mathbf{r}_1)u_{b2}(\mathbf{r}_2)\rangle = \Gamma^*(\mathbf{r}_1 - \mathbf{r}_2) \tag{11.29}$$

The expressions for complex amplitudes of the field in the speckle photograph plane formed by means of unit holes during the first and second exposures can be written

$$u_{a1}(\mathbf{r}_1) = \int_{-\infty}^{\infty} u(\boldsymbol{\xi})h_a(\mathbf{r}_1 - \boldsymbol{\xi})\, d\boldsymbol{\xi} \tag{11.30}$$

$$u_{a2}^*(\mathbf{r}_2) = \int_{-\infty}^{\infty} u(\xi) \exp[-i\psi(\xi)] h_a(\mathbf{r}_2 - \xi + \mathbf{d})\, d\xi \tag{11.31}$$

$$u_{b1}(\mathbf{r}_1) = \int_{-\infty}^{\infty} u(\xi) h_b(\mathbf{r}_1 - \xi)\, d\xi \tag{11.32}$$

$$u_{b2}(\mathbf{r}_2) = \int_{-\infty}^{\infty} u^*(\xi) \exp[i\psi(\xi)] h_b(\mathbf{r}_2 - \xi + \mathbf{d})\, d\xi \tag{11.33}$$

where $\exp[i\psi(\xi)]$ is a random phase shift caused by the irreversible change in the surface microrelief, and $\mathbf{d}$ is the vector of speckle photograph displacement between two exposures that is necessary to produce the carrier fringes.

Let us calculate, using Eqs. (11.30)–(11.33), the correlation functions entering into Eq. (11.27):

$$\langle u_{a1}(\mathbf{r}_1) u_{a2}^*(\mathbf{r}_2)\rangle = \langle \exp(-i\psi)\rangle \Gamma(\mathbf{r}_1 - \mathbf{r}_2 + \mathbf{d}) \tag{11.34}$$

$$\langle u_{b1}^*(\mathbf{r}_1) u_{b2}(\mathbf{r}_2)\rangle = \langle \exp i\psi\rangle \Gamma^*(\mathbf{r}_1 - \mathbf{r}_2 + \mathbf{d}) \tag{11.35}$$

$$\langle u_{a2}(\mathbf{r}_1) u_{a1}^*(\mathbf{r}_2)\rangle = \langle \exp i\psi\rangle \Gamma^*(\mathbf{r}_1 - \mathbf{r}_2 + \mathbf{d}) \tag{11.36}$$

$$\langle u_{b2}^*(\mathbf{r}_1) u_{b1}(\mathbf{r}_2)\rangle = \langle \exp(-i\psi)\rangle \Gamma(\mathbf{r}_1 - \mathbf{r}_2 + \mathbf{d}) \tag{11.37}$$

After substituting Eqs. (11.28), (11.29), and (11.34)–(11.37) into Eq. (11.27), we come to

$$I(\boldsymbol{\omega}) = 2\int_{-\infty}^{\infty} \{\exp[i\boldsymbol{\omega}(\mathbf{r}_1 - \mathbf{r}_2)]\}[|\Gamma(\mathbf{r}_1 - \mathbf{r}_2)|^2 + |\langle \exp i\psi\rangle|^2 |\Gamma(\mathbf{r}_1 - \mathbf{r}_2 + \mathbf{d})|^2]\, d\mathbf{r}_1\, d\mathbf{r}_2 \tag{11.38}$$

Taking into account that

$$\int_{-\infty}^{\infty} |\Gamma(\mathbf{r}_1 - \mathbf{r}_2)|^2 \exp\{i\boldsymbol{\omega}(\mathbf{r}_1 - \mathbf{r}_2)\}\, d\mathbf{r}_1\, d\mathbf{r}_2 \sim H(\boldsymbol{\omega})$$

where $H(\boldsymbol{\omega})$ is the Fourier transform of the correlation function and $\Gamma(\mathbf{r}_1 - \mathbf{r}_2) = \Gamma(\mathbf{r}_2 - \mathbf{r}_1)$, let us transform Eq. (11.38) into the following form:

$$I(\boldsymbol{\omega}) \sim H(\boldsymbol{\omega})[1 + |\langle \exp(i\psi)\rangle|^2 \cos(\boldsymbol{\omega}\mathbf{d})] \tag{11.39}$$

Equation (11.39) can be used to derive an expression for the contrast γ_{SC} of the carrier fringes in the method of speckle photography with double circular aperture,

$$\gamma_{SC} = |\langle \exp(i\psi)\rangle|^2 \tag{11.40}$$

Thus, the expression for fringe contrast in the method of speckle photography with double circular aperture (11.40) has the same form as (11.16) in the method of speckle photography.

11.2.3 Visualization of Surface Microrelief Variation Areas

The recording via correlation speckle photography of the areas where the object surface microrelief has been changed can be made using carrier fringes. Figure 11.3a shows an optical arrangement for recording speckle photographs. The object surface area under study, O, is illuminated by a plane wave W of coherent laser light. An objective lens L projects a focused image of the object surface onto the photographic plate SF. In the first exposure the speckle structure of the image of the surface in the initial state is recorded on the photographic plate. Next a force is applied to the object surface that results in an irreversible change in its microrelief, the plate SF is displaced in its plane by an amount d, and the second exposure is made. Thus, the speckle photograph will record two speckle structures of the surface image shifted with respect to one another. Note that the magnitude of d should be somewhat greater than the minimum speckle dimension.

When any part of the double-exposure speckle photograph SF obtained in this way is illuminated with an unexpanded laser beam LB (Fig. 11.3b), an interference pattern of Young's fringes of the same frequency and orientation will be observed on the screen S. At the points of the image where the surface microrelief has changed, the contrast of the Young's fringes will be less sharp (relative to the unaffected parts of the surface), or the fringes will not be discernible at all. Thus, by scanning the speckle photograph with an appropriate pitch, one can determine the region where the contrast of the Young's fringes varies and therefore where the surface microrelief has undergone a change.

The method of speckle photography also permits visualization of the entire area of surface microrelief variation against the background of carrier fringes [7]. In this case the double-exposure speckle photograph is fixed in the spatial filtration configuration shown in Fig. 11.4. A small diameter circular aperture is placed in plane A located at a distance from the Fourier plane (the focal plane of lens L). Through the aperture one observes a system of parallel straight carrier fringes, their orientation and pitch depending on the aperture position in the filtration plane and the distance from the Fourier plane to the latter. The areas where the fringes vary in contrast correspond to those where the surface microrelief undergoes a change, as occurs with holographic interferograms.

The optical arrangements discussed permit the visualization of the boundary of the surface microrelief distortion area as a set of discrete lines

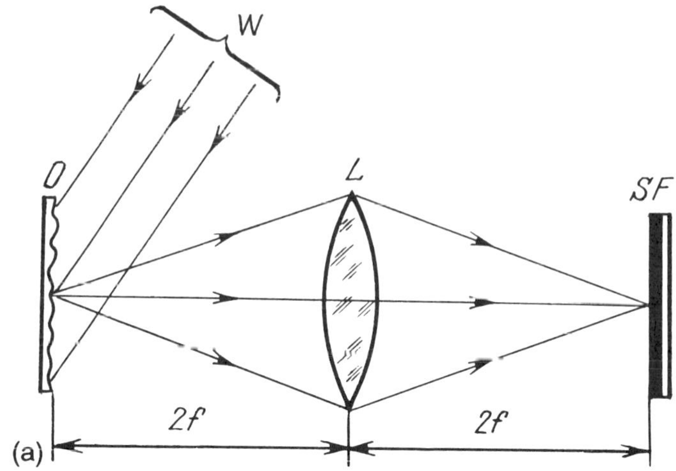

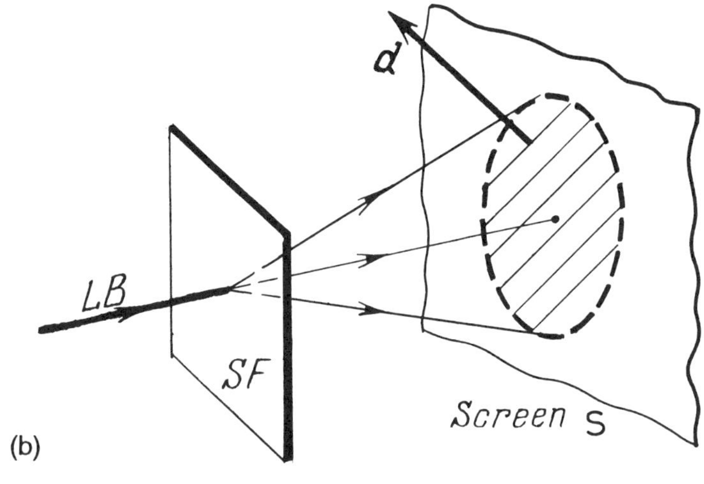

Figure 11.3 (a) Optical arrangement for speckle photograph recording; (b) observation of Young's fringes.

or points. To record the continuous boundary one can use the so-called image subtraction configurations. They provide a possibility of introducing a phase shift of π between two exposures throughout the field studied. In this case one obtains a dark-field interferogram, the surface areas with distorted microrelief being visualized as bright spots of different intensity against the dark background.

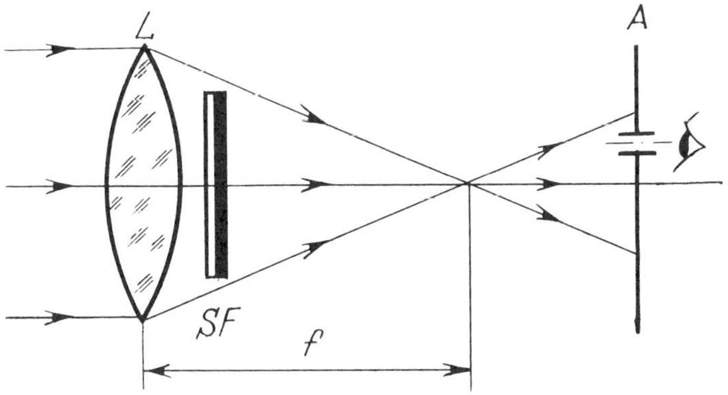

Figure 11.4 Spatial filtering outside the Fourier plane.

An arrangement for image subtraction [8] can be used to visualize the surface areas with distorted microrelief on the objects with plane surfaces. The double-exposure speckle photograph SF recorded in the optical arrangement shown in Fig. 11.3a is mounted in the optical configuration for spatial filtration (Fig. 11.5). If the photographic plate is displaced between two exposures by an amount d, the intensity distribution in the focal plane of lens L (Fourier plane) will be modulated by parallel interference fringes. If we now place in the Fourier plane a narrow slit with the axis positioned at the center of a dark fringe, the image seen through it will be dark (the case of image subtraction) except for the areas with a distorted surface

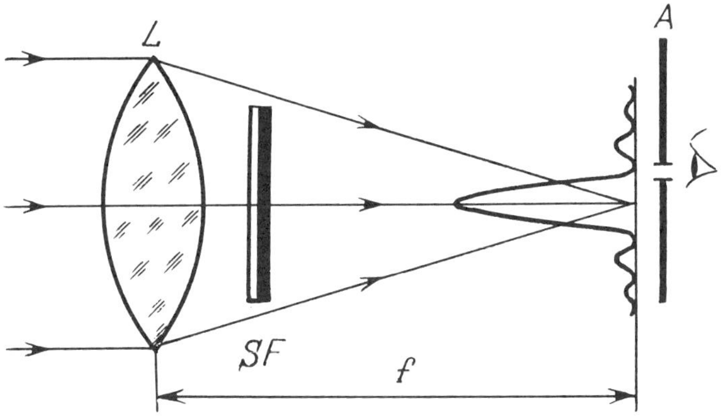

Figure 11.5 Image subtraction by means of speckle photography.

microrelief, which will be bright. The brightness results because the speckle structures corresponding to these surface areas are partially or completely decorrelated.

11.2.4 Experimental Verification of the Expression for Fringe Contrast in Correlation Speckle Photography and Correlation Holographic Interferometry

An optical arrangement (Fig. 11.6) that permits simultaneous recording of a hologram and a speckle photograph of the surface under study was used for experimental verification of Eq. (11.23). The hologram and speckle photograph are obtained by the two-exposure method in the following order. First, one records on a photographic plate SF the initial speckle structure of the surface image (first exposure of the speckle photograph). Second, a photographic plate H is introduced into the optical arrangement, and a hologram of the initial state of the surface is recorded on it (first exposure of the hologram). During this time the photographic plate SF is screened by an opaque mask A. Third, the surface of the object is subjected to a force that affects its microrelief. Fourth, the mirror M is

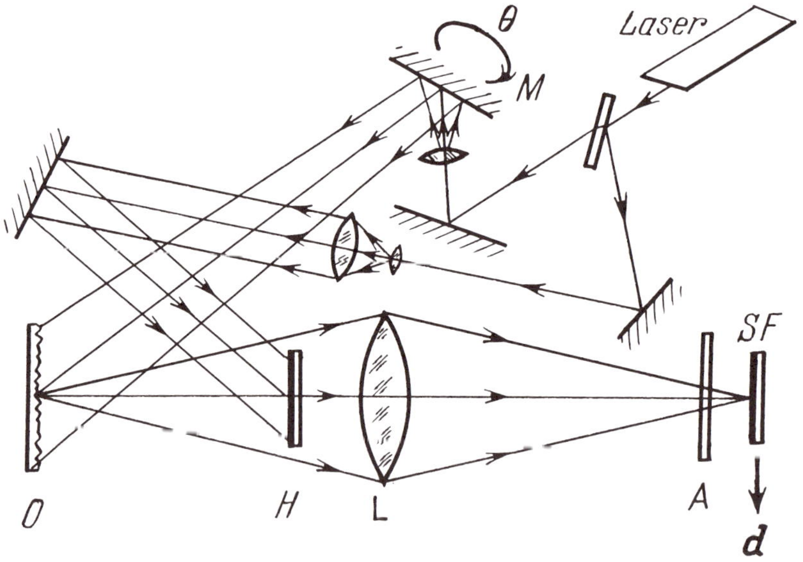

Figure 11.6 Optical arrangement for simultaneous recording by the double-exposure method of holograms and speckle photographs.

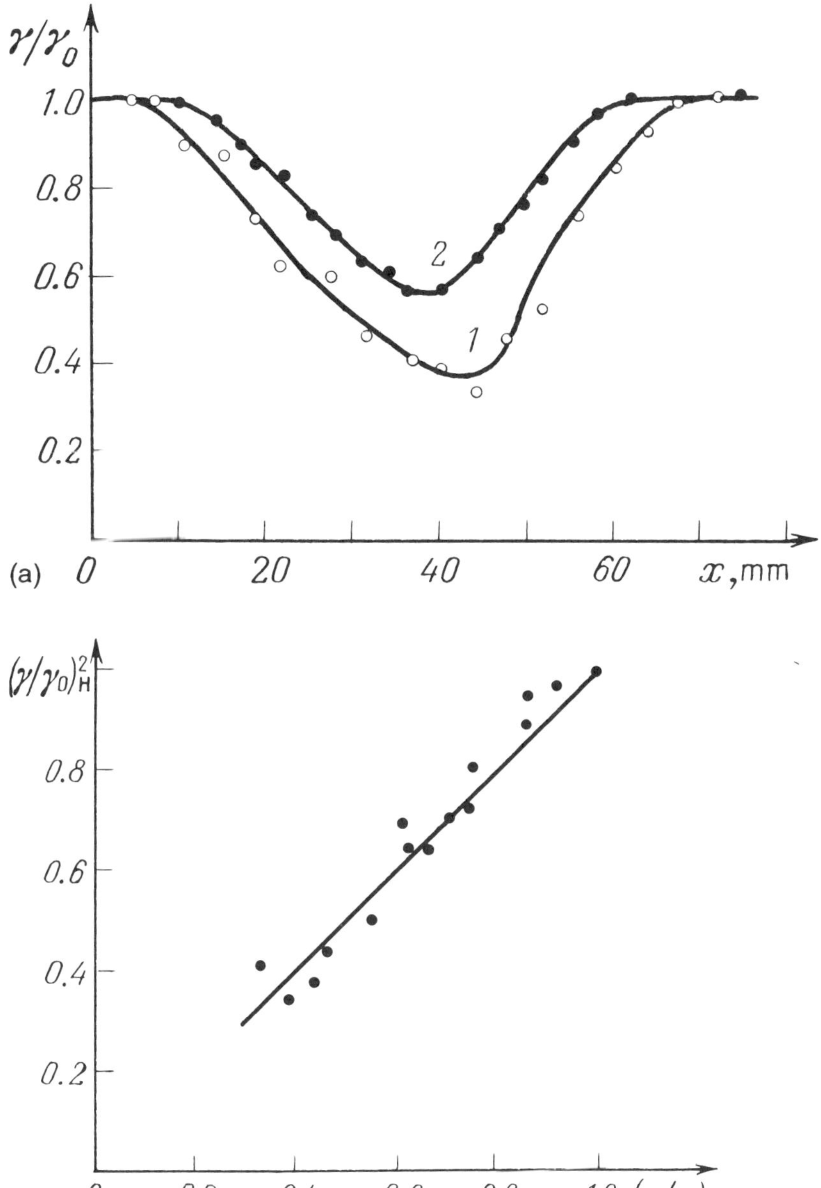

Figure 11.7 (a) Fringe contrast variation over the specimen surface after wear obtained by holographic interferometry, 2, and by speckle photography, 1. (b) Plot of $(\gamma/\gamma_0)_H^2$ versus $(\gamma/\gamma_0)_S$.

turned through a small angle θ, and a hologram of the distorted surface is recorded on plate H (second exposure of hologram).

Fifth, the photographic plate H and mask A are removed, the photographic plate SF is displaced in its plane by an amount d, and the speckle structure of the object image after the action is recorded on it (second exposure of the speckle photograph). Note that with the operations performed in this sequence, the directions of the illuminating beam during the first and second exposures of the speckle photograph differ somewhat because of the amount of rotation of the mirror M necessary to produce carrier fringes in the holographic interferogram. One could, of course, return the mirror M to its initial position before carrying out the second exposure of the speckle photograph. As shown by calculations and experiments, however, decorrelation of the speckle structures in this case is negligible, because the tilt angles of the illuminating beam necessary to produce carrier fringes in holographic interferometry do not exceed a few minutes of arc.

Thus, the same change in surface microrelief can be recorded with the preceding arrangement (Fig. 11.6) both by correlation holographic interferometry and by correlation speckle photography.

The studies were carried out on a specimen of aluminum alloy D16T with initial irregular roughness $R_a = 1.50\ \mu m$ [1]. The wear was produced by random translational motion of a plane steel bar over the central part of the specimen surface, with the peripheral parts of the specimen not subjected to wear. An abrasive was introduced into the contact zone. Figure 11.7a presents longitudinal distributions of the normalized fringe contrast γ/γ_0 (where γ_0 is the fringe contrast outside the wear zone). Curve 2 was obtained by holographic interferometry, and curve 1 by speckle photography. The same data are presented in another graph in Fig. 11.7b, with $(\gamma/\gamma_0)_S$ plotted along the horizontal axis and $(\gamma/\gamma_0)^2_H$ along the vertical axis. The straight line thus obtained passes through the origin at 45°, thus demonstrating the validity of expression (11.23) for the case of wear.

11.3 DETERMINATION OF THE CONTACT AREA

11.3.1 Evaluation of the Plastic Component of Actual Contact Surface

In the mechanics of the contact interaction of objects with rough surfaces, one deals with the actual contact surface area and the contact contour surface area. The actual, or true, contact surface area represents a sum of the areas of all contacting elements in the surface microrelief of the objects involved. The contact contour surface area is the area of the object surface bounded by the contour of the contacting regions. Obviously, the actual contact surface area is less than the contact contour surface area. In turn, the former may be represented as a sum of the elastic and plastic components. In fact, in the contact interactions of objects with rough surfaces, under loading their contacting microrelief elements undergo elastoplastic interaction. After unloading, the actual contact area decreases because of the removal of the elastic deformation.

We now return to Fig. 11.1a and denote the light intensity distribution in the plane of the speckle photograph in the first exposure by $J_1(\mathbf{r})$, and that in the second exposure by $J_2(\mathbf{r})$. Denote by $u_1(\boldsymbol{\xi})$ and $u_2(\boldsymbol{\xi})$ the complex light amplitudes at the surface of the object in the first and second exposures, respectively. Then the expressions for the complex amplitudes $v_1(\mathbf{r})$ and $v_2(\mathbf{r})$ in the plane of the speckle photograph can be written in the form [9]

$$v_1(\mathbf{r}) = \int_{F_0} u_1(\boldsymbol{\xi})h(\mathbf{r} - \boldsymbol{\xi})\, d\boldsymbol{\xi}$$

$$v_2(\mathbf{r}) = \int_{F_0} u_2(\boldsymbol{\xi})h(\mathbf{r} - \boldsymbol{\xi})\, d\boldsymbol{\xi}$$

where $h(\mathbf{r} - \boldsymbol{\xi})$ is the impulse response of the system and F_0 is the area of the object surface corresponding to the limiting resolution of the optical system used for speckle photograph recording.

The expression for $v_2(\mathbf{r})$ can now be rewritten as

$$\begin{aligned} v_2(\mathbf{r}) &= \int_{F} u_F(\boldsymbol{\xi})h(\mathbf{r} - \boldsymbol{\xi})\, d\boldsymbol{\xi} + \int_{F_0 - F} u_1(\boldsymbol{\xi} + \mathbf{d})h(\mathbf{r} - \boldsymbol{\xi})\, d\boldsymbol{\xi} \\ &= v_{2F} + v_{20} \end{aligned} \tag{11.41}$$

where u_F is the complex amplitude of the light wave reflected from the plastically deformed surface microrelief elements (it is assumed that $\langle u_F \cdot u_1 \rangle = 0$), and F is the area where microrelief has changed.

By analogy the expression for $v_1(\mathbf{r})$ can be rewritten in the form

$$\begin{aligned} v_1(\mathbf{r}) &= \int_{F} u_1(\boldsymbol{\xi})h(\mathbf{r} - \boldsymbol{\xi})\, d\boldsymbol{\xi} + \int_{F_0 - F} u_1(\boldsymbol{\xi})h(\mathbf{r} - \boldsymbol{\xi})\, d\boldsymbol{\xi} \\ &= v_{1F} + v_{10} \end{aligned} \tag{11.42}$$

Taking into account (11.41) and (11.42) for the intensities $J_1(\mathbf{r})$ and $J_2(\mathbf{r})$, we obtain the expressions

$$J_1(\mathbf{r}) = |v_1(\mathbf{r})|^2 = |v_{1F}|^2 + |v_{10}|^2 + v_{1F}v_{10}^* + v_{1F}^*v_{10} \tag{11.43}$$

and

$$J_2(\mathbf{r}) = |v_2(\mathbf{r})|^2 = |v_{2F}|^2 + |v_{20}|^2 + v_{2F}v_{20}^* + v_{2F}^*v_{20} \tag{11.44}$$

Obviously, $v_{20}(\mathbf{r}) = v_{10}(\mathbf{r} + \mathbf{d})$, where $\mathbf{d}$ is the displacement vector of the photographic plate between exposures.

According to Eq. (11.1), the intensity distribution $I_S(\boldsymbol{\omega})$ in the screen plane with a double-exposure speckle photograph illuminated by an unexpanded laser beam (Fig. 11.1b) will be

$$\begin{aligned}
I_S(\boldsymbol{\omega}) &= \int_{-\infty}^{\infty} \{\exp [i\boldsymbol{\omega}(\mathbf{r}_1 - \mathbf{r}_2)]\}\langle I(\mathbf{r}_1)I(\mathbf{r}_2)\rangle \, d\mathbf{r}_1 \, d\mathbf{r}_2 \\
&= \int_{-\infty}^{\infty} \{\exp [i\boldsymbol{\omega}(\mathbf{r}_1 - \mathbf{r}_2)]\}\langle[J_1(\mathbf{r}_1) + J_2(\mathbf{r}_1)] \\
&\quad \times [J_1(\mathbf{r}_2) + J_2(\mathbf{r}_2)]\rangle \, d\mathbf{r}_1 \, d\mathbf{r}_2 \\
&= \int_{-\infty}^{\infty} \{\exp [i\boldsymbol{\omega}(\mathbf{r}_1 - \mathbf{r}_2)]\}(\langle J_1(\mathbf{r}_1)J_1(\mathbf{r}_2)\rangle \\
&\quad + \langle J_2(\mathbf{r}_1)J_2(\mathbf{r}_2)\rangle + \langle J_1(\mathbf{r}_1)J_2(\mathbf{r}_2)\rangle \\
&\quad + \langle J_1(\mathbf{r}_2)J_2(\mathbf{r}_1)\rangle) \, d\mathbf{r}_1 \, d\mathbf{r}_2
\end{aligned} \tag{11.45}$$

where $I(\mathbf{r}) = J_1(\mathbf{r}) + J_2(\mathbf{r})$.

Using Eqs. (11.43) and (11.44), we can determine all the intensity correlation functions in Eq. (11.45),

$$\langle J_1(\mathbf{r}_1)J_1(\mathbf{r}_2)\rangle = \langle J_2(\mathbf{r}_1)J_2(\mathbf{r}_2)\rangle \sim C_1 + F_0^2|\Gamma(\mathbf{r}_1 - \mathbf{r}_2)|^2 \tag{11.46}$$

$$\langle J_1(\mathbf{r}_1)J_2(\mathbf{r}_2)\rangle = \langle J_1(\mathbf{r}_2)J_2(\mathbf{r}_1)\rangle \sim C_2 + (F_0 - F)^2|\Gamma(\mathbf{r}_1 - \mathbf{r}_2 + \mathbf{d})|^2 \tag{11.47}$$

where C_1 and C_2 are constants.

Substituting Eqs. (11.46) and (11.47) into Eq. (11.45), we obtain

$$I_S(\boldsymbol{\omega}) \sim 1 + (1 - F/F_0)^2 \cos (\boldsymbol{\omega}\mathbf{d}) \tag{11.48}$$

From Eq. (11.48) it follows that

$$\gamma_S = (1 - F/F_0)^2 \tag{11.49}$$

As shown in Ref. 10, the change in carrier fringe contrast γ_H in the method of holographic interferometry is equal to the ratio of the area F

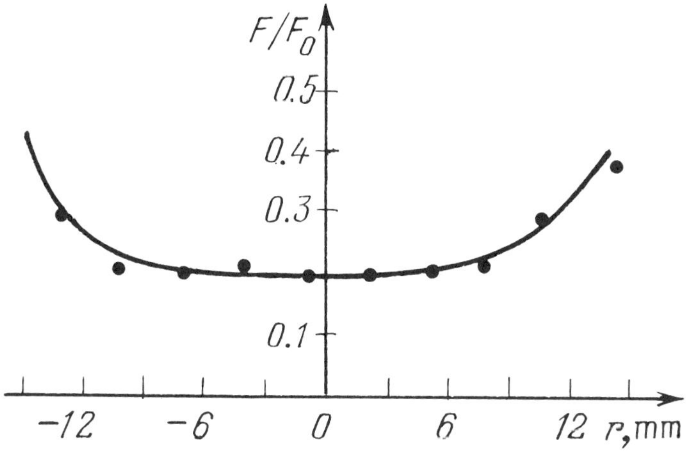

Figure 11.8 Dependence of F/F_0 on r over the surface of contact between a steel plate and the end face of a cylinder.

of the plastic component of the actual contact area to the area F_0 of the contact contour surface,

$$\gamma_H = 1 - F/F_0. \tag{11.50}$$

Comparison of Eqs. (11.49) and (11.50) confirms the formerly obtained general equation (11.23).

The dots in Fig. 11.8 show the magnitudes of F/F_0 along the contact spot diameter where the steel plate interacts under load with an end face of a cylinder. The surfaces brought into contact had the same roughness, $R_a = 1.50$ μm. The scanning laser beam diameter was equal to 1 mm.

The dependence presented in Fig. 11.8 corresponds qualitatively to the contact pressure distribution along the contact spot diameter (see Fig. 11.12).

11.3.2 Contact Contour Surface Recording

Let us record a double-exposure speckle photograph using the optical arrangement shown in Fig. 11.3a. Between exposures we bring the object under study into contact with another object under load and shift the speckle photograph in its plane. When such a speckle photograph is scanned by an unexpanded laser beam, a system of carrier fringes will be observed with the contrast over the contact area lower than over the surface areas not subjected to the contact interaction. The contact contour surface in this case corresponds to a surface area, where

$$\gamma_S < \gamma_{S0} \tag{11.51}$$

Here γ_{S0} is the fringe contrast beyond the contact area.

The determination of contact contour area with the help of Eq. (11.51) can be carried out using the arrangement for speckle photograph recording

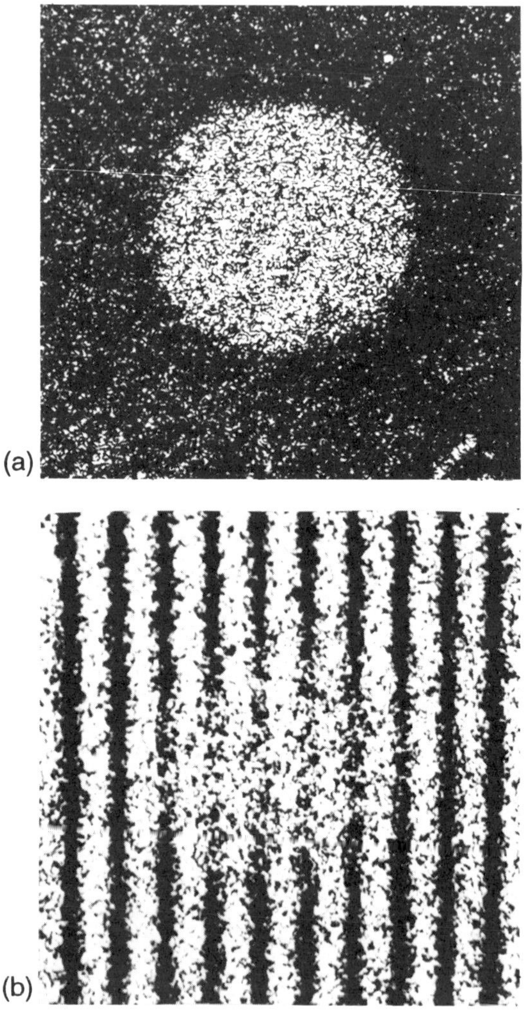

Figure 11.9 Contact surface visualization. (a) Image subtraction by speckle photography. (b) Holographic interferogram.

with a double circular aperture. In this case the laser beam transmitted through the speckle photograph does not impose on the Young's fringe pattern. The determination of the contact contour surface can be carried out most efficiently using image subtraction techniques. As an illustration, consider the problem of a contact between a steel plate and a steel ball of radius 500 mm [11]. The plate had a hardness of 39 HRC, and the ball a hardness of 44 HRC. The roughness of both contacting surfaces was R_a = 1.50 μm. Figure 11.9a presents a photograph of the steel plate obtained by means of a double-exposure speckle photograph by the image subtraction technique (see Fig. 11.5). The holographic carrier fringe interferogram shown in Fig. 11.9b was recorded simultaneously with the speckle photograph under the same load conditions. One can clearly see that whereas the carrier fringe contrast in the contact zone decreases as one approaches its center, the interference fringes do not disappear completely.

The photographs in Fig. 11.9 allow us to draw the qualitative conclusion that decorrelation of speckle structures in a double-exposure speckle photograph occurs faster than that of the waves caused by the same change in the surface microrelief but reconstructed from a double-exposure hologram.

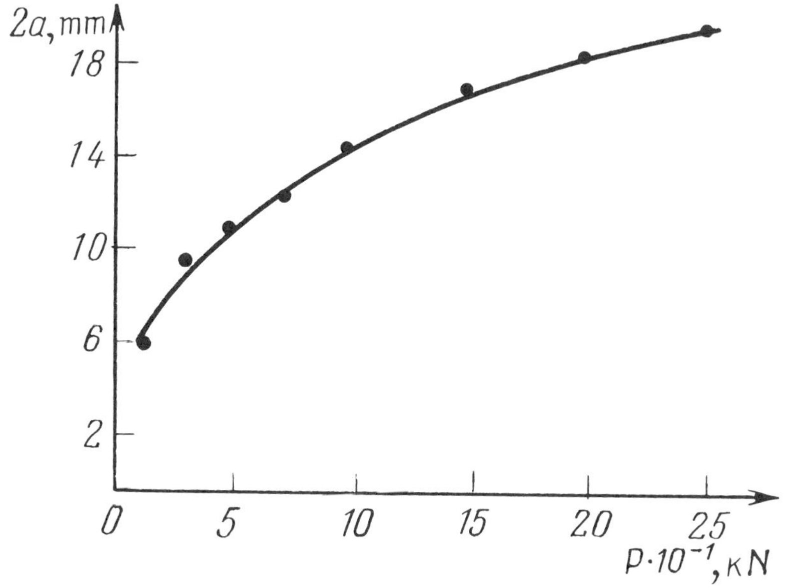

Figure 11.10 Dependence of diameter $2a$ of the contact surface on load P.

An experimental dependence of the diameter of the contact surface on the load, obtained by speckle photography, is presented by the dots in Fig. 11.10. For comparison, the solid line illustrates the theoretical dependence corresponding to Hertz's solution [12]. The good agreement between experimental data and calculated values supports the use of correlation speckle photography for measuring the contact contour surface.

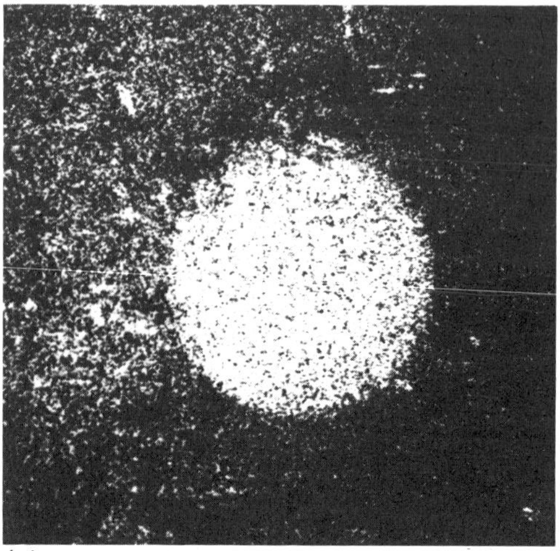

(a)

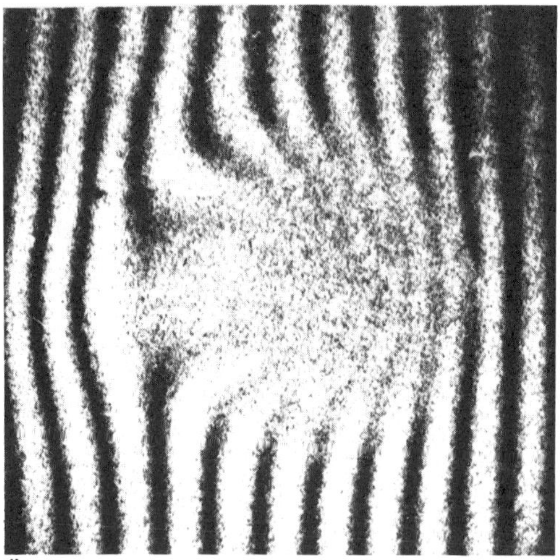

(b)

Figure 11.11 Contact surface visualized by means of one double-exposure focused image hologram. (a) Image subtraction by speckle photography; (b) carrier interference fringes.

The relative error in this case depends primarily on the speckle sizes (see Fig. 11.6a) and, as evaluations show, is not high (4%). To reduce the speckle size it is necessary to widen the slit and hence the dark interference fringes in the Fourier plane. This can be achieved by multiexposure speckle photography [13].

The high precision with which holographic interferometry can reveal the boundary of transition into the elastoplastic deformation domain in the contact between two objects [11] and the possibility of recording the continuous-contact surface boundary by speckle photography can be combined in an optical arrangement for the purpose of obtaining focused image holograms. Figure 11.11a is a photograph of the contact surface between a steel ball and a steel plate produced by speckle photography (image subtraction technique), and Fig. 11.11b is a holographic interferogram. The photographs shown in Fig. 11.11 were obtained by means of one double-exposure focused-image hologram [11]. To produce carrier fringes the object under study was turned between exposures about an axis lying in the plane of its surface, and the photographic plate was displaced in its plane. The bending of the carrier interference fringes (Fig. 11.11b) demonstrates the appearance of residual deformations in the vicinity of the contact zone, which indicates the elastoplastic nature of the contact under this load.

11.4 CONTACT PRESSURE MEASUREMENT

11.4.1 Effect of Contact Pressure on Carrier Fringe Contrast

When two objects with rough surfaces are brought into contact with each other, the plastic component of the actual contact surface will be greater in places with higher contact pressures. Hence, the carrier fringe contrast in these zones should be lower.

It is convenient to study the effect of contact pressure on fringe contrast for a plate in contact with the end face of a rigid cylinder (Fig. 11.12). The contact pressure Q can be calculated by using the expression [14]

$$Q = P(2\pi a\sqrt{a^2 - r^2})^{-1} \tag{11.52}$$

where P is the load, a is the radius of the cylinder, and r is the coordinate measured from the axis of the cylinder. It follows from (11.52) that in the central region of the contact surface there is an extended zone of almost constant pressure Q_0,

$$Q_0 - P/2F \tag{11.53}$$

where $F = \pi a^2$ is the area of the end face of the cylinder. Thus, by varying the contact pressure Q_0 and measuring the carrier fringe contrast in the

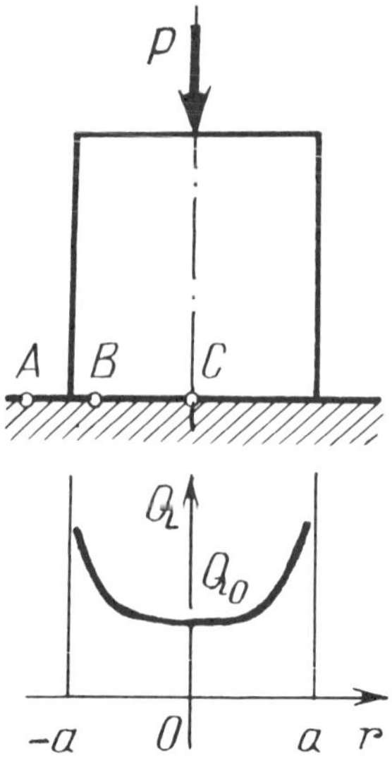

Figure 11.12 Distribution of contact pressure Q in the contact zone of a plate with the end face of a cylinder.

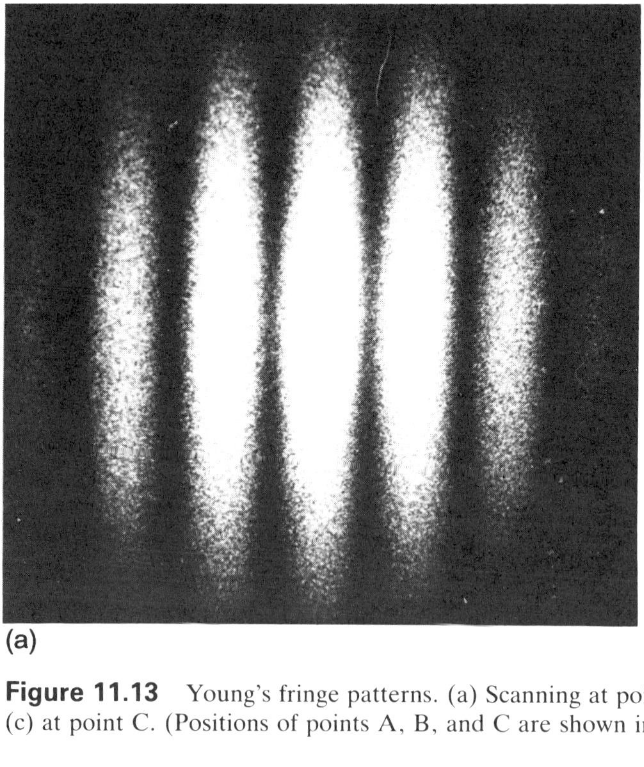

(a)

Figure 11.13 Young's fringe patterns. (a) Scanning at point A; (b) at point B; (c) at point C. (Positions of points A, B, and C are shown in Fig. 11.12.)

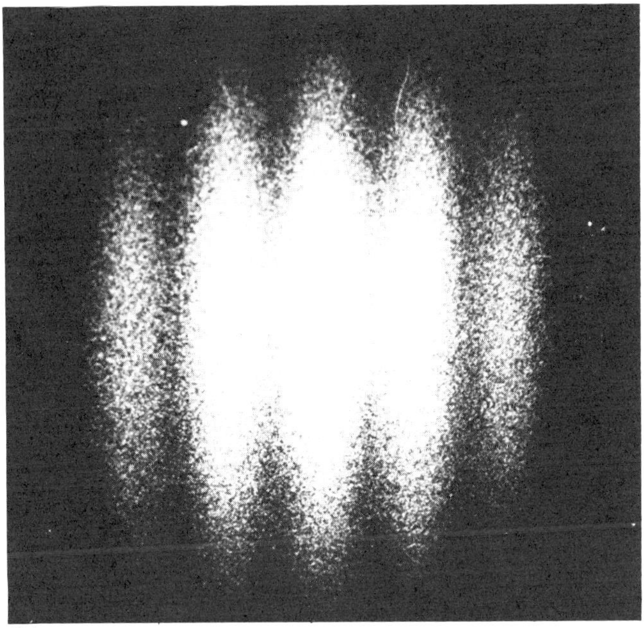

(b)

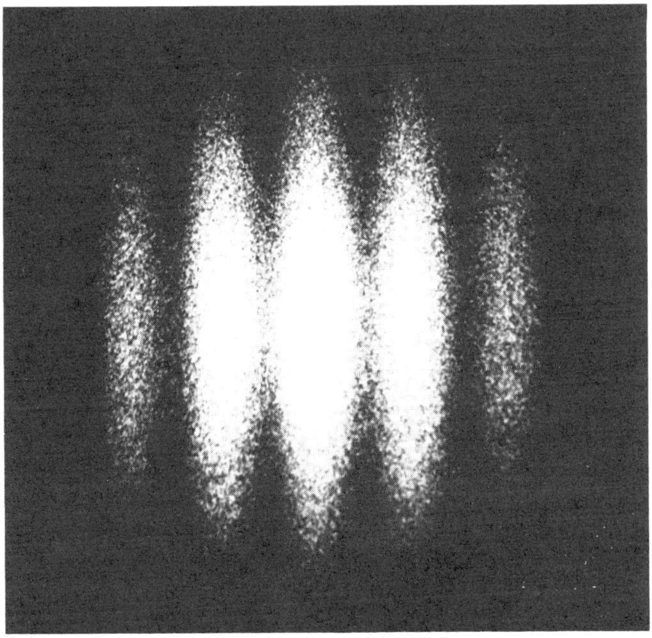

(c)

Figure 11.13 Continued.

central zone of the contact spot, one can determine the relation between Q_0 and γ.

Figure 11.13 presents three characteristic Young's fringe patterns observed when a double-exposure speckle photograph (produced with the optical arrangement of Fig. 11.3a) is illuminated with an unexpanded laser beam. The Young's fringes reveal the highest contrast when the beam passes through point A, which lies outside the contact zone (Fig. 11.12), where no change in the microrelief occurs (Fig. 11.13a). Point B lies close to the zone with the highest contact pressure (i.e., where the changes in microrelief are the largest), and therefore the fringes have a low contrast (Fig. 11.13b). Point C is located in the area of almost constant contact pressure; the corresponding Young's fringe pattern is shown in Fig. 11.13c. The contrast of these fringes is less than that of the fringes outside the contact zone but higher than the Young's fringe contrast in the zone of maximal contact pressure. This series of interferograms indicates that the change in the Young's fringe contrast is in general agreement with the change contact pressure along the radius of the contact spot.

To obtain a quantitative dependence of the fringe contrast $\Delta\gamma/\gamma_0 = (\gamma_0 - \gamma)/\gamma_0$ (where γ_0 is the fringe contrast outside the contact zone) on the contact pressure, various loads were applied to the end face of a steel cylinder of 30 mm diameter and a hardness of 45 HRC, placed either on a 39 HRC steel plate or on a plate of D16T alloy [15]. The roughness of the cylinder's end face was $R_a = 0.32$ μm, and that of the plates was within the range 0.65–2.50 μm. This range of roughness variation covers the values met most frequently in practice. Each load in the contact interaction was applied to a new area of the plate.

The Young's fringe contrast was measured for six interference fringes along a line passing through the center of the interferogram perpendicular to the fringes, and the mean value was found. The measurements were performed at five points in the central region of the area of constant contact pressure, with a subsequent averaging of the contrast values thus obtained. This treatment of the data reduces the effect of the speckles in the Young's fringe pattern on the results of the measurements.

Figure 11.14a presents some results obtained for the steel plate. Curves 1 and 2 were obtained for plate roughnesses $R_a = 2.50$ μm and $R_a = 0.65$ μm, respectively. These results imply a linear relationship between $\Delta\gamma/\gamma_0$ and the calculated pressure Q_0 [see Eq. (11.53)]. As seen from Fig. 11.14a, for a smaller degree of roughness, when the actual contact surface area is greater but the contact pressure Q_0 is the same, the plastic change of the microrelief elements is smaller than for a greater degree of roughness. In relation to this result, for smaller degrees of roughness the contact pressures can be studied over a wider range.

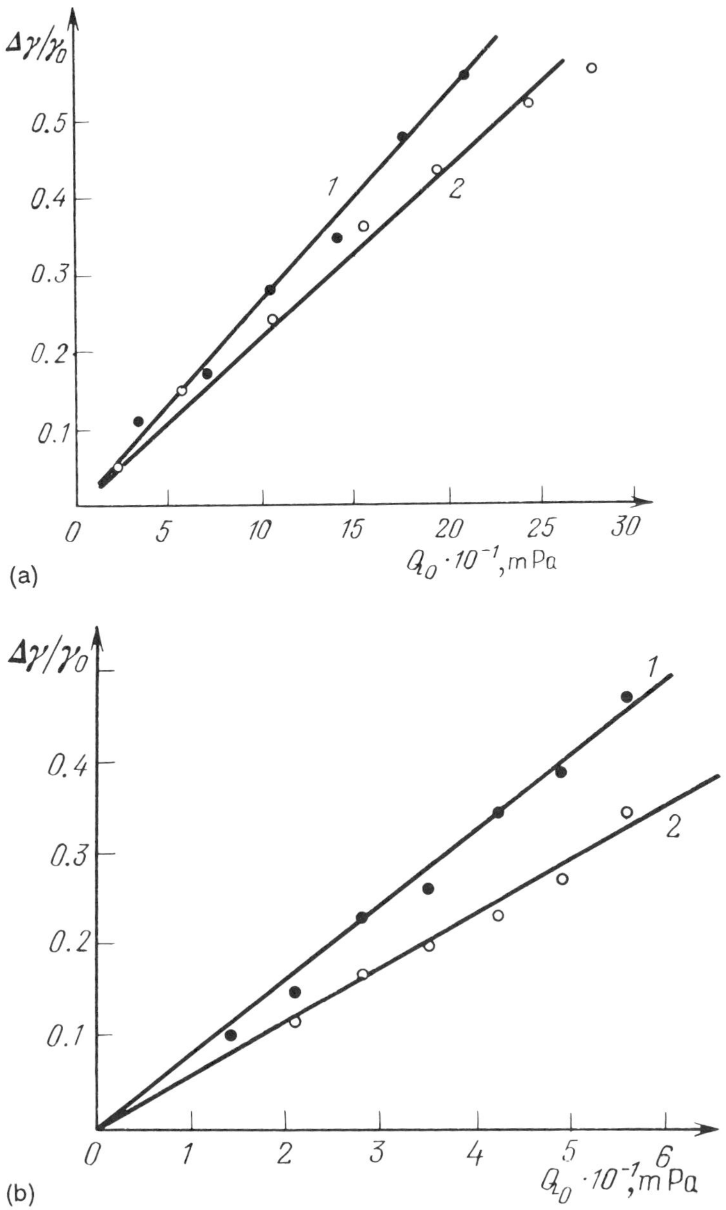

Figure 11.14 Dependence of the change in fringe contrast, $\Delta\gamma/\gamma_0$, on contact pressure Q_0. (a) Cylinder end face in contact with a steel plate. (b) Cylinder end face in contact with an aluminum alloy D16T plate.

Figure 11.14b illustrates similar measurements made for a plate of aluminum alloy D16T, which is more plastic than steel. Curves 1 and 2 correspond to a plate roughness of $R_a = 1.7$ μm and $R_a = 0.65$ μm, respectively. These relations behave in the same way as those in Fig. 11.14a.

11.4.2 Contact Pressure Measurement by Methods of Normalization and Calibration

The preceding results show that the dependence of $\Delta\gamma/\gamma_0$ on Q_0 is linear over a wide range of roughness, hardness, and mechanical properties of the objects in contact. Hence, we can assume that, in general, the following equality is satisfied [15]:

$$Q = C(\Delta\gamma/\gamma_0) \tag{11.54}$$

where C is a constant. From the condition of equilibrium (normalization) it follows that

$$P = \int_F Q\,dF = C\int_F \frac{\Delta\gamma}{\gamma_0}\,dF \tag{11.55}$$

where F is the area of the contact contour surface and P is the normal component of the load under which the objects interact. From Eq. (11.55) one can derive the constant C,

$$C = P\left(\int_F \frac{\Delta\gamma}{\gamma_0}\,dF\right)^{-1}$$

and, using Eq. (11.54), obtain an expression for the determination of the contact pressure,

$$Q = P\left(\int_F \frac{\Delta\gamma}{\gamma_0}\,dF\right)^{-1}\frac{\Delta\gamma}{\gamma_0} \tag{11.56}$$

The method of correlation speckle photography was applied by Osintsev et al. [16] to determine the contact pressures between plates made of aluminum alloy D16T and coupled by a bolt type of joint (Fig. 11.15). The plate thickness was $h = 15$ mm, and the hole diameter in the plates was $D = 15$ mm. The surfaces of the contacting plates had irregular roughnesses with $R_a = 1.5$ μm.

The dots in Fig. 11.16 show the contact pressure distribution $Q(r)$ calculated using (11.56) for a load $P = 180$ kN.

To evaluate these results, the preceding problem was solved by the finite-element technique for plate friction coefficients of 0, 0.1, and 0.2. The solid line in Fig. 11.16 was obtained by this method for a friction coefficient of 0.2, at which the experiment and calculations reveal the best fit.

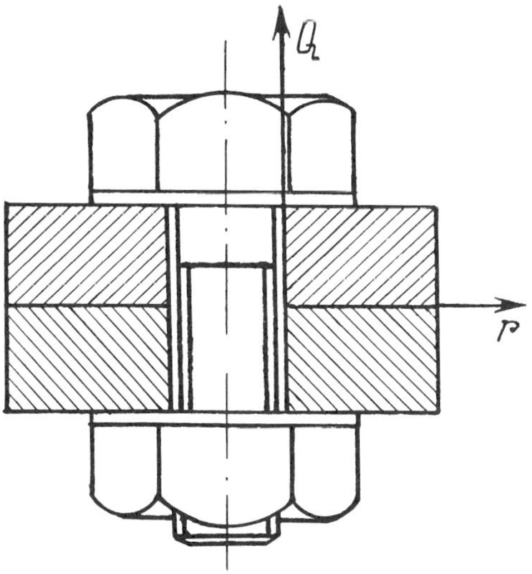

Figure 11.15 Bolt joint between two plates.

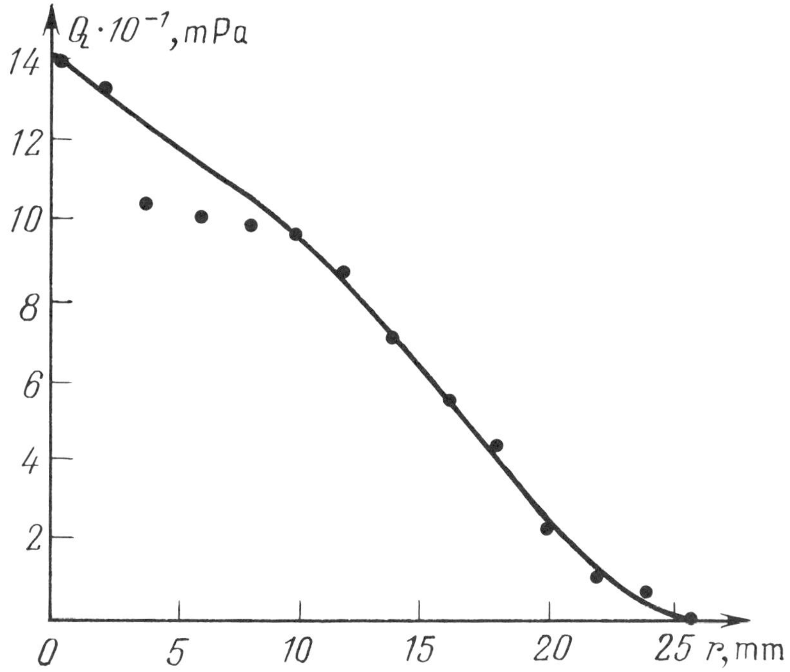

Figure 11.16 Contact pressure distribution for two plates in contact with each other.

Thus, correlation speckle photography allows correction of the calculation of contact pressures, particularly in cases where the friction coefficient cannot be determined.

Some problems, however, are met with in practice when it is impossible to derive the constant C from the condition of normalization (11.55). In this case the so-called method of calibration can be used in the following

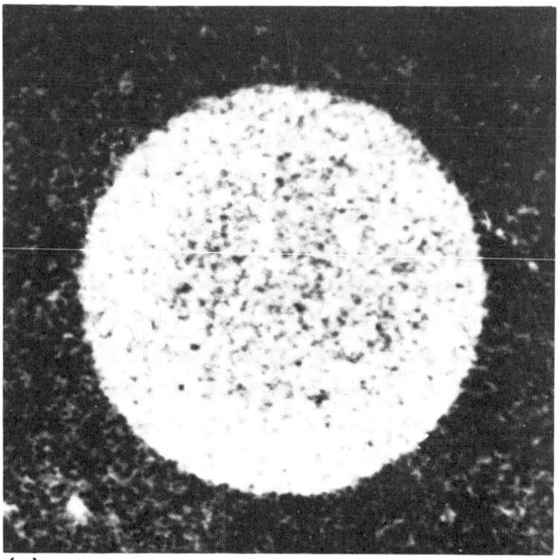

(a)

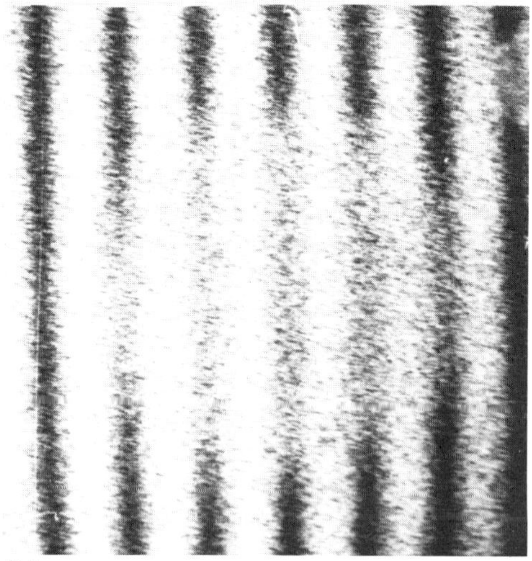

(b)

Figure 11.17 Determination of contact pressure by speckle photography (end face of cylinder in contact with plate). Patterns obtained (a) by image subtraction; (b) by spatial filtration outside the Fourier plane.

way. First, one produces a cylinder (8–30 mm in diameter and 16–60 mm in height) by the same technology and from the same material as one of the objects under study. The roughness of the cylinder's end face should also be the same as the surface of the object under study. Second, one records by means of speckle photography the several contact surfaces at different loads on the cylinder placed on the surface of the other contacting object (or a specially produced specimen). Third, one plots $(\Delta\gamma/\gamma_0)_S$ as a function of Q_0. As this dependence is linear, the tangent of the straight-line slope is equal to a constant for the conversion of the change in Young's carrier fringe contrast into contact pressure.

The experiments carried out showed an equivalence between the contact pressure measurement data obtained by methods of normalization and calibration.

The method of speckle photography permits an observation of the image of the entire contact surface, using first the image subtraction technique (Fig. 11.5). A typical pattern for the contact of the end face of a cylinder with a plate is shown in Fig. 11.17a. The uniform gray field in the central region of the imprint image corresponds to the zone of constant contact pressure (Q_0). In this case the distribution of intensity over the contact spot corresponds to that of the contact pressure.

The second technique for observing the image of the entire contact region is based on the use of spatial filtration outside the Fourier plane (Fig. 11.4). The fringe pattern thus obtained (Fig. 11.17b) is similar to the interferogram produced by holographic interferometry [17]. Both photographs shown in Fig. 11.17 were obtained by means of the same double-exposure speckle photograph.

REFERENCES

1. A. V. Osintsev, Yu. I. Ostrovsky, Yu. P. Presnyakov, and V. P. Shchepinov, Fringe contrast in correlation speckle photography and correlation holographic interferometry, *Zh. Tekhn. Fiz.*, *62*(7), 82–89 (1992).
2. J. C. Dainty, Ed., *Laser Speckle and Related Phenomena*, Springer-Verlag, New York, 1975.
3. S. M. Rytov, *Introduction to Statistical Radiophysics* (in Russian), Nauka, Moscow, 1976.
4. Yu. I. Ostrovsky, V. P. Shchepinov, and V. V. Yakovlev, *Holographic Interferometry in Experimental Mechanics*, Springer-Verlag, Berlin, 1991.
5. Yu. I. Ostrovsky and V. P. Shchepinov, Correlation holographic and speckle interferometry, *Prog. Opt.*, *30*, 87–135 (1992).
6. V. P. Shchepinov and Yu. P. Presnyakov, Correlation speckle photography with double circular aperture, *Opt. Spektr.*, (in print).
7. I. V. Volkov and I. S. Klimenko, On some peculiarities of speckle photograph recording and interpretation for objects under deformation, *Zh. Tekhn. Fiz.*, *50*, 1038–1043 (1980).

8. S. Debrus, M. Françon, and C. P. Grover, Detection of differences between two images, *Opt. Commun.*, *4*, 172–174 (1971).
9. A. V. Osintsev, Yu. P. Presnyakov, and V. P. Shchepinov, Measurement of plastic component of actual contact surface by means of correlation speckle photography, Trudy All-Union Symp. Met. Primen. Holograf. Interf. (Kuibyshev), 1990, p. 62.
10. J. T. Atkinson and M. J. Labor, Measurement of the area of real contact between, and wear of, articulating surfaces using holographic interferometry, Proc. Int. Conf. on Application of Holography and Optical Data Process, Jerusalem, 1977, p. 289.
11. A. V. Osintsev, Yu. I. Ostrovsky, V. P. Shchepinov, and V. V. Yakovlev, Contact surface determination by means of speckle photography, *Zh. Tekh. Fiz.*, *58*, 1420–1423 (1988).
12. S. P. Timoshenko and J. N. Goodier, *Theory of Elasticity*, McGraw-Hill, New York, 1970.
13. S. Debrus, M. Françon, and C. P. Grover, Detection of the difference between two images: an improved method, *Opt. Commun.*, *6*, 15–17 (1972).
14. N. I. Bezukhov, *Fundamentals of the Theory of Elasticity, Plasticity and Creep* (in Russian), Vysshaya Shkola, Moscow, 1968.
15. A. V. Osintsev, Yu. I. Ostrovsky, V. P. Shchepinov, and V. V. Yakovlev, Fringe contrast change in holographic interferometry and speckle photography for solid objects contact interaction, *Zh. Tekh. Fiz.*, *61*(8), 134–139 (1991).
16. A. V. Osintsev, Yu. I. Ostrovsky, V. P. Shchepinov, and V. V. Yakovlev, Effect of contact pressures on interference fringes contrast in speckle photography, *Zh. Tekh. Fiz.*, *62*(4), 123–128 (1992).
17. A. V. Osintsev, Yu. I. Ostrovsky, V. P. Shchepinov, and V. V. Yakovlev, Effect of contact pressures on fringe contrast in holographic interferometry, *Pis'ma Zh. Tekh. Fiz.*, *11*, 202–204 (1985).

INDEX